Space Vehicle Maneuvering, Propulsion, Dynamics and Control

Ranjan Vepa

Space Vehicle Maneuvering, Propulsion, Dynamics and Control

A Textbook for Engineers

Springer

Ranjan Vepa
School of Engineering and Materials Science
Queen Mary, University of London
London, UK

ISBN 978-3-031-65517-3 ISBN 978-3-031-65518-0 (eBook)
https://doi.org/10.1007/978-3-031-65518-0

This Springer imprint is published by the registered company Springer Nature Switzerland AG
The registered company address is: Gewerbestrasse 11, 6330 Cham, Switzerland

To Sudha

Preface

In the last decade the subject of space vehicle manoeuvring, propulsion, dynamics and control has made rapid strides. The "standard" mathematical description of orbital motion of a spacecraft has been evolving, and the book will provide students with the necessary skills to optimally synthesize spacecraft trajectories and implement advanced techniques for planning both planetary and interplanetary space missions. The underlying physics of all spacecraft propulsion technologies are described, with some detailed focus on electric propulsion, to facilitate the selection of the most appropriate propulsion systems for a specific mission.

This book is about manoeuvring a space vehicle using impulsive and continuous thrusts generated by on-board propulsion systems, the resulting dynamic trajectories of motion of the spacecraft, the control of the trajectories of motion and the tailoring of the trajectories from a departure planet to a destination planet, to suit the needs of a particular mission. The study and analysis of these topics would allow a space engineer to estimate the propulsive requirements for a specific space mission to the moon or to another planet and also return to the Earth, if so desired. Based on the propulsive requirements and depending on the trajectory the spacecraft is expected to follow in space as well as the nature of the payload the vehicle must carry and the vehicle itself, a space engineer could estimate the energy and fuel requirements for the mission. Thus the book would facilitate the design of a complete space mission.

In this book, following an introduction to the subject, a description of the space environment and its influence on the spacecraft propulsion system and its selection are discussed. This is followed by an in-depth description of Keplerian celestial mechanics, co-planar and non-planar orbital transfers involving both impulsive and continuous manoeuvers, and perturbation effects that characterize the real non-Keplerian nature of orbital motion. The use of the restricted three-body dynamics as descriptors of spacecraft motion, which has replaced the Keplerian model as the standard baseline model for studying orbital mechanics of space vehicles, the limitations of the approach in terms of orbital perturbations and an understanding of the physical source and influence of these perturbations, is discussed in some detail. The quantitative use of the rocket equation for high thrust spacecraft manoeuvres and alternative methods for low thrust trajectories are used to estimate the propulsion requirements for a mission. An introduction to space vehicle propulsion is then presented including a description of launch-vehicle dynamics, the rocket equation and metrics that are used to characterize the fuel requirements over a particular mission. The principles of the optimal synthesis of trajectories are discussed. After discussing classical on-board chemical propulsion systems and the basics of electric propulsion, the principles and features of range spacecraft thrusters are discussed to facilitate the selection of an appropriate propulsion system for meeting the requirements of a specific mission. The physics of the dynamics of plasmas is discussed in some detail. This is followed by the background on Electromagnetic theory, Antennas, Helicon waves and Radio Frequency Propulsion based on Helicon waves. Also discussed in some detail are the key drivers which determine a spacecraft design for a particular mission.

A central feature of this book is that all the material covered in the book are presented at the level of a final year graduate or a postgraduate university student. The key focus of the book is to explain the topics covered at a final year graduate or a postgraduate university student level, thus making it accessible to most practicing engineers.

I would like to thank my colleagues, and present and former students of the School of Engineering and Material Science, at Queen Mary, University of London for their support in this endeavour. A very special thank you to Prof. J. P. W. Stark for providing a copy of the book that he co-authored and edited and also for providing his lecture notes. I still remember the day when he first patiently explained the sphere of influence (SOI) concept, almost 35 years ago. That was indeed the starting point of this journey which would not have been possible without his support.

I would like to thank my wife Sudha, for her love understanding and patience. Her encouragement was a principal factor that provided the motivation to complete the project. I would also like to thank our children Sruthi, Satvi and Abhinav for their understanding during the course of this project.

London, UK
2024

Ranjan Vepa

Contents

List of Abbreviations

ABEP	Air Breathing Electric Propulsion
ADR	Active Debris Removal
AO	Atomic Oxygen
Ar	Argon
AU	Astronomical Unit
Au	Gold (Auric)
BEW	Beating Electrostatic Waves
BOL	Beginning of Life
CD	Cadmium
CH_4	Methane
CLP	Collinear Lagrange Points
CO	Carbon Monoxide
CO_2	Carbon Dioxide
CR3BP	Circular Restricted Three Body Problem
CRTBP	Circular Restricted Three Body Problem
DC	Direct Current
DCD	Discrete Circular Disc
ECR	Electron Cyclotron Resonance
EDL	Electric Double Layer
EM	Electro-magnetic
EMC	Electromagnetic Compatibility
EMF	Electromotive Force
EP	Electric Propulsion
ER3BP	Elliptic Restricted Three Body Problem
ET	Electric Thruster
EUV	Extreme Ultra Violet
Fe	Iron (Ferric)
FEEP	Field Emission Electric Propulsion
FLR	Finite-Larmor Radius
GEO	Geostationary Orbit
GIE	Gridded Ion Engine

GPS	Global Positioning System
H	Hydrogen Atom
H_2	Hydrogen molecule
H_2O	Water
HCT	Hall Current Thruster
HCW	Hill Clohessy Wiltshire
HDLT	Helicon Double Layer Thruster
He	Helium
HEO	High Earth Orbit
HET	Hall Electric Thruster
HiPEP	High Power Electric Propulsion
HPT	Helicon Plasma Thruster
HTPB	Hydroxyl-Terminated Polybutadiene
IEC	Inertial Electrostatic Confinement
IPT	IEC Plasma Thruster
IR	Infra Red
JAXA	Japan Aerospace Exploration Agency
K	Potassium
LASER	Light Amplification by Stimulated Emission of Radiation
LEO	Low Earth Orbit
LERM	Linear Equations of Relative Motion
LH2, LH	Liquid Hydrogen
LMO	Low Mars Orbit
LOX	Liquid Oxygen
LVLH	Local Vertical Local Horizontal
MASCON	Concentrated Mass
MAVEN	Mars Atmospheric and Volatile EvolutioN
MEE	Modified Equinoctial Elements
Mg	Magnesium
MHD	Magneto Hydro Dynamic
mHTX	mille Helicon Thruster Experiment
MHz	Mega Hertz
Mo	Molybdenum
MOM	Mars Orbiter Mission
MPD	Magneto Plasma Dynamics
MPDT	Magneto Plasma Dynamic Thruster
Mw	Molecular Weight
N_2	Nitrogen Molecule
Na	Sodium
NASA	National Aeronautics and Space Administration
NEO	Near Earth Orbit
NERM	Nonlinear Equations of Relative Motion
NEXT	NASA Evolutionary Xenon Thruster
Ni	Nickel

NO	Nitrous Oxide
NO_2	Nitric Dioxide
NSTAR	NASA Solar Technology Application Readiness
O_2	Oxygen Molecule
ODE	Ordinary Differential Equation
PEMP	Permanent Magnet Expanding Plasma
PIC	Particle-in-Cell
PPT	Pulsed Plasma Thruster
PV	Photo Voltaic
RAAN	Right Ascension of the Ascending Node
REF	Rotating Electric Field
RF	Radio Frequency
RFIT	Radio Frequency Ion Thruster
RH	Right Handed
RIT	Radio Frequency Ion Thruster
RMF	Rotating Magnetic Field
RMS	Root Mean Square
RTG	Radio-Isotope Thermo-Electric Generator
SOI	Sphere of Influence
SPT	Stationary Plasma Thruster
TCM	Trajectory Correction Manoeuvre
TG	Trivelpiece-Gould
Ti	Titanium
UDMH	Unsymmetrical Di-methyl Hydrazine
UV	Ultra Violet
VASIMR	VAriable Specific Impulse Magneto plasma Rocket
W	Tungsten
Z	Z particle, neutral sub-atomic particle (Boson)
Zn	Zinc

Chapter 1
Celestial Mechanics: Keplerian Orbits

1.1 Introduction: The Space Environment

Chapter 1 focusses on the background and introductory material for the book. For this reason, it includes a diverse selection of topics including elements of celestial mechanics, some background material on the environmental features around the Sun, the features of the interplanetary environment in space, the electromagnetic, thermal and vacuum environment within the Earth's sphere of influence and the local environment in the vicinity of a launch site. While these topics are admittedly quite diverse, a broad understanding of these and other related concepts is essential for a holistic understanding of the principles underpinning space vehicle manoeuvring, propulsion, dynamics and control, which is the primary aim of the book. Before embarking on a study of spacecraft dynamics and its control, while on an interplanetary mission, it is extremely important to fully comprehend and grasp not only the features of the environment in the vicinity of our planet but also the environment around the centre of our planetary system, i.e. around the Sun, the intervening region between the Sun and Earth and beyond the Earth and the limits of our planetary environment. A deeper study of these environments reveals the unique nature of our home, the planet Earth and its extraordinary features. A detailed description of the space environment and its impact on space vehicle design is discussed in some detail by Fortescue, Swinerd and Stark (2011).

One could begin with the Sun, which is the source of all energy within the solar planetary system, before considering the Earth and its relatively benign and charge-neutral nature, its unique behaviour as a magnetic dipole in space and the resulting magnetosphere around it, the key features of the intervening environment, like the vacuum, Van Allen Radiation belts in the Sun-Earth region and the region beyond the Earth's orbit around the Sun. Of particular interest are the varied influences on a spacecraft as it journeys through these distant regions.

Astronomers hypothesize the Sun and planets were born some 4.5 thousand million years ago, from a cloud of gas that collapsed under its own gravity. As the

R. Vepa, *Space Vehicle Maneuvering, Propulsion, Dynamics and Control*,
https://doi.org/10.1007/978-3-031-65518-0_1

gas cloud collapsed, its temperature and pressure increased to set off a thermonuclear reaction, which led to the birth of the Sun, and as the matter around it slowly cooled, the planets were formed. The entire solar system is part of the Milky Way galaxy, with about 100000 million stars. These are the stars any one can see in a typical night sky with the naked eye.

The Sun is a typical yellow star. At its core, a thermonuclear fusion reaction produces the energy, which keeps it at about 14 million degrees Celsius. Eventually, in about another 4.5 thousand million years, the thermonuclear fusion process will cease, producing the energy, and the star will turn into a white dwarf, which will radiate the energy remaining into space. It will then also collapse under the influence of its own gravity to about the size of the Earth.

The Sun is about 0.7 million kilometres (km) in diameter, with a mass of $\sim 2 \times 10^{30}$ kg and about 149,597,870 km from the Earth on average, as the Earth orbits the Sun in a slightly elliptic orbit around the Sun. The closest point of the Earth's approach to the Sun is 147.5 million km, indicating that the eccentricity of the Earth's orbit is 0.01671. The average distance of the Earth from the Sun is taken to be one astronomical unit (AU) and is used as a unit of distance within the planetary system of the Sun. The Sun is characterized by a sphere of influence (SOI), which encloses the planetary system and more while the Earth's SOI is about a million km. Thus, the environment around the Earth could be considered to be within 0.1 million km, while the region within 2000 km could be considered to be the near-Earth environment. Considering the Sun and its planetary system within the context of the universe, it is worth reminding ourselves that is but a mere speck within the vast expanse of the Milky Way, which itself is just one of the many galaxies that populate the universe. The primary features of the Sun and the planets are summarized in Tables 1.1 and 1.2.

Table 1.1 Features of the Sun

Diameter (at its equator)	1,392,000 km
Average mean distance from the Earth	1AU = 149,597,870.7 km
Mass of the Sun	1.9891×10^{30} kg
Ratio with the mass of the Earth	332946
Accn due to gravity (in Earth g's)	27.9
Volume ratio with the Earth	1303600
Average rotation period	25.4 Earth days
Composition	Mainly hydrogen, helium
Surface temperature	5500 °K
Core temperature	14,000,000 °K
Time taken for light to reach the Earth	8.3 min
Age of the Sun	4.5 thousand million years
Remaining life	4.5 thousand million years
Features: sunspots	Sunspot cycle (~11 years)
Other features	Flares, prominences
Distance from centre of galaxy	26,500 light-year
Speed around the galaxy	220 km/s

Table 1.2 Summary of the planets

	Mercury	Venus	Earth	Mars	Jupiter	Saturn	Uranus	Neptune	Pluto
Mean distance from the Sun (in AU)	0.39	0.72	1	1.52	5.20	9.54	19.22	30.06	39.53
Diameter @ equator (in km)	4878	12104	12756	6794	142796	120536	49946	49528	2304
Mass (in 10^{24} kg)	0.33	4.87	5.97	0.642	1898.6	568.46	86.62	102.43	0.0127
Accn. due to gravity	0.38	0.9	1	0.38	2.58	1.06	0.9	1.13	0.07
Density (g/cm^3)	5.43	5.24	5.52	3.93	1.33	0.69	1.32	1.64	2.05
Angle of rotation axis	0°	177.3	23.5	25.2	3.1	26.7	97.9	29.6	122
Length of the year	87.9 days	224.7 days	365.2422 days	686.98 days	11.86 years	29.46 years	84.07 years	164.82 years	248.6 years
Length of a day	59 days	243 days	23 h 56 mins	24 h 37 min	9 h 56 mins	10 h 14 mins	17 h 14 mins	16 h 3 mins	6 days 9 h
Moons	0	0	1	2	25	30	20	8	1
Rings	0	0	0	0	1	1000's	11	5	0
Visiting s/c	Mariner 10	Mariner 2, Venera 7, 9, Magellan		Mariner 4, Mars 2, Viking, Pathfinder, Mars 2020, Curiosity, Mars Orbiter, Maven,	Pioneer 10 & 11, Voyager 1 & 2, Galileo, Juno, New Horizons, Ulysses	Pioneer 11, Voyager 1 & 2, Cassini	Voyager 2	Voyager 2	New Horizons

(continued)

Table 1.2 (continued)

	Mercury	Venus	Earth	Mars	Jupiter	Saturn	Uranus	Neptune	Pluto
Surface temp. °C	Day: 427 Night: −179	+480 −33	Mean 22	Mean −23	Mean −150	Mean −180	Mean −-214	Mean −220	Mean −230
Atmosphere	Very little	96%CO_2 3.5% N_2	77% N_2, 21% O_2, 1% H_2O	95%CO_2 2.7% N_2, 1.6% Ar, 1.3% O_2.	82 % H_2, 17% He	93 % H_2, 7% He	85 % H_2, 15% He, 2% methane	85 % H_2, 13% He, 2% methane	Mostly methane (?)
Core	Ni, Fe	Solid Fe, Ni	Inner: solid Fe, Ni outer: Liquid Fe, Ni	Fe, Ni	Rock, ice, liquid metallic H_2, H_2, He	Rock, ice, liquid metal-lic H_2, H_2, He	Rock	Rock	Rock
Mantle	Rock	Rock	Rock	Rock			Frozen water	Frozen water	Frozen water
Crust	Rock	Rock	Rock	Rock					
Features	Moon like craters	Thick clouds of CO_2 trap-ping heat	Blue planet, covered with 71% water	Red planet, largest vol-cano, polar ice caps	Giant hurricane area, red spot	Rings of ice & rock	Extremely tilted, almost rolls along!	Fastest wind speeds, 2000 km/h	Highly elliptic orbit

Returning to the Sun, the region around the Sun has a layered structure. The inner region of the environment around the Sun is the photosphere, the middle or intermediate layer is the chromosphere and the outer region is the coronal region with a temperature exceeding 1,000,000 °K. By comparison, the temperature of the photosphere is merely 4–6000°K, while the chromosphere is 25,000 °K.

A key feature of the Sun are the coronal mass ejections, or low-density flares, which are only visible when the Sun itself is eclipsed, which in turn results in the generation and propagation of the solar wind. Solar wind, or intense clouds of high-energy particles, are produced by solar storms or the coronal mass ejections. When these high-energy particle clouds make their way to the Earth in 3–4 days, they collide with the magnetic field of the Earth and cause the magnetic field lines to change their shape. This in turn effects GPS, radio communication and navigation systems on Earth. The solar particles and solar flares are accompanied by intense electromagnetic (EM) radiation (consisting of infra-red, IR, visible 42%, ultraviolet, UV ~7%, extreme UV <0.1% (damaging), radio waves <0.1%) and are also the main energy source for the Earth. Dome of the global properties of solar flares and coronal mass ejections have been discussed by Hudson (2011). The solar particles generated by a coronal mass ejection and on their way to the Earth get trapped within the magnetosphere around it. The solar wind is a form of a plasma consisting of electrons and various ions, travelling at 300–500 km/s with a number density ~ 10 protons/cm^3. It consists mainly of 96% protons and 4% alpha particles with a mean velocity ~450 km/s and a maximum velocity ~850–900 km/s. The particle number density is ~10 /cm^3 at 1 AU with the following temperatures: 10000–100000 °K. The total radiant flux is 3.8×10^{26} Watts with a source temperature: ~6000 °K, emitting a yellow light and at a distance of 1AU. The radiation follows the "black body" radiation or Stefan-Boltzmann law with the wavelength distribution of thermal radiation being inversely proportional to the absolute temperature. A secondary feature of the surface of the Sun is the phenomenon of sunspots. Sunspots on the solar disc are small cooler areas of the Sun's photosphere. The Zurich sunspot number archive for 130 years has been studied by a number of scientists, and it is known that the sunspots appear almost periodically and could be characterised either by the ratio between the number of single spots and the number of groups or the total number of sunspots. The Zurich archive reveals a period between 10 and 11 years. Ramesh (2010) discusses the relationship between the sunspot cycle and the coronal mass ejections.

The spectrum of radiation received the Earth from the Sun as well as other radiations effects is extremely important to the Earth. The wavelength distribution of the thermal radiation from a black body is related to the temperature. This is known as Wiens law. The wavelength distribution of thermal radiation from a black body at any temperature has the same shape at any other temperature. The wavelength (in centimetres) is inversely proportional to the temperature in degress Kelvin, the proportionality constant being 0.2897.

The Stefan-Boltzmann law describes the power radiated from a black body in terms of its temperature. Specifically, the Stefan-Boltzmann law states that the total energy radiated per unit surface area of a black body across all wavelengths per unit time, also known as the black-body radiant emittance, is directly proportional to the fourth power of the black body's thermodynamic temperature T. It was with this law Josef Stefan also determined the temperature of the Sun's surface. The Sun is at the source of all our energy and is the primary source of all radiation in the solar system. Josef Stefan was also able to account for the energy flux from the Sun that is absorbed by the Earth's atmosphere. According to the Stefan-Boltzmann law,

$$\phi_{\text{out}} = \sigma\varepsilon T^4 \equiv \left\{\frac{\text{Energy}}{\text{area} \times \text{time}}\right\} = \text{Output flux} \tag{1.1}$$

In the above equation, ε is the emissivity, $=1$ for a black body and $\sigma = 5.672 \times 10^{-8}\text{W/m}^2\ ^\circ\text{K}^4$ is the Stefan-Boltzmann constant. Thus, the solar flux and input power in the vicinity of the Earth's surface can be, respectively, shown to be:

$$\phi_{\text{in}} = 1.371 \times 10^3 \left(\frac{R_\oplus}{R}\right)^2 \frac{w}{m^2}, \quad P_{\text{in}} = \phi_{\text{in}} A_{\text{proj}}. \tag{1.2}$$

In the above, $R_\oplus$ is the mean distance of the Earth's surface from the Sun, R is the distance from the Sun, where the solar flux is desired and A_{proj} is the projected surface area normal to the incident flux.

One could take a close look at the features of the Earth and other planets in the vicinity. In particular, it may be observed that Venus, Mercury and Mars, at distances of 40, 100 and 228 million km respectively, have no magnetic field compared with Earth. To be more precise, Venus has about 15/1,000,000 of Earth's magnetic field, and Mercury is about 1/100. Mars' field is also significantly lower and about ~16–40 times less than Earth's magnetic field. Thus, the Earth stands out in the planetary orbital region between the Sun and Jupiter! Moreover, there is a unique interaction between the solar wind approaching the Earth and the magnetosphere around the Earth. The magnetosphere is a magnetic bubble that protects Earth from charged particles from the sun arriving in the form of the solar wind by channelling these particles down magnetic-field lines and to the other side of our planet. The interaction between the supersonic solar wind and the terrestrial magnetic field creates the shock region, also known as the bow shock. The foreshock then forms the "upstream" of this shock region. The solar wind and the associated magnetic field, which causes the bow shock wave, is always directed away from the Sun. It results in the interaction between the charged particles in the upper layers of the terrestrial atmosphere, which are guided by the Earth's magnetic field, and the solar wind particles carrying the interplanetary magnetic field. The impact of solar winds also causes electromagnetic waves to appear as small oscillations of Earth's magnetic field. Waves with the same frequencies as waves in this foreshock region have been spotted on Earth's sun-facing side, suggesting that they can enter the magnetosphere and travel all the way to the planet's surface.

The interaction with the solar wind primarily results in the formation of the Van Allen radiation belts of "trapped" particles, which are unique to the planet Earth. James Van Allen discovered the existence of zones of energetically charged particles, which originate from within the solar wind, and are captured by and held around a planet by that planet's magnetosphere. Thus, the Van Allen radiation belts of "trapped" particles consist of two distinct groups: the outer and the inner Van Allen radiation belts. Li and Hudson (2019) illustrate the two-belt structure of the Van Allen radiation belts and also provide a schematic of the magnetosphere showing the arrival of an interplanetary shock. They also discuss the various electromagnetic wave phenomena that exist within the radiation belts. The outer belt is made up of billions of high-energy particles (electrons) that originate from the Sun and become trapped in Earth's magnetic field, an area known as the magnetosphere. The inner belt results from interactions of cosmic rays with Earth's atmosphere and contains lower-energy electrons and protons. The inner belt contains high concentrations of electrons in the range of hundreds of keV and energetic protons with energies exceeding 100 MeV, trapped by the strong (relative to the outer belts) magnetic fields in the region. Protons at lower altitudes are the result of the beta decay of neutrons created by cosmic ray collisions with nuclei of the upper atmosphere. The source of lower-energy protons is believed to be proton diffusion due to changes in the magnetic field during geomagnetic storms. The magnetic field is weak, and charged particles also get trapped in the field lines causing them to act as magnetic "mirrors". The inner Van Allen belt makes its closest approach to the surface of the Earth at the South Atlantic Anomaly. Due to the slight offset of the belts from Earth's geometric centre, the closest approach of the inner Van Allen belt to the surface of the Earth in the northern hemisphere is not symmetrically located.

Considering the Earth's magnetic field and looking closely at the Earth, the magnetic field around the Earth can be modelled as a field of a simple dipole as shown in Fig. 1.1. There are several consequences that follow, and, in particular, one can observe that the charged particles could get trapped in the field. Although there are many dipole approximations, one popular model is

$$B = B_E \left(\frac{R}{R_E} \right)^{-3} \frac{\left(4 - 3\cos^2\lambda\right)^{1/2}}{\cos^6\lambda}. \tag{1.3}$$

The field on the Earth's surface is obtained by setting $R = R_E$. In the above equation, λ is the magnetic latitude and $B_E = 0.312G$.

The Earth's local environment within about 1200 km is referred to as the Earth's atmosphere. It has a layered structure. Layers are identified according to their temperature profiles, which are relatively stable. The primary layers are as follows: the troposphere (0–11 km, upper boundary varies from 11–20 km, a region known as the tropopause); the stratosphere (20–47 km, the stratopause at 47–52 km), which is low in pressure and warm; and the mesosphere, (50–85 km). The temperature variation according to the 1976 International Standard Atmosphere model in the troposphere (0–11 km), the tropopause (11–20 km), the stratosphere (20–47 km), the

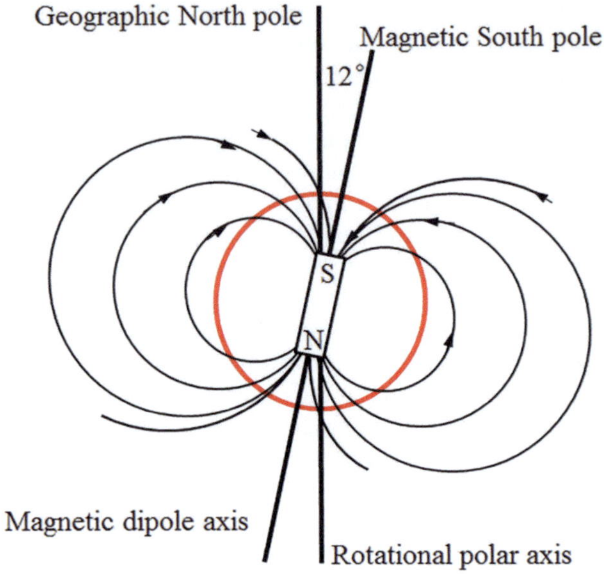

Fig. 1.1 Dipole model of the Earth's magnetic field

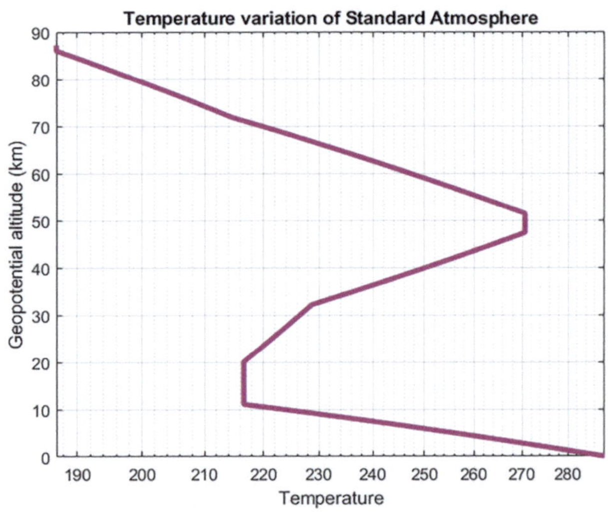

Fig. 1.2 Temperature variation according to the 1976 International Standard Atmosphere model in the troposphere (0–11 km), the tropopause (11–20 km), the stratosphere (20–47 km), the stratopause (47–52 km) and the mesosphere (52–85 km)

stratopause (47–52 km) and the mesosphere (52–85 km) is shown in Fig. 1.2. The temperature decreases with height in the mesosphere with the mesopause at its boundary. Beyond the mesopause is the thermosphere, with the upper boundary varying from 500 to 1000 km. The temperature increases with height in the thermosphere. It is bounded by the thermopause, which is followed by the exosphere. The exosphere contains most of the satellites orbiting Earth, up to 1200 km into space. The lighter particles (H, He) can escape, hence the name. It is mainly composed of hydrogen (Mw = 2), helium (4) and some heavier molecules, such as atomic oxygen

(16), nitrogen (28) and carbon dioxide (44) closer to the exobase (edge of the exosphere). The density here is too low for the domain to behave as a gas. Temperatures are constant with height and are dependent on solar activity, ranging from about 500 to 2000 K.

1.1.1 Density Behaviour in the Exosphere (Hydrostatic Model)

Two basic equations define density behaviour in the exosphere. The first relates the change in pressure to the change in the potential energy, while the second models the change in the acceleration due to gravity. Thus, the change in pressure and acceleration due to gravity are, respectively, given by

$$dp = -\rho g dz = -\frac{pM}{RT} g dz, \quad g = g_0 \frac{r_0}{(r_0 + z)^2} \tag{1.4}$$

Integrating, the first of the above equations,

$$\rho = \rho_0 \exp\left(-\frac{(z - z_0)}{h_0}\right) = \rho_0 \exp\left(-\frac{h}{h_0}\right), \quad h_0 = \frac{RT}{Mg_0} \tag{1.5}$$

In the above equation, M is the mean molecular weight (kg/kmol), while h_0 is the scale height, with low mean molecular weights at high scale heights.

1.1.2 Key Features of the Intervening Environment Between the Earth and the Sun

The Sun is the main energy source for all the planets and the environment around them. It is primarily a neutral environment characterized by ultra-high vacuum and relatively low gravity. The low-gravity environment translates to a domain of microgravity but not zero gravity, and it is important to recognize that Newton's law of gravitation still applies. Moreover, the intervening region between the planets and the Sun is also a plasma environment due to the presence of high-energy (ionizing) particles, arriving from the Sun, following coronal mass ejections in the form of cosmic rays and particles. These particles could cause high-velocity micro impacts, while micrometeorites and space debris, present everywhere in the interplanetary environment, could cause macro impacts. And finally, closer to the Earth's surface, it is important to consider the launch environment.

The neutral environment, with its ultra-high vacuum characteristic, is a feature closer to the Earth, up to 100000 km. The vacuum could be compared with a high

Table 1.3 Specific compositions of the atmospheres in the vicinity of various planets and a moon

Planet/ Moon	Composition % of primary constituents	Surface pressure (1 bar=98692.3 Pa)	Surface temperature (°K)	Temperature @ 200 km (°K)	Ionosphere (electrons/ cm^3)
Mercury	Exosphere with traces of O_2, Na, H_2, He, K.	10^{-12} bar	700 to -100	100–280	
Venus	CO_2 (96%), N_2 (3.5%)	92	750	800–1100	~10^6
Earth	N_2 (77), O_2 (21), H_2 (1)	1	285	310	~10^6
Moon	Exosphere with traces of He, Ar, Na, K	3×10^{-15}	253		0.016 (max)
Mars	CO_2 (95), Ar (1.6), N_2 (2.7)	0.006	220		~10^5
Jupiter	H_2 (89), CH_4 (0.2), He (11)	Gaseous	165[a]		~10^5
Saturn	H_2 (93), CH_4 (0.2), He (7)	Gaseous	130[a]		
Titan	N_2 (90–99), CH_2 (1–5), Ar (0–6)	1.5	95	150	~10^3
Uranus	H_2 (85), CH_4 (<1), He (15)	Gaseous	80[a]		
Neptune	H_2 (90), CH_4 (<1), He (10)	Gaseous	70[a]		
Pluto	N_2, CH_4/CO (traces only)		40		

[a]Temperature given where pressure is the same as on Earth at sea level

vacuum chamber with pressures in the region of 3×10^{-6} Pa. To put this in context, spacecraft, in a low earth orbit, is generally at altitudes of 300–1200 km from Earth's surface with pressures equal to 1×10^{-5} Pa to about 3×10^{-7} Pa. A geostationary orbit (GEO) at an altitude of about 36000 km from Earth surface is typically characterized by pressures equal to 3×10^{-7} Pa or less.

The actual composition of the atmospheres in the vicinity of various planets could be vastly different. Table 1.3 gives the specific compositions of the atmospheres in the vicinity of various planets and a moon. It clearly shows the unique features of the Earth's own atmosphere.

In order to analyse the interactions between the cosmic particles and the spacecraft, some background on the kinetic theory of gases is essential. When the molecular particles strike a surface, there is a transfer of momentum. Over time, some finite amount of momentum will be transferred to a unit area, and this results in a net force, spreading over an area which is equivalent to a distribution of pressure. The pressure could then be related to an equivalent translational kinetic energy and an equivalent thermal energy. The change in momentum is given by $p = mv \times \cos$ (angle to the surface) – ($-mv \times \cos$(angle to the surface)) and $p = 2mv \times \cos$(angle to the surface). The total pressure exerted on the surface can be found by integrating over all molecules. The pressure can then be combined with the ideal gas law to obtain the translational kinetic energy of the molecules, which could be related to the temperature (3/2)*kT.

An atmosphere exists at low earth orbit heights—hence, there is a varying drag force due to density variations. The drag force on the spacecraft results in a removal of energy from it, and consequently, the orbit will decay. Drag variations are well correlated with solar ultraviolet (UV) output and the atmospheric heating that occurs during geomagnetic storms. There are definite timescales for the variations of the density and the drag force. The rate at which these variations happen is dependent upon height and time (long and short timescales); the long-term variation is due to solar activity (higher temp due to increased UV). When there is a higher density, it is accompanied by an increased rate of loss of altitude hence the amount of propellant needed/used in thrusters to overcome drag will be more and depend on when the mission is launched. Short-term variations are due to day (in sunlight temperature is high) and night (no sunlight, low temperature) variations, and these diurnal (daily) effects have a phase lag relative to the local solar time.

Two indices are generally used to measure the solar radiation and geomagnetic activity levels. The solar radio flux at 10.7 cm (2800 MHz) is an excellent indicator of solar activity. Called the F10.7 index, it is one of the longest-running records of solar activity. The F10.7 radio emissions originate high in the chromosphere and low in the corona of the solar atmosphere. Currently, it is used as a measure for extreme UV (EUV) radiation in atmospheric models. The F10.7 index for solar flux is a measure of the solar flux emitted at a wavelength of 10.7 cm. Since the EUV radiation is absorbed in the thermosphere, it is difficult to obtain a measurement of the solar flux at these wavelengths using instruments at the Earth's surface. It has been found that the F10.7 index presents a good correlation with solar activity. The K_p index for geomagnetic storms quantifies disturbances in the horizontal component of Earth's magnetic field. It indicates the magnitude of geomagnetic disturbance on a 0–9 scale, with 0 being very quiet and 9 indicating a major geomagnetic storm. The K_p index is used in estimates of spacecraft charging, geomagnetically induced currents and satellite drag. The A_p index for geomagnetic activity is an equivalent index, which is a measure of the general, daily, average level of geomagnetic activity on the Earth for a given day. It is obtained by averaging measurements of the magnetic field variations made at different locations over 3-hour time frame. Geomagnetic storms are characterized by a sudden increase in this index.

The outer gaseous shell of a planet's atmosphere, particularly in the thermosphere, exchanges energy with the space plasma environment. The sources of energy are absorption of EUV radiation (10–200 nm), joule heating by electrical currents, particle precipitation from the magnetosphere and dissipation of upward propagating waves (tides, planetary and gravity waves), while the energy sinks are thermal conduction into the mesosphere; IR cooling by CO_2 NO and O; and related chemical reactions. The heating of the exosphere by solar radiation is largely due to the EUV radiation. However, the radiation at 10.7 cm wavelength, characterized by the F10.7 index, has historically been used in atmospheric modelling. Indeed, both radiations originate from the same region of the Sun atmosphere (and show a good correlation), and the 10.7 cm radiation can be readily measured on the ground (at the Penticton site in Canada).

1.1.3 Environmental Effects to Spacecraft

The primary effects of the space environment of a spacecraft due to the vacuum, atomic oxygen (AO) and UV radiation are outgassing and increase in the fatigue strength. The increased particulate or ionizing radiation in the vicinity of the South Atlantic Anomaly is increased radiation. The increased radiation and consequent increase in plasma could result in radiation damage to solar cells. The extreme temperatures and thermal cycling could result in a mismatch of thermal expansion coefficients in a spacecraft. The micrometeoroid/orbital debris impact could not only damage the spacecraft but also result in a loss of orbital momentum. The interactions between the molecules and the walls of spacecraft lead to the creation of drag and physical sputtering. Sputtering is a process where a gas molecule severs the chemical bonds in some materials. This can happen if the energy associated with the molecule is greater than the energy required to break the chemical bond. As already mentioned earlier at altitudes above 86 km, the main atmospheric constituent is atomic oxygen (AO). AO interacts with a wide range of materials, leading to oxidation, erosion and general degradation. Spacecraft materials are usually chosen for their thermal characteristics and low weight. Some of these are kept thin in order to minimize launch costs. Degradation due to AO may have serious consequences on the thermal characteristics of the material. Degradation due to AO, however, is of greatest concern for long-term missions. The dominant trace species in the exosphere and beyond is AO, which is not only a chemically reactive atom but usually is characterized by a high velocity (8 km/s $\equiv$ energy is 5 eV where 1 eV is the charge of an electron subjected to a potential difference of 1 volt ~1.6×10^{-19} (C) $\times$ 1 Volt $= 1.6 \times 10^{-19}$ joules). It can lead to chemically induced sputtering followed by material loss. The first example is the Space Transportation System, STS-3 (1982): 35% mass loss of Kapton from a camera thermal blanket led to a long-term programme of investigation for preparation of the International Space Station (ISS) due to concerns of structural integrity, should AO remove the binder in composite materials. NASA initiated the Long Duration Exposure Facility mission, to provide long-term data on the space environment and its effects on space systems and operations. It was launched in 1984 with a design life of six months, although it lasted for 6 years when it was retrieved by the Space Shuttle Columbia. In its 69 months in space, it provided scientific data on the long-term effects of space exposure on materials, components and systems that have benefited spacecraft designers.

The results of high vacuum influence the following: (i) Outgassing, the direct phase change from solid to vapour (sublimation), which removes material from a designed location that could condense on a cooler surface—could be serious if the

Table 1.4 Temperatures in ° K for different given sublimation rates in materials

Element	0.1 μm/year	10 μm/year	1000 μm/year
Cd	311	350	395
Zn	344	400	450
Mg	383	444	506
Au	933	1073	1223
Ti	1193	1343	1523
Mo	1653	1903	2173
W	2143	2423	2753

Based on Fortescue et al. (2011)

condensation is on optics or a solar array! It is important to note that sublimation rates are different for Cd, Zn, Mg, Au, Ti, Mo and W (Tungsten). These rates are summarized in Table 1.4.

High vacuum also influences (ii) material's strength especially when surface conditions are critical either when (a) surface cracks absorb gases or due to (b) surface oxidation. For example: Glass is ~3× stronger in vacuum than in air. Note: Fatigue strength is also influenced and is generally increased. High vacuum along with UV radiation influences iii) plastics and lubricants resulting in the evaporation of low-molecular-weight fractions (e.g. contaminants, additives, absorbed gases, etc.), resulting in a change in rigidity or strength or electrical characteristics or optical properties along with a small but significant (~3%) loss of weight. In the case of lubricants, there is a loss of volatile components leading to an absence of oxygen or an absence of water vapour. One possible solution is to use solid lubricants, e.g. molybdenum di-sulphide (MoS_2). Electric and magnetic fields can have a significant effect on motion of charged particles resulting in forces on charged particles, electrons (negative charge) and ions that generally carry a positive charge, as well as some neutral particles. An electric field (vector) results in an electrostatic force, $\boldsymbol{F}_E = q\boldsymbol{E}$. A magnetic field results in a Lorentz force: $\boldsymbol{F}_B = q(\boldsymbol{v} \times \boldsymbol{B})$. In a magnetic field, particle rotates with an angular velocity $= \boldsymbol{\omega}_c = q\boldsymbol{B}/m_p$. This is the Larmor or cyclotron frequency resulting in a velocity, $\boldsymbol{v}_{perp} = \boldsymbol{\omega}_c\, r_i$ in a circular path of radius, r_i.

The plasma environment is driven by the Sun's solar wind and contains several high-velocity charged particles. It is most easily imaged using a solar coronagraph, which is often the only way of capturing the coronal mass ejections. The plasma environment can affect a spacecraft in several ways via surface charging, internal charging and galactic cosmic ray impact, trapped radiation, solar particle and ion sputtering, O+ erosion, plasma-induced scintillation and plasma-induced wave refraction. Two primary effects of exposure to radiation are radiation damage of a spacecraft and the induced radioactivity in the spacecraft. To understand the radiation damage to materials, electrons, in the plasma, must be considered differently from protons and high Z particles (subatomic neutral particles). Electrons lead to long-term degradation in the efficiency of semiconductor devices due to the absorption of trails of light ionization tracks through the device. This directly leads to the reduced efficiency of solar cells and to increased noise in circuits that may be sensitive to noise. As already mentioned, AO (dominant species at orbital altitudes)—chemically enhanced sputtering due to chemical activity and high impact speed (8 km/s)—leads to erosion of materials such as Kapton. It can also change the properties of other polymeric materials (further protection is possible using fluorinated surface treatments). Outgassing of materials such as plastics and some other materials due to the vapour pressure of materials in high vacuum is a distinct possibility. Particular problems could arise for materials including some polymers and metals such as beryllium, with the removal of different molecular weight components (especially with large molecular weight components being removed preferentially). Damage to materials in a space environment following exposure to UV or solar radiation results in the degradation of plastic materials, primarily due to the photochemical dissociation of long-chain polymers. It results in changes to both optical properties and electrical properties and can significantly damage electronic circuits. Protons and high Z particles lead to significant local damage in a circuit with the potential for single-event upsets due to substantial charge generated in the circuit. This requires error-tolerant software and error correction codes. Generally, protection is provided by screening all electronics (e.g. cover-glass on solar cells).

The principal influence of radiation damage is due to charged particles that disrupt semiconductor structure leading to the introduction of defects. This results in a loss of photovoltaic (PV) cell efficiency—a loss of ~25% after 10 years in GEO. Thus, there is a requirement to oversize the solar array at the beginning of life (BOL) to compensate for these losses over the lifetime of the array. The deposition of charge results in several types of errors in semiconductor components, including date errors in "scientific" satellites, communication satellites, latch-up errors leading to the burn-out of a component and inadvertent switch on/off of systems. These are generally called single-event upset of circuits and require hardening of electronics if they are to be avoided. The high radiation levels often result in the growth of "whiskers", which is particularly problematic in components containing cadmium and lithium.

There are various solutions to radiation damage, including hardware solutions, the use of radiation-hardened components (often military standard), the screening of components, software solutions and error tolerance and error correction coding.

1.1.4 The Thermal Environment and Thermal Control

A key feature of the thermal environment in space is the total absence of convection of the entire range of pressures (ultra-vacuum 10^{-14} bar $< p < 10^{-17}$ bar) that one could experience for satellites in space. Direct sunlight is the main source of heating in space, and it is conveyed by radiation. The temperatures in deep space could reach as high as 3000 °K for satellites, although these temperatures are not steady and there are indeed thermal cycles and significant variations in temperature. This is due to the rapidly changing illumination conditions as well as other local conditions, such as solar eclipse, planetary exploration and other such issues. The solar flux diminishes as $1/d^2$ (very high heat fluxes for solar orbiters, very low for far away missions) with dramatically changing illumination conditions for different space probes. Atmospheric conditions on other planets during planetary exploration could vary quite significantly. A spacecraft's thermal control system is usually designed for maintaining a spacecraft's thermal environment while on a prescribed orbit. An Earth-orbiting spacecraft's thermal environment is determined primarily by the incoming radiation from the Sun (60%), the Albedo (30%, reflected radiation from the Sun) and the direct infrared radiation from the Earth or earthshine (10%). The losses are only due to the re-radiation of thermal energy by the satellite. The black body radiation curves for the Sun and the Earth reveal a range of wavelengths. However, there is very little overlap between these ranges and the peak radiation, although the curves indicating the distribution of the radiation for different wavelengths are similar. In fact, the wavelength at which the radiation is a peak for the Earth is about 20 times the corresponding peak wavelength for the Sun. The primary objective of a thermal control system is to maintain all spacecraft and payload components within their required temperature limits over the entire mission. Therefore, there is a need to set various limits on the temperatures within the spacecraft's environment. This would include operational limits, survival limits and gradient limits. It is important to recognize that the thermal control of the temperatures within the spacecraft's environment could be accomplished by active or passive means. Passive thermal control relies on conductive and radiative heat paths created by devices such as radiators and insulators, which require no electrical or mechanical input. Active thermal control relies on pumps, thermostats and moving parts and essentially requires electrical power.

1.1.5 Space Debris

Debris is generated through in-orbit activity such as launch. The study of debris generation includes the mechanisms of generation, such as the separation of stages leading to the discarding of structures in space as well as other motor casings left in orbit. Furthermore, the firing of solid rockets can cause "clag" from exhausted Al-Ox particles.

Most space operations could finally result in debris generation with several hundred end-of-life spacecraft and rocket breakups contributing to the debris in space. The continuous evolution of the obits, which sometimes require high-impact velocities, is often responsible for debris generation. Some orbits may become inoperable due to a continuous cascade of debris. International law and legislation are now in place for end-of-life de-orbiting of a spacecraft and possible re-entry into a LEO orbit and the lower atmosphere, where debris could either burn up or fall back onto the Earth's surface. Spacecraft in GEO orbit must be moved into a graveyard orbit well beyond the GEO orbit, so they do not prove to be hazardous to another GEO spacecraft within the crowded GEO orbit region. Debris in space is ever-increasing even if 20% is automatically removed! Space debris mitigation is required to minimize particular creation of debris, the deliberate deorbiting of spacecraft in LEO orbits, removal of GEO satellites from the GEO orbital domain, the venting of upper stages after operation after payload delivery to reduce explosions in space as well as disastrous effects of collisions with them. Space debris prevention measures are mandatory to preclude explosive failures of spacecraft and upper stages, which include elimination of the placement of space debris objects in outer space and removal procedures which require reducing the number and mass of objects, thereby reducing the probability and severity of on-orbit hypervelocity collisions.

There are a number of design guidelines for space debris mitigation. Spacecraft and orbital stages should be designed not to release debris during normal operations. The potential for break-ups during all phases of the mission should be minimized. Spacecraft or orbital stages that are terminating their operational phases in orbits and that pass through the LEO region should be deorbited or, where appropriate, manoeuvred into an orbit with a reduced lifetime (25 years is a reasonable lifetime limit). If a spacecraft or orbital stage is to be disposed of by re-entry into the atmosphere, it should be ensured that debris that survives to reach the surface of the Earth should not pose an undue risk. Missions should estimate and limit the probability of accidental collisions during the orbital lifetime. Collision avoidance manoeuvres for spacecraft and coordination of launch windows should be considered and incorporated in the design. This would require maintaining an additional fuel provision for these operations.

Satellites in GEO and higher orbits, which are too far away to renter the atmosphere, should be moved into a graveyard orbit well outside the region used by active satellites. The aim is to create and maintain a protected zone of a few hundred kilometres on either side of the GEO orbit.

1.1.6 The Launch Environment

The launch environment is characterized by extreme vibrations due to the existence of an intense acoustic field, shock and high temperatures. Thus, requirements are imposed, so these characteristics are maintained within limits.

The acoustic field is generally maximum at the time when the launcher is just released and when it reaches transonic speeds. It could be as high as ~160 dB for 60 s (relative to a pressure of 2×10^{-5} N/m^2). These are usually accepted as the maximum limits. The acceleration itself depends on the launch vehicle used and is of the order of 8–12 g in the case of the Saturn V/Apollo launch vehicle, under 3 g for the Shuttle and about 4 g for the Ariane family.

The shocks are created during the initiation and continuation of combustion and during the separation of stages. Solar cells and other electronics are designed to withstand 2500 g for as long as a millisecond or two. There are also extensive thermal requirements that must be met during the launch phase.

1.2 Orbital Motion: The Two-Body Problem and Kepler's Laws

The understanding of the motion of planetary bodies evolved over many centuries. The early view of planetary motion was that it was completely Earth-centred. However, Nicolaus Copernicus (1473–1543) introduced the idea that planetary motion was heliocentric, although he did believe that planets orbited the Sun in circles. Although Tycho Brahe (1546–1601) reverted to the idea of a geocentric model, he was indeed a keen observer, and he recorded his observations extensively. It was Galileo Galilei (1564–1642) who firmly believed in the heliocentric model. However, Johannes Kepler (1571–1630), a researcher who worked closely with Tycho Brahe and had access to his recorded observations, established the heliocentric model of planetary motion on a firm footing by introducing his three laws of planetary motion. Kepler's first law states that each planet revolves around the Sun in an elliptical path, with the Sun occupying one of the foci of the ellipse. Kepler was able to show that with Sun at one of the foci, every planet traces an elliptic path around the centre of the ellipse as shown in Fig. 1.3. Kepler's second law states that the straight line joining the Sun and a planet sweeps out equal areas in equal intervals of time. Kepler's remarkable third law relates the squares of the planets' orbital periods to the cubes of the semi-major axes of their orbits, which are stated to be proportional.

According to Kepler's first law, the equation for the orbit is given by

$$r(\theta) = \frac{a(1 - e^2)}{1 + e\cos(\theta)}. \tag{1.6}$$

The distance a is known as the semi-major axis and e is the eccentricity. The point of closest approach ($\theta = 0, r = a(1 - e)$) is known as the periapsis, while the point farthest from the focus ($\theta = \pi$, $r = a(1 + e)$) is the apoapsis. The distance from the centre to the focus is the product of the semi-major axis and the eccentricity ae. The

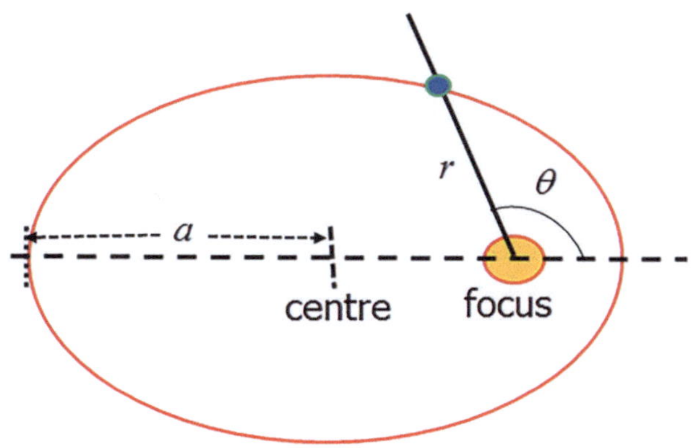

Fig. 1.3 Kepler's first law

moment of momentum of the planet $mr^2d\theta/dt$ is conserved and hence a constant. Thus, the quantity $r^2d\theta/dt$ is labelled as h and is always a constant. In fact, it can be shown that $h^2 = \mu a(1 - e^2) = \mu p$, where p is known as the semi-latus rectum of the ellipse. It can be shown that when a planet traverses an infinitesimal angle $d\theta$ in a time interval dt, the area swept by the planet is approximately equal to half of $r^2d\theta$. Consequently, the areal velocity is half of $r^2d\theta/dt$ which is nothing but $h/2$ and constant, thus proving Kepler's second law. According to Kepler's third law,

$$a^3 = \frac{G(M + m)}{4\pi^2}T^2. \tag{1.7}$$

The quantity T is the orbital period of the satellite, a is the semi-major axis, G is the gravitational constant, M is the mass of the "Sun" and m is the mass of the planet. In celestial mechanics, this law provides the timescales of the solar system. One can observe rather precisely the orbital periods of planets, and from this information, one can infer the scale of the solar system. All distances are then normalized by the Earth's orbital radius, which is defined as 1 astronomical unit (1 AU). The quantity $\mu = G(M + m)$ is a constant, uniquely associated with the central and orbiting body, and when $m < < M$, it can be approximated as $\mu = GM$.

1.2.1 Elliptic, Parabolic and Hyperbolic Orbits

Equation (1.6) is the equation of an ellipse for the eccentricity $e < 1$, the equation for a parabola for $e = 1$ and an equation for a hyperbola for $e > 1$. When $e = 0$, it reduces to path of constant $r(\theta)$ for all θ and hence traces a circular path. This aspect is illustrated in Fig. 1.4.

1.2.2 Newton's Law of Gravitation

The onus of proving Kepler's laws mathematically fell upon Sir Isaac Newton (1643–1727), and he formulated his celebrated law of gravitation and invented differential calculus for doing so. Newton's law of gravitation states that when a particle, on mass, m moves in the vicinity of celestial body, it experiences a force of attraction, which is directed towards a fixed centre of attraction, located at the centre of mass of the attracting central body. The force is mutual in the sense that the attracting central body experiences a force directed towards the centre of mass of the particle of the same magnitude.

This force acting on an orbiting body of mass m, modelled as a point mass with all its mass concentrated at a point, is given by Newton's law of gravitation as

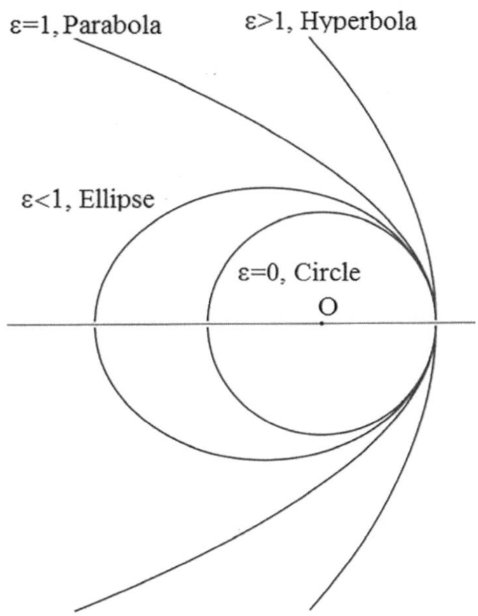

Fig. 1.4 Orbits with different eccentricities

$$F = -G\frac{mM}{r^2} \tag{1.8}$$

where M is the mass of the attracting, central body, G is the universal gravitational constant and r is the distance between the centres of masses of the two bodies. As the force is always directed to the central body, the force normal to the central direction is necessarily zero, and consequently, the momentum in this direction h is conserved. Assuming that that central, attracting body is fixed in space, the equation of motion of the orbiting body is

$$-G\frac{Mm}{r^2} = m\left(\ddot{r} - r\dot{\theta}^2\right). \tag{1.9}$$

Although Eq. (1.9) is apparently nonlinear, it can be transformed to a linear equation by using a transformation, $u = 1/r$, and the conic solution given by Eq. (1.6) follows. When expressed in terms of the potential function of the gravitational force due to a central particle $U(r)$, Eq. (1.9) may be expressed as

$$m\left(\ddot{r} - r\dot{\theta}^2\right) = \nabla\left(G\frac{Mm}{r}\right) = -\nabla U(r) \tag{1.10}$$

where $\nabla()$ is the gradient vector of the potential function. Equation (1.10) is valid even when the body is no longer a point mass, although the expression for the potential function $U(r)$ is given by a general expression.

The equation is valid. In case the central body is not fixed in space, then it can be shown that the relative position vector **r**, of the orbiting body, relative to the central body satisfies

$$\ddot{\mathbf{r}} + \left(\frac{G(M+m)}{r^2}\right)\left(\frac{\mathbf{r}}{r}\right) = 0. \tag{1.11}$$

This is the celebrated equation for two-body relative motion. Combining Newton's law of gravitation given by Eq. (1.8) and Newton's second law, the equations of motion of a system of particles in a space-fixed Cartesian frame experience only the gravitational force between the particles, given by

$$m_j \frac{d^2\mathbf{r}_j}{dt^2} = -G\sum_{k\neq j} \frac{m_j m_k\left(\mathbf{r}_j - \mathbf{r}_k\right)}{r_{jk}}, \quad j=1,2,\cdots,N. \tag{1.12}$$

1.2.3 Keplerian Model, Orbital Elements

In addition to the semi-major axis a, the eccentricity e and the angle between the direction of the periapsis and the direction of location of the orbiting body from the focus where the central body is located will henceforth be denoted as ν and known as the true anomaly; to distinguish it from the generic polar coordinate θ, Kepler defined three Euler angles, belonging to 3-1-3 sequence of Euler angles, to define the orientation of the orbit plane in space. Thus, the semi-major axis defines the size of the orbit, the eccentricity defines the shape of the orbit, the true anomaly defines the location of the orbiting body in its trajectory and the Euler angle sequence defines the orientation of the orbit plane relative to the plane fixed in space. Of the two nodes on the line of nodes on a reference plane, one is the ascending node, where the satellite crosses it going from the south to the north of the reference plane. The other is the descending node, where the satellite crosses the reference plane from the north to the south. The argument of the ascending node, also referred to as the right ascension of the ascending node (RAAN or Ω), is the first of the three Euler angles in the 3-1-3 sequence that orients the orbital plane in space. Once the orbital inclination, the second Euler angle, is defined, there are an infinite number of orbits possible in an orbital plane. Once the orbital plane is oriented in space, it is essential to orient the orbit ellipse within the orbital plane. This is done by a single angular element known as the argument of perigee. Thus, the three Euler angles uniquely determine the orbit within the orbit plane and are the argument of the ascending node (Ω) from a reference direction fixed in space and the line of nodes from the descending to the ascending node; the inclination angle of the orbit plane (i), which represents a rotation of the space-fixed plane about the line nodes; and finally, the argument of the periapsis (ω), which represents the angular position of the periapsis from the

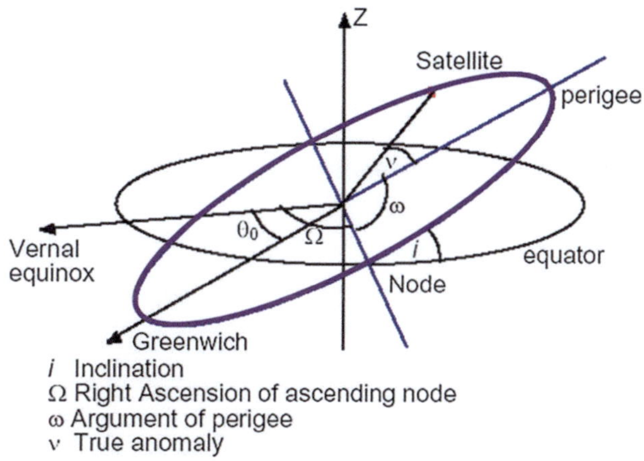

Fig. 1.5 Keplerian angular elements for a typical Earth-orbiting satellite

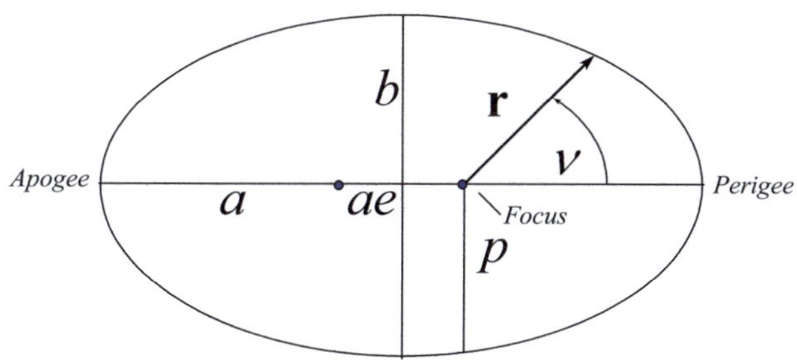

Fig. 1.6 Definition of the semi-major axis, the semi-minor axis, the eccentricity, the semi-latus rectum, the true anomaly, the apogee and the perigee

ascending node in the orbit plane. Together, the six elements, a, e, ν, Ω, i, ω, represent the *classical orbital elements* that were used by Kepler to locate the orbiting body in space and are also known as the Keplerian orbital elements. Four of these are angles which are illustrated in Fig. 1.5.

By convention, we specify the location of the ascending node by the "right ascension of ascending node" which is an angle, measured at the centre of the planet, from the vernal equinox, an apparently fixed but slowly moving reference point in the sky (known as the First Point in Aries, as it was located in the constellation Aries in the year, 150CE) where the right ascension to the ascending node is defined to be zero. It is, an angle measured in the reference plane from the vernal equinox, which is characterized by an equal length in time of the day and night, every year, on the 21st of March.

So far, the four orbital elements define the orientation and location of the orbit in the orbital plane, the orientation of the orbit ellipse in the orbital plane and the precise location of the satellite in the orbit. The size and shape of the orbit ellipse are determined by the semi-major axis and the eccentricity, which are illustrated in Fig. 1.6, which refers specifically to an earth-orbiting ellipse where the closest and farthest points in the orbit to the central body, the Earth, are respectively known as the perigee and apogee.

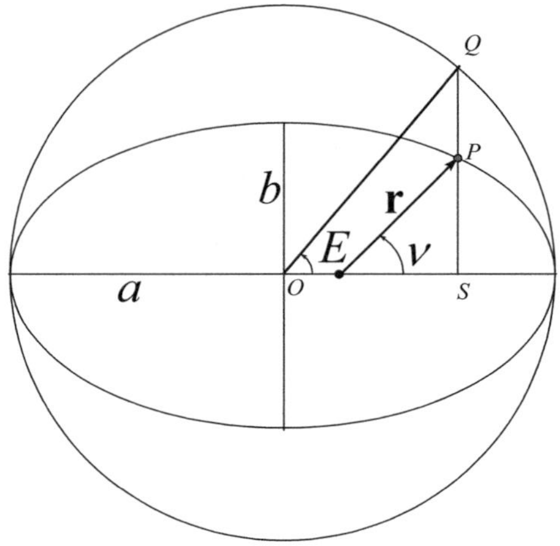

Fig. 1.7 Definition of the eccentric anomaly

The mean rate of motion or angular speed of the orbiting body is usually given in units of revolutions per day. Kepler's third law of orbital motion gives us a precise relationship between the speed of the satellite and the size of the orbit. So, by specifying the speed of the satellite or its mean rate of motion, it is possible to define the size of the orbit. The mean motion or mean anomaly is then defined as the product of the mean rate of motion and the time of passage from the periapsis or the perigee of the orbit. The mean anomaly (M) is simply an angle that marches uniformly in time from 0° to 360° during one revolution. It is defined to be 0° at the perigee or periapsis and as 180° at the apogee or apoapsis. It is related to the true anomaly (ν) and is often employed as an alternate element to ν. One of Kepler's finest contributions was to establish its relationship to the true anomaly. However, to do this, it was essential to define the eccentric anomaly, E. The definition, which uses a superscribed circle and the projection of the orbit body onto it to precisely define the eccentric anomaly, E, is illustrated in Fig. 1.7.

Kepler's equation relates the *eccentric anomaly*, E, to the *mean anomaly*, M, which is defined in terms of the time of flight from the periapsis, $t - t_p$, and the orbit period T, or the mean orbital rate n. Kepler's equation is

$$M = E - e\sin E,\ \ M = \left(t - t_p\right)\sqrt{\mu/a^3} = 2\pi\left(t - t_p\right)/T = n\left(t - t_p\right) \tag{1.13}$$

Then, the time of flight between two points on an orbit with prescribed eccentric anomalies, from E_0 to E_1, is

$$tof = (M_1 - M_0)/n = \sqrt{a^3/\mu}\{(E_1 - e\sin E_1) - (E_0 - e\sin E_0)\} \tag{1.14}$$

Using the fact that $r = a(1 - e\cos E)$, one could derive Gauss's equations relating the eccentric anomaly, E, to the true anomaly, ν,

$$\cos\nu = \frac{\cos E - e}{1 - e\cos E}, \quad \sin\nu = \frac{\sin E\sqrt{1-e^2}}{1 - e\cos E},$$
$$\cos E = \frac{\cos\nu + e}{1 + e\cos\nu}, \quad \sin E = \frac{\sin\nu\sqrt{1-e^2}}{1 + e\cos\nu}, \quad \tan\frac{E}{2} = \sqrt{\frac{1-e}{1+e}}\tan\frac{\nu}{2}. \tag{1.15}$$

1.2.4 Definition of a Keplerian Orbit

The definition and calculation of a Keplerian orbit is achieved by relating the position and velocity vectors of an orbiting body to its Keplerian elements. Recall that the equation of relative motion is given by the equations

$$\ddot{\mathbf{r}}_1 = \mathbf{F}_1/m_1, \quad \ddot{\mathbf{r}}_2 = \mathbf{F}_2/m_2, \quad \ddot{\mathbf{r}} = \ddot{\mathbf{r}}_2 - \ddot{\mathbf{r}}_1 = -\left(\frac{G(m_1+m_2)}{r^2}\right)\left(\frac{\mathbf{r}}{r}\right), \tag{1.16}$$

so

$$\ddot{\mathbf{r}} + \frac{G(m_1+m_2)}{|\mathbf{r}|^2}\frac{\mathbf{r}}{|\mathbf{r}|} = \ddot{\mathbf{r}} + \frac{\mu}{|\mathbf{r}|^2}\frac{\mathbf{r}}{|\mathbf{r}|} = \mathbf{0}, \tag{1.17}$$

where $\mu = G(m_1 + m_2) \approx Gm_2$ when m_2, the mass of the orbiting body, is very small in comparison to m_1 which is assumed to be the mass of the central body. It should be noted that in the *Système International* (SI) units, the units of μ are $Nm^2/Kg = m^3/s^2$. It then follows that

$$\mathbf{r}\times\dot{\mathbf{r}} = \mathbf{r}\times\mathbf{v} = \mathbf{h}, \mathbf{r}\times\ddot{\mathbf{r}} = \dot{\mathbf{h}} = \mathbf{0}. \tag{1.18}$$

Integrating the cross product of Eq. (1.17), with **h**, using the relations $\mathbf{r}\cdot\dot{\mathbf{r}} = r\dot{r}$ and

$$\begin{aligned}\ddot{\mathbf{r}}\times\mathbf{h} &= -\frac{\mu}{r^3}(\mathbf{r}\times\mathbf{h}) = -\frac{\mu}{r^3}(\mathbf{r}\times(\mathbf{r}\times\dot{\mathbf{r}})) = \frac{\mu}{r^3}(\dot{\mathbf{r}}(\mathbf{r}\times\mathbf{r})\text{-}\mathbf{r}(\mathbf{r}\times\dot{\mathbf{r}})) \\ &= \mu\left(\frac{\dot{\mathbf{r}}r}{r^2} - \frac{\mathbf{r}\dot{r}}{r^2}\right) = \mu\frac{d}{dt}\left(\frac{\mathbf{r}}{r}\right)\end{aligned} \tag{1.19}$$

it follows that

$$\dot{\mathbf{r}} = \mathbf{v} = \frac{\mathbf{h}}{h^2}\times\mu\left(\frac{\mathbf{r}}{|\mathbf{r}|} + \mathbf{e}\right). \tag{1.20}$$

Taking the cross product of Eq. (1.19), with **h**,

$$\mathbf{v}\times\mathbf{h}=\mu\left(\frac{\mathbf{r}}{|\mathbf{r}|}+\mathbf{e}\right)\quad\Rightarrow\quad\mathbf{e}=\frac{\mathbf{v}\times\mathbf{h}}{\mu}-\frac{\mathbf{r}}{|\mathbf{r}|}\tag{1.21}$$

But,

$$\dot{\mathbf{r}}\times\mathbf{h}=\dot{\mathbf{r}}\times(\mathbf{r}\times\dot{\mathbf{r}})=\mathbf{r}(\dot{\mathbf{r}}\times\dot{\mathbf{r}})\text{-}\dot{\mathbf{r}}(\dot{\mathbf{r}}\times\mathbf{r})=\mathbf{r}v^2-\mathbf{v}(\mathbf{v}\times\mathbf{r}).\tag{1.22}$$

Hence,

$$\mathbf{e}=\frac{1}{\mu}\left(\mathbf{r}v^2-\mathbf{v}(\mathbf{v}\times\mathbf{r})\right)-\frac{\mathbf{r}}{|\mathbf{r}|}.\tag{1.23}$$

The magnitude of the vector constant, $\mathbf{e}$, is defined as the eccentricity, e.

By applying Pythagoras theorem, it follows that the tangential component and the radial component of the moment of momentum satisfy

$$r^2v^2=h^2+(\mathbf{v}\cdot\mathbf{r})^2.\tag{1.24}$$

All the six orbital elements may now be found from the position and velocity vectors. Thus, given $\mathbf{r}$ and $\mathbf{v}$ construct the moment of the momentum vector $\mathbf{h}=\mathbf{r}\times\mathbf{v}\equiv[h_x\ h_y\ h_z]$. The inclination and the argument of the ascending node or the RAAN are, respectively, given by

$$i=\cos^{-1}(h_z/|\mathbf{h}|),\ \ \Omega=\tan^{-1}\left(-h_x/h_y\right).\tag{1.25}$$

The line nodes are given by

$$\mathbf{n}=\mathbf{k}\times\mathbf{h}=[-h_y\quad h_x\quad 0]/h,\ \ h=|\mathbf{h}|.\tag{1.26}$$

The eccentricity vector and the eccentricity are, respectively, given by

$$\mathbf{e}=\frac{\mathbf{v}\times\mathbf{h}}{\mu}-\frac{\mathbf{r}}{|\mathbf{r}|},\ \ e=|\mathbf{e}|.\tag{1.27}$$

The argument of the periapsis or the perigee is obtained from

$$\omega=\cos^{-1}\left(\frac{\mathbf{e}}{|\mathbf{e}|}\frac{\mathbf{n}}{|\mathbf{n}|}\right).\tag{1.28}$$

Thus, the true anomaly ν is obtained from

$$r = \frac{a(1-e^2)}{1+e\cos\nu} = \frac{p}{1+e\cos\nu} = \frac{\mu h}{1+e\cos\nu} \Rightarrow \nu = \cos^{-1}\left(\frac{\mu h - r}{er}\right). \tag{1.29}$$

Having obtained all the angular elements, the semi-major axis is given by

$$a = \frac{p}{1-e^2} = \frac{\mu h}{1-e^2}. \tag{1.30}$$

Thus, all the six *classical orbital elements*, a, e, ν, Ω, i, ω, are obtained.

To obtain the position and velocity vectors from the six *classical orbital elements*, a, e, ν, Ω, i, ω, the magnitude of the position vector is obtained from Eq. (1.29). The tangential velocity component is obtained from

$$v_\theta = h/r = \mu(1+e\cos\nu)/h. \tag{1.31}$$

The radial velocity component is obtained from the flight path angle ϕ by

$$v_r = v_\theta \tan\phi = (\mu/h)e\sin\nu, \quad \tan\phi = \frac{v_r}{v_\theta} = \frac{\dot{r}}{r\dot{\nu}} = \frac{e\sin\nu}{1+e\cos\nu}. \tag{1.32}$$

The magnitude of the position vector and the tangential and radial velocity components give the position and velocity of the orbiting body in a central body-centric frame of reference. These must now be transformed to space-fixed coordinates and that can be done using the fact the Keplerian orbital plane is defined by a 3-1-3 Euler angle sequence. For the 3-1-3 Euler angle sequence, the first rotation of the space-fixed frame is about the third axis by the angle Ω, the second about the first axis by the inclination angle and the third rotation is again about the third axis by the satellite's longitude $\theta = \omega + \nu$. The transformation from the space-fixed frame to the central body-centric orbit frame for the 3-1-3 Euler angle sequence is

$$\begin{bmatrix} x_o \\ y_o \\ z_o \end{bmatrix} = \begin{bmatrix} c\theta & s\theta & 0 \\ -s\theta & c\theta & 0 \\ 0 & 0 & 1 \end{bmatrix} \begin{bmatrix} 1 & 0 & 0 \\ 0 & ci & si \\ 0 & -si & ci \end{bmatrix} \begin{bmatrix} c\Omega & s\Omega & 0 \\ -s\Omega & c\Omega & 0 \\ 0 & 0 & 1 \end{bmatrix} \begin{bmatrix} x \\ y \\ z \end{bmatrix}. \tag{1.33}$$

Thus, the position vector is given by the inverse transformation, leading to the position vector which is

$$\mathbf{r} = \begin{bmatrix} c\Omega & -s\Omega & 0 \\ s\Omega & c\Omega & 0 \\ 0 & 0 & 1 \end{bmatrix} \begin{bmatrix} 1 & 0 & 0 \\ 0 & ci & -si \\ 0 & si & ci \end{bmatrix} \begin{bmatrix} c\theta & -s\theta & 0 \\ s\theta & c\theta & 0 \\ 0 & 0 & 1 \end{bmatrix} \begin{bmatrix} r \\ 0 \\ 0 \end{bmatrix}. \tag{1.34}$$

The inverse of the same transformation from the space-fixed frame to the orbit frame can be used for constructing the velocity vector. The inverse transformation leading to the velocity vector in the space-fixed frame is

$$\mathbf{v} = \begin{bmatrix} c\Omega & -s\Omega & 0 \\ s\Omega & c\Omega & 0 \\ 0 & 0 & 1 \end{bmatrix} \begin{bmatrix} 1 & 0 & 0 \\ 0 & ci & -si \\ 0 & si & ci \end{bmatrix} \begin{bmatrix} c\theta & -s\theta & 0 \\ s\theta & c\theta & 0 \\ 0 & 0 & 1 \end{bmatrix} \begin{bmatrix} v_r \\ v_\theta \\ 0 \end{bmatrix}. \tag{1.35}$$

Thus, the position and velocity vectors of the orbiting body can be uniquely determined from the six *classical orbital elements*, a, e, ν, Ω, i, ω.

The orbit determination problem has been discussed comprehensively by Bate, Mueller and White (1971), Vallado (2004) and Curtis (2005). The problem of determining the position of orbiting body in an orbit is discussed by Chobotov (2002).

1.3 Energy Considerations: The Vis-Viva Equation

It has already been indicated in Eq. (1.10) that the gravitational force can be obtained from a potential function, which is given by

$$U(r) = -m\left(\frac{\mu}{r}\right). \tag{1.36}$$

In Eq. (1.36), m is the mass of the orbiting body. Moreover, the kinetic energy of the orbiting body, which is assumed to be concentrated point mass, is given by

$$T_{KE} = \frac{m}{2}v^2. \tag{1.37}$$

As there no other forces to consider, and no non-conservative or dissipative forces acting on the orbiting body, in free space, the total energy in conserved. It follows that the total energy is conserved and one has

$$\frac{m}{2}v^2 - m\left(\frac{\mu}{r}\right) = \text{a constant}. \tag{1.38}$$

The constant in Eq. (1.38) is evaluated from the conditions at the point of closest approach or the point of farthest approach. At the point of closest approach or the periapsis, the orbiting body is at a distance equal to

$$r = a(1-e). \tag{1.39}$$

The radial velocity is zero and the tangential velocity is given by

$$v = h/r = h/(a(1-e)). \tag{1.40}$$

Since $h^2 = \mu a(1 - e^2) = \mu p$, $v^2 = h^2/(a^2(1 - e)^2) = \mu(1 + e)/(a(1 - e))$ and Eq. (1.38) reduces to

$$\frac{m}{2}v^2 - m\left(\frac{\mu}{r}\right) = \frac{m}{2}\frac{\mu}{a}\frac{1+e}{1-e} - m\frac{\mu}{a(1-e)} = -\frac{m}{2}\frac{\mu}{a}. \tag{1.41}$$

Equation (1.41) may be written as

$$\frac{v^2}{2} - \frac{\mu}{r} = -\frac{\mu}{2a}. \tag{1.42}$$

Equation (1.42) is the celebrated vis-viva equation, where the first term is the specific kinetic energy or the kinetic energy per unit mass of the orbiting body, the second is the specific potential energy and the quantity to the right is the total energy. A number of useful relations can be deduced from the vis-viva equation.

1.3.1 Escape Velocity, Circular Velocity

When the orbit is parabolic, the semi-major axis is infinite, and the orbiting body can never expect to complete the orbit around the central body. In this case, it can be deemed to have "escaped" from the gravitational field of the central body. In this case,

$$\frac{v^2}{2} - \frac{\mu}{r} = 0 \Rightarrow v = \sqrt{\frac{2\mu}{r}}. \tag{1.43}$$

Equation (1.43) is known as the escape velocity and is the minimum velocity beyond which the orbiting body will leave the gravitational field of the central body. In this case, the eccentricity $e = 1$ and $a = \infty$.

When the orbit is circular, $r = a$ and the vis-viva equation reduces to

$$\frac{v^2}{2} = \frac{\mu}{a} - \frac{\mu}{2a} = \frac{\mu}{2a} \Rightarrow v = \sqrt{\frac{\mu}{r}}. \tag{1.44}$$

Equation (1.44) is the circular velocity of the orbiting body when the orbit is indeed a circle, i.e. the eccentricity $e = 0$.

1.3.2 *Hyperbolic Orbits: The Turning Angle and the Excess Velocity*

It is also important to consider the case when the eccentricity $e > 1$. Hyperbolic orbits have a number of key features that distinguish them from elliptic orbits. Not only is the eccentricity of a hyperbolic orbit greater than unity $e > 1$ and the semi-major axis is negative $a < 0$, the energy constant in the vis-viva equation is positive, $\mathcal{E} > 0$, indicating that as the body in such an orbit is infinitely far away from the central body, it still has a finite amount of kinetic energy to continue on its trajectory. Furthermore, the hyperbolic orbit has a property that no other orbit has. A vehicle traveling the length of the orbit, while coming from some point at infinity, will arrive and then fly by through the closest approach point (periapsis) before leaving, going on to another point at infinity. The approach direction comes in from the direction of the approach velocity at infinity, denoted as $-v_\infty$, and the departure direction in the direction of the departure velocity at infinity, which is $+v_\infty$. The angle through which the vehicle turns, while approaching from one direction at infinity and departing in another direction at infinity, is called the *turning angle*. This turning angle can be determined from the properties of the hyperbola. Consequent to this behaviour, the flight path angle is not periodic, as in an elliptic orbit, but starts at either $\pm\pi/2$ and finishes at $\mp\pi/2$.

From the vis-viva equation at any point in the orbit,

$$\frac{1}{2}v^2 - \frac{\mu}{r} = -\frac{\mu}{2a} = \varepsilon \tag{1.45}$$

Along a parabolic escape trajectory, when $r \to r_\infty \to \infty$ and $v_\infty \to 0$,

$$\frac{1}{2}v^2 - \frac{\mu}{r} = \frac{1}{2}v_{\text{esc}}^2 - \frac{\mu}{r} = \frac{1}{2}v_\infty^2 - \frac{\mu}{r_\infty} = 0. \tag{1.46}$$

Hence, local escape velocity at its current position r is $v_{esc} = \sqrt{2\mu/r}$.
For a circular orbit, it has already been shown that

$$a = r, \ \ \frac{1}{2}v^2 - \frac{\mu}{a} = -\frac{\mu}{2a} = \varepsilon, \ \ v^2 = \frac{\mu}{a}. \tag{1.47}$$

Thus, one could also define a local circular velocity at its current position r, as $v_{\text{cir}} = \sqrt{\mu/r}$.

Along a hyperbolic orbit, however, $v_\infty > 0$, and it follows that

$$\frac{1}{2}v^2 - \frac{\mu}{r} = \frac{1}{2}v_\infty^2 - \frac{\mu}{r_\infty} = \frac{1}{2}v_\infty^2 \Rightarrow v_\infty^2 = v^2 - \frac{2\mu}{r}. \tag{1.48}$$

From the vis-viva equation,

$$v^2 = 2\varepsilon + \frac{2\mu}{r} \Rightarrow v^2 = 2\varepsilon + v_{\text{esc}}^2. \tag{1.49}$$

At any point very far from the central body (at infinity), as $r \to \infty$, the velocity approaches an asymptotic value. This can be shown by using the vis-viva equation $v^2 \to 2\varepsilon = v_\infty^2 - 2\mu/r_\infty = v_\infty^2$. Thus, for a hyperbolic orbit, the vis-viva equation itself is expressed as

$$v^2 = v_\infty^2 + \frac{2\mu}{r} = v_\infty^2 + v_{\text{esc}}^2 = v_\infty^2 + 2v_{\text{cir}}^2. \tag{1.50}$$

Hence, the quantity v_∞ is referred to as the *hyperbolic excess velocity* and is $v_\infty^2 = v^2 - v_{\text{esc}}^2 = 2\varepsilon = -\mu/a$. It is only a function of μ and the semi-major axis a. From the orbit equation, $r = p/(1 + e\cos\nu)$, $\cos\nu|_{r=\infty} \equiv \cos\nu_\infty$, $\cos\nu_\infty = -1/e$. But, by definition, the turning angle δ is the deflection angle of the velocity vector, as illustrated in Fig. 1.8, and is related to ν_∞, the true anomaly at infinity, by the relation $\nu_\infty = (\pi + \delta)/2$. Hence, the turning angle is only a function of the eccentricity, $\cos\nu_\infty = -\sin(\delta/2) = -1/e$. Thus, given the turning angle, the eccentricity is given by $e = 1/\sin(\delta/2)$. Finally, the velocity at the periapsis of the orbit is obtained from the vis-viva equation and the position of the orbiting body

$$\frac{1}{2}v^2 - \frac{\mu}{r} = \frac{1}{2}v_p^2 - \frac{\mu}{r_p} = \frac{1}{2}v_\infty^2 \Rightarrow v_p^2 = v_\infty^2 + \frac{2\mu}{r_p}. \tag{1.51}$$

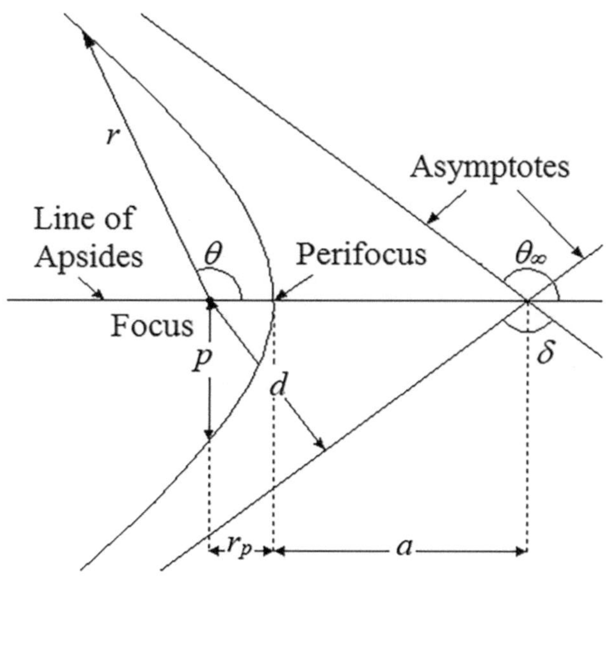

Fig. 1.8 Geometry relating turning angle to the true anomaly

Thus, the velocity at the periapsis, or the perifocus of the hyperbolic orbit, may be obtained from the hyperbolic excess velocity and the local escape velocity at the periapsis.

1.3.3 Application to Gravity Assist

The feature of a hyperbolic orbit in turning an approaching body with a velocity at infinity equal to v_∞, through a finite turning angle, which depends only on the eccentricity of the hyperbolic orbit, has an important application in turning a body approaching a planet in a hyperbolic orbit. The fact that the approaching body undergoes a finite change in its velocity vector, because of the change in the direction, can be interpreted as the velocity change provided by a planet during a fly-by due to its gravitational force, in terms of the hyperbolic excess velocity v_∞ and the turning angle δ. The velocity change due to the gravitation force or the gravity assist, during a fly-by or a swing-by, provided by a planet is

$$\Delta v = 2v_\infty \sin\left(\frac{\delta}{2}\right). \tag{1.52}$$

Thus, comparing with the local circular velocity at the perifocus of the hyperbolic orbit,

$$\frac{\Delta v}{v_{\text{cir},p}} = 2\frac{v_\infty}{v_{\text{cir},p}} \sin\left(\frac{\delta}{2}\right). \tag{1.53}$$

Since the local circular velocity is given by $v_{\text{cir},p} = \sqrt{\mu/r_p}$, it follows that

$$v_\infty^2 = -\mu/a = -r_p v_{\text{cir},p}^2/a, \;\; a = -r_p v_{\text{cir},p}^2/v_\infty^2. \tag{1.54}$$

Thus, the relationship between the eccentricity and the position of the perifocus may be expressed as

$$e = 1 - r_p/a = 1 + r_p v_\infty^2/r_p v_{\text{cir},p}^2 \equiv 1 + \psi^2, \quad \psi = v_\infty/v_{\text{cir},p}. \tag{1.55}$$

Thus, the relationship between the eccentricity and turning angle is

$$\sin(\delta/2) = 1/e = 1/(1+\psi). \tag{1.56}$$

Thus, Eq. (1.55) may be expressed as

$$\frac{\Delta v}{v_{\mathrm{cir},p}}=\frac{2}{1+\psi^2}\frac{v_\infty}{v_{\mathrm{cir},p}}=\frac{2\psi}{1+\psi^2},\quad \psi=\left(\frac{v_\infty}{v_{cir,p}}\right). \tag{1.57}$$

Maximizing the left-hand side of Eq. (1.57),

$$\begin{aligned}\frac{\partial}{\partial\psi}\left(\frac{\Delta v}{v_{\mathrm{cir},p}}\right)=\frac{\partial}{\partial\psi}\left(\frac{2\psi}{1+\psi^2}\right)&=\frac{2}{1+\psi^2}-\frac{4\psi^2}{\left(1+\psi^2\right)^2}\\&=\frac{2+2\psi^2-4\psi^2}{\left(1+\psi^2\right)^2}=\frac{2-2\psi^2}{\left(1+\psi^2\right)^2}.\end{aligned} \tag{1.58}$$

Hence, for maximizing $\Delta v/v_{\mathrm{cir},\,p}$, $\psi=1$. It follows that the maximum obtainable Δv from a gravity assist manoeuvre is given by

$$\left.\frac{\Delta v}{v_{\mathrm{cir},p}}\right|_{\max}=1\Rightarrow \Delta v|_{\max}=v_{\mathrm{cir},p},\quad \text{when}\quad v_\infty=v_{\mathrm{cir},p}. \tag{1.59}$$

1.3.4 Lambert's Problem and Its Solution

Given two points on an orbit at $\vec{r}_1$ and $\vec{r}_2$ and the included angle θ and the time-of-flight (TOF) $t_f=\Delta t$, the problem is to find the *orbital elements of the elliptic orbit* passing through the two points, as shown in Fig. 1.9 with the periapsis at P and apoapsis at A, and $\vec{v}_1$, $\vec{v}_2$, *the velocities at the two ends of the given orbital*

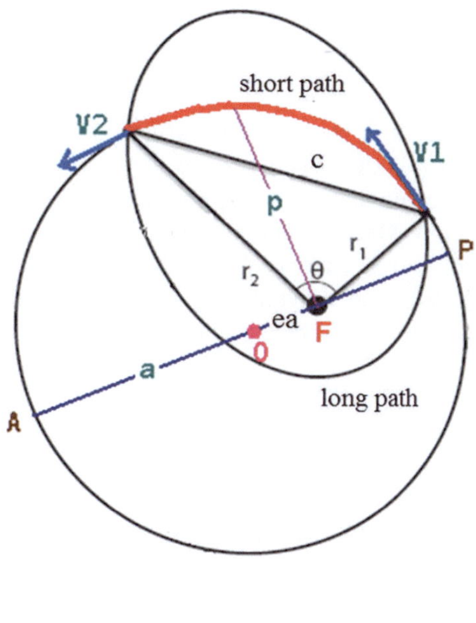

Fig. 1.9 Lambert's problem

segment. There are a family of orbits passing through the two points, but only one of them can be achieved in the given TOF.

The transfer angle for a prograde/retrograde orbit could be determined from the following:

$$\theta = \cos^{-1}\left(\frac{\vec{r}_1 \cdot \vec{r}_2}{r_1 r_2}\right), \text{if } \left(\vec{r}_1 \times \vec{r}_2\right)_k \geq 0, \theta = 2\pi \\ - \cos^{-1}\left(\frac{\vec{r}_1 \cdot \vec{r}_2}{r_1 r_2}\right), \text{if } \left(\vec{r}_1 \times \vec{r}_2\right)_k < 0, \tag{1.60}$$

for prograde orbits and

$$\theta = \cos^{-1}\left(\frac{\vec{r}_1 \cdot \vec{r}_2}{r_1 r_2}\right), \text{if } \left(\vec{r}_1 \times \vec{r}_2\right)_k < 0, \theta = 2\pi \\ - \cos^{-1}\left(\frac{\vec{r}_1 \cdot \vec{r}_2}{r_1 r_2}\right), \text{if } \left(\vec{r}_1 \times \vec{r}_2\right)_k \geq 0, \tag{1.61}$$

for retrograde orbits.

The time to traverse an arc between two points is related to the transfer orbit by *Kepler's* TOF equation and is only a function of $r_1 + r_2$, the semi-major axis, a, and

$$c = \sqrt{r_1^2 + r_2^2 - 2r_1 r_2 \cos\theta}. \tag{1.62}$$

It is given by Lambert's equation, which could be derived from Kepler's equation, and is

$$\sqrt{\frac{\mu}{a^3}}\Delta t = \alpha - \beta - (\sin\alpha - \sin\beta), \quad \sin\frac{\alpha}{2} = \pm\sqrt{\frac{s}{2a}}, \quad \sin\frac{\beta}{2} = \pm\sqrt{\frac{s-c}{2a}}, \tag{1.63}$$

where s is the semi-perimeter and satisfies $2s = r_1 + r_2 + c$ with the following: $0 \leq \alpha \leq 2\pi$, $0 \leq \beta \leq \pi$ for $\theta \leq \pi$, $0 \leq \alpha \leq 2\pi$, $-\pi \leq \beta \leq 0$ for $\theta \geq \pi$. It must be observed that there is a minimum value of $a = a_m$, which is half the semi-perimeter below which a solution does not exist as the orbiting body has insufficient energy to get across. There is also a minimum value for the eccentricity which is given by the ratio of the difference in lengths between $\vec{r}_1$ and $\vec{r}_2$ and c. Thus, Lambert's equation could be written as

$$\sqrt{a^3} = \frac{\sqrt{\mu}\Delta t}{\alpha - \beta - (\sin\alpha - \sin\beta)}, \tag{1.64}$$

and solved for a, iteratively. The methods of solution of Lambert's equation are discussed by Prussing and Conway (1993).

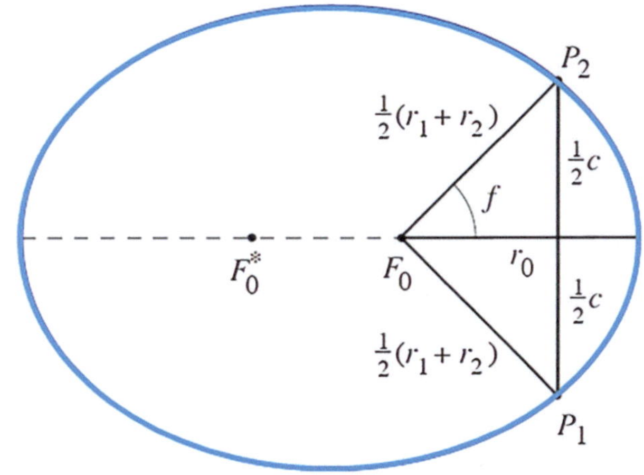

Fig. 1.10 The transformed orbit

Given the TOF formula, the orbit can be transformed to any shape desired as long as the $r_1 + r_2$, c and a are held constant. The transformed orbit is shown in Fig. 1.10. Lambert defined two non-dimensional parameters that are functions only of the problem geometry. The first is a non-dimensional parameter known as Lambert's parameter, λ,

$$\lambda = \sqrt{\frac{r_1 r_2}{s}} \cos\frac{\theta}{2} = \pm\sqrt{\frac{s-c}{s}}, \quad -1 < \lambda < 1. \tag{1.65}$$

The sign ambiguity in the square root can be resolved as $\lambda > 0$, $0 < \theta < \pi$ and $\lambda < 0$, $\pi < \theta < 2\pi$.

The second is a dimensionless TOF parameter which is

$$T = \sqrt{\frac{8\mu}{s^3}}\Delta t. \tag{1.66}$$

With the orbit transformation, the equation for the pericentric radius is

$$r_0 = a(1-e) = r_{op} \sec^2\frac{E_2 - E_1}{4}, \tag{1.67}$$

where the radius of the parabolic segment at the pericentre is

$$r_{op} = \frac{1}{4}\left(r_1 + r_2 + 2\sqrt{r_1 r_2}\cos\frac{\theta}{2}\right) = \frac{s}{4}(1+\lambda)^2. \tag{1.68}$$

The transformed orbit is now shown relative to the superscribed circle and the parabolic orbit in Fig. 1.11.

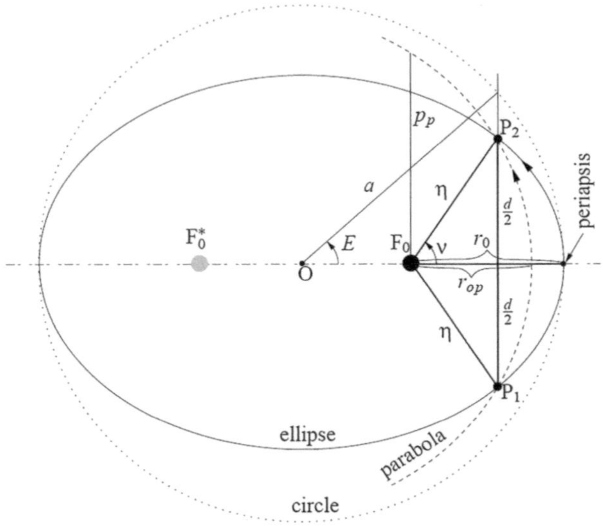

Fig. 1.11 The transformed orbit is shown relative to the superscribed circle and the parabolic orbit

Battin (1999) then introduced two new variables, l and m, which are always positive and depend only on the problem geometry as

$$l = \left(\frac{1-\lambda}{1+\lambda}\right)^2, \quad m = \frac{\mu \Delta t^2}{8 r_{op}^3} = \frac{T^2}{(1+\lambda)^6}. \tag{1.69}$$

These variables transformed Lambert's TOF equation to a cubic, which is

$$y^3 - (1-h_1)y^2 - h_2 = 0, \tag{1.70}$$

where y is related to l and m by

$$y^2 = \frac{m}{(l+x)(1+x)}, \tag{1.71}$$

and x is defined by

$$x = \sqrt{\left(\frac{1-l}{2}\right)^2 + \frac{m}{y^2}} - \frac{1+l}{2}. \tag{1.72}$$

The parameters, h_1 and h_2, are functions of l, m, x and a continued fraction, $\xi(x)$, which are defined.

The two parameters h_1 and h_2 for Battin's cubic are

$$h_1 = \frac{(l+x)^2(1+3x+\xi(x))}{(1+2x+l)\{4x+\xi(x)(3+x)\}} \quad \text{and}$$
$$h_2 = \frac{m(x-l+\xi(x))}{(1+2x+l)\{4x+\xi(x)(3+x)\}}. \tag{1.73}$$

The continued fraction function is

$$\xi(x) = \frac{8\left(\sqrt{1+x}+1\right)}{3+\frac{1}{5+\eta+\frac{\frac{9}{7}\eta}{1+\frac{c_\eta \eta}{1+\frac{c_\eta \eta}{1+\cdots}}}}}, \quad \eta = \frac{x}{\left(\sqrt{1+x}+1\right)^2}, \quad -1 < \eta < 1, \tag{1.74}$$

and

$$c_\eta = \frac{n^2}{(2n)^2 - 1}, \quad n = 4, 5, \ldots. \tag{1.75}$$

The solution of the cubic is obtained by Battin's methods (1999), by the method of Gooding (1990), by Sun's method (1979) or by the universal method discussed by Prussing and Conway (1993). The f and g equations, also discussed by Prussing and Conway (1993), are used to obtain the velocities.

1.4 Relative Motion Equations: Hill-Clohessy-Wiltshire Equations

Considering the motion of most planetary and celestial bodies, almost all of them are relative to each other. There is actually no known celestial body that moves relative to a fixed point in space. Thus, the study of the relative motion of an orbiting body relative to a nominal or reference celestial body in a circular or an elliptic orbit is particularly important. The nonlinear equations of motion of an orbiting body relative to a nominal or reference body in a circular orbit in the absence of perturbation forces are first developed. This is followed by the consideration of certain special cases as well as a decoupled representation of the linearized equations of motion

Consider a nominal planetary body in a circular orbit, which is referred to as the reference body. The body's local vertical local horizontal (LVLH) frame is defined by three Cartesian coordinates: the first centred in a nominal reference body, which is assumed to be in a circular orbit and pointing away from the direction of the gravitational force of the central body; the second being in the orbit plane and normal to the first, pointing to the direction of forward motion of the nominal reference body; and the third normal to the orbit plane and mutually perpendicular to the first two. In the LVLH frame, the position vector $\mathbf{r}_d$ of another orbiting body is given by the following:

$$\mathbf{r}_d = \mathbf{r} + \vec{\rho} = (r + x)i + yi + zk, \;\; \mathbf{r}_d \cdot \mathbf{r}_d = (r + x)^2 + y^2 + z^2, \tag{1.76}$$

where $\mathbf{r}$ is the position vector of the reference body, r is the magnitude of $\mathbf{r}$ and $\vec{\rho} = xi + yi + zk$ is the position vector of the other orbiting body relative to the reference body. The angular velocity and acceleration of the LVLH frame, which is moving with the reference body, are given by $\omega = \dot{\nu}k = (h/r^2)k$ and $\alpha = \ddot{\nu}k = -(2\dot{r}/r)\omega = -(2\dot{r}/r)(h/r^2)k$, respectively, where ν is the true anomaly and $h = |\mathbf{r} \times \dot{\mathbf{r}}|$ is the magnitude of the orbital angular momentum. From kinematics, the equation of motion for the other orbiting body in the reference body's LVLH frame is given by the following:

$$\ddot{\mathbf{r}}_d = \ddot{\mathbf{r}} + \ddot{\rho} + 2\omega \times \dot{\rho} + \omega \times (\omega \times \rho) + \alpha \times \rho. \tag{1.77}$$

Thus, in component form, one has

$$\ddot{\mathbf{r}}_d = \left(\ddot{r} + \ddot{x} - 2\dot{\nu}\dot{y} - \dot{\nu}^2 x - \ddot{f} y\right)i + \left(\ddot{y} + 2\dot{x}\dot{\nu} - \dot{\nu}^2 y + \ddot{\nu}x\right)j + \ddot{z}k. \tag{1.78}$$

Using Kepler's two-body equations of motion both for the primary and for the other orbiting body, both orbiting the Earth, $\ddot{r} = -\mu_e/r^2$ and $\ddot{\mathbf{r}}_d = -\mu_e \mathbf{r}_d / r_d^3$ respectively, the equations of motion are expressed as

$$-\mu_e \mathbf{r}_d / r_d^3 = \left(-\mu_e/r^2 + \ddot{x} - 2\dot{\nu}\dot{y} - \dot{\nu}^2 x - \ddot{\nu} y\right)i + \left(\ddot{y} + 2\dot{x}\dot{\nu} - \dot{\nu}^2 y + \ddot{\nu}x\right)j + \ddot{z}k. \tag{1.79}$$

In terms of the Cartesian coordinates, one has the following three component equations:

$$\ddot{x} - 2\dot{\nu}\dot{y} - \dot{\nu}^2 x - \ddot{\nu} y = \left(\mu_e/r^2\right) - \mu_e(r + x)/r_d^3, \tag{1.80}$$

$$\ddot{y} + 2\dot{x}\dot{\nu} - \dot{\nu}^2 y + \ddot{\nu}x = -\mu_e y/r_d^3, \tag{1.81}$$

$$\ddot{z} = -\mu_e z/r_d^3. \tag{1.82}$$

But,

$$r_d^{-3} = r^{-3}\left(1 + 2\frac{x}{r} + \frac{x^2 + y^2 + z^2}{r^2}\right)^{-3/2}. \tag{1.83}$$

Hence, the equations for the other orbiting body relative position coordinates are

$$\ddot{x} - 2\dot{\nu}\dot{y} - \dot{\nu}^2 x - \ddot{\nu} y = \frac{\mu_e r}{r^3} - \frac{\mu_e (r + x)}{r_d^3} = \frac{\mu_e r}{r^3} \times \left(1 - \left(1 + \frac{x}{r}\right)\left(1 + 2\frac{x}{r} + \frac{x^2 + y^2 + z^2}{r^2}\right)^{-3/2}\right), \tag{1.84}$$

$$\ddot{y} + 2\dot{x}\dot{\nu} - \dot{\nu}^2 y + \ddot{\nu} x = -\frac{\mu_e y}{r^3}\left(1 + 2\frac{x}{r} + \frac{x^2 + y^2 + z^2}{r^2}\right)^{-3/2}, \tag{1.85}$$

$$\ddot{z} = -\frac{\mu_e z}{r^3}\left(1 + 2\frac{x}{r} + \frac{x^2 + y^2 + z^2}{r^2}\right)^{-3/2}. \tag{1.86}$$

Linearizing Eqs. (1.84)–(1.86),

$$\ddot{x} - 2\dot{\nu}\dot{y} - \dot{\nu}^2 x - \ddot{\nu} y = 2\mu_e x / r^3, \tag{1.87}$$

$$\ddot{y} + 2\dot{x}\dot{\nu} - \dot{\nu}^2 y + \ddot{\nu} x = -\mu_e y / r^3, \tag{1.88}$$

$$\ddot{z} = -\mu_e z / r^3. \tag{1.89}$$

These equations are called the linearized equations of relative motion (LERM). Thus, the nonlinear equations of relative motion (NERM) can be expressed as

$$\ddot{x} - 2\dot{\nu}\dot{y} - \left(\dot{\nu}^2 + 2\frac{\mu_e}{r^3}\right)x - \ddot{\nu} y = \frac{\mu_e r}{r^3} \times \left(1 - 2\frac{x}{r} - \left(1 + \frac{x}{r}\right)\left(1 + 2\frac{x}{r} + \frac{x^2 + y^2 + z^2}{r^2}\right)^{-3/2}\right) \tag{1.90}$$

$$\ddot{y} + 2\dot{x}\dot{\nu} - \left(\dot{\nu}^2 - \frac{\mu_e}{r^3}\right)y + \ddot{\nu} x = \frac{\mu_e y}{r^3}\left(1 - \left(1 + 2\frac{x}{r} + \frac{x^2 + y^2 + z^2}{r^2}\right)^{-3/2}\right), \tag{1.91}$$

$$\ddot{z} + \frac{\mu_e z}{r^3} = \frac{\mu_e z}{r^3}\left(1 - \left(1 + 2\frac{x}{r} + \frac{x^2 + y^2 + z^2}{r^2}\right)^{-3/2}\right). \tag{1.92}$$

To obtain the Hill-Clohessy-Wiltshire (HCW) equations, one lets $\dot{\nu} - n = \ddot{\nu} = 0$, $\dot{\nu}^2 = \mu_e / r^3 = n^2$, where n represents the mean rate of motion of the reference body or the mean motion. To write the equations (1.90)–(1.92) in state space form one defines the velocity components in the LVLH frame by

$$\begin{bmatrix} \Delta v_x \\ \Delta v_y \\ \Delta v_z \end{bmatrix} = \begin{bmatrix} \dot{x} \\ \dot{y} \\ \dot{z} \end{bmatrix} + \boldsymbol{\omega} \times \begin{bmatrix} x \\ y \\ z \end{bmatrix} = \begin{bmatrix} \dot{x} \\ \dot{y} \\ \dot{z} \end{bmatrix} + \begin{bmatrix} 0 & -\dot{\nu} & 0 \\ \dot{\nu} & 0 & 0 \\ 0 & 0 & 0 \end{bmatrix} \begin{bmatrix} x \\ y \\ z \end{bmatrix}. \tag{1.93}$$

Thus,

$$\begin{bmatrix} \Delta\dot{v}_x \\ \Delta\dot{v}_y \\ \Delta\dot{v}_z \end{bmatrix} + \begin{bmatrix} 0 & -\dot{\nu} & 0 \\ \dot{\nu} & 0 & 0 \\ 0 & 0 & 0 \end{bmatrix} \begin{bmatrix} \Delta v_x \\ \Delta v_y \\ \Delta v_z \end{bmatrix} = \begin{bmatrix} \ddot{x} \\ \ddot{y} \\ \ddot{z} \end{bmatrix} + 2 \begin{bmatrix} 0 & -\dot{\nu} & 0 \\ \dot{\nu} & 0 & 0 \\ 0 & 0 & 0 \end{bmatrix} \begin{bmatrix} \dot{x} \\ \dot{y} \\ \dot{z} \end{bmatrix} - \begin{bmatrix} \dot{\nu}^2 & \ddot{\theta} & 0 \\ -\ddot{\nu} & \dot{\nu}^2 & 0 \\ 0 & 0 & 0 \end{bmatrix} \begin{bmatrix} x \\ y \\ z \end{bmatrix}. \tag{1.94}$$

Hence, one may write Eqs. (1.93)–(1.94) in state-space form as

$$\begin{bmatrix} \dot{x} \\ \dot{y} \\ \dot{z} \end{bmatrix} = \begin{bmatrix} \Delta v_x \\ \Delta v_y \\ \Delta v_z \end{bmatrix} - \begin{bmatrix} 0 & -\dot{\nu} & 0 \\ \dot{\nu} & 0 & 0 \\ 0 & 0 & 0 \end{bmatrix} \begin{bmatrix} x \\ y \\ z \end{bmatrix}, \tag{1.95}$$

$$\begin{bmatrix} \Delta\dot{v}_x \\ \Delta\dot{v}_y \\ \Delta\dot{v}_z \end{bmatrix} = -\begin{bmatrix} 0 & -\dot{\nu} & 0 \\ \dot{\nu} & 0 & 0 \\ 0 & 0 & 0 \end{bmatrix} \begin{bmatrix} \Delta v_x \\ \Delta v_y \\ \Delta v_z \end{bmatrix} + \frac{\mu_e}{r^3} \begin{bmatrix} 2x \\ -y \\ -z \end{bmatrix} + \frac{\mu_e}{r^3} \begin{bmatrix} rf_1 \\ yf_2 \\ zf_3 \end{bmatrix}. \tag{1.96}$$

In state-space form, the LERM are given by the equation

$$\dot{\mathbf{x}} = \mathbf{A}(t)\mathbf{x}, \tag{1.97}$$

where

$$\mathbf{x} = [x \quad y \quad z \quad \Delta v_x \quad \Delta v_y \quad \Delta v_z]^T, \tag{1.98}$$

$$\mathbf{A}(t) = \begin{bmatrix} -\begin{bmatrix} 0 & -\dot{\nu} & 0 \\ \dot{\nu} & 0 & 0 \\ 0 & 0 & 0 \end{bmatrix} & \begin{bmatrix} 1 & 0 & 0 \\ 0 & 1 & 0 \\ 0 & 0 & 1 \end{bmatrix} \\ \frac{\mu_e}{r^3} \begin{bmatrix} 2 & 0 & 0 \\ 0 & -1 & 0 \\ 0 & 0 & -1 \end{bmatrix} & -\begin{bmatrix} 0 & -\dot{\nu} & 0 \\ \dot{\nu} & 0 & 0 \\ 0 & 0 & 0 \end{bmatrix} \end{bmatrix}. \tag{1.99}$$

The matrix $\mathbf{A}(t)$ is time-varying as r, $\dot{\nu}$ and $\ddot{\nu}$ vary with time. Thus, $\mathbf{A}(t)$ is periodic with the period equal to $T = 2\pi/n$. Including the nonlinear terms on the right-hand side, the equations of motion may be expressed in state-space form as

$$\dot{\mathbf{x}} = \mathbf{A}(t)\mathbf{x} + \mathbf{B}, \quad \mathbf{B} = \left(\mu_e/r^3\right)[rf_1 \quad yf_2 \quad zf_3]^T. \tag{1.100}$$

In terms of the mean rate of motion n, the NERM are

$$\ddot{x} - 2\dot{y}n - 3n^2x = 2\dot{y}(\dot{\nu} - n) + \left(\dot{\nu}^2 + 2\frac{\mu_e}{r^3} - 3n^2\right)x + \ddot{\nu}y + \frac{\mu_e r}{r^3} \times \left(1 - 2\frac{x}{r} - \left(1 + \frac{x}{r}\right)\left(1 + 2\frac{x}{r} + \frac{x^2 + y^2 + z^2}{r^2}\right)^{-3/2}\right), \tag{1.101}$$

$$\ddot{y} + 2\dot{x}n = 2\dot{x}(n - \dot{\nu}) + \left(\dot{\nu}^2 - \frac{\mu_e}{r^3}\right)y - \ddot{\nu}x + \frac{\mu_e y}{r^3} \times \left(1 - \left(1 + 2\frac{x}{r} + \frac{x^2 + y^2 + z^2}{r^2}\right)^{-\frac{3}{2}}\right) \tag{1.102}$$

$$\ddot{z} + n^2z = \left(n^2 - \frac{\mu_e}{r^3}\right)z + \frac{\mu_e z}{r^3}\left(1 - \left(1 + 2\frac{x}{r} + \frac{x^2 + y^2 + z^2}{r^2}\right)^{-3/2}\right). \tag{1.103}$$

The Hill-Clohessy-Wiltshire (HCW) equations (Clohessy and Wiltshire, 1960) are obtained by setting the right-hand sides of Eqs. (1.101)–(1.103) to zero. The general solutions to the LVLH coordinates and their rates in the HCW equations are as follows:

$$x(t) = (\dot{x}_0/n)\sin nt - (3x_0 + 2\dot{y}_0/n)\cos nt + 2(2x_0 + \dot{y}_0/n), \tag{1.104}$$

$$y(t) = 2(\dot{x}_0/n)\cos nt + 2(3x_0 + 2\dot{y}_0/n)\sin nt - 3(2nx_0 + \dot{y}_0)t - 2(\dot{x}_0/n) + y_0, \tag{1.105}$$

$$z(t) = (\dot{z}_0/n)\sin nt + z_0\cos nt, \tag{1.106}$$

$$\dot{x}(t) = n(\dot{x}_0/n)\cos nt + n(3x_0 + 2\dot{y}_0/n)\sin nt, \tag{1.107}$$

$$\dot{y}(t) = -2n(\dot{x}_0/n)\sin nt + 2n(3x_0 + 2\dot{y}_0/n)\cos nt - 3(2nx_0 + \dot{y}_0), \tag{1.108}$$

$$\dot{z}(t) = n(\dot{z}_0/n)\cos nt - nz_0\sin nt. \tag{1.109}$$

First, it may be observed that only one of the solutions has a secular or linearly time dependent term. It may also be observed that timescales of the periodic terms are the same, but the magnitudes of the secular would depend on the initial velocity perturbations in a direction tangential to the orbit.

From the solutions given by Eqs. (1.104)–(1.109), it is observed that the secular and constant terms may be isolated by considering the combinations of states given by

$$2\dot{x}(t) - ny(t) = 2\dot{x}_0 - ny_0 + 3n(\dot{y}_0 + 2nx_0)t,$$
$$\dot{y}(t) + 2nx(t) = -3(2nx_0 + \dot{y}_0) + 4(2nx_0 + \dot{y}_0) = \dot{y}_0 + 2nx_0. \tag{1.110}$$

From the above equations, one may introduce a transformation of the variable to separate the periodic and secular parts of the solution:

$$x_r \equiv nx(t) + \frac{\dot{y}(t)}{2} = nx_0 + \frac{\dot{y}_0}{2}; \quad \Rightarrow \dot{y}(t) = 2(x_r(t) - nx(t)), \tag{1.111}$$

$$y_r \equiv ny(t) - 2\dot{x}(t) = ny_0 - 2\dot{x}_0 - 6\left(nx_0 + \frac{\dot{y}_0}{2}\right)t;$$

$$\Rightarrow \dot{x}(t) = \frac{1}{2}(ny(t) - y_r(t)). \tag{1.112}$$

Using the above definitions of the relative orbital states, x_r and y_r, the original in-plane HCW equations may be expressed as

$$\frac{1}{2}(2\ddot{x}(t) - n\dot{y}(t)) = 3n\left(nx(t) + \frac{\dot{y}(t)}{2}\right); \quad \Rightarrow \dot{y}_r(t) = -6nx_r(t), \tag{1.113}$$

$$2\left(\frac{\ddot{y}(t)}{2} + n\dot{x}(t)\right) = 0; \quad \Rightarrow \dot{x}_r(t) = 0. \tag{1.114}$$

For the periodic part, one observes that the combinations of states $2\dot{y}(t) + 3nx(t)$ and the radial relative velocity $\dot{x}(t)$ are purely periodic.

$$2\dot{y}(t) + 3nx(t) = 2\dot{y}_0 + 4nx_0 - nx(t) = -\dot{x}_0 \sin nt + (3nx_0 + 2\dot{y}_0)\cos nt, \tag{1.115}$$

$$\dot{x}(t) = \dot{x}_0 \cos nt + (3nx_0 + 2\dot{y}_0)\sin nt. \tag{1.116}$$

It follows that the amplitudes of these velocities satisfy

$$(2\dot{y}(t) + 3nx(t))^2 + \dot{x}^2(t) = \dot{x}_0^2 + (3nx_0 + 2\dot{y}_0)^2 = A^2. \tag{1.117}$$

Writing the velocities in terms of the relative orbital states, x_r and y_r, and using Eqs. (1.111) and (1.112), the amplitude relation reduces to

$$(4x_r(t) - nx(t))^2 + \frac{1}{4}(ny(t) - y_r(t))^2 = A^2. \tag{1.118}$$

Introducing the phase angle Φ as well as x_p and y_p, such that

$$A\cos\Phi = 4x_r(t) - nx(t) \equiv x_p(t), \;\; A\sin\Phi = \frac{1}{2}(ny(t) - y_r(t)) \equiv y_p(t), \tag{1.119}$$

it follows that

$$\dot{x}(t) = \frac{1}{2}(ny(t) - y_r(t)) = A\sin\Phi \equiv y_p(t), \tag{1.120}$$

$$\dot{y}(t) = 2(x_r(t) - nx(t)) = 2A\cos\Phi - 6x_r(t) = 2x_p(t) - 6x_r(t). \tag{1.121}$$

It can then be shown that oscillation amplitude A and the phase angle Φ, respectively, satisfy,

$$\dot{A} = 0, \quad \dot{\Phi} = n. \tag{1.122}$$

Hence, it also follows that one may express the position components of the in-plane state vector in terms of the secular and periodic relative orbital states, x_r, y_r, x_p, y_p, as

$$x(t) = \frac{x_r}{n} - \frac{\dot{y}(t)}{2n} = \frac{4x_r(t)}{n} - \frac{x_p(t)}{n}, \quad y(t) = \frac{y_r}{n} + \frac{2\dot{x}(t)}{n} = \frac{y_r(t)}{n} + \frac{2y_p(t)}{n}. \tag{1.123}$$

In the out-of-plane direction, the motion is purely oscillatory, and for this reason, the out-of-plane oscillation amplitude A_z and phase angle Ψ may be defined from

$$\dot{z} = A_z \cos\Psi, \quad nz = A_z \sin\Psi, \quad \tan\Psi = \frac{nz}{\dot{z}}, \quad z(t) = \frac{A_z}{n}\sin\Psi. \tag{1.124}$$

Thus, the equations governing the evolution of the out-of-plane oscillation amplitude A_z and phase angle Ψ may be shown to be

$$\dot{A}_z = 0, \quad \dot{\Psi} = n. \tag{1.125}$$

To establish a complete state-space representation of the HCW equations, one could use Eqs. (1.111), (1.112), (1.113) and (1.114) and the out-of-plane states. However, it is equally important to avoid the need for linearization. Hence, the complete set of secular and periodic relative orbital states x_r, y_r, x_p, y_p, z and $v_z = \dot{z}$ are defined by Eqs. (1.111), (1.112) and (1.119) and the out-of-plane states z and $v_z = \dot{z}$. The four linear transformation equations, for the in-plane states, may be summarized as

$$\frac{x_r}{n} \equiv x + \frac{\dot{y}}{2n}, \quad \frac{y_r}{n} \equiv y - \frac{2}{n}\dot{x}, \quad x_p \equiv 3nx + 2\dot{y} \quad \text{and} \quad y_p \equiv \dot{x}. \tag{1.126}$$

The inverse transformation equations are

$$x = \frac{4x_r}{n} - \frac{x_p}{n}, \quad y = \frac{y_r}{n} + \frac{2y_p}{n}, \quad \dot{x} = y_p, \quad \dot{y} = 2\left(x_p - 3n\frac{x_r}{n}\right). \tag{1.127}$$

The linear state-space representation of the HCW equations may be expressed in terms of the relative orbit states, x_r, y_r, x_p and y_p, as

$$\dot{x}_r = 0, \quad \dot{y}_r = -6nx_r, \quad \dot{x}_p = -ny_p, \quad \dot{y}_p = nx_p. \tag{1.128}$$

It is observed that the system of equations is decoupled into three subsystems: secular, periodic and out-of-plane subsystems. Further properties of the NERM and the LERM equations are discussed by Vepa (2019).

1.5 Real Orbits, the Three-Body Problem

So far, the application of the orbit theory to the orbit of a real Earth satellite has not been discussed. Consider, for example, the only known naturally occurring Earth satellite, the moon. The moon orbits the Earth, which in turn orbits the Sun. The Sun is the central body of the heliocentric orbits of all planets, including the Earth. One cannot ignore the influence of the Sun, in studying the orbit of an Earth satellite, unless one restricts oneself to orbits that are relatively very close to the Earth. The Keplerian two-body theory is only applicable in a small domain very close to the Earth. However, not all the results obtained for Kepler's two-body solution can be translated to the multibody case. It is generally customary to study the orbit of the moon, relative to the Sun-Earth, which provides a moving or rotating reference frame for the study of the motions of the moon, a typical Earth satellite.

The three-body problem refers to a configuration of orbiting bodies consisting of three arbitrary masses that attract one another. Newton's law of gravitation is applied, so the three bodies mutually attract each other and vectorially add up all the forces (and convert to accelerations) acting on each of the bodies. It is often possible to define the barycentre of the system and describe the motion relative to it. Generally, it is true that only numerical solutions are tractable, as there are no analytical solutions. However, numerical solutions could be chaotic, and for this and other reasons, several simplifications are made. Thus the motion of the three bodies is restricted to remain within the same plane and it is further assumed that the orbits of the primary central body and the secondary orbiting body are circles, and that the third body is comparatively of very low mass. This is the more common problem dealt with in classical celestial mechanics and is known as the "circular restricted three-body problem". Not only the Sun-Earth-Moon system can be approximated by this problem but also the Sun-Jupiter and any of its Galilean moons. Jupiter's four largest moons, Io, Europa, Ganymede and Calisto, were the first moons discovered beyond Earth and are known as the Galilean moons after Italian astronomer Galileo Galilei, who first observed them in 1610. The circular restricted three-body problem (CR3BP) is also applicable to the Sun-Jupiter and each of several asteroids, circling the Sun in orbits between Mars and Jupiter. So, the CR3BP can be treated as an archetypal problem that can not only approximate Keplerian orbits in the limit but is also applicable to several planetary configurations.

Two of the larger bodies in the CR3BP are known as the primary and secondary bodies, while the mass of the third body is neglected. The frame of reference used to model the dynamics or motion of the third body is assumed to rotate with the line joining the primary and secondary bodies. The origin of the frame is located at the barycentre of the primary and secondary bodies. The reference frame rotates about the barycentre of the primary and secondary bodies with the angular velocity of the two-body system, which can be found from Kepler's third law and is assumed to be uniform. Thus, the equation of motion of the restricted problem includes the effects of rotation and the control acceleration vector $\vec{u}_c$ and is given by

$$\begin{aligned} \left.\frac{\mathrm{d}\vec{v}}{\mathrm{d}t}\right|_B &= \left.\frac{\mathrm{d}^2\vec{r}}{\mathrm{d}t^2}\right|_R + 2\vec{\omega}\times\left.\frac{\mathrm{d}\vec{r}}{\mathrm{d}t}\right|_R + \vec{\omega}\times\left(\vec{\omega}\times\vec{r}\right) + \left.\frac{\mathrm{d}\vec{\omega}}{\mathrm{d}t}\right|_B \times\vec{r} \\ &= -G\left(\frac{M_1}{r_1^3}\vec{r}_1 + \frac{M_2}{r_2^3}\vec{r}_2\right) + \vec{u}_c, \end{aligned} \tag{1.129}$$

When the angular velocity is assumed to be constant,

$$\left.\frac{\mathrm{d}^2\vec{r}}{\mathrm{d}t^2}\right|_R = -G\left(\frac{M_1}{r_1^3}\vec{r}_1 + \frac{M_2}{r_2^3}\vec{r}_2\right) - 2\vec{\omega}\times\left.\frac{\mathrm{d}\vec{r}}{\mathrm{d}t}\right|_R - \vec{\omega}\times\left(\vec{\omega}\times\vec{r}\right) + \vec{u}_c, \tag{1.130}$$

The gravitational and centrifugal accelerations can be written as the gradient of a single potential function. Thus, one may define a new potential function (the Roche potential function), as

$$U = -G\left(\frac{M_1}{r_1} + \frac{M_2}{r_2}\right) - \frac{1}{2}\omega^2 r^2. \tag{1.131}$$

The equation of motion of the third body may be stated compactly as

$$\frac{\mathrm{d}^2\vec{r}}{\mathrm{d}t^2} = -\nabla U - 2\vec{\omega}\times\frac{\mathrm{d}\vec{r}}{\mathrm{d}t} + \vec{u}_c. \tag{1.132}$$

Given the orbital rate n of the primary and secondary bodies together in the orbital plane, from the balance of the gravitational and centrifugal forces, one has

$$M_1 n^2 d_1 = \frac{GM_1M_2}{(d_1+d_2)^2} = \frac{M_1\mu_2}{(d_1+d_2)^2}, \tag{1.133}$$

$$M_2 n^2 d_2 = \frac{GM_1M_2}{(d_1+d_2)^2} = \frac{M_2\mu_1}{(d_1+d_2)^2}. \tag{1.134}$$

In the two Eqs. (1.133) and (1.134), d_1 and d_2 are the distances of the primary and secondary bodies from the barycentre. Adding the same two equations, one obtains

$$n^2 = \frac{\mu_1 + \mu_2}{(d_1 + d_2)^3}. \tag{1.135}$$

It is convenient to assume that the distance between the primary and secondary bodies is normalized to unity; thus, $d_1 + d_2 = 1$. Furthermore, it is convenient to scale the unit of time, so $\tau = nt$. The equation of motion reduces to

$$\frac{\mathrm{d}^2 \vec{r}}{\mathrm{d}\tau^2} = -\nabla \overline{U} - 2\frac{\vec{\omega}}{n} \times \frac{\mathrm{d}\vec{r}}{\mathrm{d}\tau} + \frac{1}{n^2}\vec{u}_c, \quad \vec{\omega} = [0 \quad 0 \quad n], \quad n^2 = \mu_1 + \mu_2, \tag{1.136}$$

with

$$\overline{U} = -\frac{1}{r_1} - \frac{\mu_2}{\mu_1 + \mu_2}\left(\frac{1}{r_2} - \frac{1}{r_1}\right) - \frac{1}{2}r^2. \tag{1.137}$$

The scalar CR3BP equations are

$$\ddot{x} - 2\dot{y} - x = -(1-\mu)\left(\frac{x+\mu}{\|\mathbf{r}_1\|^3}\right) - \mu\left(\frac{x-1+\mu}{\|\mathbf{r}_2\|^3}\right) + u_x, \tag{1.138}$$

$$\ddot{y} + 2\dot{x} - y = -(1-\mu)\left(\frac{y}{\|\mathbf{r}_1\|^3}\right) - \mu\left(\frac{y}{\|\mathbf{r}_2\|^3}\right) + u_y, \tag{1.139}$$

$$\ddot{z} = -(1-\mu)\left(\frac{z}{\|\mathbf{r}_1\|^3}\right) - \mu\left(\frac{z}{\|\mathbf{r}_2\|^3}\right) + u_z, \tag{1.140}$$

where the dot represents differentiation with respect to τ, $\mathbf{r}_1^2 = (x+\mu)^2 + y^2 + z^2$, $\mathbf{r}_2^2 = (x+\mu-1)^2 + y^2 + z^2$ and in the context of the CR3BP, $\mu = \mu_2/(\mu_1 + \mu_2)$. Moreover, u_x, u_y and u_z are the components of the normalized control acceleration $\vec{u}_c/n^2$.

1.5.1 The Equilibrium Points

Under conditions of equilibrium, derivatives with respect to τ could be set to zero. Thus, the conditions of equilibrium in the absence of any control accelerations are

$$x_e = (1-\mu)\left(\frac{x_e+\mu}{\|\mathbf{r}_{1e}\|^3}\right) + \mu\left(\frac{x_e-1+\mu}{\|\mathbf{r}_{2e}\|^3}\right), \tag{1.141}$$

$$y_e = (1-\mu)\left(\frac{y_e}{\|\mathbf{r}_{1e}\|^3}\right) + \mu\left(\frac{y_e}{\|\mathbf{r}_{2e}\|^3}\right), \tag{1.142}$$

$$0 = (1-\mu)\left(\frac{z_e}{\|\mathbf{r}_{1e}\|^3}\right) + \mu\left(\frac{z_e}{\|\mathbf{r}_{2e}\|^3}\right), \tag{1.143}$$

The last equation implies that $z_e = 0$. When the equilibrium points are collinear along the line joining the primary and secondary bodies, one can additionally assume that $y_e = 0$. In this case, Eq. (1.141) reduces to

$$x_e = \frac{(1-\mu)}{(x_e+\mu)^2} + \frac{\mu}{(x_e+\mu-1)^2}. \tag{1.144}$$

Equation (1.144) is then expressed as

$$\begin{aligned} x_e = x_e &\left(\frac{(1-\mu)}{(x_e+\mu)^3} + \frac{\mu}{(x_e+\mu-1)^3}\right) \\ &+\mu(1-\mu)\left(\frac{1}{(x_e+\mu)^3} - \frac{1}{(x_e+\mu-1)^3}\right). \end{aligned} \tag{1.145}$$

Equation (1.145) could be solved numerically. In fact, there are a total of five equilibrium solutions, and of these, three are collinear, along the line joining the primary and secondary bodies. The approximate expressions for the Lagrange points can be obtained and are

$$\begin{aligned} &L_{1,2} \simeq \left(1 \mp \left(\tfrac{\mu}{3}\right)^{1/3}, 0\right), \quad L_3 \simeq -\left(1 + \frac{5}{4}\left(\frac{\mu}{3}\right), 0\right) \quad \text{and} \\ &L_{4,5} = \frac{1}{2}\left(1-2\mu, \pm\sqrt{3}\right). \end{aligned} \tag{1.146}$$

The first three are unstable while the last two exact and stable. An example to consider is the Earth-Moon system for the primary and secondary bodies. The equilibrium points are illustrated in this case in Fig. 1.12.

With reference to the Earth-Moon-satellite CR3BP, the L_1 point lies on the line joining the primary and secondary bodies and between them. A body which orbits the Earth more closely than the Moon would normally have a shorter orbital period than the Moon, but if the third body is directly between Earth and Moon, then the effect of lunar gravity is to weaken the force pulling the body towards the Earth and thus increase the orbital period of the third body. The closer the object is to the Moon, the greater is this effect. At the L_1 point, the orbital period of the object becomes exactly equal to the lunar orbital period. The L_2 point lies on the line joining the primary and secondary bodies but beyond the secondary body, which is assumed

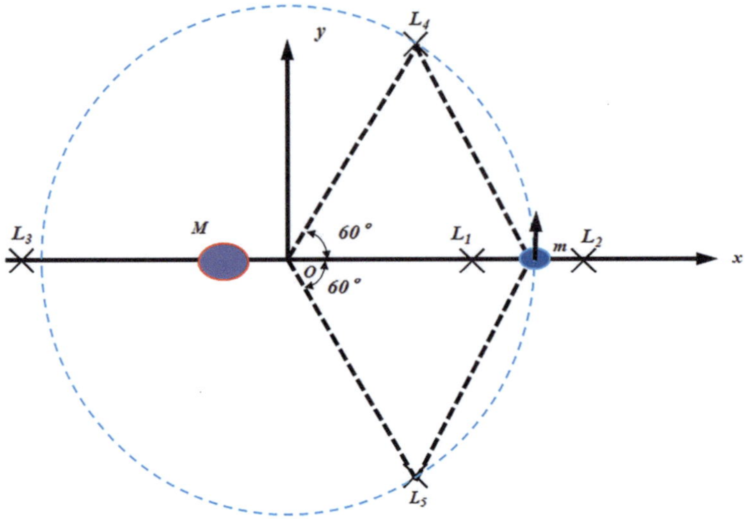

Fig. 1.12 Equilibrium points for the Earth-Moon-satellite CR3BP:L_4 and L_5 are at 60° to the horizontal

to be of lower mass than the primary. On the side of the Moon away from the Earth, the orbital period of a third body would normally be greater than that of the Moon. Lunar gravity decreases the orbital period of the body; at the L_2 point, it has the same period as the Moon. If the second body has mass M_2 and much smaller than the mass M_1 of the main body, then L_1 and L_2 are at approximately equal distances from the second body (L_1 is actually a little closer), the Hill sphere radius, which is given by $(\mu/3)^{1/3}$. The significance of the Hill sphere is explained in a later chapter. The L_3 point lies on the line defined by the two large bodies, beyond the larger of the two. In the Earth-Moon system, L_3 is on the opposite side of the Moon, a little further away from the Earth than the Moon is, where the combined pull of the Moon and Earth again causes the object to orbit with the same period as the Moon.

The L_4 and L_5 points lie at the third point of an equilateral triangle with the base of the line defined by the two bodies, such that the point is, respectively, ahead of, or behind, the body with the smaller mass in its orbit around the body with the larger mass. L_4 and L_5 are sometimes called *triangular Lagrange points* or *Trojan points*.

1.5.2 *Stability of the Lagrange Points*

The collinear Lagrange points L_1, L_2 and L_3 are unstable for arrange of values of the parameter $\mu = \mu_2/(\mu_1 + \mu_2)$, while the L_4 and L_5 points are stable. To study the stability of the Lagrange points, the nonlinear equations of motion are generally linearized for small motions about the equilibrium points. The process of doing this is instructive and will be demonstrated here for the collinear Lagrange points. The

CR3BP equations with the external, control or disturbance accelerations included, in component form, are

$$\ddot{x} - 2\dot{y} - x = -(1-\mu)\left(\frac{x+\mu}{\|\mathbf{r}_1\|^3}\right) - \mu\left(\frac{x-1+\mu}{\|\mathbf{r}_2\|^3}\right) + u_x, \tag{1.147}$$

$$\ddot{y} + 2\dot{x} - y = -(1-\mu)\left(\frac{y}{\|\mathbf{r}_1\|^3}\right) - \mu\left(\frac{y}{\|\mathbf{r}_2\|^3}\right) + u_y, \tag{1.148}$$

$$\ddot{z} = -(1-\mu)\left(\frac{z}{\|\mathbf{r}_1\|^3}\right) - \mu\left(\frac{z}{\|\mathbf{r}_2\|^3}\right) + u_z, \tag{1.149}$$

with

$$\mathbf{r}_1^2 = (x+\mu)^2 + y^2 + z^2, \quad \mathbf{r}_2^2 = (x+\mu-1)^2 + y^2 + z^2. \tag{1.150}$$

In the above, the overhead dot $(\dot{})$ represents differentiation with respect to non-dimensional time.

Let $x_0 = x_e$ be the distance of a *collinear* equilibrium point from the barycentre. Then, it can be shown that

$$\frac{x+x_0+\mu}{\left((x+x_0+\mu)^2+y^2+z^2\right)^{3/2}} \approx \frac{\left(\frac{x}{x_0+\mu}+1\right)\left(1-3\frac{x}{\mu+x_0}\right)}{(x_0+\mu)^2} = \frac{1}{(x_0+\mu)^2}\left(1-\frac{2x}{x_0+\mu}\right), \tag{1.151}$$

$$\frac{x+x_0+\mu-1}{\left((x+x_0+\mu-1)^2+y^2+z^2\right)^{3/2}} \approx \frac{1}{(x_0+\mu-1)^2}\left(1-\frac{2x}{x_0+\mu-1}\right), \tag{1.152}$$

$$\frac{y}{\left((x+x_0+\mu)^2+y^2+z^2\right)^{3/2}} \approx \frac{y}{(x_0+\mu)^3}\left(1-3\frac{x}{\mu+x_0}\right) \approx \frac{y}{(x_0+\mu)^3}, \tag{1.153}$$

$$\frac{z}{\left((x+x_0+\mu)^2+y^2+z^2\right)^{3/2}} \approx \frac{z}{(x_0+\mu)^3}\left(1-3\frac{x}{\mu+x_0}\right) \approx \frac{z}{(x_0+\mu)^3}. \tag{1.154}$$

Hence, it follows that

$$\begin{aligned}\ddot{x}-2\dot{y}-\left(1+2\frac{(1-\mu)}{(x_0+\mu)^3}+2\frac{\mu}{(x_0+\mu-1)^3}\right)x=x_0-\frac{(1-\mu)}{(x_0+\mu)^2}\\-\frac{\mu}{(x_0+\mu-1)^2}+u_x.\end{aligned} \tag{1.155}$$

Since x_0 satisfies

$$x_0-\frac{(1-\mu)}{(x_0+\mu)^2}-\frac{\mu}{(x_0+\mu-1)^2}=0, \tag{1.156}$$

it follows that

$$\begin{aligned}x_0=x_0\left(\frac{(1-\mu)}{(x_0+\mu)^3}+\frac{\mu}{(x_0+\mu-1)^3}\right)\\+\mu(1-\mu)\left(\frac{1}{(x_0+\mu)^3}-\frac{1}{(x_0+\mu-1)^3}\right).\end{aligned} \tag{1.157}$$

The preceding equation is expressed as

$$\begin{aligned}\left(\frac{(1-\mu)}{(x_0+\mu)^3}+\frac{\mu}{(x_0+\mu-1)^3}\right)=1-\frac{\mu(1-\mu)}{x_0}\\\times\left(\frac{1}{(x_0+\mu)^3}-\frac{1}{(x_0+\mu-1)^3}\right).\end{aligned} \tag{1.158}$$

If one lets:

$$B_L=\left(\frac{(1-\mu)}{(x_0+\mu)^3}+\frac{\mu}{(x_0+\mu-1)^3}\right). \tag{1.159}$$

Thus, one has

$$\ddot{x}-2\dot{y}-(2B_L+1)x=u_x,\ \ \ddot{y}+2\dot{x}+(B_L-1)y=u_y,\ \ \ddot{z}+B_Lz=u_z. \tag{1.160}$$

If one lets $x_0=1+\gamma-\mu$, i.e. γ is the distance of the co-linear equilibrium point from the centre of mass of the second body, it follows that

$$B_L = \left(\frac{(1-\mu)}{(1+\gamma)^3} + \frac{\mu}{|\gamma|^3} \right). \tag{1.161}$$

The stability of the above linear equations could be studied in terms of the single parameter, B_L, introduced by Farquhar (1968). One could easily verify the instability of the collinear equilibrium points for $B_L > 0$ and $B_L \neq 1$.

1.5.3 Application of the CR3BP and the Lagrange Points

The main application of the CR3BP is trajectory planning for an interplanetary satellite mission. For example, on a mission to Mars from the planet Earth, one could consider the Sun-Earth-satellite CR3BP for the departure phase. Somewhere midway, beyond the Hill sphere of the departure planet, the coordinates are changed, and after establishing the initial conditions, Sun-Mars-satellite CR3BP is considered till the satellite arrives at Mars and enters the Hill sphere of the arrival planet. This and other similar examples will be considered in the subsequent chapters. Such an approach should provide a reasonably accurate trajectory for any interplanetary mission and also useful in estimating the time of flight and fuel requirements for the missions under consideration. One could also plan a mission to the Moon, a mission to Jupiter and Saturn and beyond to one of their moons.

Exercises

1. (a) A key feature in the design of all space systems is the type of materials selected.
   (i) Identify the processes that high-energy particle radiations have upon electronic subsystems used on space missions. Identify the nature of the damage caused in each case and the ways in which the damage may be reduced. Materials erosion is a particular feature for those materials on the outer surface of all space vehicles. Identify the principal ways in which a spacecraft surface material becomes either eroded or changed as a result of being exposed in low earth orbit.
   (b) Identify the principal sources of space debris and the methods used to mitigate the production of space debris.
2. Identify the Keplerian orbital elements that define the orientation of a spacecraft's orbit plane.
   What other orbital elements are required to define the size and shape of the orbit? Name at least one orbital element that may be used to locate the satellite precisely on the orbit, without the need for any additional calculations.

Describe briefly all the expected perturbative influences that change these elements for a real satellite in orbit.

3. (a) Describe the structure of the Earth's upper atmosphere and identify how the dominant molecular and atomic species change with altitude.
(b) Atmospheric density at the altitude of low earth orbiting satellites varies over long and short periods of time.

    (i) Identify the appropriate timescales for these variations and describe the primary source and the physical mechanism by which atmospheric density changes.
    (ii) What radiation measurements are generally used in linking source variations with predicted atmospheric density and drag estimates?

(c) Describe the processes responsible for the formation of the Van Allen radiation belts, and identify the key impact that these belts have on spacecraft design.

4. (a) Identify the Keplerian orbital elements that define the orientation of a spacecraft's orbit plane. What other orbital elements are required to define the size and shape of the orbit? Name at least one orbital element that may be used to locate the satellite precisely on the orbit, without the need for any additional calculations. Explain carefully and distinguish between osculating orbital elements and mean orbital elements.
(b) Given the magnitude of the perturbation force acting on a satellite and the position and velocity vectors of the satellite at a particular instant, derive from first principles expressions for the three components of the perturbation force, along the radius vector, normal to the orbit plane and mutually perpendicular to the position and moment of momentum vectors.
(c) Explain the distinguishing features of a hyperbolic trajectory of satellite relative to a planet.
(d) Given only the position and velocity vectors of the satellite at a particular instant, explain without equations how you can estimate any three orbital elements of a hyperbolic orbit.

5. (a) Identify the six orbital elements used to describe an orbit and exactly what feature of the orbit each of them determines.
(b) Using the general equation for a conic, obtain expressions for the transverse or tangential velocity v_θ, and the radial velocity v_r which are the two velocity components of a satellite in a Keplerian orbit.
(c) At a particular instant of time t, in units of distance and time for which the gravitational constant μ, for the two-body problem, is equal to unity ($\mu = 1$), the following vectors define the position and velocity of an orbiting body in a certain space-fixed coordinate system:

$$\mathbf{r} = 1\mathbf{i} + 1\mathbf{j} + 0\mathbf{k} \text{ and } \mathbf{v} = 0\mathbf{i} + 0\mathbf{j} + 1\mathbf{k}.$$

Calculate the moment of momentum vector **h** and the vector defining the line of nodes, **n**, from the descending to the ascending node. Estimate a value for the orbit inclination angle and the eccentricity vector. Hence, also choose appropriate values for the argument of the ascending node and the argument of the perigee. Use the principle of conservation of energy (the vis-viva equation) to determine the semi-major axis, a. Determine also the escape velocity for this orbit when $r = 2$ and the corresponding velocity increment at that position of the orbiting body.

6. (a) What is the circular orbital velocity of the planet Earth? Given the distance unit is 1 AU and the velocity unit is 29.7859 km/s, what is the value of μ? What is the corresponding time unit?
(b) A satellite orbits the Earth at a radius of 6400 km in a circular orbit. Determine its velocity. Given the distance unit is 6400 km and the velocity unit is 7.8910 km/s, what is the value of μ? What is the corresponding time unit?
(c) A satellite is orbiting the planet Earth with $\mathbf{r} = [8250 \quad 390 \quad 6900]$km and $\mathbf{v} = [-0.7 \quad 6.6 \quad -0.6]$km/s. Obtain the flight path angle and the six classical Keplerian orbital elements.
(d) For an Earth satellite, given $r_p = 7000$ km and $r_a = 10,000$ km, find ν and r, 3600 s after perigee passage and the period of the orbit. When $\nu = 90°$, find r, v and altitude.
7. A comet makes a close approach to the Earth on a hyperbolic passage. The fly-by induces a change in the direction of the comet's velocity vector with respect to the Earth of 60° at its velocity at a distance very far away from the Earth (infinity) and with respect to it is 2000 m/s.
Starting from the first principles, determine (i) the semi-major axis of the Comet's orbit, (ii) the eccentricity, (iii) the true anomaly of the comet at infinity, (iv) the periapsis altitude and (v) the velocity at the periapsis.
Compare the velocity at infinity to the velocity of the comet at a distance of 1000000 km from the Earth.
8. At a particular instant in time, the following vectors define the position and velocity of a vehicle orbiting Earth, respectively, in a space-fixed coordinate system:

$$\mathbf{r} = (1\mathbf{i} + 1\mathbf{j} + 0\mathbf{k})/\sqrt{2}, \mathbf{v} = (-1\mathbf{i} + 1\mathbf{j} + 0\mathbf{k})/\sqrt{2}$$

The unit of time is taken to be 1000 s. The unit of length is taken as 8000 km.

(a) Determine the orbit inclination and the eccentricity vector. You may assume the value of the gravitation parameter to be $\mu = 400,000$ km^3/s^2. Hence, also choose appropriate values for the argument of the ascending node and the argument of the perigee.
(b) Calculate the true anomaly of the orbiting vehicle at a point which is 8000 km from the Earth's surface.
(c) Explain the term *Eccentric anomaly* and its use.
(d) Explain the term *vis-viva equation*. At the point where $\mathbf{r} = (1\mathbf{i} + 1\mathbf{j} + 0\mathbf{k})/\sqrt{2}$, calculate the magnitude of the velocity increment required by the orbiting vehicle, in km/s, if the vehicle is required to leave the Earth's gravitational field.

9. (a) A satellite's position is known at two points, at times t_1 and t_2 with the corresponding eccentric anomalies E_1 and E_2 and the corresponding vector positions being $\mathbf{r}_1$ and $\mathbf{r}_2$. Obtain general relations between the eccentricity of the orbit and the time of flight.
   (b) A satellite on a transfer orbit (TO) is required to be in the two positions defined by the vectors $\mathbf{r}_1 = [1 \quad 0 \quad 0]$ and $\mathbf{r}_2 = [-0.0767 \quad 1.5217 \quad 0]$. The TO's semi-major axis is 1.5648 and its eccentricity is 0.3666. Find the time of flight.
10. A satellite is approaching the planet Jupiter with a relative velocity of 10 km/s at the edge of its sphere of influence. The eccentricity of the satellite's orbit about Jupiter is 1.08. Explaining all your assumptions and calculation steps, estimate:
    (a) The semi-major axis of the orbit and the distance of closest approach to Jupiter
    (b) The excess velocity and the total energy of the spacecraft as it approaches the planet
    (c) The turning angle
    (d) The velocity increment that the satellite acquires as it swings by the planet Jupiter
    (e) The excess velocity of the satellite as it departs the planet Jupiter
11. A satellite is approaching the planet Saturn ($\mu = 3.7931 \times 10^7$ km^3/s^2) with a relative velocity of 10 km/s at the edge of its sphere of influence. The eccentricity of the satellite's orbit about Saturn is 1.08. Explaining all your assumptions and calculation steps, estimate:
    (a) The semi-major axis of the orbit and the distance of closest approach to Saturn
    (b) The excess velocity and the total energy of the spacecraft as it approaches the planet
    (c) The turning angle
    (d) The velocity increment that the satellite acquires as it swings by the planet Saturn
    (e) The excess velocity of the satellite as it departs the planet Saturn
12. The semi-major axis of the planet Jupiter was 5.2034 AU, (1 AU = 1.496 $\times$ 10^{08} km) on a particular day in the year 2000. The orbit eccentricity was 0.04839, and the true anomaly was 19.65°.
    (a) Assume the orbit is Keplerian, and calculate its position or distance from the Sun in km.
    (b) Estimate the velocity of the planet at that particular position in km/s. How does this velocity differ if the velocity of the planet is assumed to be circular with the radius of the orbit equal to the semi-major axis (for a heliocentric orbit, $\mu = 1.327 \times 10^{11}$ km^3/s^2)?

(c) Given that the Inclination was 1.31°, the longitude of ascending node was 100.56° and the argument of periapsis was 14.75°, obtain the transformation to convert the orbit position and velocity to a heliocentric Cartesian frame. Clearly explain the derivation of the transformation. Hence, transform the position, the radial component of the velocity and the tangential component of velocity to obtain the position and velocity vector of the asteroid in the heliocentric Cartesian frame.

13. Reconsider the preceding exercise no 9, and obtain the corresponding modified equinoctial elements of the planet's orbit on the given particular day in the year 2000.
14. (a) Consider two reference frames, the first frame being fixed in space with its origin located at the Earth's centre. The frame's x_i axis passes through the equatorial plane and points to a fixed point in space. The y_i axis is perpendicular to the x_i axis and in the equatorial plane, while the z_i axis is normal to the equatorial plane. The second reference frame is a moving reference frame with its origin at the Earth's centre and with its x_s axis directed along the position vector of a spacecraft orbiting the Earth. The y_s axis of the moving reference frame is perpendicular to the position vector of the spacecraft and in the orbit plane, while the z_s axis is mutually perpendicular to the first two axes.
Using the Euler angle sequence formulation and a sequence of three Euler angle rotations, obtain from first principles the transformation matrix to transform any vector in the space-fixed reference frame to the moving reference frame. Hence, obtain the angular velocity vector of a body which is stationary in the moving reference frame.
(b) Obtain and state the components of the angular velocity vector in the moving reference frame defined in part (a). Explain and justify the steps involved in your derivation.

References

Bate, R., Mueller, D., & White, J. (1971). Orbit determination from two positions and time. In *Fundamentals of astrodynamics* (1st ed.). Dover Publications, Inc.

Battin, R. (1999). Solving Lambert's problem. In *An introduction to the mathematics and methods of astrodynamics* (Revised Edition). AIAA Education Series.

Chobotov, V. A. (2002). Position and velocity as a function of time. In *Orbital mechanics* (Third Edition). AIAA Education Series.

Clohessy, W. H., & Wiltshire, R. S. (1960). Terminal guidance system for satellite rendezvous. *Journal of the Aerospace Sciences, 27*, 5, 653–658, 674.

Curtis, H. (2005). Preliminary orbit determination. In *Orbital mechanics for engineering students* (1st ed.). Elsevier Butterworth-Heinemann.

Farquhar, R. W. (1968). Linear feedback control. In *The control and use of libration-point satellites* (pp. 73–91). SUDAAR NO. 350, Department of Aeronautics and Astronautics, Stanford University.

Fortescue, P. W., Swinerd, G. S., & Stark, J. P. W. (2011). *Spacecraft systems engineering* (4th ed.). Wiley Ltd.

Gooding, R. H. (1990). A procedure for the solution of lambert's orbital boundary-value problem. In *Celestial mechanics and dynamical astronomy* (Vol. 48(2)) (pp. 145–165). Kluwer Academic Publishers.

Hudson, H. S. (2011). Global properties of solar flares. *Space Science Reviews, 158*, 5–41. https://doi.org/10.1007/s11214-010-9721-4

Li, W., & Hudson, M. K. (2019). Earth's Van Allen radiation belts: From discovery to the Van Allen Probes era. *Journal of Geophysical Research: Space Physics, 124*, 8319–8351. https://doi.org/10.1029/2018JA025940

Prussing, J. E., & Conway, B. A. (1993). Position in orbit as a function of time. In *Orbital mechanics*. Oxford University Press.

Ramesh, K. B. (2010). Coronal mass ejections and sunspots—Solar cycle perspective. *The Astrophysical Journal Letters, 712*, L77–L80. https://doi.org/10.1088/2041-8205/712/1/L77

Sun, F. T. (1979) *On the minimum time trajectory and multiple solutions of Lambert's problem.* AAS/AIAA Astrodynamics Specialist Conference, AAS-79-164.

Vallado, D. (2004). Initial orbit determination. In *Fundamentals of astrodynamics and applications* (2nd ed.). Microcosm Press.

Vepa, R. (2019). *Dynamics and control of autonomous space vehicles and robotics*. Cambridge University Press.

Chapter 2
Celestial Mechanics: Real Orbits

2.1 Real Orbits: Perturbations

The definitions of the Keplerian orbits were based on several assumptions. To summarize, these assumptions were (i) the central body is a "point" mass (it is the only body other than the spacecraft—its gravity is the only force), (ii) the moving body is a point mass of negligible mass and finally (iii) there are no other forces/torques acting. However, when considering real orbits about the Earth, it is important to recognize that (i) the Earth is not a point mass; it is not spherical nor does it have a uniform mass distribution. It is not the only body. Its gravity is not the only force as the Sun, Moon and planets exert gravitational forces too. (ii) Gravity is not the only force as there is also the atmospheric drag, solar radiation pressure and several other disturbing forces. One also needs to consider orbit manoeuvres that could take a finite time or be impulsive.

The additional perturbation specific force vector (force vector per unit mass) or more simply the perturbing acceleration acting on the satellite can be represented as a three component vector, in one of two notations as

$$\mathbf{F} = F_r \mathbf{e}_R + F_s \mathbf{e}_T + F_n \mathbf{e}_N = R\mathbf{e}_R + T\mathbf{e}_T + N\mathbf{e}_N, \tag{2.1}$$

where $\mathbf{e}_R$ is a unit vector pointing radially outward and in the orbit plane, $\mathbf{e}_T$ is a unit vector pointing tangentially in the direction of motion and in the orbit plane and $\mathbf{e}_N$ is a unit vector pointing normal to the orbit plane and mutually perpendicular to the other two unit vectors. It may be recalled that for orbits around the Earth, the radial direction is along the geocentric radius vector of the spacecraft measured positive in a direction away from the geocentre, the tangential direction is perpendicular to this radius vector measured positive in the direction of orbital motion and the normal direction is positive along the angular momentum vector of the spacecraft's orbit.

R. Vepa, *Space Vehicle Maneuvering, Propulsion, Dynamics and Control*,
https://doi.org/10.1007/978-3-031-65518-0_2

The vector equation for Keplerian motion was derived in Chap. 1 and is

$$\ddot{\mathbf{r}}+\frac{G(m_1+m_2)}{|\mathbf{r}|^2}\frac{\mathbf{r}}{|\mathbf{r}|}=\ddot{\mathbf{r}}+\frac{\mu}{|\mathbf{r}|^2}\frac{\mathbf{r}}{|\mathbf{r}|}=\mathbf{0}, \tag{1.17}$$

With the perturbation force vector,

$$\ddot{\mathbf{r}}+\frac{G(m_1+m_2)}{|\mathbf{r}|^2}\frac{\mathbf{r}}{|\mathbf{r}|}=\ddot{\mathbf{r}}+\frac{\mu}{|\mathbf{r}|^2}\frac{\mathbf{r}}{|\mathbf{r}|}=\mathbf{F}=F_r\mathbf{e}_R+F_s\mathbf{e}_T+F_n\mathbf{e}_N. \tag{2.2}$$

The unit vectors are defined as

$$\mathbf{e}_R=\mathbf{r}/|\mathbf{r}|.\mathbf{e}_N=\mathbf{r}\times\mathbf{v}/|\mathbf{r}\times\mathbf{v}| \quad \text{and} \quad \mathbf{e}_T=\mathbf{h}\times\mathbf{r}/|\mathbf{h}\times\mathbf{r}|. \tag{2.3}$$

The consequence of the existence of the additional perturbation forces is that key conservation principles, such as conservation of tangential momentum and the conservation of energy, are no longer valid but can be used to derive the perturbation equations. The vis-viva equation

$$\frac{1}{2}v^2-\frac{\mu}{r}=-\frac{\mu}{2a}=\mathcal{E}. \tag{1.45}$$

is not valid. $\mathcal{E}$ is *not* a constant, and the semi-major axis a is *not* a constant. The angular momentum $\mathbf{h}=\mathbf{r}\times\mathbf{v}$ is *not* a constant. So $h^2=\mu a(1-e^2)=\mu p$ is *not* a constant. Thus, the eccentricity, e, and the semi-latus rectum, p, are *not* constants. However, from the visa-viva equation:$a=\mu/(2\mathcal{E})$. Hence it follows that

$$\dot{a}=-\frac{\mu}{2\mathcal{E}^2}\dot{\mathcal{E}}=-a\frac{\dot{\mathcal{E}}}{\mathcal{E}}=\frac{2a^2}{\mu}\dot{\mathcal{E}}, \quad \text{or,} \quad \frac{\dot{a}}{a}=-\frac{\dot{\mathcal{E}}}{\mathcal{E}}=\frac{2a}{\mu}\dot{\mathcal{E}}. \tag{2.4}$$

Any loss of energy will cause a to decay!

From the principle of virtual work and energy, $d\mathcal{E}=\mathbf{F}\cdot d\mathbf{r}$,

$$\dot{\mathcal{E}}=\mathbf{F}\cdot\dot{\mathbf{r}}=\mathbf{F}\cdot\left(\dot{r}\mathbf{e}_R+r\dot{\theta}\mathbf{e}_T\right)=\dot{r}R+r\dot{\theta}T. \tag{2.5}$$

So $\dot{a}$ may be found, if the perturbing force components are known.

Furthermore, since $h^2=\mu a(1-e^2)=\mu p$,

$$e^2=1-\left(h^2/\mu a\right)=1+2\left(h^2/\mu^2\right)\mathcal{E}. \tag{2.6}$$

So,

$$\frac{\mu^2}{h^2\mathcal{E}}e\dot{e}=2\frac{\dot{h}}{h}+\frac{\dot{\mathcal{E}}}{\mathcal{E}}=2\frac{\dot{h}}{h}-\frac{\dot{a}}{a}, \quad \text{or,} \quad \dot{e}=\frac{h^2\mathcal{E}}{e\mu^2}\left(2\frac{\dot{h}}{h}-\frac{\dot{a}}{a}\right). \tag{2.7}$$

From Newton's second law,

$$\dot{\mathbf{h}} = \mathbf{r} \times \mathbf{F} = \mathbf{F} \cdot \left(\dot{r}\mathbf{e}_R + r\dot{\theta}\mathbf{e}_T\right) = rT\mathbf{e}_N - rN\mathbf{e}_T. \tag{2.8}$$

However, since changes in the magnitude of $\mathbf{h}$ are normal to the orbit plane, $\dot{h} = rT$; so $\dot{e}$ may be found.

To find the time rate of change of the true anomaly, recall that $r^2\dot{\nu} = h$. For a Keplerian orbit,

$$\dot{\nu} = \frac{h}{r^2} = \frac{\mu^2}{h^3}(1 + e\cos\nu)^2. \tag{2.9}$$

To find $\dot{i}$, $\dot{\omega}$ and $\dot{\Omega}$, one needs to consider the definitions of these angles and differentiate (Vallado, 2004; Curtis, 2005; Prussing & Conway, 1993)

$$i = \cos^{-1}(h_z/|\mathbf{h}|), \quad \omega = \cos^{-1}\left(\frac{\mathbf{e}}{|\mathbf{e}|}\frac{\mathbf{n}}{|\mathbf{n}|}\right), \quad \Omega = \tan^{-1}\left(-h_x/h_y\right). \tag{2.10}$$

One also needs to differentiate the line node equation, the eccentricity vector and the eccentricity:

$$\cos\nu = \frac{\mathbf{r} \cdot \mathbf{e}}{|\mathbf{r}||\mathbf{e}|}, \quad \mathbf{n} = \mathbf{k} \times \mathbf{h} = \left[-h_y \;\; h_x \;\; 0\right]/h, \quad h = |\mathbf{h}|, \quad \mathbf{e} = \frac{\mathbf{v} \times \mathbf{h}}{\mu} - \frac{\mathbf{r}}{|\mathbf{r}|}, \quad e = |\mathbf{e}|. \tag{2.11}$$

Thus, in principle, the rates of change of the Keplerian orbital elements due to the presence of known perturbing forces may be found.

2.2 Gauss Perturbation Equations

The simplicity of the two-body motion and the fact that real problems of celestial mechanics involve perturbation forces make it practical to use two-body motion as a zero approximation. The position and velocity of the orbiting body can be calculated by the standard formulas of two-body problem which are generally valid with constant elements that represent the usual solution of the two-body problem, by imposing additional constraints. Thus, we generate a set of elements that are functions of time, which coincide with the fixed elements at the instant of their computation. The velocity is independent of variations in the orbital elements (constraint in the method of variation of parameters). These are the osculating orbital elements (osculating means touch).

The equations for rates of change of the six *classical orbital elements*, $\dot{a}, \dot{e}, \dot{\nu}, \dot{\Omega}, \dot{i}, \dot{\omega}$, in terms of the perturbation force components, F_r, F_s and F_n, are known as the Gauss perturbation equations (Bate et al., 1971) and summarized below.

$$\dot{a} = \frac{da}{dt} = \frac{2a^2}{h}[F_r e \sin\nu + (1 + e\cos\nu)F_s] = \frac{2e\sin\nu}{n\sqrt{1-e^2}}F_r + \frac{2a\sqrt{1-e^2}}{nr}F_s, \tag{2.12}$$

$$\dot{e} = \frac{de}{dt} = \frac{\sqrt{1-e^2}\sin\nu}{na}F_r + \frac{\sqrt{1-e^2}}{nea^2}\left[\frac{a^2(1-e^2)}{r} - r\right]F_s, \tag{2.13}$$

$$\dot{\omega} + \dot{\Omega}\cos i = \frac{d\omega}{dt} + \frac{d\Omega}{dt}\cos i = -\frac{\sqrt{1-e^2}}{nae} \times \left[F_r\cos\nu - \left(1 + \frac{1}{1+e\cos\nu}\right)F_s\sin\nu\right], \tag{2.14}$$

$$\dot{i} = \frac{di}{dt} = \frac{r\cos\theta}{na^2\sqrt{1-e^2}}F_n = \frac{r\cos\theta}{h}F_n, \quad \theta = \omega + \nu, \tag{2.15}$$

$$\dot{\Omega} = \frac{d\Omega}{dt} = \frac{r}{na^2\sqrt{1-e^2}}\frac{\sin\theta}{\sin i}F_n = \frac{r}{h}\frac{\sin\theta}{\sin i}F_n = \frac{1}{\sin i}\frac{di}{dt}, \tag{2.16}$$

$$\dot{\nu} = \frac{d\nu}{dt} = \frac{h}{r^2} + \frac{1}{he}[F_r p\cos\nu - (p+r)F_s\sin\nu] = \frac{h}{r^2} - \left(\frac{d\omega}{dt} + \frac{d\Omega}{dt}\cos i\right), \tag{2.17}$$

or,

$$\dot{M} = \frac{dM}{dt} = n - \frac{2r}{na^2}F_r - \sqrt{1-e^2}\left(\frac{d\omega}{dt} + \frac{d\Omega}{dt}\cos i\right). \tag{2.18}$$

In the above equations, θ is the argument of the latitude.

The primary effects of the perturbation forces can be classified into two groups. In cases the integral around an orbit (averaged over orbit) of the perturbation is zero, there are no secular changes to orbital elements. When the integral is nonzero, it implies that the perturbation forces potentially lead to secular changes to orbital elements (linear increase with time) or sustained changes in the orbital elements over the period of the orbit.

2.2.1 Nonsingular Orbital Elements

Examining the above perturbation equations, it is easy to observe the singular nature of these equations for certain values of the eccentricity and the inclination. For these reasons, several alternate sets or orbital elements have been proposed. In what follows, one such element set is defined as it is extremely useful for purposes of orbit propagation. These orbital elements are known as the modified equinoctial orbital elements which are a set of orbital elements that are also useful for trajectory analysis and optimization. They are valid for circular, elliptic and hyperbolic orbits. The modified equinoctial element evolution equations (Brouke & Cefola, 1972; Walker et al., 1985) exhibit no singularity for zero eccentricity and orbital inclinations equal to 0 and 90°. However, two of the components are singular for an orbital inclination of 180°. The relationships between modified equinoctial and classical orbital elements are stated first.

$$\begin{aligned} &p=a(1-e^2),\ f=e\cos(\omega+\Omega),\ \ g=e\sin(\omega+\Omega),\\ &h=\tan(i/2)\cos\Omega,\ k=\tan(i/2)\sin\Omega,\ \ L=\omega+\Omega+\nu, \end{aligned} \tag{2.19}$$

In the above equations, L is known as the true longitude. It is easy to show that the inverse relationships are

$$a=p/\left(1-f^2-g^2\right), \tag{2.20}$$

$$e=\sqrt{f^2+g^2}, \tag{2.21}$$

$$i=2\tan^{-1}\left(\sqrt{h^2+k^2}\right)=\tan^{-1}\left(2\sqrt{h^2+k^2},1-h^2-k^2\right), \tag{2.22}$$

$$\omega=\tan^{-1}(g/f)-\tan^{-1}(k/h)=\tan^{-1}(gh-fk,fh+gk), \tag{2.23}$$

$$\Omega=\tan^{-1}(k,h),\nu=L-\omega-\Omega=L-\tan^{-1}(g/f) \tag{2.24}$$

or,

$$\theta=\omega+\nu=\tan^{-1}(h\sin L-k\cos L,h\cos L+k\sin L). \tag{2.25}$$

In these equations, an expression of the form $\tan^{-1}(y,x)$ indicates a four quadrant inverse tangent calculation. The relationships between the position and velocity vectors, respectively, in the Earth centred inertial frames of reference and the modified equinoctial elements may be shown to be given by

$$\mathbf{r}=\frac{r}{s^2}\begin{bmatrix}(1+\alpha^2)\cos L+2hk\sin L\\(1-\alpha^2)\sin L+2hk\cos L\\2h\sin L-2k\cos L\end{bmatrix}. \tag{2.26}$$

$$\mathbf{v}=\frac{1}{s^2}\sqrt{\frac{\mu}{p}}\begin{bmatrix}(1+\alpha^2)(g+\sin L)-2hk(\cos L+f)\\(1-\alpha^2)(\cos L+f)-2hk(\sin L+g)\\2h\cos L+2k\sin L+2fh+2gk\end{bmatrix}. \tag{2.27}$$

or,

$$\alpha^2 = h^2 - k^2,\ s^2 = 1 + h^2 + k^2,\ r = p/w \quad \text{and} \quad w = 1 + f\cos L + g\sin L. \tag{2.28}$$

The Gauss perturbation equations may now be expressed in terms of the modified equinoctial elements and are

$$\dot{p} = \frac{dp}{dt} = 2\sqrt{\frac{p}{\mu}}\frac{p}{w}F_s, \tag{2.29}$$

$$\dot{f} = \frac{df}{dt} = \sqrt{\frac{p}{\mu}}\left(F_r \sin L + \frac{(f + (w+1)\cos L)}{w}F_s - (h\sin L - k\cos L)\frac{g}{w}F_n\right), \tag{2.30}$$

$$\dot{g} = \frac{dg}{dt} = \sqrt{\frac{p}{\mu}}\left(-F_r\cos L + \frac{(g + (w+1)\sin L)}{w}F_s + (h\sin L - k\cos L)\frac{g}{w}F_n\right), \tag{2.31}$$

$$\dot{h} = \frac{dh}{dt} = \sqrt{\frac{p}{\mu}}\frac{s^2}{2w}F_n\cos L. \tag{2.32}$$

$$\dot{k} = \frac{dk}{dt} = \sqrt{\frac{p}{\mu}}\frac{s^2}{2w}F_n\sin L, \tag{2.33}$$

and

$$\dot{L} = \frac{dL}{dt} = \sqrt{\mu p}\frac{w^2}{p^2} + \frac{1}{w}\sqrt{\frac{p}{\mu}}(h\sin L - k\cos L)F_n. \tag{2.34}$$

Each of the orbital elements are now no longer constant throughout an orbit. But they do define an "instantaneous" Keplerian orbit called an osculating orbit (meaning a touching orbit), which is illustrated in Fig. 2.1. If the perturbing force is removed, the orbit has the orbital elements at that instant in time. For unperturbed two-body motion, there are no perturbing forces. Therefore, for two-body motion, the first five modified equinoctial orbital elements are, indeed, constant.

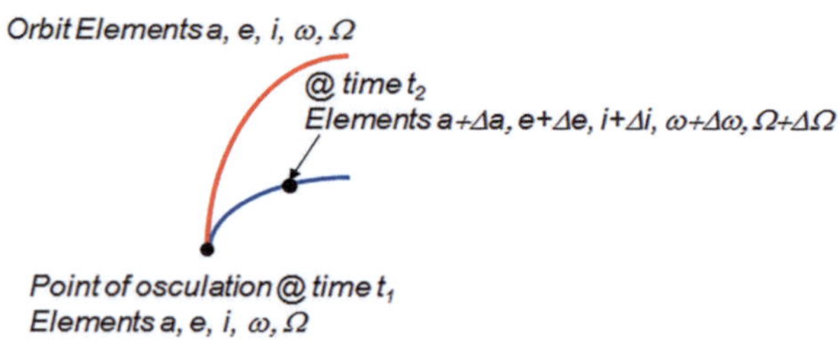

Fig. 2.1 Illustration of an osculating orbit

The special case of a circular orbit presents an interesting example to consider. A spacecraft in circular orbit about the Earth has an electrothermal propulsion unit, which provides a thrust tangentially to the orbit in the plane of the orbit. Moreover, for a circular Keplerian orbit,

$$v = \frac{\sqrt{\mu}}{\sqrt{r}}. \tag{2.35}$$

Differentiating Eq. 2.35 and applying Newton's second law of motion,

$$\dot{v} = -\frac{1}{2}\frac{\sqrt{\mu}}{\sqrt{r^3}}\dot{r} = -f. \tag{2.36}$$

A negative thrust, $-f$, must be applied to slow the satellite down for orbit raising. If one lets $r = a$ and assume the mass is almost constant, under these conditions, the rate of change in semi-major axis may be written as

$$\frac{da}{dt} = 2f\sqrt{\frac{a^3}{\mu}}. \tag{2.37}$$

where f is the thrust per unit mass acting on the satellite. If it is assumed that the thrust is constant and the satellite is initially in a circular orbit of semi-major axis r_1, then it can be shown that the decrement in the velocity from v_1 to v_2, $v_1 - v_2 = \Delta v$ required to raise the orbit from r_1 to r_2 may be determined. Integrating both sides of Eq. 2.37 from r_1 to r_2 and applying the limits,

$$\int \frac{da}{\sqrt{a^3}} = \int \frac{2}{\sqrt{\mu}} f dt \Rightarrow \frac{\sqrt{\mu}}{2}\int_{r_1}^{r_2} \frac{da}{\sqrt{a^3}} = f(t_2 - t_1) = f\Delta t = \Delta v. \tag{2.38}$$

Thus, it follows that

$$\Delta v = \frac{\sqrt{\mu}}{2}\int_{r_1}^{r_2} \frac{da}{\sqrt{a^3}} = \frac{\sqrt{\mu}}{\sqrt{r_1}} - \frac{\sqrt{\mu}}{\sqrt{r_2}} = \frac{\sqrt{\mu}}{\sqrt{r_1}}\left(1 - \frac{\sqrt{r_1}}{\sqrt{r_2}}\right) = v_1\left(1 - \frac{\sqrt{r_1}}{\sqrt{r_2}}\right). \tag{2.39}$$

From Eq. 2.39, it follows that

$$\frac{\Delta v}{v_1} = 1 - \frac{\sqrt{r_1}}{\sqrt{r_2}}, \tag{2.40}$$

where v_1 is the initial orbital velocity of the satellite. One can also estimate the time taken for orbit raising, $\Delta t = \Delta v/f$, and the mass of fuel required assuming a constant fuel consumption rate ($m_{\text{fuel}} = \dot{m}_{\text{fuel}} \Delta t$).

2.2.2 Gravitational Perturbations, Station Keeping

The primary perturbation force acting on an Earth orbiting satellite is due gravitational perturbations. There are several gravitational perturbations acting on satellite in an Earth orbit. Of these, the non-spherical nature of the planet Earth causes some significant perturbations.

2.2.3 Non-spherical Nature of Central Body-1

The gravitational potential of the Earth, the central body for all satellite orbits around it, is not only spherically symmetric but also spherically nonuniform. The principal non-spherical variation is the bulging of the Earth along the equator due to the Earth's rotation. This variation, which causes the Earth to be an oblate spheroid, is still axially symmetric.

The standard gravitational potential function of a sphere of mass m_s is $U = Gm_s/r = \mu/r$. For an oblate spheroid model of the Earth, the gravitational potential has an additional term:

$$U = G\oiiint_V \frac{dm}{r_{dm}} = \frac{\mu}{r} - \frac{\mu}{r}\sum_{k=2}^{\infty}\left(\frac{R_E}{r}\right)^k J_k P_k(\cos\theta). \tag{2.41}$$

In Eq. 2.41, R_E is the radius of the Earth's equator, J_k are constant coefficients, P_k are Legendre polynomial functions of order k and θ is the colatitude, r_{dm} is the distance of the spacecraft from an element of mass dm and r is the distance of the spacecraft from CM of the Earth (or planet). Equation 2.41 represents the zonal harmonic gravitation model for a perfect oblate spheroid. The frame of reference used is spherical polar frame, with the radius, the colatitude and the azimuth or longitude as the three coordinates. The frame is illustrated in Fig. 2.2. The Cartesian coordinates are related to the spherical polar coordinates by the relations

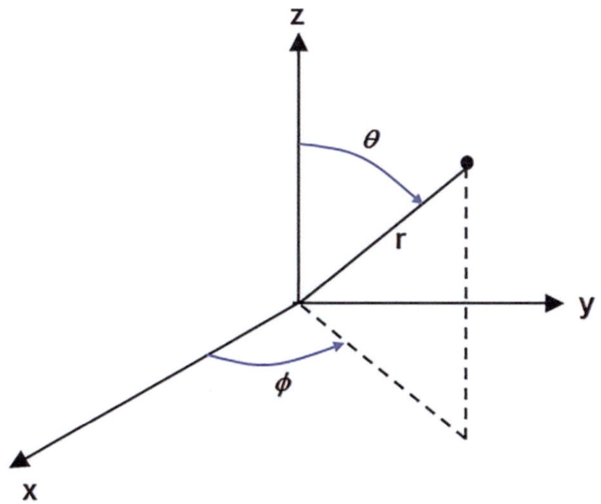

Fig. 2.2 The spherical polar coordinates

$$x = r\sin\theta\cos\phi, \quad y = r\sin\theta\sin\phi, \quad z = r\cos\theta. \tag{2.42}$$

The divergence operator is defined by

$$\nabla\cdot\mathbf{V} = \frac{1}{r^2}\frac{\partial}{\partial r}\left(r^2V_r\right) + \frac{1}{r\sin\theta}\frac{\partial}{\partial\theta}(\sin\theta V_\theta) + \frac{1}{r\sin\theta}\frac{\partial}{\partial\phi}\left(V_\phi\right). \tag{2.43}$$

The gradient operator is

$$\nabla V = \frac{\partial V}{\partial r}\boldsymbol{u}_r + \frac{1}{r}\frac{\partial V}{\partial\theta}\boldsymbol{u}_\theta + \frac{1}{r\sin\theta}\frac{\partial V}{\partial\phi}\boldsymbol{u}_\phi. \tag{2.44}$$

The gravitational acceleration is then given by the gradient of the potential. The non-spherical gravitational acceleration vector can be expressed as

$$\nabla U = \mathbf{g} = -g_r\boldsymbol{u}_r + g_\theta\boldsymbol{u}_\theta. \tag{2.44}$$

After truncating the series in the potential function, the non-spherical gravitational acceleration vector is given by

$$\begin{aligned}\mathbf{g} = -g_r\boldsymbol{u}_r + g_\theta\boldsymbol{u}_\theta = \frac{\mu}{r^2}\sum_{k=2}^{K}(k+1)\left(\frac{R_e}{r}\right)^k P_k(\cos\theta)J_k\boldsymbol{u}_r \\ - \frac{\mu\sin\theta}{r^2}\sum_{k=2}^{K}\left(\frac{R_e}{r}\right)^k P'_k(\cos\theta)J_k\boldsymbol{u}_\theta.\end{aligned} \tag{2.45}$$

The first six J_k coefficients are

$$J_2 = 1082.64 \times 10^{-6},\ J_3 = -2.69 \times 10^{-6},\ J_4 = -1.58 \times 10^{-6},$$
$$J_5 = -0.15 \times 10^{-6},\ J_6 = 0.59 \times 10^{-6},\ J_7 = -0.44 \times 10^{-6}. \tag{2.46}$$

For a zonal only Earth gravity model, the east component is identically zero. Physically, the rotation of the planet causes the equatorial bulge and this is reflected in the J_2 term. Thus, for a uniformly distributed planet, like Mars, J_2 is proportional to the ratio of centrifugal acceleration at the equator to the gravitational acceleration, $(=\omega^2 R_P^3/\mu)$. However, for a centrally concentrated planet, like Jupiter and Saturn, the ratio is lower, and consequently these planet's effective radius is smaller!

In Eq. 2.45, the Legendre polynomials P_k may be defined recursively from

$$P_0(q) = 1,\ P_1(q) = q,\ P_2(q) = (3q^2 - 1)/2,$$
$$P_3(q) = q(5q^2 - 3)/2,\ P_n = \frac{1}{2^n n!}\frac{\partial^n}{\partial q^n}(q^2 - 1)^n. \tag{2.47}$$

The zonal harmonic expansion given by Eq. 2.41 is independent of the longitude, $\phi = \lambda$, and valid for the oblate spheroid model of the Earth. There is a more general model where the coefficients also depend on the longitude, λ. The associated non-spherical gravitational field may be expressed in terms of Legendre (P_k) and associated Legendre polynomials (P_{nm}), the distance r of the satellite from the Earth's centre, the longitude λ of the satellite measured from the Greenwich meridian and positive eastward, the latitude l of the satellite taken to be positive towards north and the equatorial Earth radius, $R_e = 6378.16$ km:

$$U = \frac{\mu}{r} - \frac{\mu}{r} \times \left[\sum_{n=2}^{\infty}\left(\frac{R_e}{r}\right)^n J_n \times P_n(\sin l) - \sum_{n=2}^{\infty}\sum_{m=1}^{\infty}\left(\frac{R_e}{r}\right)^n J_{nm} \times P_{nm}(\sin l)\cos(m(\lambda - \lambda_{nm}))\right]. \tag{2.48}$$

The functions P_n and P_{nm}, Legendre and associated Legendre polynomials, respectively, are given by

$$P_n(x) = \frac{1}{2^n!} \times \frac{d^n}{dx^n}(x^2 - 1)^n,\ P_{nm}(x) = (1 - x^2)^{\frac{m}{2}} \times \frac{d^m}{dx^m} P_n(x). \tag{2.49}$$

J_n, J_{nm} and λ_{nm} are constants characterizing the Earth's mass distribution. J_n are the *zonal* harmonics which were introduced earlier. J_{nm} and λ_{nm} are associated with the *tesseral* and *sectorial* ($n = m$) harmonics related to the ellipticity of the equator which results in a 150 m difference between the equator's major and minor axes (most important: $n = 2, 3$, $m = 1, 2$). For the Earth, $J_{21} = 0$, $J_{22} = 1.86 \times 10^{-6}$, $J_{31} = 2.1061 \times 10^{-6}$ and all other J_{nm} are at most equal to 1.0×10^{-6}.

2.2.4 *Oblateness (J_2)* Perturbations

However, the principal terms in Earth's model of the gravitational potential are given by the J_2 term. The zonal gravity perturbation contribution due to J_2 is

$$U = \frac{\mu}{r} - \frac{\mu}{r}\left(\frac{R_E}{r}\right)^2 J_2 P_2(\cos\theta) = \frac{\mu}{r} - \frac{\mu}{r}\left(\frac{R_E}{r}\right)^2 J_2\left(\frac{3z^2}{2r^2} - \frac{1}{2}\right). \tag{2.50}$$

The definition of the colatitude θ and the longitude $\phi = \lambda$ of a point defining the spacecraft's location on a sphere of radius r is shown in Fig. 2.2. Furthermore, $z = r \sin \theta = r \sin i \sin (\omega + \nu)$. The perturbation force vector is

$$\mathbf{F}_{obs} = \nabla U = \left(\mathbf{e}_r \frac{\partial}{\partial r} + \mathbf{e}_z \frac{\partial}{\partial z}\right) \cdot U = -\mu R_E^2 J_2 \left(\mathbf{e}_r \frac{\partial}{\partial r} + \mathbf{e}_z \frac{\partial}{\partial z}\right) \cdot \left(\frac{3z^2}{2r^5} - \frac{1}{2r^3}\right). \tag{2.51}$$

Equation 2.51 reduces to

$$\begin{aligned}\mathbf{F}_{obs} &= -\mu R_E^2 J_2 \left(\mathbf{e}_r\left(\frac{3}{2r^4} - \frac{15z^2}{2r^6}\right) + \mathbf{e}_z\left(\frac{3z}{r^5}\right)\right) \quad \text{with} \\ \mathbf{e}_r &= \mathbf{e}_x \cos\phi \sin\theta + \mathbf{e}_y \sin\phi \sin\theta + \mathbf{e}_z \cos\theta.\end{aligned} \tag{2.52}$$

The gravitation force vector is transformed to the Cartesian coordinates and then to the planetary coordinates using the Euler angle transformation defined in Chap. 1. Thus, observing that the angles Ω, i and $\theta = \omega + \nu$ represent a 3-1-3 sequence of Euler angles, the unit vectors satisfy the relations

$$\begin{bmatrix}\mathbf{e}_R\\ \mathbf{e}_T\\ \mathbf{e}_N\end{bmatrix} = \begin{bmatrix}c\theta & s\theta & 0\\ -s\theta & c\theta & 0\\ 0 & 0 & 1\end{bmatrix}\begin{bmatrix}1 & 0 & 0\\ 0 & ci & si\\ 0 & -si & ci\end{bmatrix}\begin{bmatrix}c\Omega & s\Omega & 0\\ -s\Omega & c\Omega & 0\\ 0 & 0 & 1\end{bmatrix}\begin{bmatrix}\mathbf{e}_x\\ \mathbf{e}_y\\ \mathbf{e}_z\end{bmatrix}, \tag{2.53}$$

$$\begin{bmatrix}\mathbf{e}_x\\ \mathbf{e}_y\\ \mathbf{e}_z\end{bmatrix} = \begin{bmatrix}c\Omega & -s\Omega & 0\\ s\Omega & c\Omega & 0\\ 0 & 0 & 1\end{bmatrix}\begin{bmatrix}1 & 0 & 0\\ 0 & ci & -si\\ 0 & si & ci\end{bmatrix}\begin{bmatrix}c\theta & -s\theta & 0\\ s\theta & c\theta & 0\\ 0 & 0 & 1\end{bmatrix}\begin{bmatrix}\mathbf{e}_R\\ \mathbf{e}_T\\ \mathbf{e}_N\end{bmatrix}. \tag{2.54}$$

The expression for the perturbation force is

$$\mathbf{F}_{obs} = -\frac{3\mu R_E^2 J_2}{r^4}\left(\frac{\mathbf{e}_R}{2}\left(1 - 3\sin^2 i \sin^2\theta\right) + \sin\theta \sin i(\sin i\mathbf{e}_T + \cos i\mathbf{e}_N)\right). \tag{2.55}$$

Thus, the three components of the force vector are

$$
\begin{aligned}
R_{obs} &= -\frac{3\mu R_E^2 J_2}{2r^2 p^2}(1+e\cos(\theta-\omega))^2\left(1-3\sin^2\theta\sin^2 i\right),\\
T_{obs} &= -\frac{3\mu R_E^2 J_2}{r^2 p^2}(1+e\cos(\theta-\omega))^2\sin\theta\sin^2 i,\\
N_{obs} &= -\frac{3\mu R_E^2 J_2}{r^2 p^2}(1+e\cos(\theta-\omega))^2\sin\theta\sin i\cos i.
\end{aligned}
\tag{2.56}
$$

These components could also be expressed in terms of the modified equinoctial elements as

$$
\begin{aligned}
R_{obs} &= -\frac{3\mu R_E^2 J_2}{2r^4}\left(1-\frac{12(h\sin L-k\cos L)^2}{1+h^2+k^2}\right),\\
T_{obs} &= -\frac{12\mu R_E^2 J_2}{r^4}\left(\frac{(h\sin L-k\cos L)(h\cos L+k\sin L)}{1+h^2+k^2}\right),\\
N_{obs} &= -\frac{6\mu R_E^2 J_2}{r^4}\left(\frac{(1-h^2-k^2)(h\sin L-k\cos L)}{1+h^2+k^2}\right).
\end{aligned}
\tag{2.57}
$$

2.2.5 *Orbital Averaging, Precession of the Line of Apsides, Regression of the Line of Nodes, Secular J_2 Perturbations*

Using the Gauss perturbation equations, one could establish equations for $\dot{\Omega}$, $\dot{\omega}$ and $\dot{M}$. Of particular importance are the so-called "secular" components of these equations which grow with time. The periodic terms could be averaged over a period of the orbit. The only secular terms that remain are in $\dot{\Omega}$ (rate of change of the RAAN), $\dot{\omega}$ (rate of change of the argument of the perigee) and $\dot{M}$ (rate of change of the mean anomaly) remain. The equation for the time rate of change of the argument of the ascending node, $\dot{\Omega}$, is given by

$$
\dot{\Omega} = \frac{h}{\mu}\frac{1}{(1+e\cos(\theta-\omega))}\frac{\sin\theta}{\sin i}N. \tag{2.58}
$$

Consider the function

$$\frac{1}{(1+e\cos(\theta-\omega))^3}\frac{\sin\theta}{\sin i}N=\widetilde{N}=(1+e\cos(\theta-\omega))\sin^2\theta\cos i. \tag{2.59}$$

Averaging $\widetilde{N}$ over the orbit,

$$\begin{aligned}\overline{\widetilde{N}}&=\frac{1}{2\pi}\int_0^{2\pi}(1+e\cos(\theta-\omega))\sin^2\theta\cos i\,d\theta=\frac{\cos i}{2\pi}\int_0^{2\pi}\sin^2\theta\,d\theta\\&=\frac{\cos i}{2}.\end{aligned} \tag{2.60}$$

Further the average value of $\dot{\theta}$ is the orbital rate n. Thus,

$$\langle\dot{\Omega}\rangle_{av}=-\frac{3\mu R_E^2J_2}{p^2}\left\langle\frac{h}{\mu r^2}\right\rangle\overline{\widetilde{N}}=-\frac{R_E^2J_2}{p^2}\langle\dot{\theta}\rangle_{av}\overline{\widetilde{N}}=-\frac{3n}{2}\left(\frac{R_E}{p}\right)^2J_2\cos i. \tag{2.61}$$

The average rate of change in $\langle\dot{\Omega}\rangle_{av}$ results in the regression (as it is negative) of the line of nodes $\Delta\Omega$, due to planetary oblateness for a circular orbit of radius a, with an inclination to the equator of i (since $p=a$ and $n=2\pi$):

$$\langle\Delta\Omega\rangle_{av}=-3\pi\left(\frac{R_P}{a}\right)^2J_2\cos i\text{ rads/orbit revolution.} \tag{2.62}$$

where R_P is the mean radius of the planet. The line of nodes and line of apsides are illustrated in Fig. 2.3.

Consider equatorial bulge which results in an additional force relative to the spherical body. The force is responsible for a tangential torque on s/c, so the rate of change of momentum is no longer a constant $dh/dt\neq 0$. Consider direction of torque, which is

$$\mathbf{T}=\mathbf{r}\times\mathbf{F}_{\text{due to bulge}}. \tag{2.63}$$

For prograde orbits $(i<90°)$, the torque is a westward torque both in the southern hemisphere and in the northern hemisphere. Thus, for prograde orbits, the orbit plane rotates westward (Ω positive east). This westward motion is called the regression of the line of nodes ($\Delta\Omega$ is negative). The regression rate is proportional to the inclination. When the inclination angle is zero, the regression rate is a maximum. It is minimum and equal to zero for polar orbits.

It is therefore conceivable that for a certain value of the inclination angle, the regression rate is synchronous with the orbital motion of the Sun. Sun-synchronous orbits are required to orient the solar panels towards the Sun, continually. The regression of the line of nodes in an orbit circular about the Earth may be approximated by the expression given by Eq. 2.62. It is the time taken by the Earth to rotate through one revolution with respect to the vernal equinox. The orbital period of the

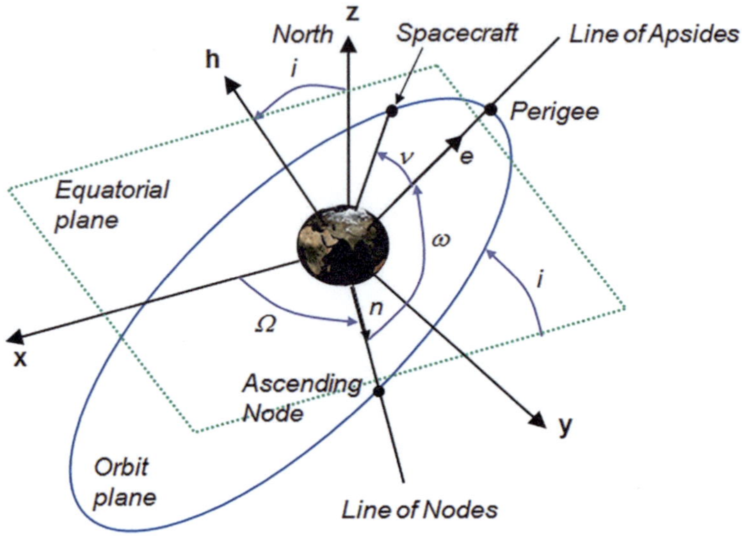

Fig. 2.3 Lines of nodes and apsides

Earth around the Sun, in a heliocentric orbit of radius r_e, for one revolution, and a satellite orbital period, in a circular geocentric orbit of radius r_c, around the Earth are, respectively, given by

$$T_s = \frac{2\pi}{n_s} = 2\pi\sqrt{\frac{r_e^3}{\mu_{\text{sun}}}},\ \ T_e = \frac{2\pi}{n_e} = 2\pi\sqrt{\frac{r_c^3}{\mu_{\text{earth}}}}. \tag{2.64}$$

The regression angle rate per second, for the planet Earth, is

$$\Delta\dot{\Omega} = -\frac{3\pi}{T_e}\left(\frac{R_P}{r_c}\right)^2 J_2 \cos i = -\frac{3}{2}R_E^2 J_2 \cos i\sqrt{\frac{GM_e}{r_c^7}}. \tag{2.65}$$

For a Sun-synchronous prograde orbit, $\left|\Delta\dot{\Omega}\right| = 1/T_s$. Thus,

$$\left|\Delta\dot{\Omega}\right| = -\frac{3}{2}R_E^2 J_2 \cos i\sqrt{\frac{GM_e}{r_c^7}} = -\sqrt{\frac{GM_s}{r_e^3}}. \tag{2.66}$$

Thus, the inclination angle for Sun-synchronous orbits must satisfy

$$\cos i = -\frac{2}{3J_2}\sqrt{\frac{M_s}{M_e}}\sqrt{\frac{r_c^7}{R_E^4 r_e^3}} = -k_{ss}\sqrt{r_c^7}. \tag{2.67}$$

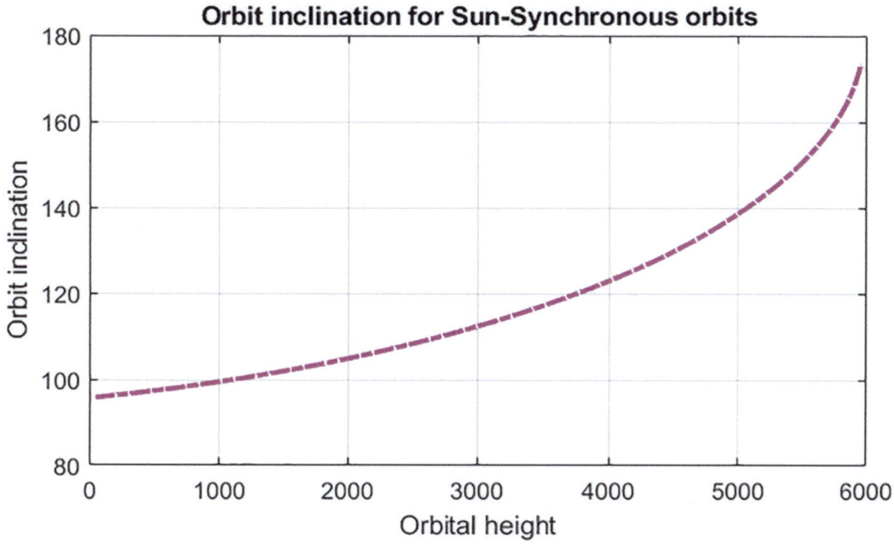

Fig. 2.4 Variation of the orbit inclination (in °) for various orbital heights (in km)

In Eq. 2.67, k_{ss} is a constant. The variation of the orbit inclination for various orbital heights is shown in Fig. 2.4.

A Sun-synchronous orbit precesses through one complete revolution/year, so it always maintains the same orientation with respect to the Sun. A Sun-synchronous orbit is achieved by having the orbital plane precess or rotate, about an axis normal to the Earth's orbital plane (ecliptic), approximately 1° eastward each day with respect to the celestial sphere to keep pace with the Earth's orbit around the Sun. A Sun-synchronous orbit can place a satellite in constant sunlight, which allows the solar panels to work continually. However, a Sun-synchronous orbit is only feasible for orbital heights up to about 5974 km. For low Earth orbits (LEO), the inclination is just over 90°. For orbital heights closer to 5974 km, the inclination is near 180°. An example of a Sun-synchronous orbit is shown in Fig. 2.5. The design of such orbits is considered in some detail by Boain (2004).

The mean motion of the argument of the perigee is given by

$$\begin{aligned}\frac{d\omega}{dt}+\frac{d\Omega}{dt}\cos i = & -\frac{\sqrt{1-e^2}}{nae}\\ & \times\left[R\cos(\theta-\omega)-\left(1+\frac{1}{1+e\cos(\theta-\omega)}\right)T\sin(\theta-\omega)\right].\end{aligned} \tag{2.68}$$

The averaging of $\frac{d\omega}{dt}+\frac{d\Omega}{dt}\cos i$ may be done just as before. Hence,

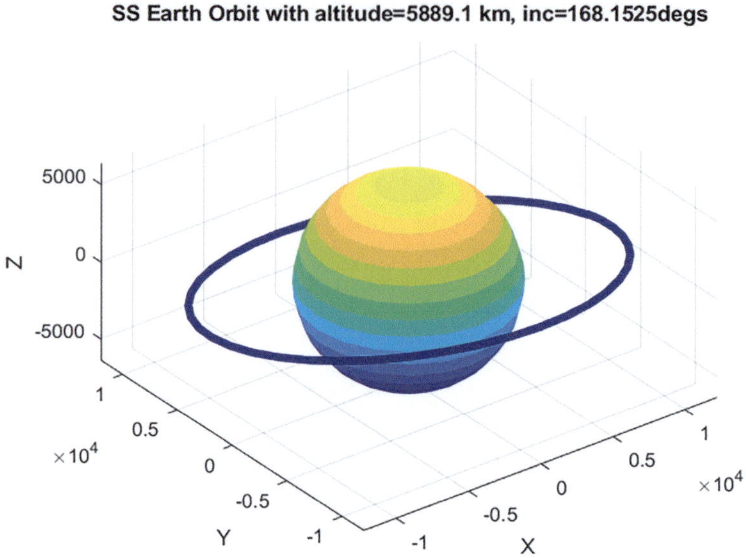

Fig. 2.5 A typical example of a Sun-synchronous orbit

$$\langle\dot{\omega}\rangle_{av} = \frac{3n}{2}\left(\frac{R_E}{p}\right)^2 J_2\left(2 - \frac{5}{2}\sin^2 i\right). \tag{2.69}$$

The mean motion of the argument of the perigee is known as the precession of the line of apsides, the line joining the perigee with the apogee. Clearly when

$$\sin^2 i = 4/5, \ \ \langle\dot{\omega}\rangle_{av} = 0. \tag{2.70}$$

The Russian Molniya orbits for Earth satellites are good examples of orbits that ensure the precession of the line of apsides is absent. They are highly elliptical orbits with an inclination of 63.43°, an argument of perigee of 270° and an orbital period of approximately half a sidereal day as shown in Fig. 2.6. There are innovative uses for the Molniya orbit.

The other important effect of the J_2 term is the increase in the mean anomaly. Thus, one may summarize the mean effects of the J_2 term as

$$\langle\dot{\Omega}\rangle_{av} = -\frac{3n}{2}\left(\frac{R_E}{p}\right)^2 J_2\cos i, \ \ \langle\dot{\omega}\rangle_{av} = \frac{3n}{2}\left(\frac{R_E}{p}\right)^2 J_2\left(2 - \frac{5}{2}\sin^2 i\right),$$

$$\langle\dot{M}\rangle_{av} = \frac{3n}{2}\left(\frac{R_E}{p}\right)^2 J_2\sqrt{1-e^2}\left(1 - \frac{3}{2}\sin^2 i\right). \tag{2.71}$$

Fig. 2.6 Example of a Molniya orbit

2.2.6 *Stability of a Spacecraft in a Geostationary Orbit: Station Keeping*

Ideally the final manoeuvres performed after launch would place a satellite into a required geostationary orbit. However, there are two drift effects that disturb the spacecraft's equilibrium position in the orbit. The first is a north-south drift due to a changing inclination of the orbit, as a consequence of the perturbations of the Moon and the Sun. These are due to the gravitational effects of the Moon and the Sun on an Earth orbiting satellite. The second is the east-west drift due to the changing longitude of the orbit. This is due to variations in the Earth's gravitational field, mainly the J_{22} term in the gravitational potential function of the Earth. This results in four equilibrium points on the orbit, two stable and two unstable. Retaining the most relevant terms in the gravitation perturbation, the gravitation potential is dominated by the J_{22} term. This is often referred to as the triaxiality perturbation. As a consequence, the longitude dependence of the perturbing force becomes approximately sinusoidal with four equilibrium points, along the orbit in the equatorial plane. Two of these are stable equilibrium points since any small longitude deviation from them would induce a drift back towards them. The other two points are unstable. Although theoretically supposed to be symmetrically located, they are marginally nonsymmetric. The spacecraft will drift towards the stable points and away from the unstable points. The persistent gravitational triaxiality perturbations

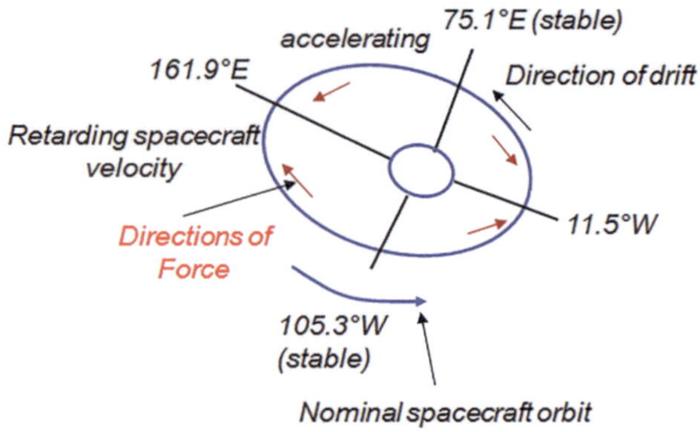

Fig. 2.7 Equilibrium points of a geostationary satellite along the orbit in the equatorial plane

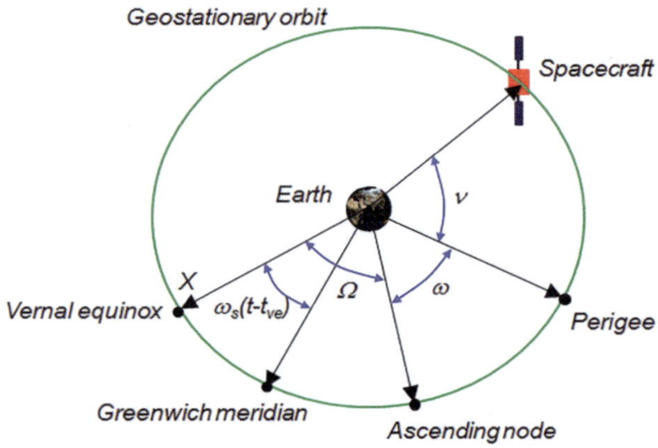

Fig. 2.8 Definition of the satellite's longitude

create two stable and two unstable equilibrium points at 75.1°E and 105.3°W and at 161.9°E and 11.5°W, along the geostationary orbit, respectively. Due to the perpetual existence of these perturbations, the nominal location of a geostationary satellite relative to Earth changes as the satellites drift away from the unstable equilibrium points towards the stable equilibrium points. To maintain a satellite on a geostationary orbit location, within its specified longitudinal and latitudinal bounds, periodic corrective impulses must be applied (satellite station keeping), with satellites nearer the unstable equilibrium points requiring more vigorous station keeping. The net longitudinal drift is always towards the stable equilibrium points which are located at 75.1°E and 105.3°W, as illustrated in Fig. 2.7.

The Earth satellite's longitude is defined from the Greenwich meridian to the satellite's position as shown in Fig. 2.8. Every time the longitude reaches the maximum allowable value, the E-W thruster provides an impulse changing the sign of longitude velocity. This burn brings the GEO satellite to reach the other edge of the E-W dead band, with zero longitude drift acceleration, and the cycle repeats itself. As the cycle is then repeated, and the GEO satellite is maintained within the dead band limits, as required. This is known as the single sided orbit control. The satellite longitude is measured from the Greenwich meridian to the satellite true location:

$$\lambda = \Omega + \omega + \nu - \omega_e(t - t_{ve}). \tag{2.72}$$

The Gauss perturbation equations are written in terms of six synchronous elements: the semi-major axis or the drift rate, the longitude and two components of the eccentricity vector and the inclination. When linearized, the first two equations, in terms of a perturbation potential *R*, decouple from the last four, and only these are used.

$$\frac{da}{dt} = \frac{2}{na}\frac{\partial R}{\partial \lambda}, \quad \frac{d\lambda}{dt} = n - \omega_e - \frac{2}{na}\frac{\partial R}{\partial a}. \tag{2.73}$$

The equations may be solved and it can be shown that

$$\dot{\lambda} = -18 J_{22}\omega_e \left(\frac{R_E}{a_{geo}}\right)^2 \sin^2(2(\lambda - \lambda_{22})). \tag{2.74}$$

With $k \approx 0.9 < 1$, the Δv for station keeping is

$$\Delta v = 4\pi k \sqrt{\Delta\lambda \left|\ddot{\lambda}_{\max}\right|}. \tag{2.75}$$

Generally, $\left|\ddot{\lambda}_{\max}\right|$ is of the order of 0.2°/day/day and typically $\Delta v_{\max}$ is of the order of 2 *m/s* per year.

2.2.7 Third-Body Perturbations-2

If one lets $\mu = G(m_\oplus + m_S) \approx Gm_\oplus$, when $m_S < < m_\oplus$, then in the case of *j* third bodies, one may derive the following relationship for the satellite's acceleration which may be expressed as

$$\ddot{\vec{r}}_{\oplus \to S} = -\mu \frac{\vec{r}_{\oplus \to S}}{r^3_{\oplus \to S}} + \sum_{j=1}^{J} Gm_{B3j} \left(\frac{\vec{r}_{\oplus \to B3j} - \vec{r}_{\oplus \to S}}{\left|\vec{r}_{\oplus \to B3j} - \vec{r}_{\oplus \to S}\right|^3} - \frac{\vec{r}_{\oplus \to B3j}}{r^3_{\oplus \to B3j}} \right). \tag{2.76}$$

In Eq. 2.76, the subscripts $\oplus$, $B3_j$ and S refer to the Earth, a third body and the satellite, respectively, and the subscript $\oplus \rightarrow S$ indicates a vector that originates at the Earth and ends at the satellite. The first term in preceding equation is the familiar two-body equation of motion, while the second term arises from third-body perturbations. With $\mathbf{r} = \vec{r}_{\oplus \rightarrow S}$, the equation is expressed as

$$\ddot{\mathbf{r}} = -\mu \frac{\mathbf{r}}{|\mathbf{r}|^3} + \mathbf{f}_{3S}\left(\vec{r}_{\oplus \rightarrow B3j}, \vec{r}_{\oplus \rightarrow S}\right) = -\mu \frac{\mathbf{r}}{|\mathbf{r}|^3} + \mathbf{f}_{3S}\left(\vec{r}_{\oplus \rightarrow B3j}, \vec{r}_{\oplus \rightarrow S}\right). \tag{2.77}$$

with the perturbation force vector

$$\begin{aligned} \begin{bmatrix} F^{3rd}_{x,p} \\ F^{3rd}_{y,p} \\ F^{3rd}_{z,p} \end{bmatrix} &= \mathbf{f}_{3S}\left(\vec{r}_{\oplus \rightarrow B3j}, \vec{r}_{\oplus \rightarrow S}\right) \\ &= \sum_{j=1}^{J} Gm_{B3j} \left(\frac{\vec{r}_{\oplus \rightarrow B3j} - \vec{r}_{\oplus \rightarrow S}}{\left|\vec{r}_{\oplus \rightarrow B3j} - \vec{r}_{\oplus \rightarrow S}\right|^3} - \frac{\vec{r}_{\oplus \rightarrow B3j}}{r^3_{\oplus \rightarrow B3j}} \right). \end{aligned} \tag{2.78}$$

2.2.8 *Third-Body Effects of the Moon and the Sun (Lunisolar)*

When applied to the Sun (slow) and the Moon (fast), their influence on a spacecraft is about the same magnitude as the J_2 term in a geostationary orbit. The lunar perturbation effects are typically twice the solar perturbation effects. Key secular effects are similar in form to variations in Ω (RAAN), ω and M (mean anomaly) which remain. They are similar in form to the J_2 term, but smaller in magnitude. While the J_2 term causes the precession of the orbit normal (ON) about the Earth's N-S axis, solar perturbations cause precession of the ON about the ecliptic plane's $N - S$ axis. A key periodic term that dominates is the variations in the inclination i which can vary from $0° < i < 15°$. Typically, $\Delta v_{\max}$ is of the order of 40–50 *m/s* per year.

2.3 Non-gravitational Perturbations

There are several important non-gravitational perturbations with primary contributions due to aerodynamic drag forces, solar wind and radiation pressure, magnetic forces, forces and moments exerted by the impact of space debris and control forces and moments.

2.3.1 Aerodynamic Drag-3

With certain assumptions, which are valid for satellites in low Earth orbits, the effect of aerodynamic drag may be assessed. The primary influence of aerodynamic drag is to retard the satellite. The main assumption is that the aerodynamic drag force acts in a direction opposite the velocity of the satellite. When the orbit is nearly circular, the normal and radial force components are $N = R = 0$, while only the tangential force is nonzero and given by

$$T = -(1/2)\rho F^2 v^2 (C_D A/m). \tag{2.79}$$

where ρ is the local atmospheric density, F is the ratio of the satellite's relative velocity with respect to the Earth and to its orbital velocity, v is the satellite orbital velocity, C_D is the drag coefficient, m is the satellite's mass, and A is the projected area of the satellite normal to the flow. The quantity in the brackets, $C_D A/m$, is known as the ballistic coefficient.

For a near-circular orbit, the equation for the time rate of change of the semi-major axis $\dot{a}$ is given by the first of Gauss's perturbation Eq. (2.12) and is

$$\frac{\dot{a}}{a} = 2\sqrt{\frac{a}{\mu(1-e^2)}}(R \times e \sin \nu + T \times e \cos \nu + T). \tag{2.80}$$

For a circular orbit from the vis-viva equation (1.42),

$$\frac{v^2}{2} = \frac{\mu}{a} - \frac{\mu}{2a} = \frac{\mu}{2a}. \tag{2.81}$$

Thus, the Eq. (2.80) reduces to

$$\dot{a} = 2a^2 T/h = -\sqrt{\mu a}\rho F^2 (C_D A/m). \tag{2.82}$$

Assuming that the atmospheric density, ρ, is a constant, and rearranging and integrating Eq. (2.82),

$$\int_{a_i}^{a_f} a^{-1/2} da = -\sqrt{\mu}\rho F^2 (C_D A/m)(t_f - t_i) = \sqrt{a_f} - \sqrt{a_i}. \tag{2.83}$$

While the orientation of the orbit plane is unaffected by the drag, the semi-major axis changes as a result of the continual orbital decay, causing the spacecraft to crash into the Earth unless appropriate corrective forces are applied. The decay of the tangential momentum may be assessed by the application of Newton's law, and this results in

$$m\frac{dV}{dt} = \frac{1}{2}\rho V^2 A C_D = m \times \frac{1}{2}\rho V^2 \frac{AC_D}{m} = mqB^*. \tag{2.84}$$

If the variation of density with height and the dynamics of the semi-major axis is, respectively, given by

$$\rho = \rho_{SL} \exp(-h/h_{\text{scale}}), \quad \dot{a} = -\sqrt{\mu a}\rho F^2 B^*, \tag{2.85}$$

the dynamics of the semi-major axis reduces to

$$\dot{a} = -\sqrt{\mu a} F^2 B^* \rho_{SL} \exp(-(a - r_E)/h_{\text{scale}}). \tag{2.86}$$

Integrating Eq. 2.86 results in

$$t = -\int_{a_0}^{a} \exp((a - r_E)/h_{\text{scale}}) da / \left(F^2 B^* \rho_{SL} \sqrt{\mu a}\right). \tag{2.87}$$

It follows that

$$t_{\text{decay}} = \{\exp(h_0/h_{\text{scale}}) - 1\} h_{\text{scale}} / \left(F^2 B^* \rho_{SL} \sqrt{\mu r_E}\right). \tag{2.88}$$

Generally, the aerodynamic drag is highest at perigee of an orbit where the velocity is highest. Atmospheric drag has the tendency to "circularize" the orbit due to larger change in radius at the apogee and smaller change in radius at the perigee. So, the trajectory spirals down continuously, until the orbit is almost circular at zero height. Finally the trajectory is a continuous spiral as it approaches the Earth's surface.

In the case of noncircular orbit, the radial, tangential and normal perturbations due to aerodynamic drag are given by the following three expressions:

$$R = -(1/2)\rho F^2 v v_r (C_D A/m), \quad T = -(1/2)\rho F^2 v v_t (C_D A/m), \quad N = 0. \tag{2.89}$$

In Eq. 2.89, in terms of the modified equinoctial elements,

$$v_r = (\mu/p)(f\sin L - g\cos L), \quad v_t = (\mu/p)(1 + f\cos L + g\sin L), \tag{2.90}$$

Gauss's equations in terms of the modified equinoctial elements may now be numerically integrated to predict the trajectory of the satellite in the presence of atmospheric drag.

2.3.2 *Prediction of the Drag Coefficient*

A reasonably accurate method of predicting the drag coefficient is essential to predict the drag force, which is highly variable. There are three methods available: (i) hypersonic gas flow (modified Newtonian flow), (ii) free molecular flow theory and (iii) the hyperthermal approximation (Newtonian). The methods allow one to establish bounds on the drag force and the drag coefficient.

To begin with, a consistent model of the atmospheric density is adopted for most planets:

$$\rho/\rho_0 = \exp(-(h - h_0)/h_{\max}), \quad 0 \le h \le h_0, \tag{2.91}$$

where $h_{\max}$ is a scale height. The model constants differ for all planets. There are many other models that have been proposed based on Eq. 2.91 which defines a basic hydrostatic model and which seeks to incorporate the nonuniformities in time and space. The models that go further are the Harris and Priester model which model the diurnal density bulge and a host of other models proposed by Jacchia and his co-workers to account for temporal as well as geomagnetic and latitudinal variations. These models have been summarized by Montenbruck and Gill (2005). To calculate drag, using one of these density models, one needs to be able to evaluate the drag coefficient. However, flow is not a continuum as in traditional fluid mechanics.

Classical Newtonian theory, which is the basis for the *modified Newtonian approach*, is sometimes called impact theory because it is derived by assuming a column of air moving normal to the surface brought to a complete stop relative to the surface.

Consider a fluid approaching an inclined surface as shown in Fig. 2.9. In this situation, fluid particles loose normal momentum but conserve tangential momentum. Let V_∞ be velocity of fluid approaching an inclined surface with angle θ, and then the velocity normal to the surface and the mass flux blocked by the surface are, respectively, given by

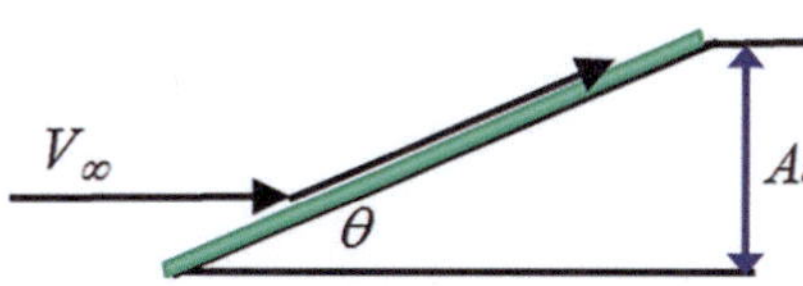

Fig. 2.9 Fluid flow approaching an inclined surface

$$V_n = V_\infty \sin\theta, \quad q_n = \rho_\infty V_\infty \sin\theta. \tag{2.92}$$

The momentum lost by the fluid per unit time, is given by,

$$p_n = \rho_\infty A V_\infty \sin\theta \times V_\infty \sin\theta = \rho_\infty A V_\infty^2 \sin^2\theta. \tag{2.93}$$

The momentum lost per unit time is converted to a force acting on the plate. Hence, the force and pressure acting on the plate are, respectively, given by

$$F_n = \rho_\infty A V_\infty^2 \sin^2\theta, \quad \Delta P_n = F_n/A = \rho_\infty V_\infty^2 \sin^2\theta. \tag{2.94}$$

Impulse-momentum considerations then readily lead to

$$P - P_\infty = \Delta P_n = F_n/A = \rho_\infty V_\infty^2 \sin^2\theta. \tag{2.95}$$

If a pressure coefficient is defined as $c_p = 2(P - P_\infty)/\rho_\infty V_\infty^2$, it follows that

$$c_p = 2(P - P_\infty)/\rho_\infty V_\infty^2 = 2\sin^2\theta. \tag{2.96}$$

Modifications to the above have been proposed where the "2" is replaced by $c_{p,\max}$. Thus,

$$c_p = c_{p,\max} \sin^2\theta. \tag{2.97}$$

To further modify the expression for the pressure coefficient, first the plate is assumed to be normal to the flow. The streamline impacting the stagnation point on the plate passes through a normal shock before being isentropically compressed to zero speed. Rayleigh's pitot formula is used to estimate $c_{p,\max}$. Using the Rayleigh's pitot formula,

$$P_{\text{stag}} = P_\infty \left(\frac{(\gamma+1)M_\infty^2}{2}\right)^{\gamma/(\gamma-1)} \left(\frac{\gamma+1}{2\gamma M_\infty^2 - (\gamma-1)}\right)^{1/(\gamma-1)}. \tag{2.98}$$

where from compressible fluid dynamics theory, P_{stag} is equal to the stagnation pressure and γ is the ratio of specific heats. Integrating the pressure over the surface, let

$$C_D = F_D / \left(0.5\rho V_\infty^2 A_{\text{ref}}\right). \tag{2.99}$$

An expression for the coefficient of drag C_D is given by

$$C_D = 2\left\{\left(\frac{(\gamma+1)^2 M_\infty^2}{4\gamma M_\infty^2 - 2(\gamma-1)}\right)^{\gamma/(\gamma-1)}\left(\frac{(2M_\infty^2-1)}{\gamma M_\infty^2} + \frac{2(1-M_\infty^2)}{(\gamma+1)\gamma M_\infty^2}\right) - \frac{1}{\gamma M_\infty^2}\right\}. \tag{2.100}$$

In hypersonic flow, it is important to set M_∞ to ∞. As free stream Mach number, $M_\infty \to \infty$

$$C_D = 4\left\{\left(\frac{(\gamma+1)^2}{4\gamma}\right)^{\gamma/(\gamma-1)}\left(\frac{1}{\gamma+1}\right)\right\}. \tag{2.101}$$

For diatomic gases at very low temperature and pressure, the ratio of specific heats is $\gamma = 5/3$. Hence,

$$C_D = \left(\frac{16}{15}\right)^{1.6}\left(\frac{3}{2}\right) = 1.6632. \tag{2.102}$$

For monoatomic gases, $\gamma = 1.4$ and

$$C_D = \left(\frac{5.76}{5.6}\right)^{3.5}\left(\frac{4}{2.4}\right) = 1.8394. \tag{2.103}$$

In the upper reaches of the atmosphere, the flow is considerably rarefied, and it is then possible to make some simplifying assumptions to model the flow. The molecular mean free path is the average distance travelled by a molecule before encountering another. The effect of reflected particles is neglected and the flow is undisturbed by the presence of the body. Hence, one may assume there are no shock waves. The atmospheric density is small enough, and one may assume that the molecular mean free path is large compared with the dimensions of a typical spacecraft. The incoming flow of molecules colliding with the spacecraft's outer surfaces may be treated as independent particles and separately from the flow of molecules moving away from those surfaces. This simplified model of the aerodynamic flow is referred to as *free molecular flow*. In this case, consider the mean free path λ of the flow. For altitudes of about 200 km, the mean free path λ is about 240 m. For altitudes of about 800 km, the mean free path λ is about 1400 m. If we define the Knudsen number as $K_n = \lambda/d$, where $d =$ characteristic dimension in the flow, when $K_n < < 1$, the flow is a continuum flow; when $K_n \approx 1$, the flow is in transition; and when $K_n > > 1$, the flow is a free molecular flow. In the presence of free molecular flow, Cook (1965) has established a formula for the drag coefficient for a flat plate, given by

$$C_D = 2\left(1 + \frac{2}{3}\sqrt{1-\alpha}\sin\phi\right), \quad \alpha = \frac{3.6\mu}{(1+\mu)^2}. \tag{2.104}$$

Typically, $\mu = 0.6$, where μ is the ratio of the (mean) mass of the incident gas atom to the mass of the surface atom and ϕ is the angle of incidence of the plate to the flow. The drag coefficient is

$$C_D = 2.527, \quad C_{D_{\max}} = 2.6667. \tag{2.105}$$

When $K_n \approx 1$, the flow is in transition and it tends to slip on a surface and $V_{\text{flow}} \neq 0$ at the surface. Also, $T_{\text{flow}} \neq T_{\text{surface}}$. This may be characterized by considering the energy exchange or momentum exchange between the flow species and the surface. In this case, it is essential to define an energy accommodation coefficient:

$$a_{\text{acc}} = (T_{\text{flow}} - T_{\text{reflected}})/(T_{\text{flow}} - T_{\text{surface}})$$

Hence, $a_{\text{acc}} = 1$ for "perfect" accommodation. The coefficient of drag is expressed as

$$C_D = 2a_{\text{acc}}\left(1 + \frac{2}{3}\sqrt{1-\alpha}\sin\phi\right). \tag{2.106}$$

An alternate approach to the study of the estimation of the drag force is based on the *hyperthermal approximation* although the results are similar to the modified Newtonian approach. Consider the flow past a body as shown in Fig. 2.10.

It is assumed that the molecules come to rest on impact and that there is negligible transverse velocity. This is the hyperthermal approximation. Hence, the volume swept out in $dV_s = SVdL$ where S = projected area of the spacecraft. Assuming further that the N atoms/m^3 each has mass of m kg, the momentum increment Δp to the surface of the body is

$$\Delta p = NmSVdl \Rightarrow \Delta p/\Delta t \rightarrow NmSVdl/dt = \text{drag force} = \rho SV^2. \tag{2.107}$$

Hence, the coefficient of drag can be expressed as

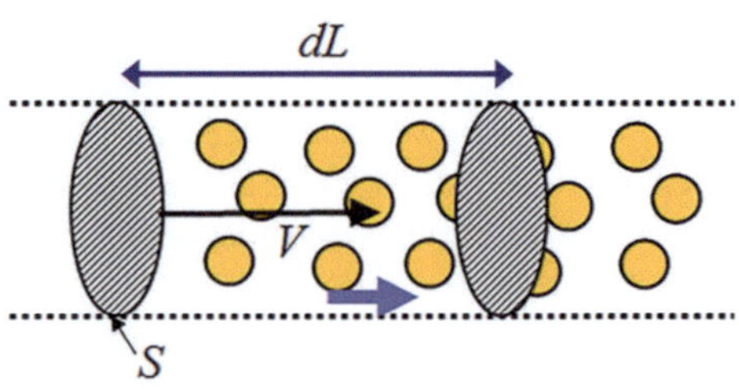

Fig. 2.10 Flow past a body: the hyperthermal approximation

$$C_D = \frac{f_{\text{drag}}}{(1/2)\rho S V^2} = 2. \tag{2.108}$$

The result is the same as that obtained from the Newtonian approximation. To fully comprehend the validity of the hyper-thermal approximation, it may be noted that if specular reflection (atom reflected from surface) is present with only a reversal of normal component of velocity, the coefficient of drag would be $C_D = 4$. Typically accepted value for a sphere $C_D = 2.2$. In general, the drag coefficient is a function of surface texture, surface temp, flow constituents, flow temp and the flow velocity. Moreover, hyperthermal approximation is not fully valid at the spacecraft's velocities which could be too high!

2.4 Solar Radiation Pressure and Wind

2.4.1 *Photonic Pressure-4*

An Earth satellite is exposed to solar radiation and experiences a net force that arises from the cumulative effect of the small impact forces generated by photons as they are absorbed or reflected by the spacecraft's surface. Thus, the acceleration imparted to satellite by the exposure to solar radiation depends predominantly on its mass and surface area, quite unlike the gravitational perturbations. The magnitude of the solar radiation pressure is determined by the solar flux, given by

$$\Phi = \Delta E/(A\Delta t). \tag{2.109}$$

When a single photon strikes a solid surface, part of is momentum is absorbed by the surface resulting in solar radiation pressure (SRP). The total energy, ΔE, that is provided to an area, A, and normal to it by all of the N impinging photons, in a time interval Δt, is

$$\Delta E = \sum_N E_\nu = \sum_N hf, \tag{2.110}$$

where f is the wave frequency in cycles/sec and h is Planck's constant given by $h = 6.62607004 \times 10^{-34}\ m^2\ kg/s$.

A single photon of energy, E_ν, carries an impulse, given by its momentum, $p = E_\nu/c$, where c is the velocity of light. Thus, summing up, the total impulse of an absorbing body that is exposed to the Sun is

$$\Delta p = \Delta E/c = \Phi A \Delta t/c. \tag{2.111}$$

The time interval over which this happens is Δt. The force experienced by the satellite is

$$F = \Delta p/\Delta t = \Delta E/c\Delta t = \Phi A/c. \tag{2.112}$$

It is proportional to the product of the photonic pressure (Φ/c) and the area of exposure. The acceleration imparted to the spacecraft of mass m_{sc} is

$$a = F/m_{sc} = \Phi A/m_{sc}c. \tag{2.113}$$

At a distance of one AU, the solar flux provided by the Sun to the Earth and the corresponding photonic pressure can be measured and are, respectively, given by

$$\Phi = 1.367\text{Watts/m}^2,\ \ P = \Phi/c = 4.56 \times 10^{-6}\text{N/m}^2. \tag{2.114}$$

In deriving the above relation, we have assumed that the surface area is normal to the incoming radiation, without including the effects of reflections, eclipses and shadows. Given the solar radiation flux at the Earth's surface, the acceleration the spacecraft experiences due to SRP is estimated from $a = PAc_r/m_{sc}$, where c_r is the coefficient of reflectivity. The coefficient of reflectivity depends on the optical properties for the spacecraft's surface and its geometry. For a flat, highly reflective surface facing the Sun, $c_r \approx 2$, while for a fully absorbent surface, $c_r = 1$.

One can include the effects of an area being at an angle to the incident ration by considering Fig. 2.11. If one expresses the SRP as, $P = \Phi/c \equiv I/c$, where the incident flux is $I = I_0$, and the corresponding SRP is, $P = I_0/c$. The corresponding incident and reflected forces are

$$F_0 = A(I_0\cos\theta/c)(\cos\theta\boldsymbol{i} + \sin\theta\boldsymbol{j}),\ \ F_R = RA(I_0\cos\theta/c)(\cos\theta\boldsymbol{i} - \sin\theta\boldsymbol{j}). \tag{2.115}$$

The total force is

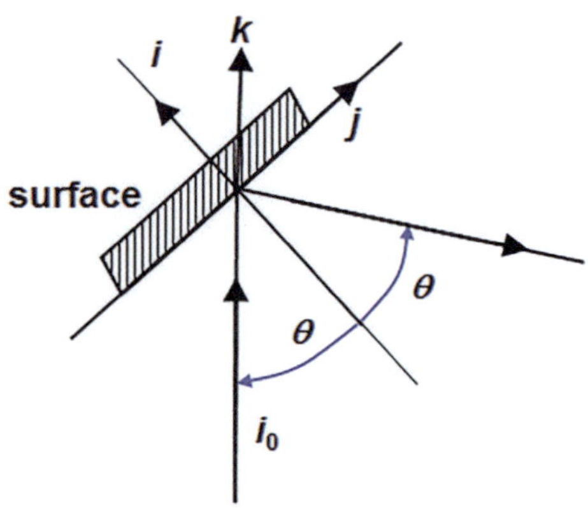

Fig. 2.11 Geometry of the incident SRP and the corresponding flux

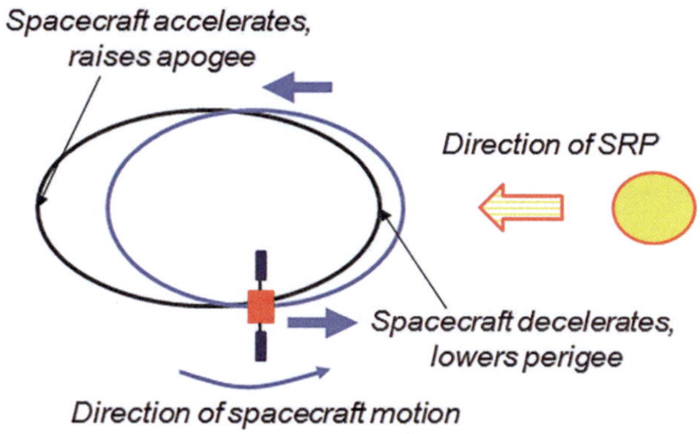

Fig. 2.12 Influence of the SRP on an Earth satellite's orbit

$$F = A(I_0\cos\theta/c)(\cos\theta\boldsymbol{i} + \sin\theta\boldsymbol{j}) + RA(I_0\cos\theta/c)(\cos\theta\boldsymbol{i} - \sin\theta\boldsymbol{j}). \quad (2.116)$$

For a perfect reflector, $R = 1$, or for a perfect absorber, $R = 0$, and one has, respectively,

$$F = 2A(I_0 cos\theta/c)cos\theta\boldsymbol{i}, \quad F = A(I_0\cos\theta/c)(\cos\theta\boldsymbol{i} + \sin\theta\boldsymbol{j}). \quad (2.117)$$

The direction of the total force on a spacecraft due to the SRP is determined by the three components of the force and has an important consequence in increasing the eccentricity of the orbit. This is illustrated in Fig. 2.12.

When photons strike a surface, they could be partially absorbed or be reflected either in a mirrorlike or specular manner, each in a specific direction, or be reflected diffusively in many directions. Whether photons of differing wavelengths are absorbed or reflected either specularly or diffusively depends to a large extent on the nature of the surface they strike, the directions of the photons relative to the surface, the texture of the surface as well as properties which determine the absorbent energy. Thus, typically following Marshall and Luthcke (1994), Vallado (2004) and Zardaín et al. (2020), a spacecraft is modelled as a collection of N plates, and the total acceleration due to the SRP which could be adsorbed or reflected either specularly or diffusively is given by

$$a = -(P/m_{\mathrm{sc}}) \sum_{k=1}^{N} A_k \cos\theta_k \left(\left(1 - c_{sr}^k\right)\boldsymbol{r}_s + 2\left(c_{sr}^k \cos\theta_k + c_{dr}^k/3\right)\boldsymbol{n}_k \right) H(\theta_k). \quad (2.118)$$

In Eq. 2.115, A_k is the area of the kth plate, $\boldsymbol{n}_k$ is the normal vector to the plate and defines its orientation, $\boldsymbol{r}_s$ is the normalized direction of the Sun from the spacecraft,c_{sr}^k and c_{dr}^k are its specular and diffusive reflectivity coefficients, respectively, and $\cos\theta_k = \boldsymbol{r}_s \cdot \boldsymbol{n}_k$ is the scalar product of the two unit vectors. The function

$H(\theta_k)$ is to ensure that the side of the plate is accounted for, so $H(\theta_k) = 0$ if $\cos\theta_k < 0$ and 1 elsewhere.

The extension of Eq. 2.113 to the case including the effects of eclipses and shadows is discussed by Montenbruck and Gill (2005).

Wetterer et al. (2014), Kubo-oka and Sengoku (1999) and List et al. (2015) have discussed modelling and estimation of SRP in some detail. Some novel applications of the SRP have been proposed by Lücking et al. (2012). In fact, it is now recognized that the geometry (topology) of a spacecraft could be optimized to minimize the effects of the perturbing forces.

2.4.2 Infrared Radiation + Albedo + Solar Wind-5, 6

One could compare the solar flux around the Earth's surface (1 AU from the Sun), $\Phi = 1.367$ Watts/m^2 with ALBEDO, the radiation from the Earth itself: $\Phi_E = 0.3$ Watts/m^2 which is about 22%. The SRP is $P = 4.56 \times 10^{-6}$N/m^2, while the corresponding Earth's albedo pressure is $P_E = 1.00 \times 10^{-6}$N/m^2. So generally to include the effects of both Earth's albedo and the solar wind pressure, the flux and pressure are empirically increased by about 30%. However, there do exist sophisticated finite element based methods not only to estimate the SRP but also the effects of both Earth's albedo and the solar wind pressure.

2.5 Magnetic Forces-7, Meteors and Space Debris Impact-8

A magnetic force, a force of attraction or repulsion, generally exists between electrically charged particles because of their motion. While electrostatic forces exist among stationary electric charges, both electric and magnetic forces are generated in the presence of moving electric charges. The magnetic force between two moving charges may be described as the effect exerted upon either charge by the magnetic field due to the other. Thus, the magnetic force on the second particle is proportional to its charge, the magnitude of its velocity, the magnitude of the magnetic field produced by the first moving charge and the direction of the path of the second particle relative to the direction of the magnetic field. The force is zero if the second charge is travelling in the same direction of the magnetic field and is greatest if it travels across the magnetic field. The magnetic force on a moving charge is exerted in a direction at a right angle to the plane formed by the direction of its velocity and the direction of the surrounding magnetic field.

Extending the argument further, both moving charges result in magnetic fields, and the force of attraction or repulsion is due to the interaction of the fields. The magnetic fields are a store of intrinsic potential energy. The interaction of the fields converts some of the potential energy to kinetic energy. Placing two unlike magnetic poles, the sources (north poles) or sinks (south poles) of the fields, together causes them to attract. When you try to place two like poles together (north to north or south to south), they will repel each other. The alignment of the fields due to the internal crystalline structure results in either a permanent or a temporary magnetic field. All magnetic materials have net intrinsic magnetization which results in a net field and in the generation of a force when that material is placed in a magnetic field such as the Earth's magnetic field. The forces acting unsymmetrically on a body can result in a net torque. The forces and torques can adversely influence the motion of a satellite in an orbit as well as its attitude when in motion. If one treats a spacecraft as a charged object, the electromagnetic force of the Earth's magnetic field on the charged object or the Lorentz force could be used to balance other perturbation effects such as the SRP as described by Yousef et al. (2022).

Impact forces due the collisions with space debris as well as meteors and other celestial objects acting unsymmetrically on a body can also result in a net torque. These impact forces and torques too can adversely influence the motion of a satellite in an orbit as well as its attitude when in motion.

2.6 Control Forces-9

Control forces are normally designed to apply a specific force on the satellite to affect a specific change to the orbit. These and their effects will be discussed in a later chapter.

On could summarize the comparative magnitudes of perturbation forces acting on a satellite which are illustrated Fig. 2.13. Third-body effects of the Sun and the Moon increase as one moves away from the Earth. The drag forces decrease rapidly as the satellite's altitude increases.

Exercises

1. (a) Identify the two principal assumptions for an orbit to be predicted by Kepler's laws.

   (b) Identify and describe the effects of the four dominant perturbations acting on a geostationary satellite, resulting in the orbit for such a satellite being non-Keplerian.

   (c) Obtain an approximate expression for the apsidal precession per orbit revolution, in terms of the semi-major axis, the orbit inclination, the orbit eccentricity, the Earth oblateness parameter J_2, the magnitude of the second zonal harmonic for the planet and the mean radius of the Earth.

   Explain what apsidal precession is, physically, and identify the required inclination for zero apsidal precession for a satellite in Earth orbit, when where J_2 has the value of 10^{-3}.

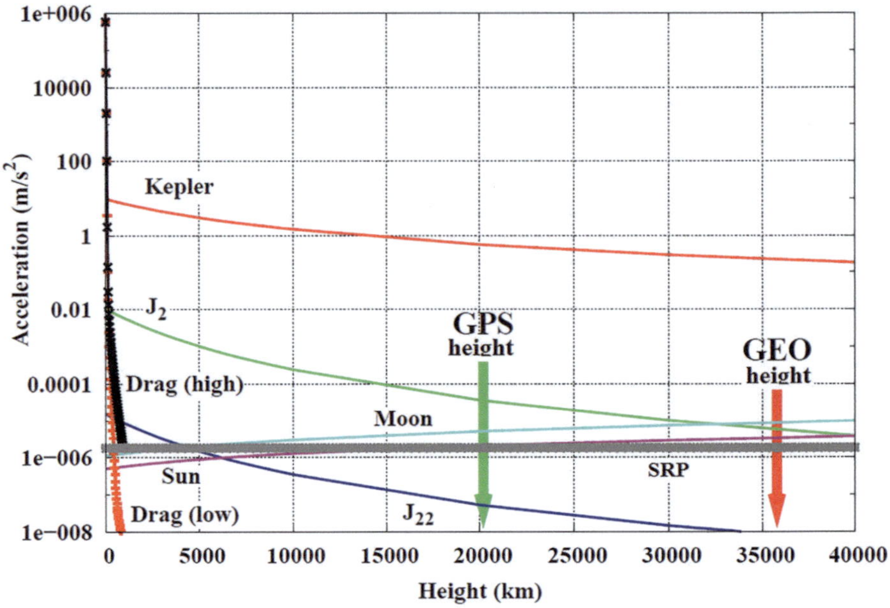

Fig. 2.13 Comparative magnitudes of perturbation accelerations acting on a satellite in Earth orbit with a ballistic coefficient,$B^* = 50$, and with high and low drag coefficients

(d) A satellite having a zero apsidal precession has an orbit period of 12 hours with an eccentricity of 0.8. This satellite may be viewed from a specific location on the ground once the satellite's altitude is above 6000 km. Calculate the fraction of the orbit period for which the satellite will be above this altitude. Hence, indicate why such an orbit may be useful.

2. (a) Describe briefly all the perturbative influences that change the classical orbital elements for a real satellite in orbit. Discuss in detail the effects of these perturbative influences upon the motion in the following specific cases, indicating any useful purposes.

   (i) A remote sensing spacecraft in near polar orbit, at an altitude of 700 km.
   A "Molniya" orbit communications satellite having a 12-hour period.
   (b) Obtain, from first principles, an approximation to the regression of the line of nodes $\Delta\Omega$, per orbit revolution, due to planetary oblateness for a circular orbit of radius a, with an inclination to the equator of i, when R_P is the mean radius of the planet and J_2 is the magnitude of the second zonal harmonic for the planet.

   (i) Consider a planet of mass M_P in circular orbit r_c about the Sun, whose mass is M_{sun}. A satellite in circular orbit of radius a about the planet is required to be in Sun-synchronous orbit. Obtain a relationship between the satellite's orbit radius and its inclination, in terms of the ratio of the planet mass to the mass

Table 2.1 Earth, Mars and Jupiter Data

	Earth	Mars	Jupiter
Mass	5.97×10^{24} kg	0.642×10^{24} kg	1898×10^{24} kg
Planetary mean radius	6378 km	3396 km	71,492 km
Length of planet day	23 h 56 min	24 h 36 min	9 h 54 min
Mean distance to the Sun	1 Au	1.52 Au	5.2 Au

of the Sun, the mean radius of the planet and the planet's orbital radius, and J_2 is the magnitude of the second zonal harmonic for the planet.

(ii) The values for J_2 for Mars and Jupiter are, respectively, given by 1.956×10^{-3} and 14.733×10^{-3}, whereas for the Earth, it is 1.083×10^{-3}. Explain why this is so with reference to the data in Table 2.1.

3. (a) Describe briefly all the perturbative influences that change the classical orbital elements for a real satellite in orbit.
   (b) Explain the physical reasons for the flattening of the planets at the poles. Why is it that certain planets are more flattened while others are less flattened?
   (c) Explain the consequences of the flattening effect on the motion of the satellites in orbit about planets.
   (d) Name one particular orbit that exploits the flattening effect and describe its features.
   (e) Explain how the other gravitational perturbations may influence the motion of an Earth orbiting satellite, and identify the manoeuvres that must be performed to compensate the motion and maintain a steady desired orbit. Give at least two examples of such manoeuvres.

4. The transformation of a position vector in an Earth centred space fixed frame to an Earth centred frame fixed in the orbit plane can be described by a sequence of Euler angles.

(a) Identify and explain the selection of the most appropriate Euler angle sequence and the angles assuming that the motion of the satellite is described by a set of Keplerian orbital elements.
(b) Hence or otherwise determine the transformation, as the product of three elementary rotational transformations, to transform the unit vectors in the orbiting frame to an Earth fixed frame.
(c) Explain what you understand by the term "first point of Aries"? What is its relationship to the vernal equinox?
(d) Satellites in low Earth orbit (LEO) are subject to aerodynamic drag. Briefly explain why such satellites do not generate shock waves.
(e) Hence show, with appropriate approximations, that the drag coefficient CD for a satellite is expected to lie in the regime, $2 < CD < 4$.

5. (a) Given that the mean radius of the Earth is 6378.1 km, the density of air is 2.5411×10^{-10} kg/m^3 at an altitude of 200 km and the gravitation constant (GM) for the Earth is 398,600 km^3/s/s, sketch the variation of a satellite's velocity, if it is orbiting the Earth in a nearly circular orbit as its altitude varies from 200 to 800 km.
(b) On a separate sketch, show the variations of the drag force acting on the satellite, for the same variations of altitude. You may find it convenient to use a log scale for the drag force. You may assume the projected cross-sectional area of the satellite normal to the velocity to be 0.080 m^2 and the drag coefficient to be 2. The atmospheric density in kg/m^3 is a function of the altitude h in km and is approximated for the given range of altitudes by

$$\rho = \exp\left[0.129 \times (h/100)^2 - 2.92997 \times (h/100) - 16.92580\right].$$

(c) Hence find the ratio of the maximum drag force to the minimum drag force as the altitude varies over the range 200–800 km.

6. A satellite in LEO is subject to a drag force. The relationship between the rate of change of semi-major axis a and drag may be approximated by the following expression:

$$da/dt = -2f_{\text{drag}}\sqrt{a^3/\mu_E}$$

Here f_{drag} is the drag force acceleration given by $f_{\text{drag}} = 0.5C_D\rho v^2 A/m$, ρ is the atmospheric density, V is the satellite velocity, A is the projected area of the satellite perpendicular to the velocity vector and m is the mass of the satellite.

(a) Show that if the satellite orbit remains circular throughout, then the change in semi-major axis per orbit Δa_{rev} is given by the equation

$$\Delta a_{\text{rev}} \approx -2\pi C_D(A/M)\rho a^2.$$

(b) An Earth observation satellite is in a circular orbit at a height of 800 km. The mission requires that the orbit period is maintained to an accuracy of 1 second. Calculate the number of orbits between successive orbit maintenance operations to carry out the mission.

References

Bate, R., Mueller, D., & White, J. (1971). Orbit determination from two positions and time, Chapter 5. In *Fundamentals of astrodynamics* (1st ed.). Dover Publications.

Boain, R. J. (2004). A-B-Cs of sun-synchronous orbit mission design. In *14th AAS/AIAA space flight mechanics conference, Maui, Hawaii*, Feb 8–12, 2004.

Brouke, R. A., & Cefola, P. J. (1972). On the equinoctial orbital elements. *Celestial Mechanics, 5*, 303–310.

Cook, G. E. (1965). Satellite drag coefficients. *Planetary and Space Science, 13*, 929–946.

Curtis, H. (2005). Preliminary orbit determination, Chapter 5. In *Orbital mechanics for engineering students* (1st ed.). Elsevier Butterworth-Heinemann, Linacre House, Jordan Hill, Oxford.

Kubo-oka, T., & Sengoku, A. (1999). Solar radiation pressure model for the relay satellite of SELENE. *Earth, Planets and Space, 51*, 979–986.

List, M., Bremer, S., Rievers, B., & Selig, H. (2015). Modelling of solar radiation pressure effects: Parameter analysis for the MICROSCOPE mission. *International Journal of Aerospace Engineering, 2015*, ID 928206 (1- 14). https://doi.org/10.1155/2015/928206

Lücking, C., Colombo, C., & McInnes, C. R. (2012). A passive satellite deorbiting strategy for medium earth orbit using solar radiation pressure and the J2 effect. *Acta Astronautica, 77*, 197–206. https://doi.org/10.1016/j.actaastro.2012.03.026

Marshall, J. A., & Luthcke, S. B. (1994). Modeling radiation forces acting on Topex/Poseidon for precision orbit determination. *Journal of Spacecraft and Rockets, 31*(1), 99–105.

Montenbruck, O., & Gill, E. (2005). *Satellite orbits: Models, methods and applications*. Springer.

Prussing, J. E., & Conway, B. A. (1993). Position in orbit as a function of time, Chapter 2. In *Orbital Mechanics*. Oxford University Press.

Vallado, D. (2004). *Fundamentals of Astrodynamics and Applications* (2nd ed.). Microcosm Press.

Walker, M. J. H., Ireland, B., & Owens, J. (1985). A set of modified equinoctial orbital elements. *Celestial Mechanics, 36*, 409–419.

Wetterer, C. J., Linares, R., Crassidis, J. L., Kelecy, T. M., Ziebart, M. K., Jah, M. K., & Cefola, P. J. (2014). Refining space object radiation pressure modeling with bidirectional reflectance distribution functions. *Journal of Guidance, Control, and Dynamics, 37*(1), 185–196.

Yousef, M. A., El-Saftawy, M. I., & Mostafa, A. (2022). Balancing the effects of solar radiation pressure on the orbital elements of a spacecraft using Lorentz force. *Scientific Reports, 12*, 15819. https://doi.org/10.1038/s41598-022-20166-y

Zardaín, L., Farrés, A., & Puig, A. (2020). High-fidelity modeling and visualizing of solar radiation pressure: A framework for high-fidelity analysis, AAS 20–481, AAS/AIAA Astrodynamics Specialist Virtual Conference, Lake Tahoe, http://www.ub.edu/wai/wp-content/uploads/2021/07/AAS-20-481.pdf

Chapter 3
Space Vehicle Manoeuvring

3.1 Principles of Space Vehicle Manoeuvring: Impulse Momentum—Basic Problem

Consider Newton's second law as applied to a single particle of mass m, and assume a near constant mass:

$$m\dot{v} = F. \tag{3.1}$$

Integrate both sides over a small time interval:

$$\begin{aligned}\int_{t_1}^{t_2} \dot{v}dt = v_2 - v_1 = \Delta v = \int_{t_1}^{t_2} (F/m)dt \approx (F/m) \int_{t_1}^{t_2} dt = (F/m)\Delta t \text{ or } \Delta v \\ = (F/m)\Delta t = I/m.\end{aligned} \tag{3.2}$$

To specify the manoeuvring force, we need to only specify Δv and change the velocity vector. The typical Δv required for completing various manoeuvres and space missions are summarized in Table 3.1.

Consider Newton's second law as applied to a single particle of mass m and subjected to a thrust force:

$$m\dot{v} = F = \dot{m}_p v_e. \tag{3.3}$$

Integrate the right side of Eq. 3.3 over a small time interval:

R. Vepa, *Space Vehicle Maneuvering, Propulsion, Dynamics and Control*,
https://doi.org/10.1007/978-3-031-65518-0_3

Table 3.1 Typical Δv for various manoeuvres and missions

Satellite displacement	**Δv in m/s per year**
Drag compensation at 500 km height	30
Attitude control	2–6
GEO orbit station keeping	50
Orbit transfers	Δv in km/s
Earth to LEO	9.7
LEO to GEO	4.3
LEO to Moon	5.9
LEO to L_1	3.7
LEO to L_2	3.4
LEO to Mars	5.3
Interplanetary transfers	Δv in km/s
Nearby planets	5–8
Far off planets	8–15
100–1000 AU	100
10,000 AU	1000
Interstellar domain	20–30,000

$$\begin{aligned}\int_{t_0}^{t_f} F dt = \int_{t_0}^{t_f} \dot{m}_p v_e dt = v_e \int_{t_0}^{t_f} \dot{m}_p dt = m_p v_e \\ \equiv I \text{ or} v_e = \left(I/m_p\right), I_{sp} \triangleq m_p v_e / m_p g_0 = I/m_p g_0.\end{aligned} \tag{3.4}$$

The maximum manoeuvring impulse is directly related to the specific impulse I_{sp} and the propellant mass m_p.

3.2 Estimating Δv Requirements for Orbit Changes

The basic Δvs required for some simple manoeuvres are now derived. These include the laws of cosines and sines, the Δv required for manoeuvres like rotating the line of apsides, orbit raising of circular and elliptic orbits and other transfers required to synthesize complex manoeuvres.

3.2.1 Velocity Components, Law of Cosines, Law of Sines

A fundamental in altering the velocity vector both in terms of its magnitude and in its direction is the law of cosines. Consider altering a velocity vector from v_i to v_f both in magnitude and direction (Fig. 3.1).

Then, initial and final velocity vectors satisfy

Fig. 3.1 Law of cosines

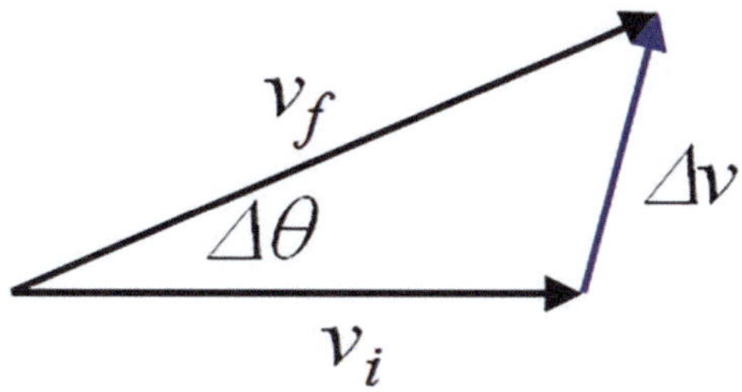

Fig. 3.2 Law of sines

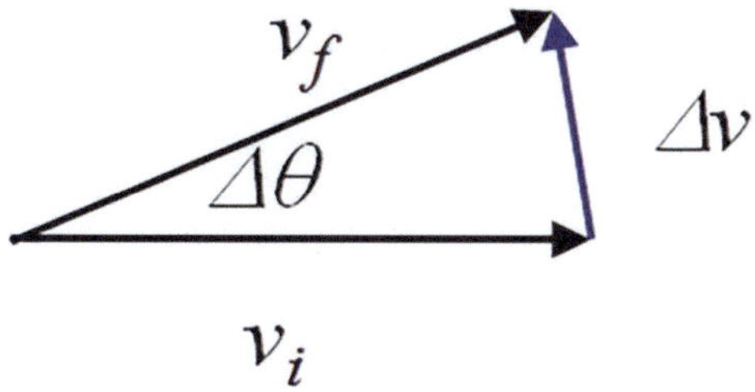

$$\Delta v^2 = v_f^2 + v_i^2 - 2v_f v_i \cos \Delta\theta. \tag{3.5}$$

The above law is known as the law of cosines.

Consider altering a velocity vector from v_i to v_f only in direction, so $v_f^2 = v_i^2$. This is a special case of the law of cosines and is illustrated in Fig. 3.2. Then,

$$\Delta v^2 = v_i^2 + v_i^2 - 2v_i v_i \cos \Delta\theta = v_i^2 \left(\sqrt{2(1 - \cos \Delta\theta)} \right)^2 = 4v_i^2 (\sin(\Delta\theta/2))^2. \tag{3.6}$$

Hence,

$$\Delta v = 2v_i \sin(\Delta\theta/2). \tag{3.7}$$

The above law is known as the law of sines.

3.2.2 Orbit Plane Inclination Change

One could consider some applications of the basic laws. Consider rotating an orbit plane about the line of nodes. This involves changing the inclination angle of the orbit plane. To understand this manoeuvre, it is essential to first understand the two components of the velocity of the satellite. The velocity of the satellite has two components in the orbit plane, a radial component, v_r, and a component tangential to the orbit, v_θ. The tangential component can be found from the orbital momentum,

$h = rv_\theta$. So, $v_\theta = h/r$. To find the radial velocity component v_r, consider the equation of the conic:

$$r = \left(h^2/\mu\right)/(1 + e\cos\nu). \tag{3.8}$$

Two important relations are important to note and these are

$$p = a\left(1 - e^2\right), r(1 + e\cos\nu) = h^2/\mu = p \tag{3.9}$$

Furthermore, from the definition of $h = r^2\dot{\nu}$,

$$\dot{\nu} = h/r^2 = (1 + e\cos\nu)^2\sqrt{\mu/p^3}. \tag{3.10}$$

Thus, it follows that

$$v_\theta = h/r = \mu(1 + e\cos\nu)/h, \dot{r}(1 + e\cos\nu) - er\dot{\nu}\sin\nu = 0. \tag{3.11}$$

It follows that, given the flight path angle ϕ,

$$\tan\phi = \frac{v_r}{v_\theta} = \frac{\dot{r}}{r\dot{\nu}} = \frac{e\sin\nu}{1 + e\cos\nu}. \tag{3.12}$$

Furthermore,

$$v_r = v_\theta\tan\phi = (\mu/h)e\sin\nu. \tag{3.13}$$

Returning to the question of changing the inclination angle must be either increased or decreased (or a negative increase). From the law of cosines,

$$\Delta v^2 = v_i^2 + v_f^2 - 2v_i v_f\cos\Delta\theta_{if}. \tag{3.14}$$

For a plane change or a change in the inclination, only the transverse or tangential component of the velocity vector needs to change direction. Hence, $v_i = v_f = v_\theta$ and $\Delta\theta_{if} = \Delta i$. No change is necessary to the radial component v_r.

Hence, it follows that

$$\Delta v = v_\theta\sqrt{2(1 - \cos\Delta i)} = 2v_\theta\sin(\Delta i/2) = 2(h/r)\sin(\Delta i/2). \tag{3.15}$$

The direction of the velocity increment is normal to the orbit plane. The impulse must be applied at a node.

To summarize, for a plane change or a change in the inclination, the orbit plane must be rotated about the line of nodes. The direction of the impulse, on the satellite, is normal to the orbit plane. The impulse must be applied at a node.

3.2.3 Multiple Orbital Element Change

Often it is required to change two or more orbital elements of an orbiting spacecraft, to change the orientation of a spacecraft. Such manoeuvres require rotating the orbit on the surface of a sphere. Thus, the relations between the four orbital angular elements could reduce to spherical trigonometric relations.

Consider two orbits around a sphere. If, as an example shown in Fig. 3.3, it is required to change both the RAAN (the argument of the ascending node) and also the inclination simultaneously, the simultaneously change in both the RAAN and the inclination requires the solution of a spherical trigonometric relation. The relationships involve the angles at the corners of spherical triangles. Spherical triangles arise when the sides of the triangles themselves are angles as are the angles between the sides of the triangle. In the example shown, it is required to change the RAAN must be incremented by $\Delta\Omega$ while the inclination also changes from i to i_f.

The corresponding relationships in plane and spherical trigonometry are summarized in Fig. 3.4.

Reconsider the example where one wishes to simultaneously change in both the RAAN and the inclination. The problem is diagrammatically restated in Fig. 3.5.

As we need to add Δv at the intersection of orbits, to simultaneously change both inclination and RAAN, it is essential to consider the spherical triangle shown in

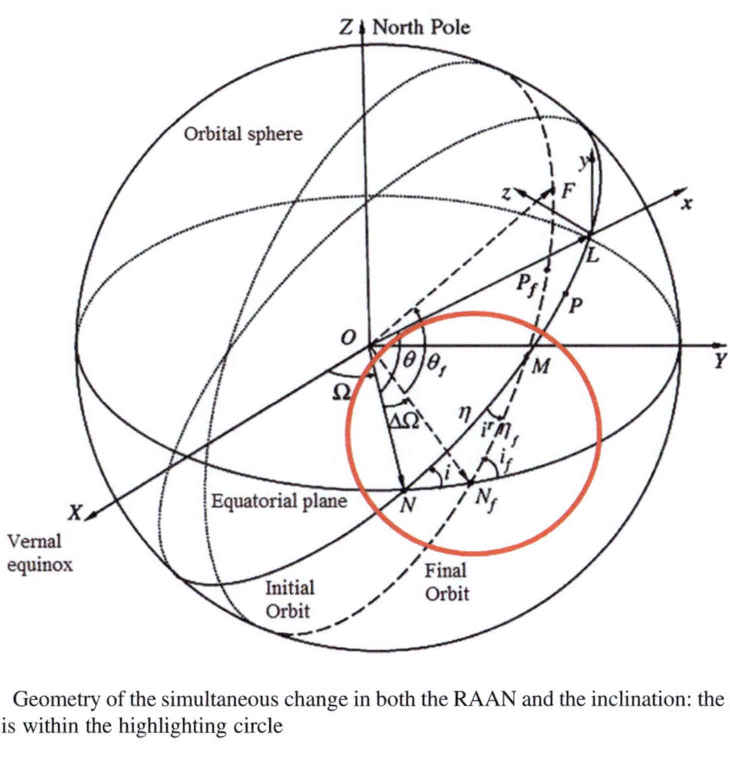

Fig. 3.3 Geometry of the simultaneous change in both the RAAN and the inclination: the spherical triangle is within the highlighting circle

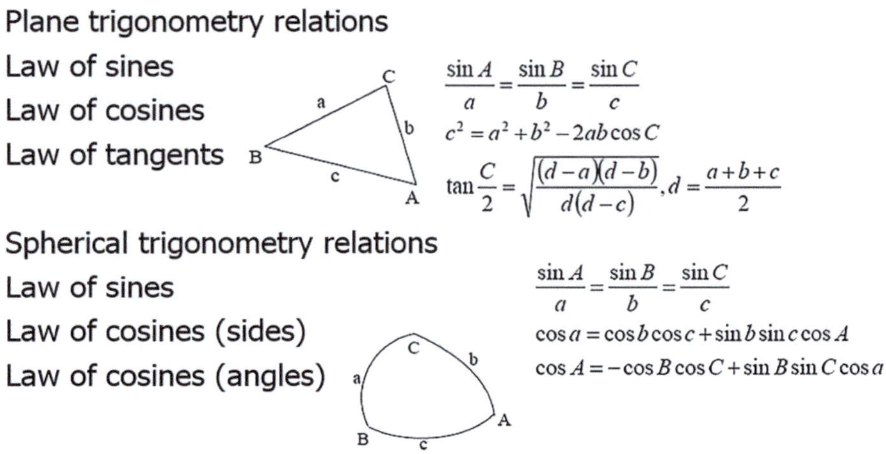

Fig. 3.4 Corresponding relationships in plane and spherical trigonometry

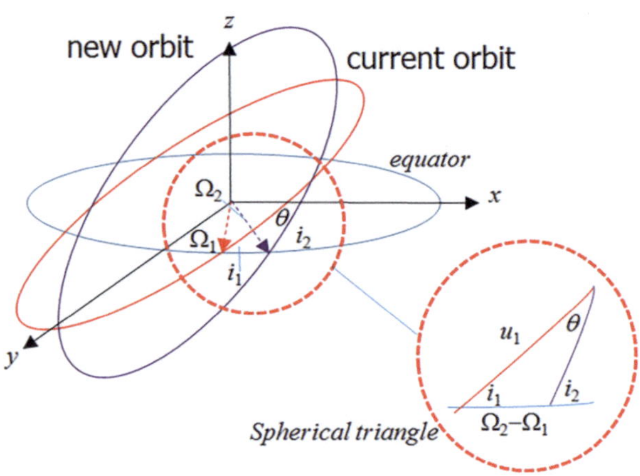

Fig. 3.5 Orbit parameters requiring simultaneous change in both the RAAN and the inclination

Fig. 3.5 and use laws of cosines, for both the angles and sides. Referring to Fig. 3.5, it is required that changes in the inclination, θ, and the angular position, u_1, where the change has to be implemented, have to be estimated. The initial inclination and RAAN are, respectively, given by i_1, Ω_1, while the final inclination and RAAN are, respectively, given by i_2, Ω_2. The law of cosines as applied to the angles and the law of cosines as applied to the sides, respectively, give

$$\begin{aligned}\cos\theta &= \cos i_1 \cos i_2 + \sin i_1 \sin i_2 \cos(\Omega_2 - \Omega_1)\\ \cos(\pi - i_2) &= \cos(\pi - i_1)\cos\theta + \sin(\pi - i_1)\sin\theta\cos u_1.\end{aligned} \tag{3.16}$$

The first of the two equations in 3.16 gives the change in the inclination, θ, while the second of two equations in 3.16 is solved to give the angular position u_1. For the position of the intersection point of the two orbits, along the orbit, the following equation must be solved:

$$\cos i_2 = \cos i_1 \cos\theta - \sin i_1 \sin\theta \cos u_1. \tag{3.17}$$

To find the Δv to affect the plane change, one may use the law of sines, as one needs to only rotate the tangential component of the orbiting satellite's velocity.

Other examples of simple Keplerian manoeuvres are now considered. Several of these manoeuvres are used to perform complex operations during a typical space mission (Fortescue et al., 2011).

3.3 Rotating the Line of Apsides

Rotating the line of apsides, which is the line joining the periapsis and apoapsis, is indeed a classic problem. Sometimes one needs to rotate the line of apsides without changing the inclination of the orbit plane (coplanar change). The velocity vector must be rotated in the orbit plane without changing the size and shape of the orbit. A simple rotation by an angle $\Delta\omega$ of the line of apsides without altering size and shape of the orbit is obtained by means of one impulsive burn at either point where the initial and final ellipses intersect on the common bisector of the angle $\Delta\omega$. Energy and angular momentum conservation imply that the magnitude and tangential component of the velocity remain unchanged. The manoeuvre is illustrated in Fig. 3.6.

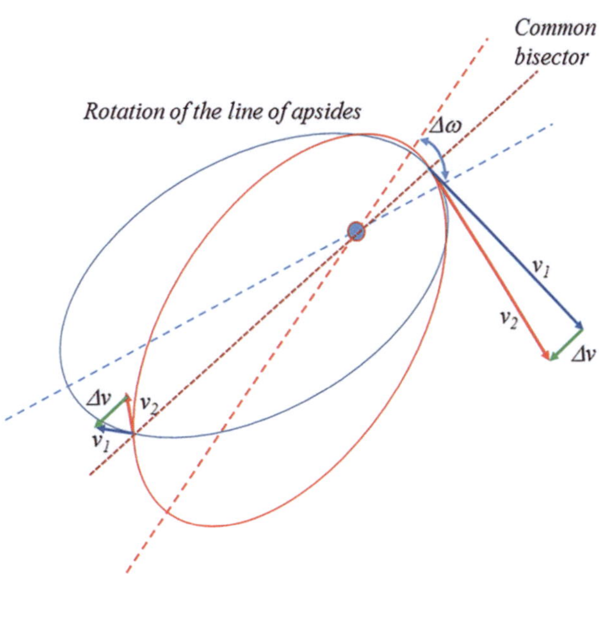

Fig. 3.6 Diagram of the manoeuvre to rotate the line of apsides indicating the initial and final velocities and the impulsive change in velocity required

The impulse may be applied at either of the two intersection points of the two orbits, but only one of these will give a minimum impulsive change in the velocity. Moreover, only the direction of radial component of the velocity needs to change at the intersection point. Hence, using the fact that $h^2 = \mu p = \mu a(1 - e^2)$, it follows that

$$\Delta v = 2v_r = 2(\mu/h)e|\sin\nu| = 2e|\sin(\Delta\omega/2)|\left(\sqrt{\mu/a(1-e^2)}\right). \tag{3.18}$$

3.3.1 Other Simple Keplerian Orbit Manoeuvres

An important manoeuvre is providing a velocity increment that would take the spacecraft from a circular orbit to a hyperbolic orbit with prescribed velocity at infinity. Using the vis-viva equation, the periapsis velocity of the hyperbolic orbit is obtained from

$$\frac{1}{2}V^2 - \frac{\mu}{r} = \frac{1}{2}v_p^2 - \frac{\mu}{r_p} = \frac{1}{2}v_\infty^2 \Rightarrow v_p^2 = v_\infty^2 + \frac{2\mu}{r_p} = v_\infty^2 + 2v_{\text{circ}}^2. \tag{3.19}$$

Thus, the velocity change for going from a circular to a hyperbolic orbit at its periapsis is given by

$$\Delta v = v_p - v_{circ} = v_{circ}\left\{\sqrt{2 + \left(\frac{v_\infty}{v_{circ}}\right)^2} - 1\right\}. \tag{3.20}$$

3.3.2 Circular to Elliptic and Vice Versa, Raising Apoapsis or Periapsis

A classic manoeuvre is incrementing the velocity to go from circular to an elliptic orbit with a given eccentricity and vice versa. Consider the velocity of a body at the periapsis in an elliptic orbit:

$$v_p = h/r_p = \sqrt{\mu p}/r_p = \sqrt{\mu a(1-e^2)}/r_p = \sqrt{\mu(1+e)}/\sqrt{r_p}. \tag{3.21}$$

The corresponding circular velocity at the same location as at the periapsis of the elliptic orbit is

$$v_c = \sqrt{\mu}/\sqrt{r_p}. \tag{3.22}$$

Thus, the required velocity increment is

$$\Delta v = \sqrt{\frac{\mu}{r_p}}\{\sqrt{1+e}-1\}. \tag{3.23}$$

Orbit raising from a circular orbit of one radius to another circular orbit with a larger radius is achieved by a velocity decrement given by

$$\Delta v = v_{circ2} - v_{circ1} = \sqrt{\frac{\mu}{r_1}}\left\{\sqrt{\frac{r_1}{r_2}}-1\right\}. \tag{3.24}$$

A velocity increment going from an elliptic orbit to a circular orbit at the apoapsis is given by

$$\Delta v = \sqrt{\frac{\mu}{r_a}}\left\{1-\sqrt{1-e}\right\}. \tag{3.25}$$

A Δv at the periapsis of an elliptic orbit will raise its apoapsis and vice versa. The Δv required in each of these two cases are, respectively, given by

$$\Delta v_p = \sqrt{\mu}\left\{\sqrt{\frac{2}{r_p}-\frac{1}{a_2}}-\sqrt{\frac{2}{r_p}-\frac{1}{a_1}}\right\}, \Delta v_a = \sqrt{\mu}\left\{\sqrt{\frac{2}{r_a}-\frac{1}{a_2}}-\sqrt{\frac{2}{r_a}-\frac{1}{a_1}}\right\}. \tag{3.26}$$

On could apply a single impulse with two components in the r and θ directions to change both a and e in the orbit plane. The velocity increments in the radial and tangential directions are, respectively, given by

$$\begin{aligned} \Delta v_r &= \sqrt{\mu}\left\{e_2\sqrt{\frac{1}{a_2\left(1-e_2^2\right)}}-e_1\sqrt{\frac{1}{a_1\left(1-e_1^2\right)}}\right\}\sin\nu, \\ \Delta v_\theta &= \sqrt{\mu}\left\{\sqrt{\frac{1}{a_2\left(1-e_2^2\right)}}-\sqrt{\frac{1}{a_1\left(1-e_1^2\right)}}\right\}(1+e\cos\nu). \end{aligned} \tag{3.27}$$

There are indeed many limitations with single impulse manoeuvres. The Δv required for completing the manoeuvre could not only be very large, and there is no room for any optimization. To be able to optimize, we need at least two or more impulses. Ideally continuous thrusting is best for optimization as continuous optimal control is feasible and may be applied in practice.

3.4 Two-Impulse Manoeuvres (Hohmann Transfer), Examples

One could combine two or more single impulse manoeuvres to implement a multi-impulse manoeuvre. The classic example of such a manoeuvre is a two-impulse Hohmann transfer. It could be implemented as a coplanar transfer from one circular or nearly circular orbit to another circular or nearly circular orbit of a different radius from the first. It is illustrated in Fig. 3.7. The semi-major axis of the transfer ellipse is the mean of the radii of the lower and higher circular orbits. Thus,

$$a_{tr} = \frac{r_2 + r_1}{2}. \tag{3.28}$$

3.4.1 *Hohmann Transfer: Coplanar Elliptic Transfer*

Hohmann transfer is minimum energy transfer along an elliptic transfer orbit with its periapsis at the same radius as the low coplanar orbit and its apoapsis at the same radius as the high circular orbit. The transfer to the elliptic orbit from the low orbit is affected by giving an impulse at the periapsis, and the transfer from the elliptic orbit to the higher orbit is affected by giving an impulse at the apoapsis of the transfer orbit. It is also possible to transfer from a high orbit to a lower orbit along an elliptic

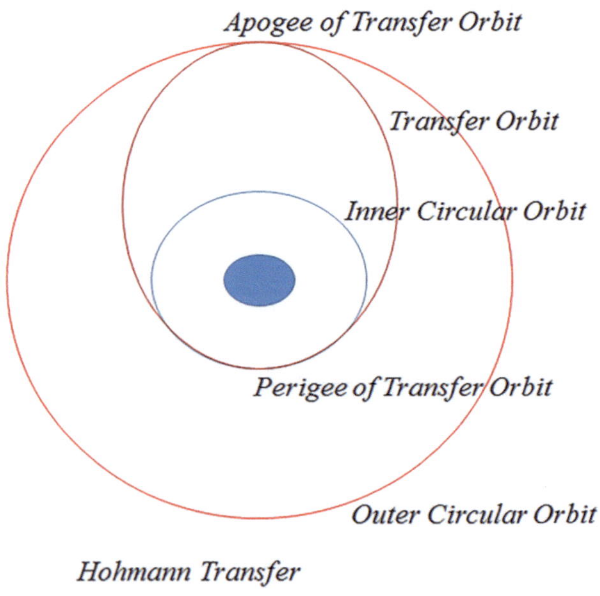

Fig. 3.7 Coplanar Hohmann transfer

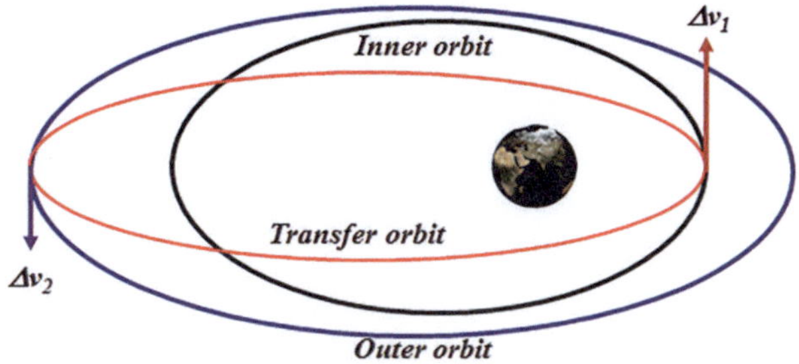

Fig. 3.8 Coplanar Hohmann transfer for coaxial elliptic initial and final orbits

transfer orbit, by providing the impulse first at apoapsis and later at the periapsis of the transfer orbit. A Δv at the periapsis of the transfer orbit will raise it to its apoapsis and vice versa. A further Δv at the apoapsis in the low to high orbit transfer and periapsis in the high to low orbit transfer simply transfer the satellite into the desired orbit. For simplicity, one may assume that the initial and final orbits are circular. Then the first impulse transfers the satellite from its initial circular orbit to the elliptic transfer orbit, while the second impulse transfers it from the elliptic transfer orbit to the desired final circular orbit. The $\Delta v's$ required for the two manoeuvres are

$$\begin{aligned}
\Delta v_p &= \sqrt{\mu}\left\{\sqrt{\frac{2}{r_1}-\frac{1}{a_{tr}}}-\sqrt{\frac{1}{r_1}}\right\} = \frac{\sqrt{\mu}}{\sqrt{r_1}}\left\{\sqrt{\frac{2r_2}{(r_1+r_2)}}-1\right\}, \\
\Delta v_a &= \sqrt{\mu}\left\{\sqrt{\frac{1}{r_2}}-\sqrt{\frac{2}{r_2}-\frac{1}{a_{tr}}}\right\} = \frac{\sqrt{\mu}}{\sqrt{r_2}}\left\{1-\sqrt{\frac{2r_1}{(r_1+r_2)}}\right\} \\
a_{tr} &= \frac{r_2+r_1}{2},
\end{aligned} \tag{3.29}$$

The Hohmann transfer is also energy optimal for coaxial elliptic initial and final orbits as shown in Fig. 3.8, while the transfer time is a maximum. However, one must check if initial impulse must be done at periapsis or apoapsis of the transfer ellipse, so it is a minimum. The transfer ellipse is shown in Fig. 3.8, from the periapsis of the inner elliptic orbit to the apoapsis of the outer elliptic orbit. A Δv at the periapse of the transfer orbit will raise it to its apoapse. The $\Delta v's$ required for the two manoeuvres are

$$\begin{aligned}
\Delta v_{p1} &= \sqrt{\mu}\left\{\sqrt{\frac{2}{r_{tr,p}}-\frac{1}{a_{tr}}}-\sqrt{\frac{2}{r_{1p}}-\frac{1}{a_1}}\right\},\\
\Delta v_{a2} &= \sqrt{\mu}\left\{\sqrt{\frac{2}{r_{2a}}-\frac{1}{a_2}}-\sqrt{\frac{2}{r_{tr,a}}-\frac{1}{a_{tr}}}\right\},\\
a_{tr} &= \frac{r_{2a}+r_{1p}}{2},\\
e_{tr} &= \frac{r_{2a}-r_{1p}}{r_{2a}+r_{1p}},\\
r_{tr,p} &= a_{tr}(1-e_{tr}) = r_{1p},\\
r_{tr,a} &= a_{tr}(1+e_{tr}) = r_{2a},\\
\Delta v &= |\Delta v_{p1}| + |\Delta v_{a2}|.
\end{aligned} \tag{3.30}$$

The transfer time is half the period of the transfer orbit and given by

$$T_{tr} = \pi\sqrt{a_{tr}^3/\mu}. \tag{3.31}$$

3.4.2 Examples of Coplanar Elliptic Hohmann Transfers

It is required to calculate $\Delta v's$ to transfer a satellite from (i) a circular parking orbit at a radius of 6500 km to a co-planar Hohmann transfer orbit to eventually transfer to a geo-stationary orbit at a radius of 4.2165×10^4 km; (ii) from the transfer orbit into a co-planar circular orbit with the same radius as the geostationary orbit and (iii) change the inclination of the orbit by 28°. Using Eqs. 3.29 and 3.15,

$$\Delta v_p = 2.4776\text{km/s}, \Delta v_a = 1.4855\text{km/s}, \Delta v_i = 1.4876\text{km/s}. \tag{3.32}$$

If the three manoeuvres are performed sequentially, the total $\Delta v = 5.4507$km/s. However, one could change the inclination of the transfer orbit before circularizing the orbit by retarding the spacecraft. At the apogee of the transfer orbit, the velocity of the spacecraft in the transfer orbit is $v_a = \left(\sqrt{\mu/r_2}\right)\left\{\sqrt{r_1/a_{tr}}\right\} = 1.5891\text{km/s}$. Hence, to change the inclination prior to circularizing is $\Delta v_i = 0.7689$km/s. Performing the circularization after the change in the inclination, the total $\Delta v = 4.7320$km/s. To transfer directly from the apogee of the transfer orbit to the geostationary orbit at an inclination of 28° can be estimated by using the cosine law and the fact that at the geostationary orbit, the circular velocity is $v_{\text{geo}} = 3.0746$km/s. Applying the cosine law to transfer directly from the apogee of the transfer orbit to the geostationary orbit at an inclination of 28°, the $\Delta v_i = 1.8305$km/s. The total Δv required is $\Delta v = \Delta v_p + 1.8305 = 4.3080$km/s, resulting in a saving of 1.1427 km/s!

In the next example, s spacecraft is at a distance of 7000 km from the Earth's centre and is in a circular orbit. It is desired to put it in a circular orbit at a radius of 20,000 km. Calculate the total amount of Δv required to transfer to the target orbit using a Hohmann transfer.

In the first step, the velocity in the parking and the target orbit is found. If the standard gravitational parameter of the Earth, the variable $r = a_{\text{park}} = 7000$ km, and the target orbit has a radius $r = a_{\text{target}} = 20000$ km, the velocities in these orbits are found. Thus, the velocities in the parking and target orbits are $v_{\text{park}} = 7.546$ km/s, $v_{\text{target}} = 4.464$ km/s). Next, the orbital characteristics of the transfer orbit are found. First, the semi-major axis and eccentricity of the transfer orbit are obtained. To do this, the average of the semi-major axes of the target orbit and the parking orbit is found. The semi-major axis of the transfer orbit a_{tr} is $a_{tr} = (a_{\text{target}} + a_{\text{park}})/2 = (20000 + 7000)/2 = 13500$ km. The eccentricity of the transfer orbit is $e_{tr} = (a_{\text{target}} + a_{\text{park}})/2a_{tr} = (20000 - 7000)/2a_{tr} = 0.4815$. Now, the velocity at periapsis of this transfer orbit is found. In this case $r = a_{\text{park}} = 7000$ km, $a = a_{tr} = 13500$ km. Vis-viva equation gives $v = \sqrt{(2\mu/r) - (\mu/a)}$. Substitute into the vis-viva equation to get $v_{\text{transfer\_peri}} = \sqrt{398600.442}\sqrt{10000/(3500 \times 13500)} = 9.185$ km/s. Then, the dv of the first manoeuvre at the periapsis of the transfer orbit is calculated. $\Delta v_1 = v_{\text{transfer \_ peri}} - v_{\text{park}} = 9.185 - 7.546 = 1.639$ km/s. This first burn will put our spacecraft into the transfer orbit.

Next, we need to calculate the speed at the transfer orbit's apoapsis. For this calculation, $r = a_{\text{target}} = 20000$km, $a = a_{tr} = 13500$ km. Substitute into the vis-viva equation to get $v_{\text{transfer\_apo}} = \sqrt{398600.442}\sqrt{3500/(10000 \times 13500)} = 3.215$ km/s. To calculate the Δv for the insertion burn, and the total Δv, the Δv for the insertion burn is $\Delta v_2 = v_{\text{target}} - v_{\text{transfer \_ apo}} = 4.464 - 3.215 = 1.249$ km/s. So the total Δv is $\Delta v = \Delta v_1 + \Delta v_2 = 2.888$ km/s.

Consider a three impulse transfer: there are two transfer orbits, an intermediate and a final transfer orbit. So, the transfer takes place in two stages. In stage 1, the transfer is from the parking (initial) to the intermediate transfer orbit. In the second stage, the transfer is from the intermediate transfer orbit to the final or target circular orbit. It is required that the total $\Delta v's$ are compared, if the intermediate transfer orbit is anywhere between the parking and the final target orbit. The situation in this three impulse transfer is illustrated in Fig. 3.9.

The plot of the total Δv required versus the radius of the intermediate transfer orbit is shown in Fig. 3.10. The calculations are left as an exercise for the reader. It is seen that the Δv for a Hohmann transfer (two impulse) is always less!

3.4.3 Hohmann Transfer: Noncoplanar Elliptic Transfers

The two-impulse noncoplanar orbit transfer for coaxial elliptic initial and final orbits without making any constraining assumptions as far as the orbits are concerned is now considered and is based on Vepa (2019). The initial and final orbit will not be

Fig. 3.9 Three impulse transfer

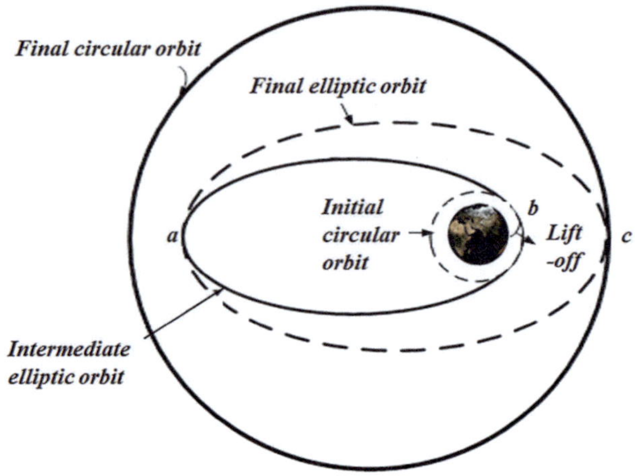

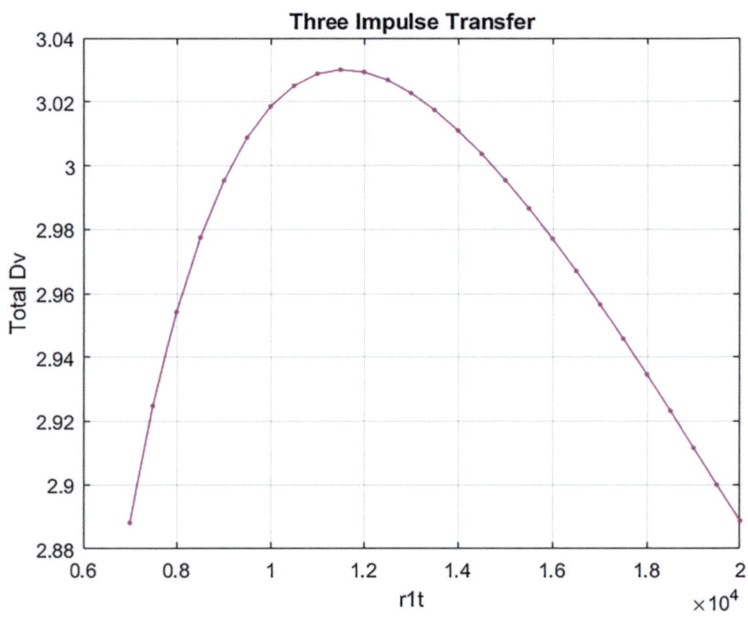

Fig. 3.10 Plot of the total Δv required for a three impulse transfer

assumed to be in the same plane. In fact, we will assume that they are inclined to each other at an angle θ. The velocity of the satellite at the periapsis of the initial orbit is

$$v_{1,p} = \sqrt{\frac{\mu}{a_i}}\sqrt{\frac{1+e_i}{1-e_i}}. \tag{3.32}$$

The velocity of the satellite at the apoapse of the final orbit is

$$v_{2,a} = \sqrt{\frac{\mu}{a_f}}\sqrt{\frac{1-e_f}{1+e_f}}. \tag{3.33}$$

For the transfer orbit, the velocity of the satellite at the periapsis of the orbit is

$$v_{t,p} = \sqrt{\frac{\mu}{a_t}}\sqrt{\frac{1+e_t}{1-e_t}}. \tag{3.34}$$

It is assumed that the semi-major axis and eccentricity of the transfer orbit are a_t and e_t, respectively. For the transfer orbit, the velocity of the satellite at the apoapsis of the orbit is

$$v_{t,a} = \sqrt{\frac{\mu}{a_t}}\sqrt{\frac{1+e_t}{1-e_t}}. \tag{3.35}$$

The major axis of the transfer orbit is assumed to be the sum of the periapsis distance of the initial orbit and the periapsis distance of the final orbit. Thus,

$$2a_t = a_f\left(1+e_f\right) + a_i(1-e_i). \tag{3.36}$$

Thus, the eccentricity may be expressed as

$$e_t = \left(a_f\left(1+e_f\right) - a_i(1-e_i)\right)/2a_t. \tag{3.37}$$

If it is assumed that the transfer orbit and the initial and final orbits are coplanar, the total Δv is given by

$$\Delta v = \left|v_{t,p} - v_{1,p}\right| + \left|v_{2,a} - v_{t,a}\right|. \tag{3.38}$$

However, if one assumes that the total plane change required is θ, one can assume that the plane change achieved after the first impulse is α and the plane change achieved after the second impulse is $\theta - \alpha$. Of course, one is seeking the minimum total Δv, and so one could minimize the total Δv with respect to α and determine what proportion of the plane change must be achieved after the first impulse. With the change in planes, one has the relations

$$\Delta v_p^2 = v_{t,p}^2 + v_{1,p}^2 - 2v_{t,p}v_{1,p}\cos\alpha, \tag{3.39}$$

$$\Delta v_a^2 = v_{2,a}^2 + v_{t,a}^2 - 2v_{2,a}v_{t,a}\cos(\theta - \alpha). \tag{3.40}$$

The total Δv is,

$$\Delta v = \mid \Delta v_p \mid + \mid \Delta v_a \mid . \tag{3.41}$$

Thus to minimize Δv and a necessary condition is that,

$$\frac{\partial \Delta v}{\partial \alpha} = 0. \tag{3.42}$$

Following Kamel and Soliman (2011), this reduces to the solution of sixth order polynomial in $x = \cos\alpha$. Kamel and Soliman (2011) write the polynomial as

$$\begin{gathered}\left(E_1^2 - E_5^2\right) + 2(E_1E_2 - E_5E_6)x + \left(2E_1E_3 - 2E_5E_7 + E_2^2 + E_5^2 - E_6^2\right)x^2 + \\ +2(E_1E_4 + E_2E_3 + E_5E_6 - E_6E_7)x^3 + \left(2E_2E_4 + 2E_5E_7 + E_3^2 + E_6^2 - E_7^2\right)x^4 + \\ +2(E_3E_4 + E_6E_7)x^5 + \left(E_4^2 + E_7^2\right)x^6 = 0.\end{gathered} \tag{3.43}$$

where the coefficients E_i are defined as

$$\begin{aligned}
E_1 &= v_{1,p}^2 v_{t,p}^2\left(v_{2,a}^2 + v_{t,a}^2\right) - v_{2,a}^2 v_{t,a}^2\left(v_{1,p}^2 + v_{t,p}^2\right)\cos^2\theta, \\
E_2 &= 2v_{1,p}v_{t,p}v_{2,a}^2 v_{t,a}^2\cos^2\theta - 2v_{1,p}^2 v_{t,p}^2 v_{2,a}v_{t,a}\cos\theta, \\
E_3 &= -v_{1,p}^2 v_{t,p}^2\left(v_{2,a}^2 + v_{t,a}^2\right) + v_{2,a}^2 v_{t,a}^2\left(v_{1,p}^2 + v_{t,p}^2\right)\left(\cos^2\theta - \sin^2\theta\right), \\
E_4 &= 2v_{1,p}^2 v_{t,p}^2 v_{2,a}v_{t,a}\cos\theta - 2v_{1,p}v_{t,p}v_{2,a}^2 v_{t,a}^2\left(\cos^2\theta - \sin^2\theta\right), \\
E_5 &= 2v_{1,p}^2 v_{t,p}^2 v_{2,a}v_{t,a}\sin\theta, \\
E_6 &= -2\cos\theta\sin\theta v_{2,a}^2 v_{t,a}^2\left(v_{1,p}^2 + v_{t,p}^2\right), \\
E_7 &= 4\cos\theta\sin\theta v_{1,p}v_{t,p}v_{2,a}^2 v_{t,a}^2 - 2\sin\theta v_{1,p}^2 v_{t,p}^2 v_{2,a}v_{t,a}.
\end{aligned} \tag{3.44}$$

Once α is determined, the complete plane change is accomplished in two stages to minimize the total Δv required. A part of the plane change given by the angle α is accomplished at the time of the application of the first impulse, while the remaining change given by $\theta - \alpha$ is accomplished during the application of the second impulse.

There are a number of special cases one could consider. For example, in a simple plane change, it is only necessary to change the direction velocity vector, which is done by providing an impulse at the ascending or descending nodes. It can be shown that in this case, $\Delta v = 2v_{\text{node}} \sin(\theta/2)$, where v_{node} is the nodal velocity. In another case, for a low inclination orbit, transfer to a zero inclination orbit is obtained by applying the impulse, satisfying the relation given by the law of cosines, $\Delta v^2 = v_1^2 + v_2^2 - 2v_1 v_2 \cos(\theta)$. Other cases are discussed by Edelbaum (1961) and Oleson (1993).

Generally, a plane change requires a considerable amount of fuel expenditure and must be carefully evaluated. Approximate computations can be very misleading.

3.5 Oberth's and Edelbaum's Multi-Impulse Manoeuvres

Consider a spacecraft orbiting a planet at a given orbital height H in a circular orbit. (Assume the radius of the planet is known). At a point P in the orbit, its thrusters are fired to bring it down to an orbit with a periapsis height closer to the planet's surface. A Hohmann transfer orbit is initiated. (i) A certain Δv or velocity increment (first velocity increment) is required to bring it down to the required orbital height after the first burn. This is given by

$$\Delta v_1 = \sqrt{\mu/r_a}\left\{1 - \sqrt{1-e}\right\}. \tag{3.45}$$

(ii) Once the spacecraft reaches the lower altitude, its thrusters are fired, so the spacecraft escapes from the planet's gravitational influence, along a hyperbolic orbit with a given excess velocity. This is the two impulse or Oberth's two-burn manoeuvre (Oberth, 1972) and is illustrated in Fig. 3.11. It is making use of the planet's gravity.

A second velocity increment is required for the second burn, given by

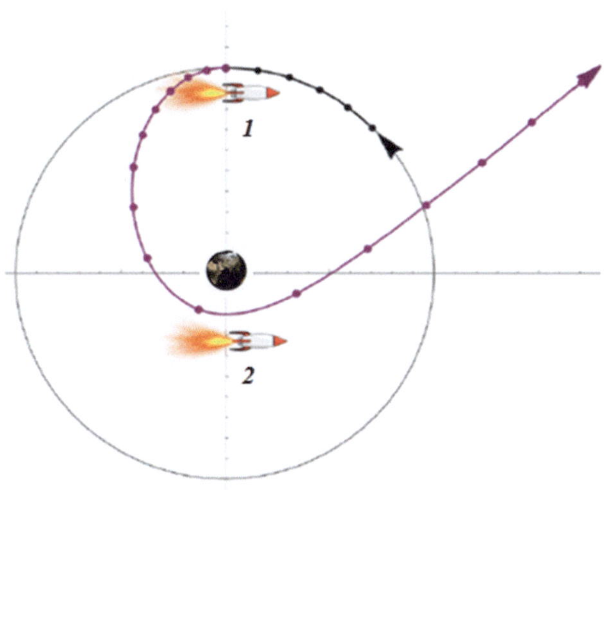

Fig. 3.11 Oberth's two-burn manoeuvre

$$\Delta v_2 = V - v_p = \sqrt{v_\infty^2 + v_{esc}^2} - v_p. \tag{3.46}$$

One can then obtain the total velocity increment required by adding the two increments, which gives

$$\Delta v = |\Delta v_1| + |\Delta v_2|. \tag{3.47}$$

(iii) If instead of this two-burn manoeuvre, the spacecraft thrusters could be fired only once at the point P on the circular orbit, it just escapes (from point P) from the planet's gravity along a hyperbolic orbit with the given excess velocity, without having to come closer to the planet. This is a direct manoeuvre or transfer from a

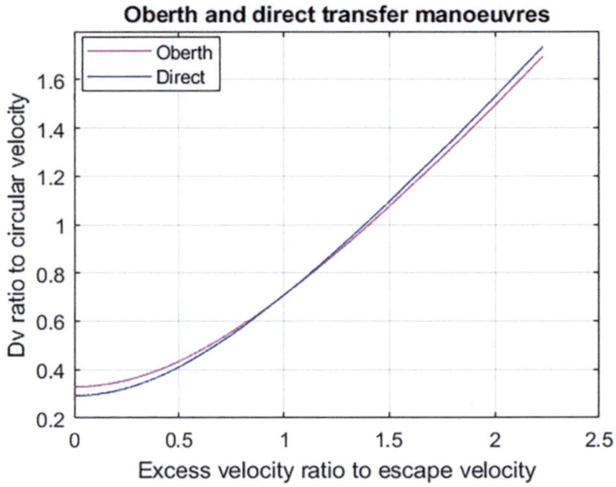

Fig. 3.12 Comparing the Oberth two-burn manoeuvre with a direct one-burn manoeuvre

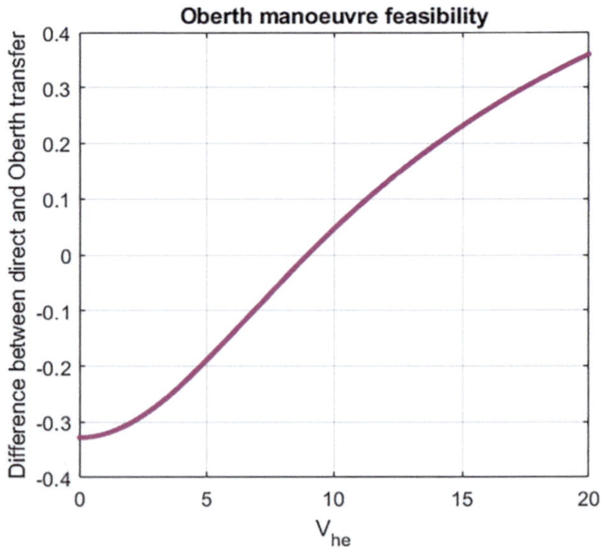

Fig. 3.13 Difference in the required Δv, shown as a plot with the desired hyperbolic excess velocity

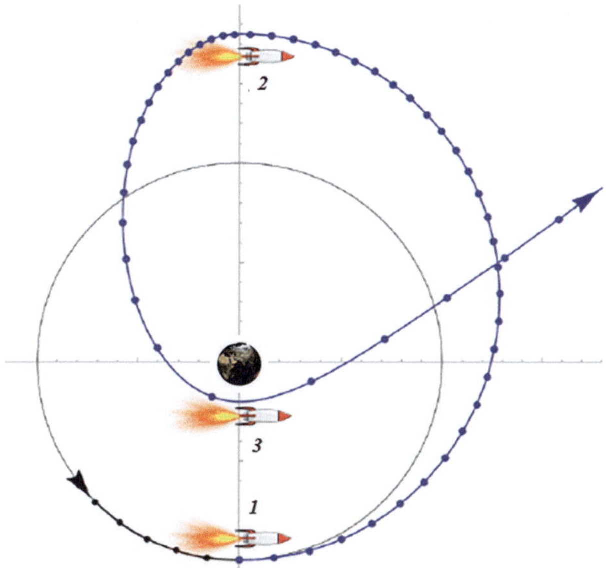

Fig. 3.14 Edelbaum's three impulse manoeuvre

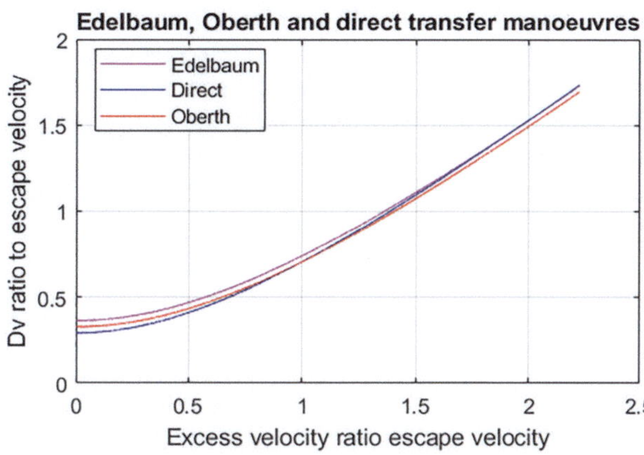

Fig. 3.15 Comparing the $\Delta v's$ for Edelbaum's, Oberth's and direct transfers

circular to the desired hyperbolic orbit. One could compare the fuel required or the ratio of the required Δv to the escape velocity, for Oberth's two-burn manoeuvre, with the corresponding ratio for the one-burn direct manoeuvre. The two manoeuvres are compared in Fig. 3.12. Figure 3.12 shows that the Oberth transfer is advantageous if the excess velocity exceeds the escape velocity at the point P. In Fig. 3.13, the difference in the required Δv is shown as a plot with the desired hyperbolic excess velocity.

Edelbaum (1959, 1967) proposed a three burn manoeuvre extending the work of Oberth. In the Edelbaum three burn manoeuvre, the spacecraft is first lifted to a high

Earth orbit in the first burn, before executing a further two-burn manoeuvre like Oberth's manoeuvre. The idea is by lifting the satellite up, the satellite gains potential energy which is then used to speed up the satellite to a lower altitude when executing Oberth's manoeuvre as shown in Fig. 3.14.

One could compare the Δv requirements for Edelbaum's three impulse manoeuvre with Oberth's and the direct manoeuvres. The comparison is shown in Fig. 3.15.

It may be observed that in terms of the total Δv required, Edelbaum's manoeuvre always requires more fuel and is eventually asymptotic to a direct transfer and always more expensive than Oberth's manoeuvre. However, it may prove to be extremely useful for meeting timing constraints as well as other requirements.

3.6 Application of Lambert's Problem to Space Vehicle Manoeuvring

In fact, timing constraints are extremely important, and in this context, one must mention the application of the solutions to Lambert's problem to space vehicle manoeuvring. The problem has been discussed at length by Prussing and Conway (1993). Here presented only is a brief summary. Lambert's Eq. (1.63) may be used to determine the terminal velocity vectors given a segment of the orbit, with only its initial and final position vectors $\vec{r}_1$ and $\vec{r}_2$, specified as well as the time of flight. The procedure involves calculating the parabolic transfer time between the two given points which is given by an equation named after Leonhard Euler. It was obtained independently and earlier than Lambert but could also be obtained by taking the limit of Lambert's equation by letting $a \rightarrow \infty$. The result is

$$\sqrt{\mu}\Delta t_p = \sqrt{\mu} \times t_p = \frac{\sqrt{2}}{3}\left(s^{3/2} - \operatorname{sgn}(\sin\theta)(s-c)^{3/2}\right), \tag{3.48}$$

where c is the chord given Eq. (1.62), θ is the transfer angle defined by Eqs. 1.60 and 1.61 and $\operatorname{sgn}(x)$ is the signum function which is 1 when x is positive and -1 otherwise. For a real elliptic orbit solution to exist, the specified transfer time must exceed the parabolic transfer time, t_p. The next step is to calculate the time t_m for the minimum energy arc, given by

$$\sqrt{\mu}t_m = \sqrt{s^3/8}(\alpha_m - \beta_m + \sin\beta_m), \alpha_m = \pi, \sin\frac{\beta_m}{2} = \sqrt{\frac{s-c}{s}}. \tag{3.49}$$

It is now feasible to compute the correct values for both α and β. Then Lambert's equation may be solved for the semi-major axis using Battin's algorithm or any of the other algorithms mentioned in Chap. 1. Once the semi-major axis is calculated, the unit vectors in the directions of $\vec{r}_1$, $\vec{r}_2$ and $\left(\vec{r}_2 - \vec{r}_1\right)/c$ are obtained from

$$e_1 = \vec{r}_1/|\vec{r}_1|, e_2 = \vec{r}_2/|\vec{r}_2|, e_c = \left(\vec{r}_2 - \vec{r}_1\right)/c. \quad (3.50)$$

The velocity vector at the initial point is given by

$$v_1 = \sqrt{\mu/(4a)}\{(\cot(\beta/2) + \cot(\alpha/2))e_c + (\cot(\beta/2) - \cot(\alpha/2))e_1\}. \quad (3.51)$$

A similar formula defines the velocity at the end point of the orbit segment which is

$$v_2 = \sqrt{\mu/(4a)}\{(\cot(\beta/2) + \cot(\alpha/2))e_c - (\cot(\beta/2) - \cot(\alpha/2))e_2\}. \quad (3.52)$$

The details may be found in by Prussing and Conway (1993).

Exercises

1. It is desired to change the inclination of the elliptic orbit of an Earth orbiting satellite to the equatorial plane by applying an impulsive force to the satellite.
   (a) Explain why the impulse must be applied to the satellite at one of the nodes.
   (b) Obtain an expression for the Δv required in terms of the orbital position, the orbital angular momentum and the required change in inclination.
   (c) It is desired to change the shape of the satellite's orbit from its elliptic shape to a circular orbit. Obtain an expression for the velocity increment required for changing the orbit from an elliptic to a circular orbit at the apogee, given the position of the apogee and the eccentricity of the elliptic orbit.
   (d) Derive an expression for the hyperbolic excess velocity for a satellite in a Keplerian orbit. Using the visa-viva equation, obtain an expression for the velocity increment required for changing the orbit from a circular to a hyperbolic orbit in terms of the hyperbolic excess velocity.
2. (a) Explain with a diagram the meaning of the term "coplanar Hohmann transfer."
   (b) Obtain an expression for the total Δv required for implementing a coplanar Hohmann transfer when the orbital elements of the initial and final orbits are known and their eccentricities are equal to zero.
   (c) Obtain an expression for the transfer time for the transfer in part (b).
   (d) Modify the expressions derived in part (b) when the eccentricities of the initial and final orbits are not equal to zero.
3. Explain, with the aid of a diagram, how only the rotation of the orbit plane can be achieved. Hence, obtain an expression for estimating the Δv requirement for changing the inclination of a spacecraft's orbit by applying a single impulse. State the point along the orbit where this impulse must be applied as well as the direction in which it must be applied.

4. (a) Obtain an expression for the velocity change provided by a planet during a flyby due to its gravitational force, in terms of both the hyperbolic excess velocity v_∞ and the turning angle δ.
   (b) Hence or otherwise, obtain an expression for the velocity change provided by a planet during a flyby, in terms of the hyperbolic excess velocity v_∞, the surface circular velocity around the planet v_s, the planet's surface radius r_s and the radius of curvature at the periapsis of the hyperbolic orbit, r_p.
   (c) Determine the relationship between the hyperbolic excess velocity v_∞, the surface circular velocity around the planet v_s, the planet's surface radius r_s and the radius of curvature at the periapsis of the hyperbolic orbit that gives the maximum possible magnitude for the velocity change provided by a planet during a flyby.
   (d) Determine the maximum possible magnitude for the velocity change provided by a planet during a flyby in terms of the surface circular velocity around the planet v_s, the planet's surface radius r_s and the radius of curvature at the periapsis of the hyperbolic orbit, r_p.

5. A spacecraft is orbiting Earth at a height of 2500 km in a circular orbit. At a point P in the orbit, its thrusters are fired to bring it down to an orbit with a perigee height above the Earth of 200 km. A Hohman transfer orbit is used.

   (a) Explaining all your calculation steps, estimate the Δv required to bring it down to the required orbital height.
   (b) Once the spacecraft reaches an altitude of 200 km, its thrusters are fired, so the spacecraft just escapes from Earth's gravity.
   Giving all your calculation steps, calculate the required for the second burn as well as the total Δv required.
   (c) Explaining the formulas used, estimate the Δv required, if instead of this two-burn manoeuvre, the spacecraft's thrusters are fired only once at the point P, so it just escapes from Earth's gravity? (Assume the radius of the Earth is 6400 km).

6. A satellite is approaching the planet Venus ($\mu = 3.2486 \times 10^5 \text{km}^3/\text{s}^2$) with a relative velocity of 5 km/s at the edge of its sphere of influence. The eccentricity of the satellite's orbit about Venus is 1.5.

   (a) Explaining all your assumptions and calculation steps, estimate (i) the semi-major axis of the orbit and the distance of closest approach to Venus, (ii) the excess velocity and the total energy of the spacecraft as it approaches the planet, (iii) the turning angle, (iv) the velocity increment the satellite acquires as it swings by the planet Venus and (v) the velocity of the satellite at its closest position to the planet Venus.

   If the spacecraft is provided with a velocity increment of 2 km/s at its closest position to the planet Venus, determine the excess velocity of the spacecraft as it departs the planet.

(b) A spacecraft is orbiting Venus at a height of 2000 km in a circular orbit. At a point P in the orbit, its thrusters are fired to bring it down to an orbit with a perigee height above the Venus of 200 km. A Hohmann transfer orbit is used.

(i) Explaining all your calculation steps, estimate the Δv required to bring it down to the required orbital height.

(ii) Once the spacecraft reaches an altitude of 200 km, its thrusters are fired, so the spacecraft just escapes from Venus's gravity.

Giving all your calculation steps, calculate the Δv required for the second burn as well as the total Δv required.

(iii) Explaining the formulas used, estimate the Δv required, if instead of this two-burn manoeuvre, the spacecraft's thrusters are fired only once at the point P, so it just escapes from Venus's gravity? (Assume the radius of the Venus is 6052 km). How does this compare with the results of parts (i) and (ii)?

References

Edelbaum, T. N. (1959). Some extensions of the Hohmann transfer manoeuver. *Journal of the American Rocket Society, 29*, 864–865.

Edelbaum, T. N. (1961). Propulsion requirements for controllable satellites. *ARS Journal, 31*, 1079–1089.

Edelbaum, T. N. (1967). How many impulses? *Astronautics and Aeronautics, 5*, 64–69.

Fortescue, P. W., Swinerd, G. S., & Stark, J. P. W. (2011). *Spacecraft systems engineering* (4th ed.). Wiley.

Kamel, O. M., & Soliman, A. S. (2011). On the generalized Hohmann transfer with plane change using energy concepts Part I. *Mechanics and Mechanical Engineering, 15*(2), 183–191.

Oberth, H. (1972). *Ways to spaceflight*. Translation of "Wege zur Raumschiffahrt," R. Oldenbourg Verlag, Munich- Berlin, 1929, NASA TT F-622.

Oleson, S. R. (1993). *An analytical optimization of electric propulsion orbit transfers* (NASA Contractor Report, NASA CR 191129). NASA Lewis Research Center.

Prussing, J. E., & Conway, B. A. (1993). *Orbital mechanics*. Oxford University Press.

Vepa, R. (2019). *Dynamics and control of autonomous space vehicles and robotics*. Cambridge University Press.

Chapter 4
Interplanetary Trajectories

4.1 Interplanetary Trajectories

One of the most important applications of the study of space vehicle manoeuvring is to the design interplanetary trajectories (Deutsch, 1963; Fortescue et al., 2011; Prussing & Conway, 1993). Interplanetary travel involves a minimum of three phases: leaving the departure planet, travelling to the domain of the destination planet and then arriving at the desired terminal orbit or position. Thus, the three phases in the interplanetary travel to another planet involve the escape from the departing planet's gravitational field in a hyperbolic orbit, travel, cruise or coast in heliocentric or Sun centred elliptic orbit from the departing planet's escape region to within the destination planet's gravitational field and eventually enter the destination planet's gravitational field in a hyperbolic orbit under the influence of the destination planet's gravitational field only. The positions and velocities are continuous as the spacecraft leaves the departing planet's gravitation field, traverses in the solar gravitational field and finally enters the destination planet gravitational field. So the first task is clearly to identify the domains of the departure planet, the destination planet and the intervening domain where the space vehicle is expected to travel, cruise or coast in heliocentric or Sun centred elliptic orbit. The basic methodology generally adopted to the design of interplanetary trajectories is the patched conic method. In the patched conic method which involves a hyperbolic exit, an elliptic transfer and a hyperbolic entry as well as the continuity of the velocity all along the trajectory, additionally in each of primary body's fields during each of the three phases, only the dominating body be it the departing planet, Sun, the central body of the planetary system or the destination planet is assumed to exist. Thus, in each of the three phases when the space vehicle is under the primary influence of either the departing planet, the Sun or the destination planet, there are only two bodies to consider, and in each of the three domains, there is only one body to consider in addition to the vehicle itself. Thus, in each of the three domains, all the orbit designs and calculations can be done on the basis of Keplerian orbits while ensuring that the

R. Vepa, *Space Vehicle Maneuvering, Propulsion, Dynamics and Control*,
https://doi.org/10.1007/978-3-031-65518-0_4

trajectories are continuous with continuous positions and velocities. The assumption that only the central body is assumed to exist in each of the three phases and that all other perturbations are ignored results in a method that is good for estimating the total velocity increment, Δv, requirements but is not very accurate in predicting the total time of flight.

4.2 Sphere of Influence (SOI)

Thus, a very useful approach in solving the interplanetary trajectory problem is to consider the influence on the spacecraft of one central body at a time, leading to a known two-body problem throughout. Consider, for instance, the departure from the planet Earth, so the only influence on vehicle during departure is assumed to be the Earth's, followed by the Sun centred transfer orbit, where only influence is that of the Sun and finally the spacecraft's arrival orbit under the target planet's gravitational influence only. To facilitate the application of the two-body problem in each phase of the flight, an essential concept is the concept of the sphere of influence (SOI), which is an important concept used in the planning of the interplanetary trajectories using the patched conic method. In the patched conic method, when a spacecraft is moving away from a large central body and towards a smaller central body, the orbiting spacecraft is expected to switch from a large body centric conic approximation to the small body centric conic approximation and vice versa. When the satellite travels from the smaller central body, it is expected to escape from the smaller body, at some point, and is considered to be "at infinity" from it. The question then arises as to how such a point could be defined. The answer to the above question is assumed to be the surface on which this perturbing third body acceleration of the larger body is equal in magnitude to the gravitational acceleration of the smaller body. This surface then defines a region of primary influence of the smaller body, as within this region, the gravitational acceleration of the smaller body would dominate. A further assumption is then made and this region is restricted to the largest sphere within the region.

Considering the line joining the two bodies and the distance between them being r_d, relative to the smaller body, the distance of point at which perturbing third body acceleration of the larger body is equal in magnitude to the gravitational acceleration of the smaller body is

$$r_{\text{soi}} \approx r_d(\mu/\mu_d)^{2/5}. \tag{4.1}$$

The derivation of Eq. 4.1 follows.

The SOI is determined by the ratio of the gravitational accelerations produced on the spacecraft by one body considered as the "central body" (M_c, a_c) and the other body considered as the "disturbing body" (M_d, a_d). The ratio is computed twice, switching the roles of the "central body" and the "disturbing body" between the two

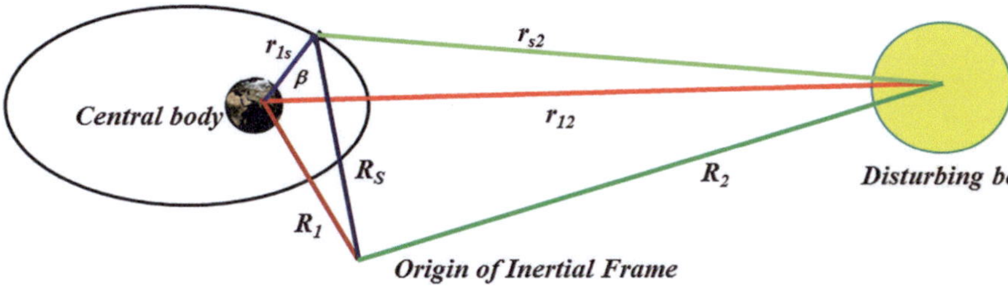

Fig. 4.1 Computation of the relative acceleration between a central and a disturbing body

bodies. The position on the line joining the two bodies, for which the two ratios are equal, constitutes the boundary of the SOI. So, it is required to compute the ratio of the disturbing body's acceleration to the central body's acceleration. The problem is illustrated diagrammatically in Fig. 4.1. Considering all the three bodies in Fig. 4.1, one could apply Newton's second law to each body, for motion with respect to the inertial frame as shown in the diagram. Considering the first, the satellite

$$m_s\ddot{\mathbf{R}}_S = G\frac{m_s m_1}{r_{1s}^3}\mathbf{r}_{1s} - G\frac{m_s m_2}{r_{s2}^3}\mathbf{r}_{s2}, \mathbf{r}_{1s} = \mathbf{R}_1 - \mathbf{R}_S, \mathbf{r}_{s2} = \mathbf{R}_s - \mathbf{R}_2. \tag{4.2}$$

Considering the motion of the central body,

$$m_1\ddot{\mathbf{R}}_1 = -G\frac{m_s m_1}{r_{1s}^3}\mathbf{r}_{1s} - G\frac{m_1 m_2}{r_{12}^3}\mathbf{r}_{12}, \mathbf{r}_{12} = \mathbf{R}_1 - \mathbf{R}_2. \tag{4.3}$$

Finally considering the disturbing body,

$$m_2\ddot{\mathbf{R}}_2 = G\frac{m_2 m_1}{r_{12}^3}\mathbf{r}_{12} + G\frac{m_s m_2}{r_{s2}^3}\mathbf{r}_{s2}. \tag{4.4}$$

Dividing each of the above equations by the respective masses,

$$\begin{aligned}
&\ddot{\mathbf{R}}_S = G\frac{m_1}{r_{1s}^3}\mathbf{r}_{1s} - G\frac{m_2}{r_{s2}^3}\mathbf{r}_{s2}, \quad \ddot{\mathbf{r}}_{1s} = \ddot{\mathbf{R}}_1 - \ddot{\mathbf{R}}_S \;\; \ddot{\mathbf{r}}_{s2} = \ddot{\mathbf{R}}_S - \ddot{\mathbf{R}}_2 = -\ddot{\mathbf{r}}_{2s},\\
&\ddot{\mathbf{R}}_1 = -G\frac{m_s}{r_{1s}^3}\mathbf{r}_{1s} - G\frac{m_2}{r_{12}^3}\mathbf{r}_{12}, \quad \ddot{\mathbf{r}}_{12} = \ddot{\mathbf{R}}_1 - \ddot{\mathbf{R}}_2,\\
&\ddot{\mathbf{R}}_2 = G\frac{m_1}{r_{12}^3}\mathbf{r}_{12} + G\frac{m_s}{r_{s2}^3}\mathbf{r}_{s2}
\end{aligned} \tag{4.5}$$

The equations for relative motion are

$$\begin{aligned}\ddot{\mathbf{r}}_{1s} &= -G\frac{m_s+m_1}{r_{1s}^3}\mathbf{r}_{1s} - Gm_2\left(\frac{\mathbf{r}_{12}}{r_{12}^3} - \frac{\mathbf{r}_{s2}}{r_{s2}^3}\right),\\ \ddot{\mathbf{r}}_{s2} &= -G\frac{m_2+m_s}{r_{s2}^3}\mathbf{r}_{s2} - Gm_1\left(\frac{\mathbf{r}_{12}}{r_{12}^3} - \frac{\mathbf{r}_{1s}}{r_{1s}^3}\right),\\ \ddot{\mathbf{r}}_{12} &= -G\frac{m_1+m_2}{r_{12}^3}\mathbf{r}_{12} - Gm_s\left(\frac{\mathbf{r}_{1s}}{r_{1s}^3} - \frac{\mathbf{r}_{s2}}{r_{s2}^3}\right).\end{aligned} \tag{4.6}$$

If $m_s < < m_1, m_2$, and since, $\mathbf{r}_{s2} = \mathbf{r}_{12} - \mathbf{r}_{1s}$, from body 1's perspective,

$$\begin{aligned}\ddot{\mathbf{r}}_{1s} = -G\frac{m_s+m_1}{r_{1s}^3}\mathbf{r}_{1s} - Gm_2\left(\frac{\mathbf{r}_{12}}{r_{12}^3} - \frac{\mathbf{r}_{s2}}{r_{s2}^3}\right) \simeq -Gm_1\frac{\mathbf{r}_{1s}}{r_{1s}^3}\\ - Gm_2\left(\frac{\mathbf{r}_{12}}{r_{12}^3} - \frac{\mathbf{r}_{12}-\mathbf{r}_{1s}}{(|\mathbf{r}_{12}-\mathbf{r}_{1s}|)^3}\right).\end{aligned} \tag{4.7}$$

From body 2's perspective,

$$\begin{aligned}\ddot{\mathbf{r}}_{s2} = -G\frac{m_2+m_s}{r_{s2}^3}\mathbf{r}_{s2} - Gm_1\left(\frac{\mathbf{r}_{12}}{r_{12}^3} - \frac{\mathbf{r}_{1s}}{r_{1s}^3}\right) \simeq -Gm_2\frac{(\mathbf{r}_{12}-\mathbf{r}_{1s})}{(|\mathbf{r}_{12}-\mathbf{r}_{1s}|)^3}\\ - Gm_1\left(\frac{\mathbf{r}_{12}}{r_{12}^3} - \frac{\mathbf{r}_{1s}}{r_{1s}^3}\right).\end{aligned} \tag{4.8}$$

In both Eqs. 4.7 and 4.8, the first term on the right is the central body's contribution, while the second term on the right is the contribution from the disturbing body. However, considering just the motion of spacecraft with respect to body 1 in Eq. (4.7), first term on the right is due to the central body 1 while the second term due to the disturbing body. The second term due to the disturbing body consists of two contributions, with latter related to the spacecraft. Considering this contribution, it could be simplified and approximated as

$$\frac{\mathbf{r}_{12}-\mathbf{r}_{1s}}{(|\mathbf{r}_{12}-\mathbf{r}_{1s}|)^3} \simeq \frac{\mathbf{r}_{12}|1-\mathbf{r}_{1s}/\mathbf{r}_{12}|}{r_{12}^3(|1-\mathbf{r}_{1s}/\mathbf{r}_{12}|)^3} = \frac{\mathbf{r}_{12}|1-\mathbf{r}_{1s}/\mathbf{r}_{12}|^{-2}}{r_{12}^3} \simeq \frac{\mathbf{r}_{12}|1+2\mathbf{r}_{1s}/\mathbf{r}_{12}|}{r_{12}^3}. \tag{4.9}$$

It follows that

$$\frac{\mathbf{r}_{12}}{r_{12}^3} - \frac{\mathbf{r}_{12}-\mathbf{r}_{1s}}{(|\mathbf{r}_{12}-\mathbf{r}_{1s}|)^3} = \frac{\mathbf{r}_{12}}{r_{12}^3} - \frac{\mathbf{r}_{12}}{r_{12}^3}|1+2\mathbf{r}_{1s}/\mathbf{r}_{12}| = -2\left|\frac{\mathbf{r}_{1s}}{r_{12}^3}\right|. \tag{4.10}$$

The ratio of the contribution to the acceleration due to the disturbing body to that of the central body, from the perspective of body 1, is

$$A_{r1} = 2\frac{m_2}{m_1}\frac{\mathbf{r}_{1s}}{r_{12}^3} \times \frac{r_{1s}^3}{\mathbf{r}_{1s}} = 2\frac{m_2}{m_1}\frac{r_{1s}^3}{r_{12}^3}. \tag{4.11}$$

From body 2's perspective, the ratio of two contributing accelerations can be shown to be

$$A_{r2} = \frac{m_1}{m_2}\frac{r_{12}^2 - r_{1s}^2}{r_{1s}^2 r_{12}^2}(|\mathbf{r}_{12} - \mathbf{r}_{1s}|)^2 \approx \frac{m_1}{m_2}\frac{r_{12}^2}{r_{1s}^2}. \tag{4.12}$$

Equating the two ratios, $A_{r1} = A_{r2}$ gives,

$$A_{r1} = 2\frac{m_2}{m_1}\frac{r_{1s}^3}{r_{12}^3} = A_{r2} = \frac{m_1}{m_2}\frac{r_{12}^2}{r_{1s}^2}. \tag{4.13}$$

Rearranging Eq. 4.13 yields

$$\left(\frac{m_1}{m_2}\right)^2 = 2\frac{r_{1s}^5}{r_{12}^5}. \tag{4.14}$$

Equation 4.14 is approximated and expressed as

$$r_{1s} = 2^{-0.2} r_{12}\left(\frac{m_1}{m_2}\right)^{2/5} = 0.8706 r_{12}\left(\frac{m_1}{m_2}\right)^{2/5} \approx r_{12}\left(\frac{m_1}{m_2}\right)^{2/5}. \tag{4.15}$$

Equation 4.15 is the final result for the diameter of the SOI.

4.3 Patched Conic Method

The travel from one planet to another is considered to be made up of three phases. The three phases in the interplanetary travel to another planet are the escape from the departing planet's gravitational field in a hyperbolic orbit, followed by the travel or coast in a heliocentric or Sun centred elliptic orbit from the departing planet's outer escape region to within the destination planet's gravitational field or the destination planet's capture region and terminating with the final phase of travel to enter the destination planet's gravitational field in a hyperbolic orbit exclusively under the influence of its gravitational field only.

The positions and velocities are required to be continuous as the spacecraft leaves the departing planet's gravitation field, traverses across the two planet's outer regions in the solar gravitational field and finally enters the destination planet's gravitational field. In the patched conic method additionally (hyperbolic exit of departure, elliptic transfer and hyperbolic entry on arrival as well as velocity continuity) in each primary body's field, only the dominating body (be it the departing planet, the Sun or the destination planet) is assumed to exert an influence of the spacecraft. Thus, only the central body is assumed to exist in each of the three phases apart from the spacecraft. All other perturbations are ignored. While the method is good for estimating the total Δv requirements, it is not very accurate in predicting the total time of flight.

Thus, a very useful approach in solving the problem of the design of an interplanetary trajectory is to consider the influence on the spacecraft of one central body at a time. As a consequence in each domain, the problem of trajectory design reduces the solution to the well-known two-body problem, followed by the synthesis of the trajectories to meet the position and velocity continuity requirements. Considering the departure from Earth, only the Earth is assumed to influence the vehicle during departure, and only the Sun is assumed to influence the vehicle as it executes a Sun centred transfer orbit, while on arrival at the destination planet, only the target planet's gravitational field influences the space vehicle. Thus, it is customary to divide the interplanetary trajectory into three different phases and patch the orbits together, which is the basis of the patched conic approximation. The solution for the trajectories in the three phases is obtained in the following order, with phase 2, when the Sun centred transfer orbit (solar gravity dominates) is obtained first using a Hohmann transfer, followed by phase 1, when the departure orbit from Earth (Earth gravity dominates) is synthesized next (direct or from a parking orbit), and finally phase 3 when the arrival orbit at the destination planet (destination planet gravity dominates) is determined last.

4.3.1 Interplanetary Trajectories

Another important approximation that is made, as the spacecraft departs from or arrives at the boundary of the SOI, is that the SOI's boundary is considered to be at infinity relative to the dominating body within the SOI, for purposes of computing the velocities long the hyperbolic trajectories and for departing from or arriving at the boundary of the SOI. This facilitates the computation of an upper bound to the total velocity requirement in that phase. This is illustrated in Fig. 4.2. The examples in a later section illustrate the computation.

4.3.2 Interplanetary Trajectories-Heliocentric Orbit Phase

In the heliocentric phase, the interplanetary (IP) transfer orbit from the edge of the departing planet's SOI to the outer boundary of the arrival planet's SOI, which is an

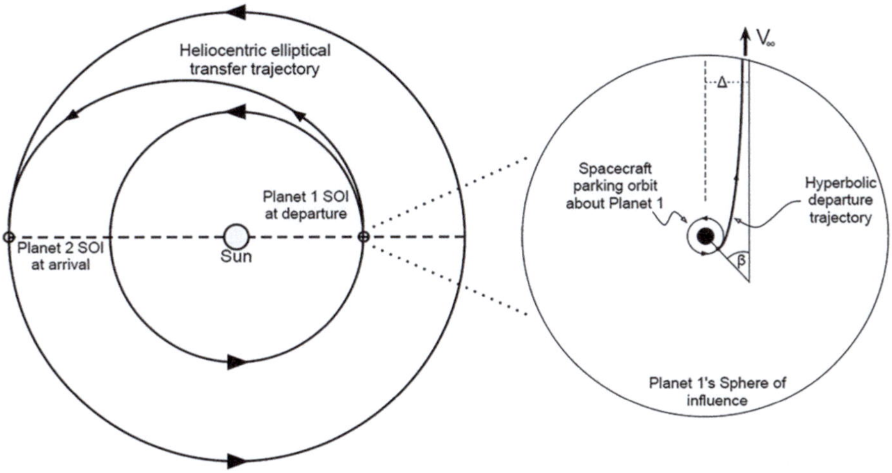

Fig. 4.2 Illustration of the departure and transfer trajectories

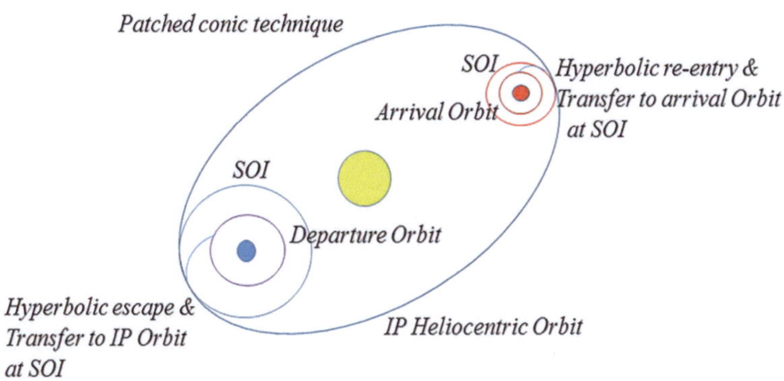

Fig. 4.3 Illustration of the heliocentric, IP elliptic transfer orbit around the Sun, involving a Hohmann transfer

elliptic transfer orbit around the Sun, involving a Hohmann transfer, is illustrated in Fig. 4.3.

4.3.3 Planetary Escape, Planetary Approach and Re-entry

It is important to consider the departure orbit during planetary escape quite closely. At the patch point, heliocentric transfer ellipse (which ignores the planet) is considered to be at the SOI of the planet itself. Relative to the planet, this is at infinity. The situation at the boundary of the departing planet's SOI is illustrated in Fig. 4.4.

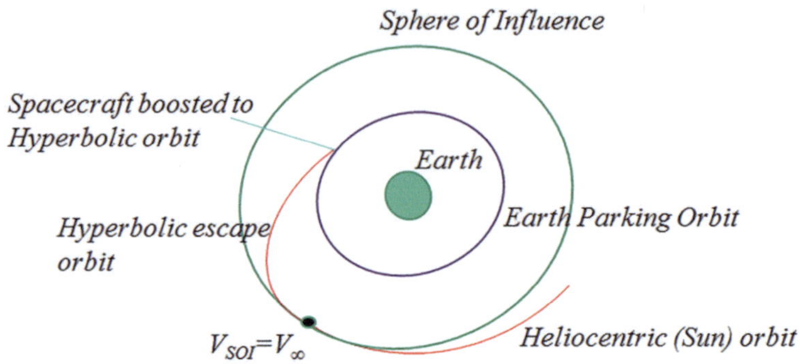

Fig. 4.4 Illustration of the departure orbit at the patch point

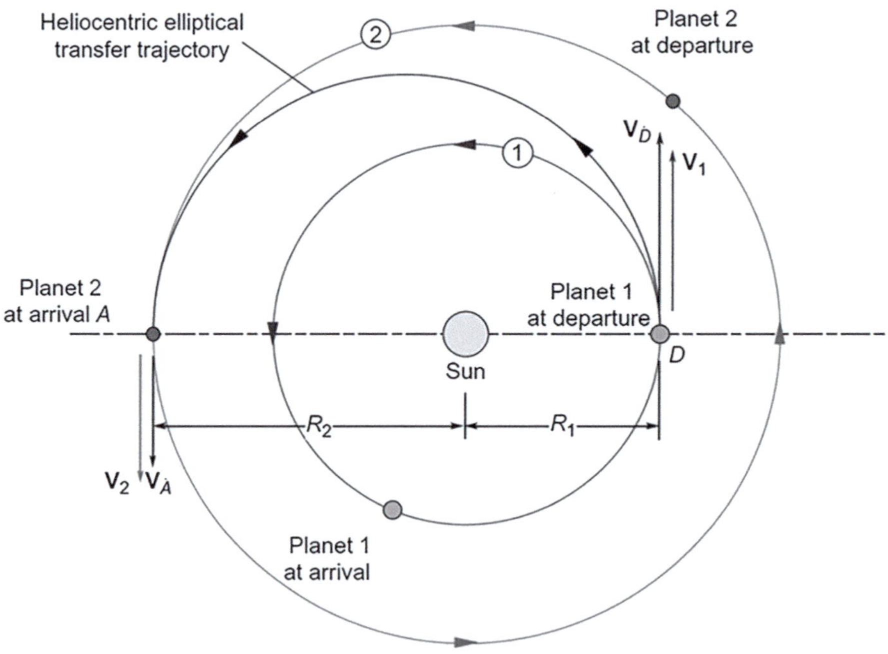

Fig. 4.5 Illustration of the heliocentric elliptic transfer orbit to the destination planet

At the patch point at the edge of the SOI of the departure planet, the spacecraft enters the Sun's gravitational field and is placed onto the transfer orbit—it continues along the transfer orbit to the destination planet. This is illustrated in Fig. 4.5. On arrival at the destination planet's SOI, the spacecraft enters the destination planet's gravitational field. From the vis-viva equation at any point on the heliocentric orbit,

$$\frac{1}{2}v^2 - \frac{\mu}{r} = -\frac{\mu}{2a} = \varepsilon. \tag{4.16}$$

To obtain the semi-major axis of the transfer ellipse, assume the radius of Earth and the destination planet are known; then, it follows that

$$a_{tr} = (R_E + R_D)/2. \tag{4.17}$$

Thus, the velocities at the perigee v_p and at the apogee v_a of the heliocentric orbit are estimated.

Along a hyperbolic orbit,

$$\frac{1}{2}v^2 - \frac{\mu}{r} = \frac{1}{2}v_\infty^2 - \frac{\mu}{r_\infty} = \frac{1}{2}v_\infty^2 = \varepsilon, 2\varepsilon = v_\infty^2 = v^2 - \frac{2\mu}{r} = v^2 - v_{\text{esc}}^2. \tag{4.18}$$

At any point very far from the central body (at infinity), as $r \to \infty$,

$$v^2 \to 2\varepsilon = v_\infty^2 - \frac{2\mu}{r_\infty} = v_\infty^2. \tag{4.19}$$

Hence, the hyperbolic excess velocity satisfies

$$v_\infty^2 = v^2 - v_{\text{esc}}^2 = 2\varepsilon = -\mu/a. \tag{4.20}$$

The satellite's required velocity at the patch point is obtained from the perigee velocity v_p of the heliocentric orbit and the Earth's circular velocity v_E, and this determines the hyperbolic excess velocity as

$$v_{\text{SOI}} = v_p - v_E = v_\infty. \tag{4.21}$$

At the departure planet's SOI, the velocity at the periapsis of the transfer orbit, relative to Earth (the departure planet), is used to determine the hyperbolic excess velocity. Thus, if one assumes a hyperbolic departure orbit at the departure planet (Earth), given the velocity at the parking orbit radius, the required Δv for going from the parking orbit to the hyperbolic departure orbit may be estimated. On departure from the Earth's SOI, the spacecraft enters the transfer orbit. An example of a transfer orbit to Mars is illustrated in Fig. 4.6.

Arrival at the destination planet is similar to departure. If one assumes a hyperbolic arrival orbit at an outer planet, at the arrival planet's SOI, the velocity on the transfer orbit, relative to Mars (the arrival planet), is obtained from the velocity of the spacecraft at the apogee of the heliocentric orbit, v_a, relative to the circular velocity of the destination planet v_D:

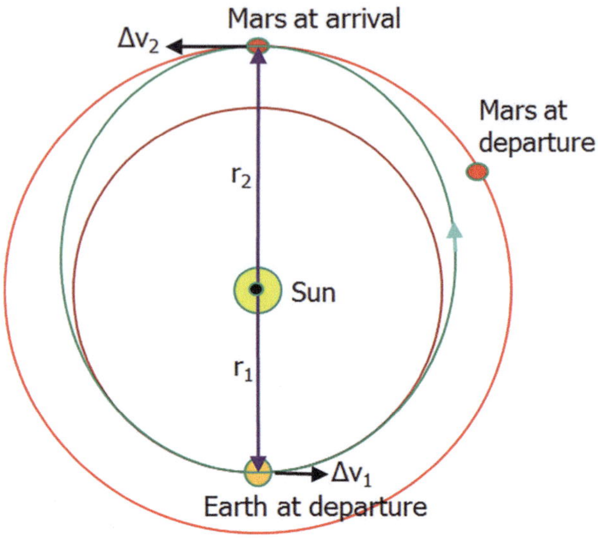

Fig. 4.6 Illustration of a Hohmann transfer orbit to Mars

$$v_{\mathrm{SOI}} = v_a - v_D = v_\infty. \tag{4.22}$$

At the patch point, the spacecraft moves into the gravitational field of the destination planet and continues along a hyperbolic orbit. The required hyperbolic arrival velocity with respect to the destination planet and the Δv required at the SOI for changing into a circular parking orbit at the desired radius may be estimated. A typical example illustrates the calculations in a later section.

4.3.4 Phasing and Launch Timing

The timing of the launch is extremely important that the arrival of spacecraft at the patch point of the destination planet's SOI synchronizes with the arrival of the destination planet at its designated location. The planet must be at the expected destination point so the spacecraft can rendezvous with it. The situation is illustrated in the phasing diagram in Fig. 4.7.

The required phase angle for flight to Mars from Earth is given by

$$\phi_{\mathrm{phase}} = \phi_f - \omega_{\mathrm{Mars}} \times t_{\mathrm{tof}}. \tag{4.23}$$

In the time $t_1 - t_0$, the satellite has travelled from the perigee to the apogee of the Hohmann transfer orbit, which is π radians. Hence, the initial phase angle between the satellite at launch and Mars should be

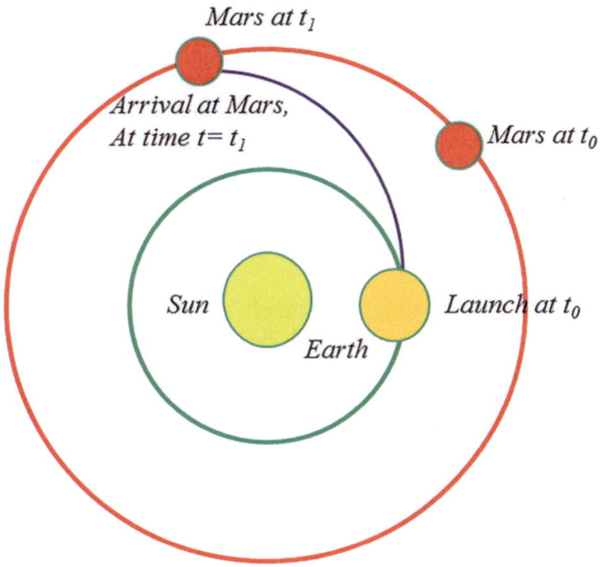

Fig. 4.7 Phasing diagram for rendezvous with the destination planet

$$\phi_{\text{phase}} = \pi - \omega_{\text{Mars}} \times t_{\text{tof}}. \tag{4.24}$$

In Eq. 4.24, ω_{Mars} is the orbital speed of Mars. If the initial phase angle of Earth is known, the wait time for launch may be estimated.

The synodic period plays a key role in determining the wait time for a launch as well as for the lay-by and return from the destination planet, if a return is contemplated. Simply stated the synodic period, is the elapsed time between two successive conjunctions of the same type. A conjunction is a configuration of two planets orbiting the Sun, when they are both in a line (or at some fixed angle to each other). The conjunction is said to be an inferior conjunction when both planets are on the same side of the Sun and thus on the same end. The conjunction is said to be a superior conjunction when both planets are on the opposite sides of the Sun and thus on the opposite ends. Given the periods of the two planets, A and B, P_A and P_B, respectively, the synodic period is given by

$$S_n = \frac{P_A P_B}{P_A - P_B} = \frac{1}{(1/P_B) - (1/P_A)}. \tag{4.25}$$

Another factor that may need to be considered is the phenomenon of orbital resonance. Orbital resonance is said to occur when

$$m/(nP_A) = (1/P_B) - (1/P_A). \tag{4.26}$$

where m and n are integers. Orbital resonance has recently been observed in many planets, especially the heavy outer planets with several Moons orbiting them. These are also known as mean motion resonances. Orbital resonance can be the source of both stability and chaos and can significantly amplify small gravitational perturbations. Another example of another resonance (spin-orbit) is the fact that the same face of the Moon is always pointing to the Earth.

Given the orbital speeds of Mars and Earth, ω_{Mars} and ω_{Earth}, respectively, the wait time for launch is given by

$$\text{Wait time} = \left(\phi_{\text{phase}} - \phi_{\text{initial}}\right) / \left(\omega_{\text{Mars}} - \omega_{\text{Earth}}\right). \tag{4.27}$$

If the opportunity for launch is missed, the next opportunity for launch will depend on the synodic period, since it is the amount of time required for a destination planet to return to the same point in its orbit relative to the departure planet, Earth, and the central body, the Sun. As an example, the synodic period for an Earth-Mars transfer is

$$S_{\text{me}} = \frac{2\pi}{\omega_E - \omega_M} = \frac{2\pi}{0.0196 - 0.0105} 10^5 = 2.188\text{yrs}. \tag{4.28}$$

After arrival at the destination planet, the waiting time to be back to Earth is the minimum lay over time. It is a fraction of the synodic period. The relative position of the planet Earth with respect to destination planet must be right for launch from the destination planet, so the planet Earth is at the correct position for the spacecraft to rendezvous with it, when it returns. For an Earth-Mars transfer, with a time of flight of 9 months, the minimum lay over time is given by

$$t_{\text{lay-over}} = \frac{2\pi}{\omega_E - \omega_M} \frac{\omega_E}{\pi} \left(1 - tof_{\text{yrs}}\right) = 0.989\text{yrs}. \tag{4.29}$$

The steps involved in the design of a trajectory for an Earth to a destination planet transfer may now be summarized.

The objective of the transfer is to orbit the destination planet. The final destination planet orbit should be posigrade at a given altitude.

Step 1. Manoeuvre into a suitable parking orbit. Assume the satellite, the destination planet and the Earth orbit Sun in the same plane (if the destination planet orbits Sun in a plane different from that of the Earth orbit, the spacecraft orbit may not be in the ecliptic plane, after launch).

Step 2. Need to transition to ecliptic plane (may need to change both the inclination and the RAAN).

Step 3. Departure and transfer at the apoapsis or periapsis of the transfer orbit, depending on whether the destination planet is inner planet or an outer planet (for Earth departure at Earth SOI). Design the Hohmann transfer from Earth to the destination planet. Recall that the Earth is orbiting the Sun with a fixed velocity, and calculate the Earth's circular velocity around the Sun and estimate relative hyperbolic excess velocity.

For the hyperbolic departure velocity from the Earth, estimate the additional velocity (magnitude, direction) required by spacecraft for Earth departure from a circular orbit, including the desired periapsis or apoapsis radius and semi-major axis of hyperbolic departure orbit at Earth's SOI.

Consider the timing of the departure and determine the wait time for departure, the time of flight and the lay over time at the destination planet.

Step 4. For arrival at the destination planet, estimate the circular velocity of the destination planet, the radius of the circular arrival orbit and the spacecraft's velocity in the arrival orbit.

Estimate also the hyperbolic excess velocity on arrival at the destination planet (for use with the vis-viva equation) as well as the required hyperbolic arrival velocity at the destination planet.

Calculate the Δv to reduce the velocity to that required at the parking orbit radius of the destination plane as well as the desired periapsis radius and semi-major axis of hyperbolic arrival orbit at the destination planet's SOI.

4.4 Application Examples, Moon, Mars (Or Other Outer Planet, E.G. Jupiter), Venus (Inner Planet)

4.4.1 Lunar Trajectory

(a) It is required to design an interplanetary orbit of a Moon probe that will take it from a circular parking orbit about the Earth to a circumlunar orbit. This example is based on Hu (2015). The radius of the parking circular orbit = 6697.1 km. The following are required:

  (i) Calculate the velocity of the circular orbit about the Earth.
  (ii) At a certain point on the circular orbit, the Moon probe is given a velocity impulse, to transfer the spacecraft to an elliptic orbit with the semi-major axis = 285,000 km. Calculate the velocity at the perigee of the transfer orbit and the Δv required.
  (iii) Calculate the distance of the moon probe on the transfer orbit at the patch point, the velocity of the probe and the flight path angle at this point, given that the angle of arrival at the patch point relative to the Earth moon axis is $\lambda_1 = 30°$. Hence, find the angle γ_1, in radians, made by the position vector at this point to the Earth-Moon axis.
  (iv) Estimate the time of flight from the perigee of the transfer orbit to the patch point on the transfer orbit.
  (v) Determine the phase angle (in radians) at the instant of departure from the parking orbit of the Earth-Moon axis relative to the probe's position vector, if the patch point must be on the SOI of the moon. You may assume the Draconian period of the Moon to be 27.2122 days.

(vi) Consider the patch point on the Moon's SOI, and estimate the velocity of the probe with respect to the Moon's centre and its direction (in radians) relative to the radius vector of the patch point to the Moon.

(vii) Hence, calculate the minimum altitude attained by the Moon probe relative to the Moon's surface. If the probe is expected to be in circumlunar flight at this point, calculate the required Δv. Hence, state the total Δv for the mission and the time of flight.

(b) Assume that the semi-major axis of the transfer ellipse varies from 160,000 km to 285,000 km, and estimate the minimum total Δv for the mission. Comment on the Δv change when the parking orbit is raised to 6800 km.

(c) The solutions are as follows: the preliminary assumed data is the Earth radius $r_e = 6378.145$km, corresponding circular velocity of the spacecraft around the Earth at zero altitude $= v_e = \sqrt{\mu_e/r_e} = 7.904$ km/s, $\mu_e = 3.986012 \times 10^5$ km^3/s^2; $\mu_{\text{moon}} = \mu_e/81.3$; the radius of the Moon, $r_{\text{moon}} = 1738$km, mean velocity of Moon with respect to the Earth's centre $= v_m = 1.018$ km/s, distance of the Moon's centre from the Earth's centre, $d_{\text{moon}} = 384400$km, the Moon's SOI radius, $r_{\text{soi}} = 66300$ km and Mean radius of the Earth orbit around the Sun, $r_E = 1\text{AU} = 149597870700 = 149.6 \times 10^6$ km. The problem is illustrated in Fig. 4.8.

The transfer orbit to the Moon is obtained from the following:

The radius of the parking circular orbit around the Earth is given to be 6697.1 km.

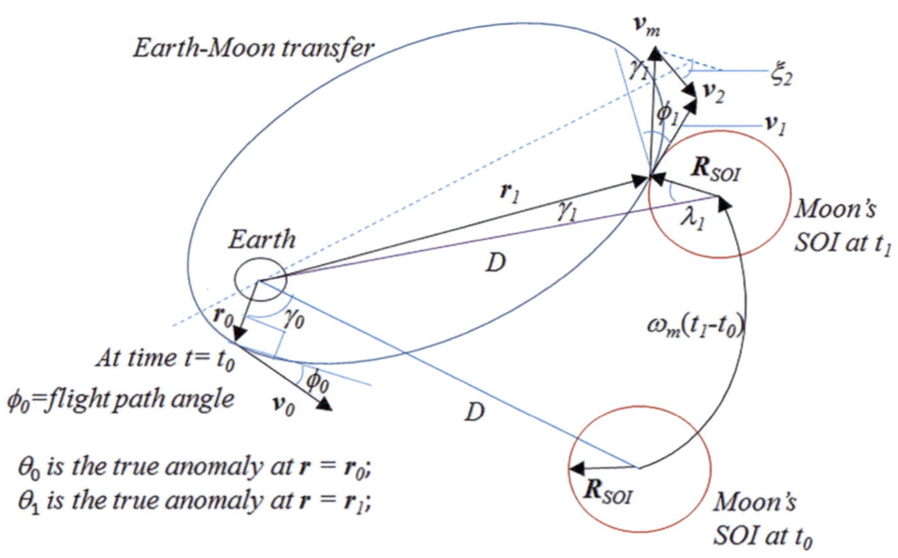

Fig. 4.8 Earth-Moon transfer orbit design

(i) Calculate the velocity of the circular orbit, $v_c = \sqrt{\mu_e/r_c} = 7.7148\ \mathrm{km/s}$.
(ii) At a certain point on the circular orbit, the Moon probe is given a velocity impulse to transfer the spacecraft to an elliptic orbit with the semi-major axis = 285,000 km.

Calculate the velocity at the perigee of the transfer orbit and the Δv required.
Distance of the initial point (perigee) in the transfer orbit = 6697.1 km.
From vis-viva equation, with $\mu = \mu_e = 3.986012 \times 10^5 km^3/s^2$.

$$\frac{1}{2}v_0^2 - \frac{\mu}{r_0} = -\frac{\mu}{2a} = \varepsilon = -0.6993, v_0 = 10.8462\ \mathrm{km/s}. \tag{4.30}$$

Velocity at that point on the transfer orbit $= v_0 =$ 10.8462 km/s, $\Delta v = 3.1314$ km/s.

(iii) Calculate the distance of the Moon probe on the transfer orbit at the patch point, the velocity of the probe and the flight path angle at this point, given that the angle of arrival at the patch point relative to the Earth moon axis is $\lambda_1 = 30°$. Hence, find the angle γ_1, in radians, made by the position vector at this point to the Earth-Moon axis.

Flight path angle $= \phi_0 = 0$, arrival angle relative to the Earth-moon axis $= \lambda_1 = 30°$. $r_0 = 6697.1$ km. $h = r_0 v_0 \cos\phi_0 = 72637.00\ \mathrm{km^2/\sec}$;

$$\begin{aligned} r_1 &= \sqrt{d_{\mathrm{moon}}^2 + r_{\mathrm{soi}}^2 - 2d_{\mathrm{moon}} r_{\mathrm{soi}} \cos\lambda_1} = 3.2866\mathrm{e} + 05\ \mathrm{km}, \\ v_1^2 &= -0.6993 \times 2 + 2\mu/r_1 = 1.0134 km/s; h/r_1 v_1 = \cos\phi_1, \phi_1 = 1.3510\ \mathrm{rads}; \\ \sin\gamma_1 &= (r_{soi}/r_1)\sin\lambda_1, \lambda_1 = 0.1010\ \mathrm{rads}. \end{aligned} \tag{4.31}$$

(iv) Estimate the time of flight from the perigee of the transfer orbit to the patch point on the transfer orbit, given that

$$\mathrm{tof}_{\mathrm{moon}} = \sqrt{\frac{a^3}{\mu}}\{(E_1 - e\sin E_1) - (E_0 - e\sin E_0)\}. \tag{4.32}$$

The parameters of the trajectory are $p = h^2/\mu = 1.3237\mathrm{e} + 04\ km$, $a = \mu_e/(2\mathcal{E}) = 2.8500\mathrm{e} + 05\ km$, $e = \sqrt{1 - p/ae} = 0.9765$.

$$r = p/(1 + e\cos\theta); r_0 = p/(1 + e\cos\theta_0); r_1 = p/(1 + e\cos\theta_1). \tag{4.33}$$

The true anomalies satisfy

$$\cos\theta_0 = \frac{p - r_0}{er_0} = 1; \theta_0 = 0; \; \cos\theta_1 = \frac{p - r_1}{er_1} = -0.9828; \theta_1 = 2.9560.$$

(v) The time of flight is given by

$$tof_{\text{moon}} = \sqrt{\frac{a^3}{\mu}}\{(E_1 - e\sin E_1) - (E_0 - e\sin E_0)\}. \tag{4.34}$$

In the above, the eccentric anomalies satisfy $\cos E_0 = \frac{e+\cos\theta_0}{1+e\cos\theta_0} = 1$; $E_0 = 0$; $\cos E_1 = \frac{e+\cos\theta_1}{1+e\cos\theta_1} = \frac{e-0.9828}{1-e^*0.9828} = -0.1563, E_1 = 1.7282$ rads.

Hence, the time of flight to the circumlunar orbit is $\text{tof}_{\text{moon}} = 1.8395\text{e} + 05$ secs $= 51.0971$ hrs.

(vi) Determine, at the instant of departure from the parking orbit, the phase angle (in radians) of the Earth-Moon axis relative to the probe's position vector, if the patch point must be on the SOI of the Moon.

$$\begin{aligned} \text{Phase angle at departure} = \gamma_0 &= \theta_1 - \theta_0 - \gamma_1 - \omega_m \text{tof}_{\text{moon}} \\ &= \theta_1 - \theta_0 - \gamma_1 - 2\pi\,\text{tof}_{\text{moon}}/T_D, \end{aligned} \tag{4.35}$$

where the Draconic or nordical period $T_D = 27.21222$ days, is the time between the Moon's passages through the same node, or intersection of its orbit with the ecliptic, the apparent pathway of the Sun.

For the initial point on the transfer orbit, the phase angle of the Moon should be

$$\begin{aligned} \gamma_0 = \theta_1 - \theta_0 - \gamma_1 - 2\pi\frac{\text{tof}_{\text{moon}}}{T_D} &= 2.9560 - 0 - 0.1010 \\ - 2\pi\frac{51.0971}{27.2122\times 24} &= 2.8550 - 0.4916 = 2.3634. \end{aligned} \tag{4.36}$$

(vii) Consider the patch point on the Moon's SOI, and estimate the velocity of the probe with respect to the Moon's centre and its direction (in radians) relative to the radius vector of the patch point to the Moon.

$$\begin{aligned} v_2 = &\sqrt{v_1^2 + v_m^2 - 2v_1 v_m \cos(\phi_1 - \gamma_1)} = 1.1885 \text{ km/s}, \\ v_2 \sin\xi_2 = &\; v_m \cos\lambda_1 - v_1 \cos(\lambda_1 - \phi_1 + \gamma_1) \Rightarrow \xi_2 = 0.1045. \end{aligned} \tag{4.37}$$

(viii) Hence, calculate the minimum altitude attained by the Moon probe relative to the Moon's surface. If the probe is expected to be in circumlunar flight at this point, calculate the required Δv.

$$h = r_2 v_2 \sin \xi_2 = 8215.5 \text{ km}^2/\text{s}. \frac{1}{2} v_2^2 - \frac{\mu_m}{r_{\text{soi\_moon}}} = -\frac{\mu_m}{2a_m} = \varepsilon = 0.6324,$$
$$p = h^2/\mu_m = 13783 \text{ km},$$
$$e = \sqrt{1 - p/a_m} = \sqrt{1 + 2\varepsilon (h/\mu_m)^2} = 2.1354, r_p = p/(1+e) = 4396 \text{ km}. \tag{4.38}$$

Probe's altitude at this point = 4396–1738 = 2658 km. For a circular orbit of the Moon, at the perigee,

$$v_{cm} = \sqrt{\mu_m/r_c} = 1.0554 \text{ km/ sec}. \tag{4.39}$$

Hence, the Δv required to transfer to a circumlunar orbit is obtained from

$$\Delta v = \sqrt{(v_2 \sin \xi_2 - v_{\text{cm}})^2 + v_2^2 \cos^2 \xi_2} = 1.5049 \text{ km/ sec}. \tag{4.40}$$

The total Δv for the mission is =3.1314 + 1.5049 = 4.6363 km/s. The time of flight to the circumlunar orbit is $\text{tof}_{\text{moon}} = 51.0971$ hrs.

b) The minimum total Δv is obtained when, the semi-major axis of the transfer ellipse is 167677.9 km and the Δv is, 3.6453 km/s. (In this case, $\theta_1 = \pi$ rads.) The calculations are done using a spreadsheet program.

When the parking orbit is raised to 6800 km, the total Δv for the mission, with the semi-major axis of the transfer ellipse equal to 285,000 km, is =3.1065 + 1.5111 = 4.6176 km/s. The minimum total Δv is obtained when the semi-major axis of the transfer ellipse is 167729.31 km and is 3.6186 km/s. However, this does not include the Δv required for raising the orbit from 6697.1 km to 6800 km. When this is included the additional Δv required is,

$$\Delta v = v_0 \left(1 - \sqrt{r_1/r_2}\right) = 7.7148 \times 0.01513 = 0.1167 km/s. \tag{4.41}$$

The minimum Δv would therefore be obtained with the parking orbit at the lowest altitude. Moreover, the additional Δv required to lower the altitude by the same ratio of radii, at the lunar end, would be almost nine times lower.

The reader may refer to Li et al. (2015) for recent developments in the synthesis of lunar trajectories.

4.4.2 *Trajectory Design to an Outer Planet: Mars*

A space mission to Mars is to be designed where the satellite is expected to survey the surface of the planet for potential future landing sites. The spacecraft is expected to leave from a circular parking Earth orbit of radius 6700 km and is expected to arrive at Mars and transfer to circular orbit of radius 3850 km. It may be assumed that Earth's orbit with radius 150×10^6 km is circular and Mar's orbit with radius 228×10^6 km is coplanar. The gravitational parameters for Earth and Mars are $\mu_{\text{earth}} = 3.986012 \times 10^5\ \text{km}^3/\text{s}^2$, $\mu_{\text{mars}} = 4.284 \times 10^4\ \text{km}^3/\text{s}^2$, respectively.

Determine the initial estimates of the total change in the velocity required, the transfer time as well as the planet phasing requirements. Assume that the current phase angle of Earth relative to Mars is $\phi_{\text{initial}} = \pi/4$.

If it is planned to bring the satellite back to Earth orbit, estimate the total time for the mission, including the lay over time in Mars.

The solutions are as given below.

Consider in the first instance the Sun centred transfer region. Considering the Hohmann transfer from Earth to Mars,

$$a_{tr} = \frac{R_E + R_M}{2} = \frac{150 + 228}{2} \times 10^6 = 189 \times 10^6\ km. \tag{4.42}$$

For the heliocentric orbit,

$$\varepsilon = -\frac{\mu_{\text{sun}}}{2a_{tr}} = -\frac{1300}{378} \times 10^2 = -343.92\ (\text{km/s})^2. \tag{4.43}$$

The spacecraft's velocity at the perigee of the transfer ellipse (near Earth) satisfies

$$v_{e,tr}^2 = 2\left(\frac{\mu}{R_E} + \varepsilon\right). \tag{4.44}$$

Hence, the spacecraft's velocity at the perigee of the transfer ellipse is

$$\begin{aligned} v_{e,tr} &= \sqrt{2\left(\frac{\mu_{\text{sun}}}{R_E} + \varepsilon\right)} = \sqrt{2\left(\frac{1300}{1.5} - 343.92\right)} = \sqrt{1733.33 - 687.84} \\ &= 32.3340\ \text{km/s}. \end{aligned} \tag{4.45}$$

The spacecraft's velocity at the apogee of the transfer ellipse (near Mars) satisfies

$$v_{tr,m}^2 = 2\left(\frac{\mu}{R_M} + \varepsilon\right). \tag{4.46}$$

Hence, the spacecraft's velocity at the apogee of the transfer ellipse is

$$v_{tr,m} = \sqrt{2\left(\frac{\mu_{\text{sun}}}{R_M} + \varepsilon\right)} = \sqrt{2\left(\frac{1300}{2.28} - 343.92\right)} = \sqrt{1140.4 - 687.84} \\ = 21.2723 \ \text{km/s}. \tag{4.47}$$

The transfer time is half the period of the transfer ellipse and consequently

$$\text{tof} = \pi\sqrt{a_{tr}^3/\mu_{\text{sun}}} = \pi\sqrt{\left(189\times 10^6\right)^3/13\times 10^{10}} = 2264\times 10^4 \ s \\ = 0.7174\text{yrs} = 8.61 \ \text{mths}. \tag{4.48}$$

Now consider the Earth centred departure region. The spacecraft must leave the Earth's SOI with a velocity, so it follows a hyperbolic trajectory with a hyperbolic excess velocity equal to $v_\infty = v_{e,\,tr} - v_E$, where v_E is the velocity of the Earth around the Sun in its actual trajectory, which is almost circular, so $a = R_E$.

The velocity of the Earth around the Sun

$$v_E = \sqrt{2\left(\frac{\mu_{\text{sun}}}{R_E} + \varepsilon_{\text{Earth}}\right)}, \varepsilon_{\text{Earth}} = -\frac{\mu_{\text{sun}}}{2R_E} = -\frac{1300\times 10^8}{300\times 10^6} \\ = -433.3334\left(\frac{\text{km}}{\text{s}}\right)^2, v_E = \sqrt{2\left(\frac{\mu_{\text{sun}}}{R_E} + \varepsilon_{\text{Earth}}\right)} = 29.4392. \tag{4.49}$$

At the patch point, the perigee of the heliocentric transfer ellipse (which ignores the planet) is considered to be at the SOI of the planet itself. Relative to the planet, this is infinite. The required hyperbolic excess velocity of spacecraft for Earth departure, relative to the Earth, is

$$v_{\text{SOI}} = v_\infty = v_{e,tr} - v_E = (32.3340 - 29.4392)\text{km/s.} = 2.8048 \ \text{km/s}. \tag{4.50}$$

The required hyperbolic departure velocity with respect to the Earth

$$v_h = \sqrt{2\left(\frac{\mu_e}{r_{ce}} + \varepsilon_\infty\right)}, \quad \varepsilon_\infty = v_\infty^2/2 = 2.8048^2/2 = 3.9335. \tag{4.51}$$

Hence,

$$v_h = \sqrt{2\left(\frac{\mu_e}{r_{ce}} + \varepsilon_\infty\right)} = \sqrt{\frac{2\times 4\times 10^5}{6700} + 7.867} = 11.2814 \quad \text{km/s}. \tag{4.52}$$

For the parking orbit with respect to the Earth, the circular velocity is

$$v_{ce} = \sqrt{\frac{\mu_e}{r_{ce}}} = \sqrt{\frac{4 \times 10^5}{6700}} = 7.7267 \text{ km/s.} \tag{4.53}$$

The required Δv is given by

$$\Delta v = v_h - v_{ce} = 11.2814 - 7.7267 \text{ km/s} = 3.5547 \text{ km/s.} \tag{4.54}$$

Now consider arrival at Mars (centred at Mars). The spacecraft must enter the Mars's SOI with a velocity, so it follows a hyperbolic trajectory with a hyperbolic excess velocity equal to $v_{\infty m} = v_M - v_{tr,\, m}$, where v_M is the velocity of Mars around the Sun in its actual trajectory, which is almost circular, so $a = R_M$.

The planet Mars's velocity around the Sun

$$\begin{aligned} v_M &= \sqrt{2\left(\frac{\mu_{\text{sun}}}{R_M} + \varepsilon_{\text{Mars}}\right)}, \\ \varepsilon_{\text{Mars}} &= -\frac{\mu_{\text{sun}}}{2R_M} = -\frac{1300 \times 10^8}{456 \times 10^6} = -285.0877\left(\frac{\text{km}}{\text{s}}\right)^2, \\ v_M &= \sqrt{2\left(\frac{\mu_{\text{sun}}}{R_M} + \varepsilon_{\text{Mars}}\right)} = 23.8783 \text{km/s.} \end{aligned} \tag{4.55}$$

At the patch point, the apogee of the heliocentric transfer ellipse (which ignores the planet) is considered to be at the SOI of the planet Mars. Relative to the planet, this is infinite.

The required hyperbolic excess velocity for Mars arrival, relative to Mars, is

$$v_{\text{SOI}} = v_{\infty m} = v_M - v_{tr,m} = (23.8783 - 21.2723)\,\text{km/s} = 2.6060 \text{ km/s.} \tag{4.56}$$

The required hyperbolic arrival velocity with respect to Mars

$$\begin{aligned} v_{hm} &= \sqrt{2\left(\frac{\mu_m}{r_{\text{cm}}} + \varepsilon_{\infty M}\right)}, \varepsilon_{\infty M} = v_{\infty m}^2/2 = 2.6060^2/2 = 3.3956, \\ v_{hm} &= \sqrt{2\left(\frac{\mu_m}{r_{\text{cm}}} + \varepsilon_{M\infty}\right)} = \sqrt{\frac{2 \times 4.284 \times 10^4}{3850} + 6.7912} = 5.3894 \text{km/s.} \end{aligned} \tag{4.57}$$

For the parking orbit with respect to Mars

$$v_{ce} = \sqrt{\frac{\mu_m}{r_{cm}}} = \sqrt{\frac{4.284 \times 10^4}{3850}} = 3.3358 km/s. \tag{4.58}$$

The required Δv is given by

$$\Delta v = v_{hm} - v_{cm} = 5.3894 - 3.3358 km/s = 2.0536 km/s. \quad (4.59)$$

For the entire mission,

$$\Delta v = 3.5547 + 2.0536 = 5.6083 km/s. \quad (4.60)$$

The total time of flight is tof = 8.61 mths.

Consider the planet phasing aspect. The angular velocities of the departure and arrival planets are

$$\omega_E = \sqrt{\frac{\mu_{\text{sun}}}{R_E^3}} = \sqrt{\frac{1.3 \times 10^{11}}{3375 \times 10^{21}}} = 0.0196 \times 10^{-5} \text{ rads}/s. \quad (4.61)$$

$$\omega_M = \sqrt{\frac{\mu_{\text{sun}}}{R_M^3}} = \sqrt{\frac{1.3 \times 10^{11}}{118520 \times 10^{21}}} = 0.0105 \times 10^{-5} \text{rads/s}. \quad (4.62)$$

The required lead angle is

$$\alpha_{\text{lead}} = \omega_M \times \text{tof} = 2.3711 \text{ rads}. \quad (4.63)$$

The required phase angle is $\phi_{\text{phase}} = \pi - \alpha_{\text{lead}} = 0.7705$ rads. The current phase angle of Earth relative to Mars = $\phi_{\text{initial}} = \pi/4$. Hence, the wait time is given by

$$\text{Wait time} = \frac{\phi_{\text{phase}} - \phi_{\text{initial}}}{\omega_M - \omega_E} = 1.6355 \times 10^5 = 163550 \text{ s} = 1.8929 \text{ days}. \quad (4.64)$$

If the opportunity for launch is missed, the next opportunity will involve a wait time. This the synodic period. The synodic period for Mars is

$$\text{synodic period} = \frac{2\pi}{\omega_E - \omega_M} = \frac{2\pi}{0.0196 - 0.0105} 10^5 = 2.188 \text{ yrs}. \quad (4.65)$$

After arrival on Mars, the waiting time to be back to Earth is the minimum lay over time.

$$t_{\text{lay-over}} = \frac{2\pi}{\omega_E - \omega_M} \frac{\omega_E}{\pi} \left(1 - \text{tof}_{\text{yrs}}\right) = 1.2169 \text{ yrs}. \quad (4.66)$$

Thus, the total time for travel to Mars on this mission and back is 2.652 yrs.

An example of Hohmann transfer and the associated trajectory correction manoeuvres (TCM) on a journey to Mars is the Mars Atmosphere and Volatile Environment Mission (MAVEN), launched by NASA into a Hohmann transfer orbit with periapsis at Earth's orbit and apoapsis at the distance of the orbit of Mars. The spacecraft travelled more than 180 degrees around the Sun in its transfer orbit, which required 10 months for Mars orbit insertion in September 2014. TCM-1 imparted a velocity increment Δv, of about 4.8 m/s to the spacecraft, as expected. TCM-1 occurred on Dec. 3, 2013. On Feb. 26, 2014, mission controllers performed a successful second TCM, also known as TCM-2. This burn lasted approximately 19 seconds and imparted a change in velocity of 68.8 cms per second. The TCM was to adjust the "aim point" of MAVEN at its closest approach when it arrives at Mars, so that it can properly enter orbit around the planet. Post-manoeuvre data review showed that TCM-1 and TCM-2 went according to plan. TCM-3 was scheduled for July 23 but was cancelled as it was deemed to be unnecessary. TCM-4 was also cancelled. MAVEN was at a distance of 21.2 million kms from Earth and 103.7 million kms from Mars on Feb. 27, 2014. The velocity was 29.44 km/s as it moved around the Sun. MAVEN was inserted into Mars orbit on September 21, 2014. TCM-5 was performed prior to insertion which was a 30-sec settling burn manoeuvre with the six TCM engines. The orbit insertion began with six thruster engines firing briefly to damp out deviations in pointing. Then, the six TCM engines quickly ignite and burn for 33 mins to slow the craft using half its fuel, allowing it to be captured in an elliptical orbit with a period of 35 h. Six smaller manoeuvres will be performed later to bring the highest and lowest points of the orbit to the altitudes desired for the science orbit. At its closest point, MAVEN was flying in the upper atmosphere, about 150 kms above the surface of Mars. MAVEN's orbit from Earth to Mars is illustrated in Fig. 4.9, showing also the points where the TCMs were executed.

India's Mars Orbiter Mission (MOM) celebrated 100 days and was 100 million kilometres out from Mars on June 16, prior to the crucial Mars orbit insertion engine firing with the rendezvous on September 24, 2014. MOM was cruising right behind NASA's MAVEN orbiter which celebrated 100 days out from Mars on Friday the

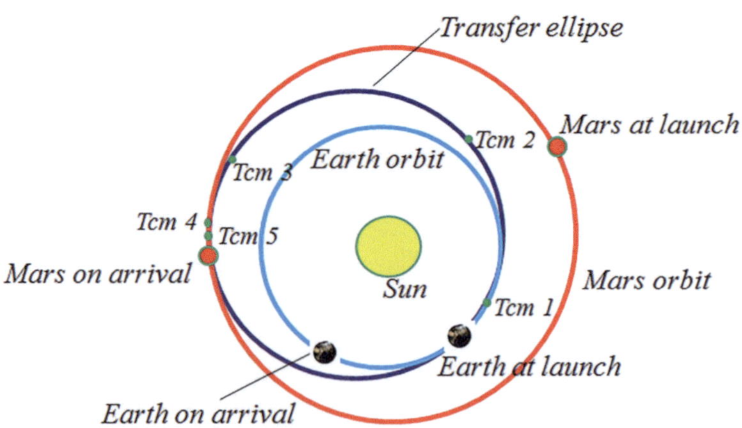

Fig. 4.9 MAVEN's orbit from Earth to Mars

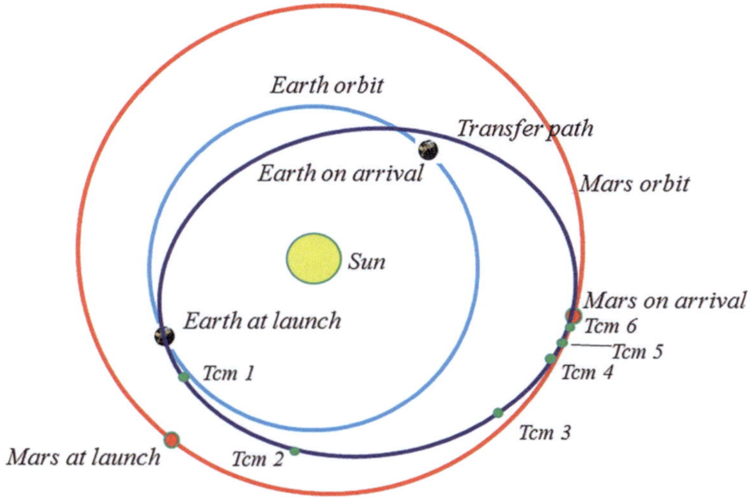

Fig. 4.10 The MARS 2020 mission's orbit from Earth to Mars

13th June. MAVEN arrived at Mars about 48 h ahead of MOM on September 21, 2014!

MAVEN, MOM, MARS 2020, and other missions to Mars have established some standards and routine manoeuvres including typical trajectories. Trajectory correction manoeuvres and orbit insertion manoeuvres as well as the associated timing schedules have also been tested. Hohmann transfer orbits as well as shorter optimal orbits have emerged as standards. The MARS 2020 mission, which transported the Perseverance rover to Mars, indicates the feasibility of such shorter optimal paths which completed the journey to Mars in under 175 days. The MARS 2020 mission's orbit from Earth to Mars is illustrated in Fig. 4.10, indicating also the points where the six TCMs were executed along the optimal path from the Earth to Mars.

Parvathi and Ramanan (2017) have proposed an improvement to the patched conic method. In this approach initially, the position vectors of the target planets at the departure and arrival epochs, respectively, are connected using the heliocentric Lambert conic. A Lambert conic is essentially Keplerian as it is based on the two-body model. The heliocentric velocities of the spacecraft in the transfer conic at the target point and at the SOI are assumed to be same. A systematic iterative approach is used to update the hyperbolic orbits at the patch points by iterating on the excess velocity at the SOI boundary till the required excess velocity is achieved. The method is a promising extension to the traditional patched conic technique.

An alternate approach to Mars is based on the concept of ballistic capture. Instead of launching the spacecraft on a transfer orbit to be eventually captured by Mars by entering into its sphere of influence, the spacecraft is launched into a Martian orbit ahead of the planet and into its orbit, so the planet eventually catches up with the spacecraft and captures it. The spacecraft does not use any fuel after it is launched into the Martian orbit. Mako and Szenkovits (2004), Circi and Teofilatto (2005) and

Prado and Neto (2006) have studied the ballistic capture of spacecraft and celestial bodies by planets of the solar system by considering the elliptic restricted three body model. Using the elliptic restricted three body model, Circi and Teofilatto (2005) have shown the existence of a band of eccentricity values of the planet satisfying a quadratic relationship for the capture to be feasible and have derived an analytical relation to determine the limiting value of the satellite capture eccentricity as a function of the pericentric distance and planet's true anomaly. Thus, the satellite approaches the planet in a hyperbolic orbit, and following the gravitational swing-by, the satellite is captured by the planet and follows an elliptic trajectory. Hyeraci and Topputo (2010) present a systematic method to design ballistic capture orbits upon arrival at a planet in interplanetary transfers. The methodology is also discussed by Luo and Topputo (2015) and by Manzi and Topputo (2021). However, ballistic capture in the restricted three body problem is a temporary phenomenon as generally after a few revolutions, the captured satellite escapes from the planet's gravity field. Yet it provides a window of opportunity, when the spacecraft can be made to descend into a lower orbit about the target planet.

4.4.3 Trajectory Design to an Inner Planet: Venus

One assumes that a change in inclination is not required. The first step in designing a suitable trajectory and estimating the required Δv to an inner planet such as Venus is to design the Hohmann transfer from a circular orbit around the departure planet Earth to a circular orbit around the destination planet, Venus. The velocities at the periapsis of the transfer ellipse at Venus and the apoapsis of the transfer ellipse at Earth, along the heliocentric transfer ellipse, given that the Earth-Sun distance, r_e, is 1 AU (1 AU = 149,598,262 km) and Venus-Sun distance, r_v, is 0.723 AU, are, respectively, given by

$$v_p = v_1^+ = \sqrt{\frac{2\mu_{\text{sun}}}{r_e}\frac{r_e^2}{r_v(r_e + r_v)}} = 42.1274 \times 0.8936 = 37.6463\text{km/s}. \tag{4.67}$$

$$v_a = v_2^+ = \sqrt{\frac{2\mu_{\text{sun}}}{r_e}\frac{r_v r_e}{r_e(r_e + r_v)}} = 42.1274 \times 0.6478 = 27.29\text{km/s}. \tag{4.68}$$

The Hohmann transfer to Venus is illustrated in Fig. 4.11.

Considering the transfer at the apoapsis, the required velocity at the apoapsis is in the heliocentric frame and must be specified. To do this, the motion of the Earth and the Earth's circular velocity are considered first.

The circular velocity of the Earth is given by

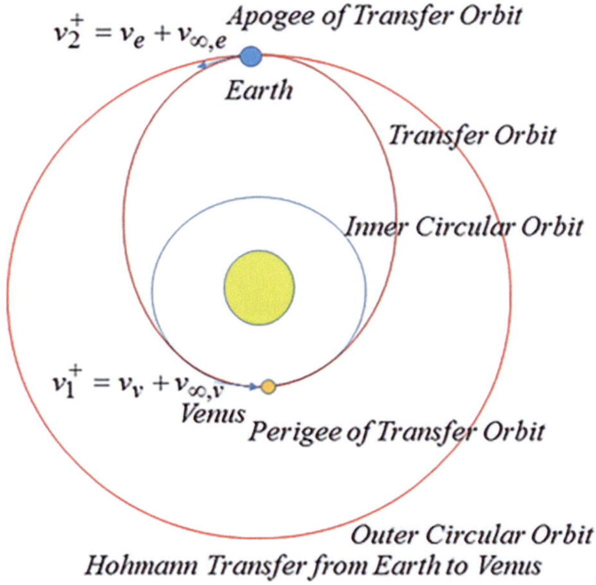

Fig. 4.11 Hohmann transfer from Earth to Venus

$$v_e = \sqrt{\mu_{\text{sun}}/r_e} = 29.7886 \ \text{km/s}. \tag{4.69}$$

For hyperbolic escape from Earth, we need to find the hyperbolic excess velocity, $v_{\infty,\ e}$, which satisfies the relation

$$v_2^+ = 27.29 = v_e + v_{\infty,e} = 29.7886 + v_{\infty,e}. \tag{4.70}$$

It follows that

$$v_{\infty,e} = \ -2.5\text{km/ sec} \Rightarrow \varepsilon_\infty = v_{\infty,e}^2/2 = 6.25/2 = 3.125. \tag{4.71}$$

The vis-viva (energy) equation may now be applied. The hyperbolic excess velocity is

$$v_\infty^2 = v^2 - v_{\text{esc}}^2 = v^2 - 2v_{\text{circ}}^2 = 2\varepsilon_\infty \Rightarrow v = \sqrt{2\left(v_{\text{circ}}^2 + \varepsilon_\infty\right)}, v_{\text{circ}} = v_{\text{parking\_orbit}} = v_{po}. \tag{4.72}$$

The required hyperbolic departure velocity from Earth is given by

$$v_h = \sqrt{2(\mu_e/r_e) + 2\varepsilon_\infty}. \tag{4.73}$$

Given that the radius of parking orbit (Earth) is 6578 km, the circular velocity in the parking orbit is

$$v_{po} = \sqrt{\mu_e / r_e} = 7.7843 \text{km/s}. \tag{4.74}$$

From the vis-viva equation,

$$\begin{aligned} 2\varepsilon_\infty &= v_h^2 - v_{esc}^2 = v_h^2 - 2v_{po}^2 = 6.25, \\ \Rightarrow v_h^2 &= 6.25 + 2v_{po}^2 = 6.25 + 2 \times 60.5959 = 127.4419, \\ \Rightarrow v_h &= \sqrt{127.4419} = 11.289 km/s. \end{aligned} \tag{4.75}$$

The required Δv to transfer the spacecraft from the parking orbit to the hyperbolic transfer orbit with the specified hyperbolic excess velocity is

$$\Delta v = v_h - v_{po} = 11.289 - 7.7843 = 3.5 \text{ km/s}. \tag{4.76}$$

As one is launching to an inner planet, there are a few additional but key questions to consider. One needs to decide if the spacecraft has to approach the destination planet from the light or the dark side, and this implies that it is important when the spacecraft is launched into orbit. The fact that the Earth rotates counterclockwise about the Sun must be considered carefully. The spacecraft is typically in an orbit that is counterclockwise about the Earth. Thus, one needs to consider the direction of the hyperbolic departure velocity. Consider the configuration in Fig. 4.12.

In the configuration shown in Fig. 4.12, the Δv occurs on the dark side with $\Delta v > 0$. It is suitable only for missions to the outer planets. For missions to inner planets, $\Delta v < 0$, and this implies that Δv occurs on the light side. The gravitational swing-by relations and the turning angle, discussed in Chap. 1, Section 1.3.3, must

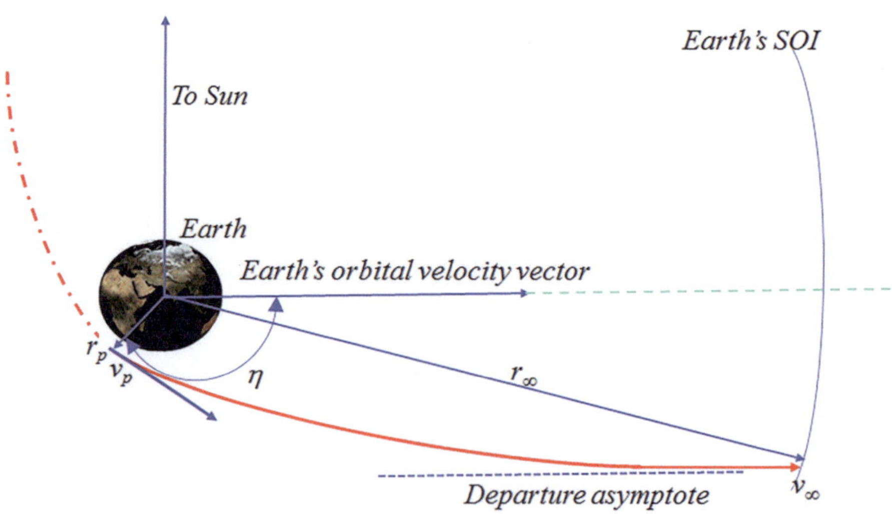

Fig. 4.12 Configuration that results in a positive Δv on departure

now be used. The Δv (positive/negative) should occur at half the turning angle $\delta/2$, before noon or midnight, where the turning angle for a gravitational swing past depends on the eccentricity and is

$$\delta = 2\sin^{-1}(1/e). \tag{4.77}$$

To find the eccentricity, we need to find the semi-major axis of the hyperbolic departure orbit, using the vis-viva equation for energy conservation. Thus,

$$\varepsilon_\infty = \frac{v_{\infty,e}^2}{2} = -\frac{\mu_e}{2a} = \frac{6.25}{2} \Rightarrow a = -63776\ \text{km}. \tag{4.78}$$

Since the transfer takes place at the perigee of the hyperbolic departure orbit, the eccentricity and the turning angle may be, respectively, estimated from

$$r_p = a(1-e) = r_c \Rightarrow e = 1.1 \Rightarrow \delta = 130.76^\circ. \tag{4.79}$$

This means that the departure must be at about 4.36 h before midnight. The situation is illustrated in Fig. 4.13. Considering the transfer at the periapsis, the required velocity at the periapsis is in the heliocentric frame and must be specified. To do this, the motion of Venus and Venus's circular velocity are considered first.

The circular velocity of Venus is given by

$$\begin{aligned} v_v &= \sqrt{\mu_{\text{sun}}/r_v} = \sqrt{\mu_{\text{sun}}/r_e} \times \sqrt{r_e/r_v} = 29.7886\sqrt{r_e/r_v} \\ &= 29.7886/\sqrt{0.723} = 35.02\ \text{km/s}. \end{aligned} \tag{4.80}$$

The radius of the circular arrival orbit (=6052 + 500 km) = 6552 km. Thus, the spacecraft's velocity in the arrival orbit must be

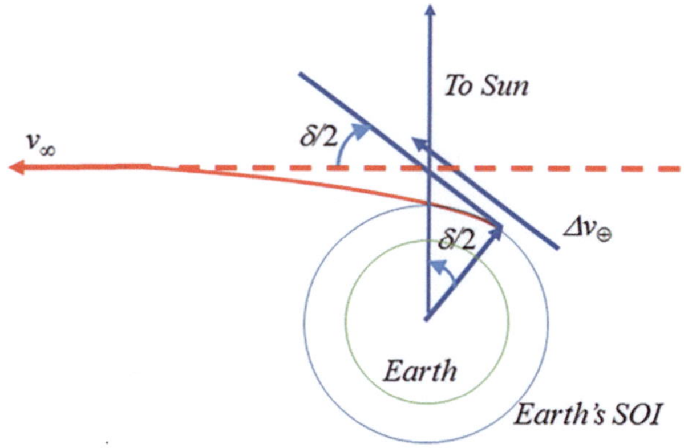

Fig. 4.13 Turning angle relationship for hyperbolic departure

$$v_{ao} = \sqrt{\mu_{\text{venus}}/r_{ao}} = 7.0414 km/s. \tag{4.81}$$

For hyperbolic entry on arrival at Venus, we need to find the hyperbolic excess velocity, $v_{\infty,\ v}$, which satisfies the relation

$$v_1^+ - v_v = v_{\infty,v} = 37.6463 - 35.02 = 2.62 \ \text{km/s}. \tag{4.82}$$

It follows that

$$v_{\infty,v} = 2.62\text{km/ sec} \Rightarrow \varepsilon_\infty = v_{\infty,v}^2/2 = 6.8644/2 = 3.4322. \tag{4.83}$$

From the radius of the arrival orbit (= 6552 km), from the vis-viva equation,

$$\begin{aligned} v_{ha}^2 &= 6.86 + 2v_{ao}^2 = 6.86 + 2 \times 7.0414^2 = 106.0233 \\ &\Rightarrow v_{ha} = \sqrt{106.0233} = 10.297 \ \text{km/s}. \end{aligned} \tag{4.84}$$

Thus to circularize the orbit, the velocity must be reduced velocity by:

$$\Delta v = 10.297 - 7.0414 = 3.2556 \ \text{km/s}. \tag{4.84}$$

However, since the surface atmospheric density on the planet Venus is 53 times that on Earth and the wind velocity around the planet Venus is high, there is indeed a very high drag force which could be exploited to make use of aerodynamic braking to slow the spacecraft down as needed. It is important to note that, as a consequence of the fact that $v_{\infty,\ v} > 0$, the spacecraft will approach Venus from behind. This means that the spacecraft is catching up to planet and is akin to the spacecraft arriving at Venus via a back door! The Hohmann transfer to Venus was illustrated in Fig. 4.11. For the counterclockwise orbit, we want to approach Venus on the dark side. Since the planet's radius (Venus radius) is 6052 km, $\mu_{\text{venus}} = 0.815\mu_{\text{earth}}$, and the distance to the Sun is 0.723 AU, there is a need to adjust conditions near arrival by executing trajectory correction manoeuvres. To find the semi-major axis of the hyperbolic arrival orbit at Venus,

$$\varepsilon_\infty = v_{\infty,v}^2/2 = -\mu_{\text{venus}}/2a = 6.8644/2 \Rightarrow a = -47325 \ \text{km}. \tag{4.85}$$

The radius of arrival orbit (Venus) = 6552 km. The desired periapsis radius at Venus is 6552 km. Thus, the eccentricity and the turning angle may be, respectively, estimated from

$$r_p = a(1 - e) = r_c \Rightarrow e = 1.1384 \Rightarrow \delta = 122.9^\circ. \tag{4.86}$$

The conditions on arrival must be adjusted to achieve the desired periapsis radius or orbit eccentricity (at patch point) as illustrated in Fig. 4.14.

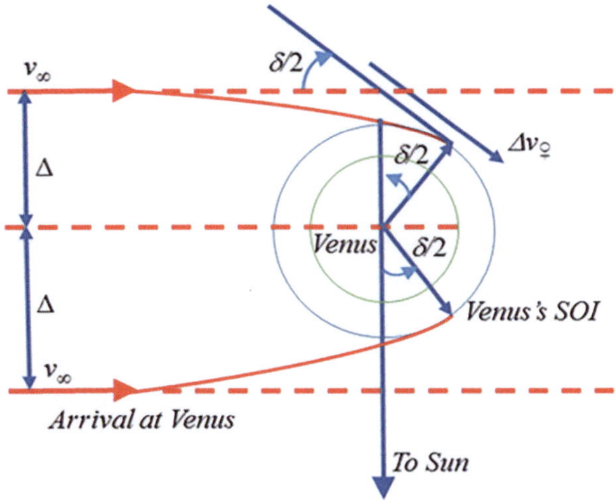

Fig. 4.14 Turning angle relationship for hyperbolic arrival at Venus

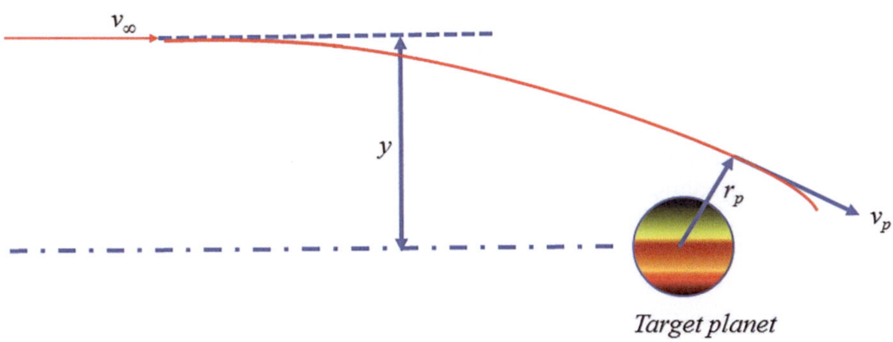

Fig. 4.15 Hyperbolic arrival at Venus

Since the angular momentum can be exactly controlled through target radius, which is given by, $\Delta = y$, the angular momentum and the semi latus rectum for arrival at Venus are, respectively, related by

$$h_v = v_{\infty,v}\Delta, p = h_v^2/\mu_{\text{venus}} = a\left(1 - e^2\right) = 14011 \text{ km}. \tag{4.87}$$

Thus, the target radius is obtained from

$$\Delta = h_v/v_{\infty,v} = \sqrt{\mu_v p}/v_{\infty,v} = 25750 \text{ km}. \tag{4.88}$$

The situation is illustrated in Fig. 4.15.

In a real design, one must also estimate the time of flight or transfer time given by $\pi\sqrt{a_{tr}^3/\mu_{\text{sun}}}$, the phasing requirements of the planet, the actual phase angle, the wait time for the launch, the synodic period and the lay over in Venus for a return flight as was done in the preceding example.

Exercises

1. You are planning a spacecraft mission to Venus and must evaluate the feasibility of using a Hohmann transfer to execute the mission.

   (a) Explain briefly what is meant by the term sphere of influence.
   (b) Draw a diagram, showing the interplanetary transfer to Venus, assuming the planetary orbits are coplanar, circular and concentric.
   (c) Include on your diagram the relevant planetary orbits, the transfer orbit and all the relevant distances and Heliocentric velocities.
   (d) Explain briefly the assumptions made when applying the patched conic method to determine interplanetary Δv requirements.
   (e) Assuming the spacecraft is in an initial circular low Earth orbit with an altitude of 200 km, determine the Δv required to inject the spacecraft into its interplanetary transfer orbit.

2. (a) What are the three key features of the *patched conic method* for interplanetary mission planning? What are the limitations of the method?
   (b) A scientific satellite is to tour the outer planets. The initial phase of this mission requires the satellite to arrive at an arrival orbit around Venus. The satellite is therefore sent on a Hohmann transfer from Earth to Venus. Data for this mission is given in Table 4.1.

      (i) Outline the steps and calculations involved in estimating the total Δv for the mission paying attention to the departure, transfer and arrival phases of the mission and the hyperbolic excess, departure and arrival velocities.
      (ii) If it is assumed that the motions of Earth and Venus are coplanar and concentric circular orbits, calculate the magnitude of the velocity change required to inject the spacecraft into an interplanetary orbit. You may assume that the circular parking orbit around the Earth at the instant of

Table 4.1 Data for a Hohmann transfer from Earth to Venus

Earth/Venus orbit planes	Coplanar
Earth orbit radius	Circular, 150×10^6 km = 1Au
Venus orbit radius	Circular, 0.723Au
Venus planet radius	6000 km
Sun gravitational parameter	1.3×10^{11} km^3/s^2
Venus gravitational parameter	3.25×10^5 km^3/s^2

departure is at an altitude of 500 km from the surface of the Earth and in the Earth's equatorial plane.

(iii) Using an appropriate approach, estimate the difference in the required Δv if the difference in the inclination between the orbits of the two planets is not ignored. Assume that the orbits of the two not planets are circular. How is this difference estimated if the orbital eccentricities of the two planetary orbits are also not ignored?

(c) The sphere of influence around a gravitational body may be written in the form $r \approx r_d(\mu/\mu_d)^{2/5}$.

Explain the meaning of sphere of influence (SOI) with reference to this equation. State the criterion that is used to derive this equation and the two principal assumptions that are made.

3. (a) A mission to Mars may be defined in terms of three phases: Earth escape phase, interplanetary coast phase and Mars injection phase. With the aid of a diagram, outline the steps involved in estimating the total Δv for the mission paying attention to the departure, transfer and arrival phases of the mission and the hyperbolic excess, departure and arrival velocities.

(b) Give the name associated with the minimum energy transfer between two coplanar circular orbits, and describe, with the aid of a diagram, the key features of this orbit.

(c) A space mission to Mars is to be designed, where the satellite is expected to survey the surface of the planet for potential future landing sites. The spacecraft is expected to leave from a circular parking Earth orbit of radius 6700 km and is expected to arrive at Mars and transfer to circular orbit of radius 3850 km. It may be assumed that Earth's orbit with radius 150×10^6 km and Mars' orbit with radius 228×10^6 km are both circular and coplanar. The gravitation constant (GM) for Earth is 398,600 $km^3/s/s$.

$$\left(\mu_{\text{Sun}} = 1327 \times 10^8 \text{km}^3/\text{s}^2, \mu_{\text{Mars}} = 0.1075 \mu_{\text{Earth}}\right)$$

Determine the estimates of the total change in the velocity required over the mission.

4. (a) A spacecraft is approaching the planet Mercury ($\mu = 2.2032 \times 10^4 \text{km}^3/s^2$) with a relative velocity of 10 km/s at the edge of its sphere of influence. The eccentricity of the spacecraft's orbit about Mercury is 1.5. Explaining all your assumptions and calculation steps, estimate the semi-major axis of the orbit and the distance of closest approach to Mercury.

(b) Estimate the excess velocity and the total energy as the spacecraft in part a) approaches the planet Mercury.

(c) Estimate the turning angle as the spacecraft in part a) orbits the planet.

(d) Estimate the velocity increment the spacecraft in part a) acquires, as it swings by the planet Mercury.

(e) Estimate the excess velocity of the spacecraft in part a), as it departs the planet Mercury.
(f) Identify at least two disadvantages of geostationary satellites for remote sensing applications.
(g) It is desired to launch a Sun-synchronous satellite. Explain the key features of such a satellite.

5. The initial position and velocity of the asteroid 2010 TK7 obtained from the initial classical orbital elements are reproduced in Table 4.2. The orbital elements of the Earth for the same day (August 27, 2011) are obtained from the JPL Horizons website and also shown in Table 4.2.

   From the orbital elements in the table, the initial position and velocity vector of the asteroid and Earth may be obtained. The asteroid's orbit can also be checked with the actual JPL HORIZONS data over the time frame of the simulation. High precision ephemerides for solar system bodies are available on-line using JPL's HORIZONS system (https://ssd.jpl.nasa.gov/horizons.cgi (accessed on November 19, 2023)). The JPL HORIZONS on-line solar system data and ephemeris computation service provide access to key solar system data and flexible production of highly accurate ephemerides for solar system objects. HORIZONS is provided by the Solar System Dynamics Group of the Jet Propulsion Laboratory.

   (a) Using the data provided in the table, propagate the orbits of 2010 TK7 and the planet Earth.
   (b) Identify the ascending and descending nodes of the asteroid orbit.
   (c) Synthesize a coplanar Hohmann transfer orbit so as to transfer a spacecraft in circular Earth orbit at an orbital height of 400 km above the equator to the asteroid's orbit at both the ascending and descending nodes. Then alter the inclination of the orbit so the spacecraft is in the same orbit plane as the asteroid. Estimate the total velocity increment required.
   (d) Synthesize a nonplanar Hohmann transfer orbit so as to transfer a spacecraft in circular Earth orbit at an orbital height of 400 km above the equator to the asteroid's orbit at both the ascending and descending nodes, so the inclination of the orbit to the spacecraft is in the same orbit plane as the asteroid. Estimate the velocity increment required and compare with result in part c).

Table 4.2 Orbital elements defining the initial position of 2010 TK7 and Earth as on JD2455800.5

	a(in au)	e	i	Ω	ω	M
2010 TK7	1.00037	0.190818	20.880°	96.539°	45.846°	217.329°
Earth	1.000806	0.01610329	0.0007°	243.305°	217.926°	233.474°

References

Circi, C., & Teofilatto, P. (2005). Effect of planetary eccentricity on ballistic capture in the solar system. *Celestial Mechanics and Dynamical Astronomy, 93*(1–4), 69–86.

Deutsch, R. (1963). *Orbital dynamics of space vehicles*. Prentice-Hall.

Fortescue, P. W., Swinerd, G. S., & Stark, J. P. W. (2011). *Spacecraft systems engineering* (4th ed.). Wiley.

Hu, W. (2015). *Fundamental spacecraft dynamics and control*. Wiley. ISBN: 978-1-118-75353-8.

Hyeraci, N., & Topputo, F. (2010). Method to design ballistic capture in the elliptic restricted three-body problem. *Journal of Guidance, Control, and Dynamics, 33*(6), 1814–1823.

Li, J., Gong, S., & Wang, X. (2015). Analytical design methods for determining Moon-to-Earth trajectories. *Aerospace Science and Technology, 40*, 138–149.

Luo, Z. F., & Topputo, F. (2015). Analysis of ballistic capture in Sun–planet models. *Advances in Space Research, 56*(6), 1030–1041.

Mako, Z., & Szenkovits, F. (2004). Capture in the circular and elliptic restricted three body problem. *Celestial Mechanics and Dynamical Astronomy, 90*, 51–58.

Manzi, M., & Topputo, F. (2021). A flow-informed strategy for ballistic capture orbit generation. *Celestial Mechanics and Dynamical Astronomy, 133*(11–12), 54. https://doi.org/10.1007/s10569-021-10048-2

Parvathi, S. P., & Ramanan, R. V. (2017). Direct transfer trajectory design options for interplanetary orbiter missions using an iterative patched conic method. *Advances in Space Research, 59*, 1763–1774.

Prado, A., & Neto, E. V. (2006). Study of the gravitational capture in the elliptical restricted three-body problem. *The Journal of the Astronautical Sciences, 54*(3–4), 567–582.

Prussing, J. E., & Conway, B. A. (1993). *Orbital mechanics*. Oxford University Press.

Chapter 5
Space Vehicle Mission Planning

5.1 Mission Planning and Space Propulsion Systems

Mission planning (Fortescue et al., 2011) involves the planning of the sequence of events starting at prelaunch and extending to end of life. This includes all events and operations prior to launch, followed by the launch phase, orbit transfers, on-station operations and finally the decommissioning of the spacecraft. A key aspect is the identification of all the critical features of the mission, the objectives of the mission and the definition of the mission requirements, including the orbits, orbit transfers and the propulsion and other subsystems that are adequate to achieve the objectives of the mission. The design guide to orbital flight by Jensen et al. (1962) lists practically all the formula for designing an elliptic trajectory and their variations as well as a host of manoeuvres along with the analysis of trajectory design problems that would be needed by flight performance engineer to establish a feasible conceptual trajectory and mission design.

Rocket propulsion is used to launch orbital vehicles from an Earth based launch pad into an orbit in space. These rocket powered launch vehicles take off vertically and then begin to progressively lean over to follow a gravity turn trajectory. By the time the fuel is completely expended, just like a projectile, the rocket powered vehicle turns to a horizontal orientation and continues to coast in the orbit. Rocket propulsion systems used in this manner, to launch a vehicle into orbit, involve a number of systems working together synergistically and are referred to as primary propulsion systems. These launch vehicles or the rockets as they are loosely referred to are primarily used to get the spacecraft into orbit and develop a thrust in the range of 60–1000 kN during a burn time spanning 100–500 s. While chemical rocket propulsion engines have a proven track record for primary thrust generation, alternate propulsion systems have also been developed for orbit and attitude control and for station keeping manoeuvres.

Another application of rocket propulsion systems is as secondary propulsion systems to correct or control the attitude and orbit of a spacecraft. The thrust

R. Vepa, *Space Vehicle Maneuvering, Propulsion, Dynamics and Control*,
https://doi.org/10.1007/978-3-031-65518-0_5

developed by the propulsion systems for such secondary applications is generally much smaller and of the order 10^{-3} to 10 N over pulsed time frame between 10^{-2} and 10 s or with low but continuous thrust as in the case electric propulsors. After the nominal orbit is established by a launch vehicle, a secondary propulsion system is used for final orbit acquisition. Some examples of such systems are a solid propellant apogee kick motor, generating a typical thrust of 69 kN over a burn time of 60 s; a liquid bipropellant motor, generating a typical thrust of thrust of 400 N; and over a burn time of 1 h, an electric thrusters such a Hall current thruster, generating a thrust of 0.02–0.04 N over a burn time that extends from hours to several days. Other examples of electric thrusters are gridded ion thrusters used for attitude control or station keeping and generating thrusts of the order of micro-Newtons.

5.2 Classification of Space Propulsion Systems

A broad classification of rocket propulsion systems is shown in Fig. 5.1.

5.2.1 Fundamental Concepts and Metrics

There are a number fundamental concepts and metrics that play a key role in understanding the relative performance of rocket propulsion systems so they can be properly differentiated and considered suitable for different generic classes of

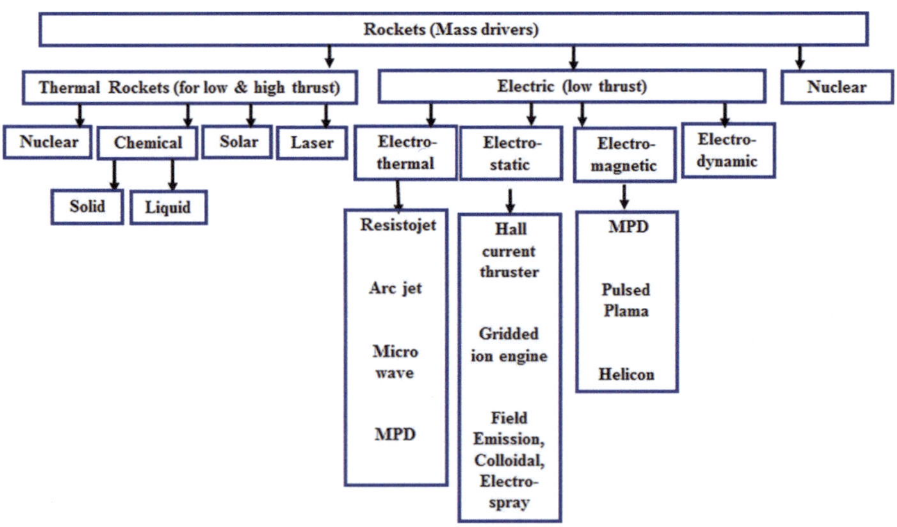

Fig. 5.1 Broad classification of rocket propulsion systems, where MPD refers to magneto-plasma dynamic thrusters

archetypal applications. Key among these features are the concepts of total impulse, specific impulse, thrust generated, specific power, propulsive efficiency, the velocity increment Δv, the burn time and the rocket equation relating the Δv increment to the mass ratio of the initial and current mass of the rocket, which will be introduced in this section.

Consider Newton's second law as applied to a single particle of mass m.

$$m\dot{v} = F. \tag{5.1}$$

Integrating both sides of Eq. 5.1 over a small time interval $\Delta t = t_2 - t_1$,

$$m \int_{t_1}^{t_2} \dot{v} dt = m(v_2 - v_1) = m\Delta v = \int_{t_1}^{t_2} F dt = I. \tag{5.2}$$

The time integral of the momentum is defined as the total impulse; let F be the thrust force acting on a vehicle, modelled as a point mass of mass m, and let $\Delta t = t_2 - t_1$ be the burn time. Equation 5.2 defines the total impulse generated by the rocket's thrust force over the given burn time. The total mechanical impulse imparted to the rocket per unit weight is given by the ratio of the impulse due to the flow of exhaust gases across the nozzle at the exit of the rocket and the weight of an equivalent mass of fuel. It is given by

$$I_{\text{mech}} = \frac{c_e \dot{m} \Delta t}{g \dot{m} \Delta t} = \frac{c_e dm}{g dm} = \frac{c_e}{g}. \tag{5.3}$$

The specific impulse is defined as the maximum possible value of the mechanical impulse per unit weight of the fuel and is

$$I_{sp} = (c_e)_{\max}/g = c_{e-\max}/g. \tag{5.4}$$

The units of the specific impulse are indeed in seconds and it is not a nondimensional quantity. It is a characteristic of the fuel employed in the rocket to generate the thrust. It is representative of the maximum possible thrust that the rocket engine can deliver with a particular fuel. For a modern high thrust chemical rocket, it is about 500 s and about 850–1000 s for a thermonuclear rocket. For the V-2 type WWII rocket engine, it is about 200 s. A reasonable assumption for most current chemical rocket launchers is about 300–400 s. It is about 2000–6000 s for electric ion or plasma thrusters.

To derive a relation for the thrust developed by a rocket engine, the principle of momentum conservation is considered. Consider a typical mass expulsion system. In Fig. 5.2, the initial mass and velocity prior to mass expulsion are $m + \Delta m$ and v, respectively, while after the mass, Δm is expelled with a velocity v_e, and the final mass and velocity of the system are m and $v + \Delta v$, respectively.

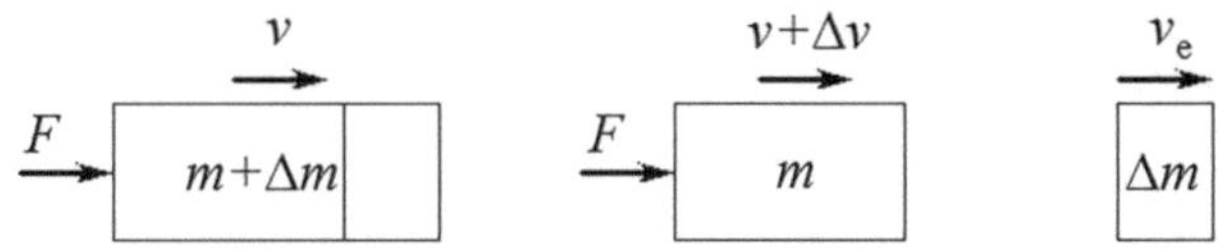

Fig. 5.2 Initial and final mass configurations of a mass expulsion system

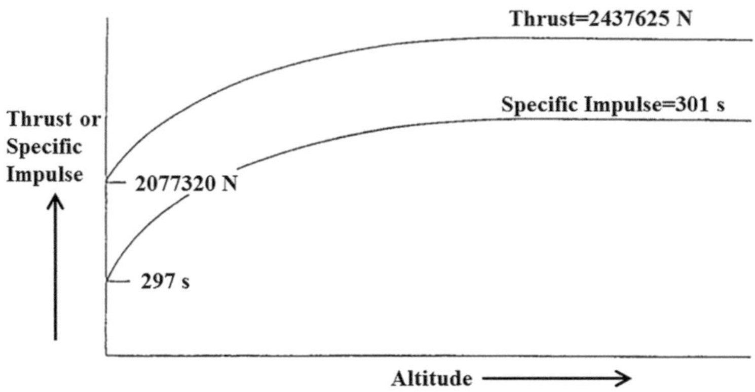

Fig. 5.3 Growth of thrust and specific impulse in a Titan IV Stage 1 rocket from sea level to maximum altitude

Referring to Fig. 5.2, the initial momentum vector and the final momentum vector to first order are, respectively,

$$\mathbf{p} = (m + \Delta m)\mathbf{v}, \mathbf{p} + \Delta\mathbf{p} = m(\mathbf{v} + \Delta\mathbf{v}) + \Delta m\mathbf{v}_e, \Delta\mathbf{p} = m\Delta\mathbf{v} + \Delta m(\mathbf{v}_e - \mathbf{v}). \quad (5.5)$$

Hence, the thrust is

$$\mathbf{T} = m\mathbf{a} = m\frac{d\mathbf{v}}{dt} = -(\mathbf{v}_e - \mathbf{v})\frac{dm}{dt} \equiv -\mathbf{c}_e\frac{dm}{dt}, F = |\mathbf{T}| = |\mathbf{c}_e||\dot{m}_f|, \quad (5.6)$$

where c_e is the relative velocity of the expelled mass relative to the main mass and $\dot{m}_f = -dm/dt$ is the fuel rate. It is a maximum when $\mathbf{v} = 0$, when c_e is the absolute velocity of the exhaust. In Eq. 5.6, $\mathbf{T}$ is the maximum thrust generated.

Figure 5.3 illustrates the typical growth of thrust and specific impulse in a Titan IV Stage 1 rocket as it reaches an altitude of 494 km in a burn time of 194 seconds, from sea level.

One may adapt the thrust equation developed for a mass expulsion system to a real rocket illustrated in Fig. 5.4.

A rocket is subjected to an aerodynamic drag and force due to the average static pressure across the nozzle exit plane as it ascends in the atmosphere. It is assumed that the rocket is closed body, wherein the fuel is burnt within a combustor and expelled to the atmosphere via a nozzle. Applying Newton's law, the equation of vertical motion is

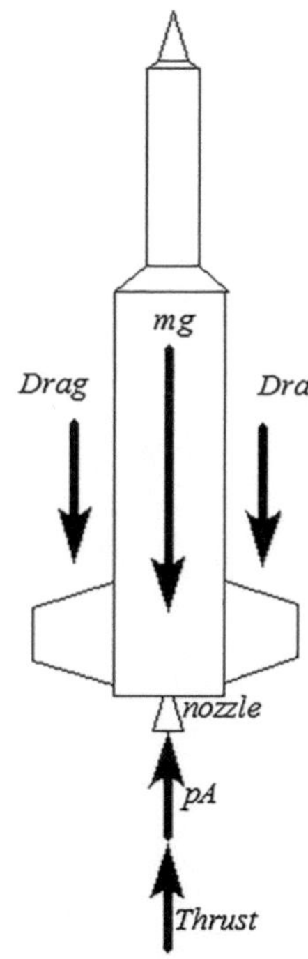

Fig. 5.4 Forces acting on a rocket as it ascends in the atmosphere

$$m_0\dot{v} = F = T + (p_e - p_a)A - D - m_0 g, T = c_e \dot{m}_f, \tag{5.7}$$

where D is the total drag and c_e is the nozzle exit velocity. One may now define the kinetic power and the specific power of the rocket.

Considering the average kinetic power P_T of a rocket, it is given by $P_T = (1/2)Fv$. When sum of the drag and weight add up to zero and the ambient pressure is equal to the atmospheric pressure so $p_e = p_a$, with $D + mg = 0$, it follows that $m_0\dot{v} = F = T = c_e\dot{m}_f$. If the acceleration $\dot{v}$ is expressed in terms of $g's$, so, $\dot{v} = ng_0$, where g_0 is the acceleration due to gravity on the surface of the Earth:

$$F = T = m_0 ng, P_T = (1/2)Fv = (1/2)m_0 \alpha g_0 v, \tag{5.8}$$

The specific power is then defined as the total kinetic power per unit mass of the rocket and is given by

$$P_T/m_0 = (1/2)\alpha g_0 v, \tag{5.9}$$

The total maximum power gained due to the thrust generated is also given by $P_T = Fv = c_e\dot{m}_f v$. In what follows, it is assumed that the velocities are scalar quantities. Thus the absolute velocity of the exhaust is $v_e = v - c_e$. The power lost due to the motion and exhaust is

$$P_e = (1/2)\dot{m}_f(v - c_e)^2, \tag{5.10}$$

Thus the total vehicle power, which is the sum of the kinetic and the exhaust power, is

$$P_e + P_T = (1/2)\dot{m}_f(v - c_e)^2 + vc_e\dot{m}_f = (1/2)\dot{m}_f\left(v^2 + c_e^2\right), \tag{5.11}$$

One can then define the propulsive efficiency as

$$\begin{aligned}\eta_p &= Fv/\left(Fv + (1/2)\dot{m}(v - c_e)^2\right) = 2Fv/\dot{m}_f\left(v^2 + c_e^2\right) = 2c_ev/\left(v^2 + c_e^2\right)\\ &= 2(v/c_e)/\left(1 + (v/c_e)^2\right).\end{aligned} \tag{5.12}$$

This may be compared with the velocity ratio which is

$$v/c_e = \left(1 \pm \sqrt{1 - \eta_p^2}\right)/\eta_p \equiv \eta_v, \tag{5.13}$$

The propulsive efficiency determines how much of the kinetic energy of the exhaust jet is useful for a vehicle travelling at a given speed. The propulsive efficiency is a maximum when $c_e = v$, i.e. when exhaust gases stand still in space. Also when $c_e = v$,

$$\eta_p = 2(v/c_e)/\left(1 + (v/c_e)^2\right) = 1, \eta_v = \left(1 \pm \sqrt{1 - \eta_p^2}\right)/\eta_p = 1. \tag{5.14}$$

The variation of the propulsive efficiency η_p, with the velocity ratio η_v, is shown in Fig. 5.5.

Since the total vehicle power is given by Eq. 5.11 and the kinetic power required to maintain the motion is $(1/2)\dot{m}v^2$, the power required for thrust generation or the jet power is

Fig. 5.5 Variation of the propulsive efficiency η_p, with the velocity ratio η_v

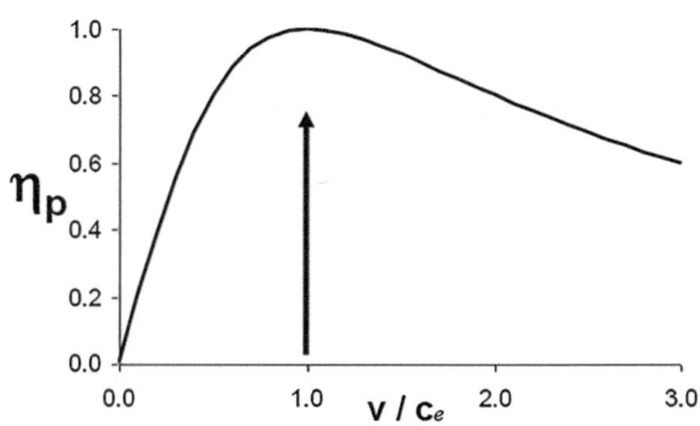

$$P_e|_{v=0} = (1/2)\dot{m}_f c_e^2 = \eta_{th}\dot{E} = \eta_{th}P_{in}. \tag{5.15}$$

Equation 5.15 gives the total static power used by at the thrust head and is the static power lost to the exhaust and used to generate thrust. If the input power to thrust head is given by P_{in}, the thrust head power conversion efficiency is

$$\eta_{th} = \frac{P_e}{P_{in}} = \frac{(1/2)\dot{m}_f c_e^2}{P_{in}} = \frac{(1/2)\dot{m}_e c_e^2}{P_{in}} = \frac{(1/2)Tc_e}{P_{in}} = \left(\frac{1}{2}gI_{sp}\right)\frac{T}{P_{in}}. \tag{5.16}$$

Applying Newton's law the equation of vertical motion as in Eq. 5.7 and assuming also that the sum of the drag and weight add up to zero and the ambient pressure is equal to the atmospheric pressure so $p_e = p_a$, with $D + mg = 0$, the equation of motion may be expressed as

$$\dot{v} = \frac{m\dot{v}}{m} = \frac{T}{m} = -c_e\frac{d\ln m}{dt}. \tag{5.17}$$

Integrating both sides of Eq. 5.17 yields

$$c_e \ln\left(\frac{m}{m_0}\right) = v_0 - v = -\Delta v. \tag{5.18}$$

Hence, it follows that

$$\frac{m}{m_0} = \exp\left(-\frac{\Delta v}{c_e}\right). \tag{5.19}$$

Equation 5.19 is the basic equation, known as the rocket equation which relates the Δv increment to the mass ratio of the initial and current mass of the rocket. Moreover, at the end of the fuel burn,

$$\frac{m_{\text{Empty}}}{m_0} = \exp\left(-\frac{\Delta v_{\text{Total}}}{c_e}\right). \tag{5.20}$$

Equation 5.20 relates the total Δv increment or the total velocity increment to the mass ratio of the initial and final mass of the rocket, when all of the fuel is used up. A typical example illustrates its use. Consider a solid booster rocket on the Space Shuttle, with an I_{sp} of 260 s. The range of the mass ratio could vary between $2 < m_{\text{Empty}}/m_0 < 10$. The exhaust velocity is $c_e = 260 \times 9.81 = 2.55$km/s giving a maximum velocity increment of $\Delta v_{\max} = 5.9$km/s. For a mission which involves (1) launch from ground (2) into a circular parking orbit at a height of 185 km and (3) deployed to final altitude, the required $\Delta v = 7.9$km/s. So single-stage rocket to parking orbit is inadequate! For a rocket with a liquid propellant, e.g. liquid O_2/ liquid H_2, with an $I_{sp} \sim 450$ s, gives a $\Delta v_{\max} = 10.1$km/s which is suitable for the mission.

5.3 Launch Vehicles

Chemical rockets consist, to varying degrees of complexity, of (1) a propellant supply and feed system, (2) a combustion chamber, and (3) an exhaust nozzle. Broadly they can be classed into two types, liquid propellant systems, which could be cold gas, bipropellant or monopropellant types and solid propellant system. However, launch vehicles generally require high thrust. As they take off vertically with low velocity, the specific impulse is low. Launch vehicles are generally propelled either by solid fuel rocket motors or by liquid propellant powered thrusters. The propellant of a solid rocket motor is stored entirely within the combustion chamber in the form of one or more shaped blocks, called grains. The main characteristic of a solid propellant rocket is its simplicity. The burning times may be from a few seconds or fractions of a second to as long as 1 or 2 min. Once the rocket has ignited, the combustion generally proceeds until all the propellant is burnt; the thrust program is fixed. The thrust generated is proportional to the burning area. Rocket grain patterns affect thrust profile, and certain standard grain patterns result in predetermined and associated thrust profiles. It is possible, with suitable grain designs, to obtain a thrust which increases (progressive burning), remains constant (neutral), or decreases (regressive) with burning time. In constant-geometry nozzles, thrust is more or less proportional to the chamber pressure. Thus, the propellant chamber must sustain high pressure and temperature. The exhaust gases are environmentally unfriendly.

Liquid propulsion thruster systems used by rocket launchers use either monopropellants or bipropellants. The fuel and oxidizer are either pressure fed or pump fed. There are several alternative propellants such as hydrazine. A variety of oxidizers are used such as hydrogen peroxide and nitrous oxide (dinitrogen monoxide). Liquid oxygen (LOX) is used exclusively with fuels like liquid hydrogen (LH_2) or kerosene or methane. The fuel and oxidizer could be supplied externally and ignited or could be hypergolic (self) ignition. The fuel and oxidizer are stored at ambient temperatures and pressures or in cryogenic conditions in pressurized tanks.

5.3.1 Secondary Propulsion Systems (Chemical and Electric)

While launch vehicles are propelled by primary propulsion systems, characterized by high thrust and low specific impulse, spacecraft in orbit use secondary propulsion systems, characterized by low thrust and high specific impulse for final orbit acquisition from the nominal orbit established by the launch vehicle and for orbit and attitude control. Typically, for example, a solid propellant based apogee kick motor, used for final orbit corrections, has a thrust of about 69 kN over a burn time of 60 s, while a typical liquid bipropellant motor for the same application has a thrust of

400 N over a burn time of 1 h. A typical electric Hall current thruster has a thrust of less than 100 milli Newtons operating from hours to days. Typical applications of the electric thruster are for attitude and orbit control and station keeping operations over the time frame of the mission.

5.4 Rocket Equation, Thrust and Specific Impulse

Whether one is dealing with a primary or secondary propulsion system, the most important characteristic of space propulsion thruster is the velocity increment Δv it could deliver, once it is in an orbit. So, it is important to develop a relationship between the velocity increment Δv delivered by a space propulsion thruster and the relationship it bears to other parameters of the vehicle. This can be done, in principle, by considering Eq. 5.20. The rocket equation for the velocity increment Δv is best rewritten in terms of certain new parameters. The mass ratio at burnout of rocket can be defined in terms of variable R defined as ratio of the initial mass of the rocket m_0 to the mass at burn out. Thus if the mass of the payload is, m_p, the expended mass of the propellant and oxidizer or fuel is, m_f, and the mass of the remaining structure is, m_s, then the ratio of the initial mass of the rocket m_0, to the mass at burn out or empty mass is,

$$R = m_0/m_E \equiv \left(m_p + m_s + m_f\right)/\left(m_p + m_s\right). \tag{5.21}$$

i.e. the ratio of the initial mass of the rocket m_0, to the empty mass or mass at burnout, m_E. Thus, Eq. 5.20 may be expressed as

$$m_{Empty}/m_0 = m_E/m_0 = 1/R = \exp(-\Delta v_{Total}/c_e) \equiv \exp(-\Delta v_m/c_e). \tag{5.22}$$

It follows that

$$\Delta v_m = -c_e \ln\left(\left(m_p + m_s + m_f\right)/\left(m_p + m_s\right)\right) = -c_e \ln R^{-1} \text{ or } \Delta v_m = c_e \ln R. \tag{5.23}$$

In the preceding analysis, it was assumed that the external forces acting on the rocket $F_{ext} \equiv 0$. In Eq. 5.7, when sum of the drag and weight do not add up to 0 and the ambient pressure is equal to the atmospheric pressure so $p_e = p_a$, with $D + mg \neq 0$, it follows that $m_0 \dot{v} = F = T - D - m_0 g = c_e \dot{m}_f - D - m_0 g$. When F_{ext} is directly proportional to the mass of the rocket and can be expressed as $F_{ext} = F - T = -m_0 g_e$ where g_e is a constant and is the equivalent acceleration due to gravity given by the sum of the acceleration due to both the gravity force and the drag force, the solution for the velocity increment must be corrected and expressed as

$$\Delta v_m = c_e \ln R - g_e t, g_e = g + D/m_0. \tag{5.24}$$

The reduction due to the last term is the gravity loss, which could be up to 35%, and represents the decrease in the velocity change in the presence of gravity and drag. If one wishes to maximize the velocity increment Δv delivered by a space propulsion thruster, it is not only important to choose a fuel with the maximum possible c_e but also simultaneously maximize the parameter R and minimize g_e. Clearly it helps to discard any unnecessary mass on board the rocket before all the fuel is expended, before burnout.

5.5 Staging

The need to maximize the velocity increment Δv delivered by a space propulsion thruster leads us naturally to the concept of staging. It is observed that the initial mass of the rocket m_0 is the sum of three contributions: the mass of the payload, m_p, the expended mass of the propellant and oxidizer or fuel, m_f, and the mass of the remaining structure, m_s. Thus, ignoring the gravity loss of the maximum velocity increment over the flight of a single-stage rocket, assuming that the payload is not jettison during the flight, is given by

$$\Delta v_m = c_e \ln R = c_e \ln\left(\frac{m_p + m_s + m_f}{m_p + m_s}\right) = c_e \ln\left(1 + \frac{m_f}{m_p + m_s}\right). \tag{5.25}$$

It is seen that this expression can be made as large as possible by reducing the mass of the structure, m_s. It is also true that Δv_mc can be made as large as possible by jettisoning as much of the mass that contributes to the mass of the structure during the flight. This is the principle of *multistaging*. Typically the rocket's flight progresses in two or three stages. In order to facilitate the optimization of the number of stages that are earmarked for jettisoning before burnout, a number of new parameters are defined. A payload mass ratio is defined as

$$p = \frac{m_p}{m_p + m_s + m_f}. \tag{5.26}$$

A structure mass coefficient is defined as

$$s = \frac{m_s}{m_s + m_f}. \tag{5.27}$$

The following important relations should be noted:

$$(1-p)s = \frac{m_s}{m_p + m_s + m_f}, 1-p = \frac{m_s + m_f}{m_p + m_s + m_f}, 1-s = \frac{m_e}{m_s + m_e}, (1-p)(1-s) = \frac{m_f}{m_p + m_s + m_f}. \tag{5.28}$$

Then it follows that

$$1-p = \frac{m_s + m_f}{m_p + m_s + m_f}, (1-p)s = \frac{m_s}{m_p + m_s + m_f}, (1-p)s + p = \frac{m_s + m_p}{m_p + m_s + m_f} = \frac{1}{R}. \tag{5.29}$$

Hence,

$$R = \frac{1}{(1-p)s + p}. \tag{5.30}$$

Also,

$$(1-p)s = \frac{1}{R} - p = \frac{1-pR}{R}, s = \frac{(1-pR)}{(1-p)R}, (1-p)s + p = s + p(1-s) = \frac{1}{R}, p = \frac{1-sR}{R(1-s)}. \tag{5.31}$$

Several American textbooks (Prussing & Conway, 1993) use a slightly different parameterization, defining a payload coefficient, usually in terms of the expended fuel ($m_e = m_f$) as

$$\lambda = \frac{m_p}{m_s + m_f} = \frac{p}{1-p}. \tag{5.32}$$

It follows that

$$p = \frac{\lambda}{\lambda + 1}, 1-p = \frac{1}{\lambda + 1}, \lambda = \frac{1-sR}{R-1}, R = \frac{1+\lambda}{s+\lambda}. \tag{5.33}$$

Thus, the maximum velocity increment over the flight of a single-stage rocket, assuming that the payload is not jettison during the flight, is given by

$$\Delta v_m = -c_e \ln R^{-1} = -c_e \ln(p + (1-p)s) = c_e \ln\left(\frac{1+\lambda}{s+\lambda}\right). \tag{5.34}$$

It may be observed that since, the maximum burn rate of fuel early in the flight or reducing the mass of the structure during the flight, gives the minimum kinetic energy to the exhaust, a multi-stage rocket provides the potential to optimize the performance and to maximise Δv_m, although this overlooks issues associated with air breathing engines as well as cost considerations. The possibility of maximizing Δv_m in a multistage rocket is considered in the next section, based on Eq. 5.34.

5.5.1 *Multistaging: Maximizing Δv*

Consider a multistage rocket with n stages. For the first stage, if the initial mass of the stage is m_{01}, the expended mass of the propellant and oxidizer or fuel is m_{f1}, and the mass of the structure of the first stage, which could be jettisoned after the burnout of the fuel in the first stage, is m_{s1}. The payload of first stage is the mass of the rest of the rocket. Thus, the mass of the payload is

$$m_{p1} = \sum_{i=2}^{n} m_{0i}. \tag{5.35}$$

For the second stage, the initial mass of the stage m_{02}, the expended mass of the propellant and oxidizer or fuel is m_{f2}, and the mass of the structure of the second stage, which could be jettisoned after the burnout of the fuel in the second stage, is m_{s2}. The mass of the payload for the second stage is

$$m_{p2} = \sum_{i=3}^{n} m_{0i}. \tag{5.36}$$

Hence, in each stage, a structure mass ratio is defined as

$$s_i = m_{si}/\left(m_{si} + m_{fi}\right). \tag{5.37}$$

The payload mass ratio is defined as

$$p_i = m_{0i+1}/m_{0i}. \tag{5.38}$$

The payload mass ratio for the last stage is defined as

$$p_n = m_p/m_{0n}. \tag{5.39}$$

Hence, total payload ratio is

$$p_{tot} = m_p/m_{01} = \prod_{i=1}^{n} p_i. \tag{5.40}$$

For a multistage rocket, it needs to sum up the Δv_i for each stage. But for each stage,

$$\frac{m_{s_i} + m_{p_i}}{m_{p_i} + m_{s_i} + m_{f_i}} = (1 - p_i)s_i + p_i = \frac{1}{R_i}. \tag{5.41}$$

The total velocity increment over the flight of a multistage rocket is

$$\Delta v_m = -c_e \sum_{i=1}^{n} \ln((1 - p_i)s_i + p_i) = -c_e \sum_i \ln R_i^{-1} = -c_e \ln\left(\prod_i R_i^{-1}\right). \tag{5.42}$$

In Eq. 5.42,

$$R_i = ((1 - p_i)s_i + p_i)^{-1}, R_i^{-1} = (1 - p_i)s_i + p_i. \tag{5.43}$$

It is essential that the total velocity increment Δv_m is maximized. Thus, one needs to maximize Δv_m defined by Eq. 5.42, subject to

$$p_{\text{tot}} = m_p/m_{01} = \prod_{i=1}^{n} p_i. \tag{5.44}$$

Thus, it is tacitly assumed the payload remains unchanged. The constrain is then expressed as

$$\ln p_{\text{tot}} - \sum_{i=1}^{n} \ln p_i = 0. \tag{5.45}$$

The problem is solved by the Lagrange multiplier method. In this method, the cost functional to be maximized, given by Eq. 5.42, is augmented by adding to it a multiple of the constrain equation given by Eq. 5.45. The multiplying factor, known as the Lagrange multiplier, is an unknown variable. The augmented cost functional is now maximized with respect to not only the original variables but also the Lagrange multiplier. Thus, the augmented cost functional is given by

$$f(p_i, \lambda) = \Delta v_m + \lambda\left(\ln p_{\text{tot}} - \sum_{i=1}^{n} \ln p_i\right). \tag{5.46}$$

Assuming that for the ith stage $c_e = c_i$, the augmented cost function is given by

$$f(p_i,\lambda) = -\sum_{i=1}^{n} c_i \ln((1-p_i)s_i + p_i) + \lambda\left(\ln p_{\text{tot}} - \sum_{i=1}^{n} \ln p_i\right). \tag{5.47}$$

To minimize the augmented cost functional, the partial derivatives of the functional with respect to each of the variables p_i and the Lagrange multiplier λ are each set to zero. Thus, the conditions for maximizing the augment cost functional with respect to each of the variables p_i and the Lagrange multiplier λ are

$$\frac{\partial}{\partial p_i} f(p_i,\lambda) = 0, \ \frac{\partial}{\partial \lambda} f(p_i,\lambda) = 0. \tag{5.48}$$

The second of the two equations in Eq. 5.48 yields the constraint Eq. 5.45. The first of the two equations in Eq. 5.48 yields

$$\frac{\partial}{\partial p_i} f(p_i,\lambda) = -c_i(1-s_i)((1-p_i)s_i + p_i)^{-1} + \frac{\lambda}{p_i} = 0. \tag{5.49}$$

Equation 5.49 may be rewritten as

$$\frac{\lambda}{p_i} = \frac{c_i(1-s_i)}{(1-p_i)s_i + p_i}. \tag{5.50}$$

Thus, it follows from Eq. 5.50 that

$$\frac{\lambda}{p_i} = \frac{c_i(1-s_i)}{(1-p_i)s_i + p_i}. \tag{5.51}$$

It follows that

$$\begin{aligned} \frac{p_i}{\lambda} &= \frac{(1-p_i)s_i + p_i}{c_i(1-s_i)} = \frac{s_i}{c_i(1-s_i)} + p_i \frac{1-s_i}{c_i(1-s_i)} \ \text{ and } \ p_i\left(\frac{1}{\lambda} - \frac{1}{c_i}\right) \\ &= \frac{s_i}{c_i(1-s_i)}. \end{aligned} \tag{5.52}$$

Hence,

$$p_i\left(\frac{1}{\lambda} - \frac{1}{c_i}\right) = \frac{s_i}{c_i(1-s_i)} = p_i \frac{c_i - \lambda}{c_i \lambda}. \tag{5.53}$$

Thus, for any i,

$$p_i = \frac{\lambda s_i}{(c_i - \lambda)(1 - s_i)} \text{ and } \lambda = p_i \frac{(c_i - \lambda)(1 - s_i)}{s_i}. \tag{5.54}$$

The optimum value of the Lagrange multiplier is

$$\lambda = \lambda_{\text{opt}} = \frac{c_i(1 - s_i)}{(1 - p_i)s_i + p_i} p_i = \frac{(c_i - \lambda)(1 - s_i)}{s_i} p_i. \tag{5.55}$$

Furthermore,

$$p_{tot} = \prod_{i=1}^{n} p_i = \lambda^n \prod_{i=1}^{n} \frac{s_i}{(c_i - \lambda)(1 - s_i)}. \tag{5.56}$$

The condition for max Δv_m is solved numerically for $\lambda = \lambda_{\text{opt}}$.
Observe that Δv_m is given by

$$\Delta v_m = -c_e \sum_{i=1}^{n} \ln((1 - p_i)s_i + p_i). \tag{5.57}$$

Furthermore, when $\lambda = \lambda_{\text{opt}}$,

$$(1 - p_i)s_i + p_i = \frac{c_i(1 - s_i)}{\lambda} p_i \text{ and } p_i = \frac{\lambda s_i}{(c_i - \lambda)(1 - s_i)}. \tag{5.58}$$

Multiplying the two gives

$$(1 - p_i)s_i + p_i = \frac{c_i s_i}{(c_i - \lambda)}. \tag{5.59}$$

Consequently, when $\lambda = \lambda_{\text{opt}}$,

$$\Delta v_{\max} = -\sum_{i=1}^{n} c_i \ln((1 - s_i)p_i + s_i) = -\sum_{i=1}^{n} c_i\{\ln(c_i s_i) - \ln(c_i - \lambda)\}. \tag{5.60}$$

The numerical algorithm for estimating the maximum velocity increment is as follows:

(i) An initial set of $c_i, s_i, p_{\text{tot}} = \prod_{i=1}^{n_s} p_i$ are given.
(ii) The number stages, n_s are also given.
(iii) The goal is to find and optimum set of p_i.
(iv) Assume an initial set of $p_i = (p_{\text{tot}})^{1/n_s}$; set $\Delta v_0 = 0$.

(v) Calculate $\lambda_i = \frac{c_i(1-s_i)}{(1-p_i)s_i + p_i} p_i$, $\lambda = \lambda_{\text{opt}} = \left(\prod_{i=1}^{n_s} \lambda_i \right)^{1/n_s}$.

(vi) Loop.

(vii) Calculate $\Delta v_{\max}$ using Eq. 5.57.

(viii) If $\Delta v_{\max} <= \Delta v_0$, stop; else, if $\Delta v_{\max} > \Delta v_0$, $\Delta v_0 = \Delta v_{\max}$.

(ix) Increment λ by a small step; $\lambda = \lambda + \varepsilon$. Update p_i using the second of Eq. 5.58.

(x) Repeat loop.

As an example, consider a three stage rocket.

5.5.2 *Application Example: Three Stage Rocket*

In this example, it is assumed that all the n stages have the same parameters. Thus,

$$
\begin{aligned}
c_e &= c_1 = c_2 = \cdots = c_{n-1} = c_n, s_1 = s_2 = \cdots = s_{n-1} = s_n = s_m, p_1 = p_2 \\
&= \cdots = p_{n-1} = p_n = p_m.
\end{aligned} \tag{5.61}
$$

It follows that

$$
p_{\text{tot}} = (p_m)^n = \lambda^n \left(\frac{s_m}{(c_e - \lambda)(1 - s_m)} \right)^n. \tag{5.62}
$$

The ratio of the total velocity increment to c_e over the flight of a multistage rocket is

$$
\Delta v_{\max} / c_e = -\ln \left((1 - s_m) p_{\text{tot}}^{1/n} + s_m \right)^n. \tag{5.63}
$$

Thus, $\Delta v_{\max}$ is only a function of n, c_e, s_m, and p_{tot}. Figure 5.6 illustrates the numerical solution for $\Delta v_{\max}/c_e$ for increasing values of n, for $s_m = 0.1$, as function of the payload ratio p_{tot}.

As $n \to \infty$, the ratio of the total velocity increment to c_e, $\Delta v_{\max}/c_e$, reduces to

$$
\Delta v_{\max} / c_e = -\ln((1 - s_m)^n p_{\text{tot}}) \to -\ln(p_{\text{tot}}). \tag{5.64}
$$

The inefficiency due to the presence of the structure is reduced to a minimum as $n \to \infty$.

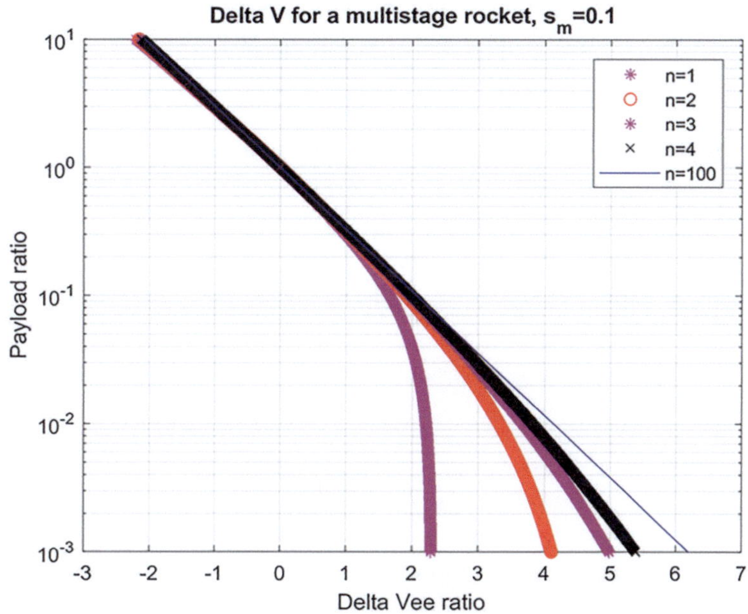

Fig. 5.6 Solution for $\Delta v_{\max}/c_e$ for increasing values of n, for $s_m = 0.1$, as function of the payload ratio p_{tot}

5.6 Maximizing Range, With Drag and Gravity Loss in the Boost Phase

It is important to be able to assess the time to burnout, velocity and altitude in the boost phase. This will assist one to maximize the altitude during the boost phase. Let the terminal velocity attained during the boost phase be defined as v_T. Assuming constant thrust and combining the total thrust and weight, the rocket equation is expressed as

$$m\dot{v} = T' - D, T' = kv_T^2. \tag{5.65}$$

Expressing the drag as $D = kv^2$, the equation of motion is written as

$$m\dot{v} = k\left(v_T^2 - v^2\right). \tag{5.66}$$

Thus, the time to burnout is

$$t = \frac{m}{k}\int_{0}^{v_t}\frac{dv}{(v_T^2 - v^2)} = \frac{m}{2kv_T}\ln\frac{v_T + v_t}{v_T - v_t}. \tag{5.67}$$

The velocity at any time t and the height at burnout (at time t) are, respectively, given by

$$v_t = \frac{1 - \exp\left(-\frac{2kv_T t}{m}\right)}{1 + \exp\left(-\frac{2kv_T t}{m}\right)}v_T, h_B = \frac{m}{k}\int_{0}^{v_t}\frac{vdv}{(v_T^2 - v^2)} = \frac{m}{2k}\ln\frac{v_T^2}{v_T^2 - v_t^2}. \tag{5.68}$$

In the coast phase, there is no propulsive thrust $T = 0$ but only the gravity force and it can be shown that

$$h_C = \frac{m_C}{2k}\ln\frac{(m_C g/k) + v_t^2}{m_C g/k}. \tag{5.69}$$

The time of coast from burnout to when the velocity is zero (at the apogee of the impending orbit) is

$$t_C = \frac{\sqrt{m_C g/k}}{g}\int_{v_t}^{0}\frac{dv}{((m_C g/k) + v^2)} = \frac{m_C}{kg}\arctan\left(\frac{v_t}{\sqrt{m_C g/k}}\right). \tag{5.70}$$

For a constant burning rate, $\dot{m}$ is a constant. Consequently, the time to burn is $t_b = (m_0 - m_E)/\dot{m}$ and the velocity increment is given by

$$\Delta v = v - v_0 = -c_e \ln\left(\frac{m_0 - \dot{m}t}{m_0}\right) - g_e t. \tag{5.71}$$

Integrating both sides (since $\int \ln(x)dx = x\ln(x) - x$),

$$h_b = \frac{m_0}{\dot{m}}c_e\left(\frac{m_0 - \dot{m}t}{m_0}\right)\left\{\ln\left(\frac{m_0 - \dot{m}t}{m_0}\right) - 1\right\} - g_e\frac{t^2}{2}. \tag{5.72}$$

If one sets $t = t_b$ with $m_E = m_0 - \dot{m}t_b$,

$$h_b = \frac{m_0}{\dot{m}}c_e\left\{\left(\frac{m_E}{m_0}\right)\ln\left(\frac{m_E}{m_0}\right) - \left(\frac{m_E}{m_0}\right) + 1\right\} - g_e\frac{t_b^2}{2}. \tag{5.73}$$

In terms of the mass ratio, $R = m_0/m_E$:

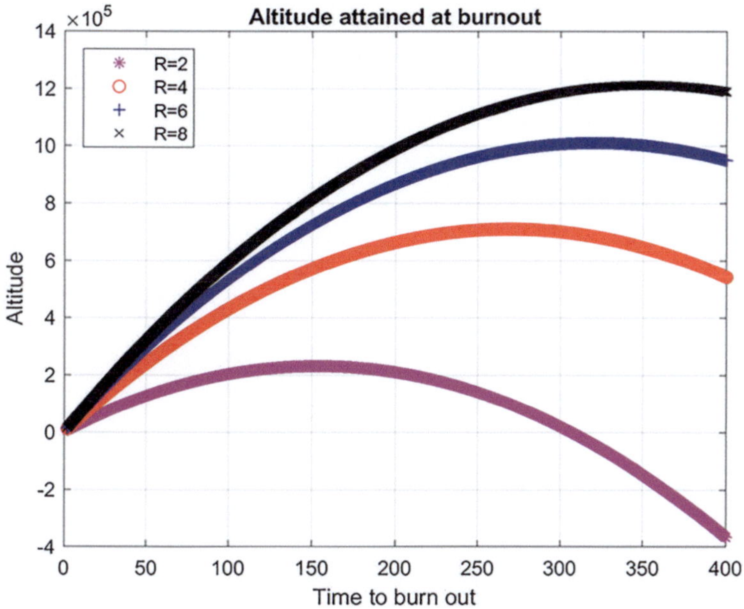

Fig. 5.7 Plots of the altitude attained at burnout for different times to burnout for increasing R

$$h_b = \frac{m_0}{\dot{m}} c_e \left\{ 1 - \frac{1}{R}(\ln R + 1) \right\} - g_e \frac{t_b^2}{2} = \frac{t_b g_0 R}{R-1} I_{sp} \left\{ 1 - \frac{1}{R}(\ln R + 1) \right\} - g_e \frac{t_b^2}{2}. \tag{5.74}$$

The function in the brackets {.} is the range or altitude function (one could assume $g_e = g$). The peak altitude attained at burnout depends on the mass ratio R. For a fixed burn time one can choose the mass ratio R to maximize the altitude. Plots of the altitude attained at burnout for different times to burnout are shown in Fig. 5.7.

For maximum altitude,

$$\frac{dh_b}{dt_b} = \frac{R g_0}{R-1} I_{sp} \left\{ 1 - \frac{1}{R}(\ln R + 1) \right\} - g_e t_b = 0. \tag{5.75}$$

Thus, the ratio of the time to burnout to the specific impulse (the burnout time ratio) is given by

$$\frac{t_b}{I_{sp}} = \frac{R}{R-1} \left\{ 1 - \frac{1}{R}(\ln R + 1) \right\}. \tag{5.76}$$

A plot of the mass ratio R for increasing burnout time ratio is shown in Fig. 5.8.

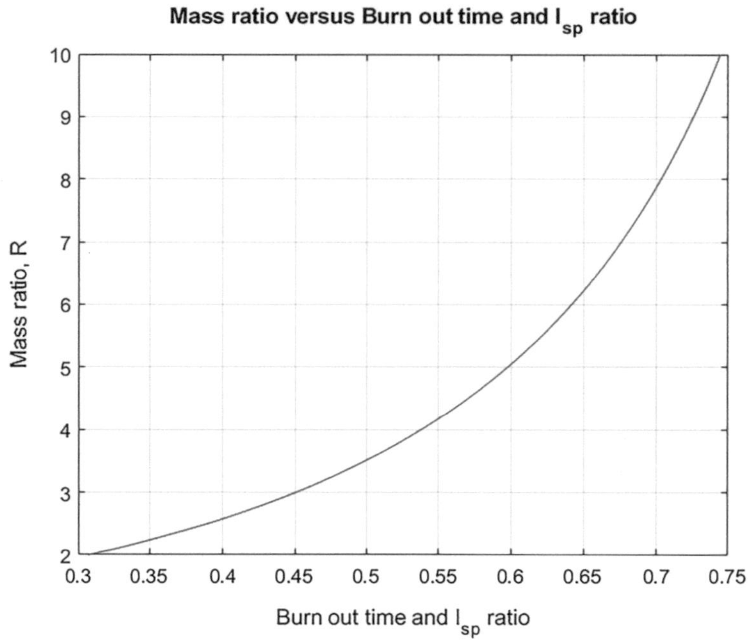

Fig. 5.8 Plot of the mass ratio R for increasing burnout time ratio

5.7 Electric Propulsion, Maximizing Δv, Payload Mass Ratio

Examining rocket Eq. 5.3 closely, one can conclude that in order to reduce the propellant consumption or, in other words, in order to deliver a significant payload mass to its destination, the exhaust velocity must at least be comparable to, or better, larger than, the velocity increment. A high exhaust speed while drastically reducing the propellant fraction translates into cost reduction or improvement of the mission characteristics, e.g. larger payload or increased duration. Chemical thrusters, whether liquid or solid, are fundamentally limited by the energy per unit of mass stored in the propellant, in such a way that v_e cannot exceed 5–6 km/s. This limitation can be circumvented by changing the energy conversion mechanism: the energy source must be de-linked from the propellant (mass), $I_{sp} \rightarrow \infty$. In electric propulsion, as in the case of nuclear propulsion an "external" energy source is used to energize/accelerate the vehicle. Electric propulsion allows for high v_e, making missions possible with large Δv which would otherwise be unaffordable with chemical thrusters. The thrust, T_{ni}, from any nonionized propellant, can be estimated from the "random" flux across a surface area, A, for an isotropic thermal distribution of temperature T_0 (typically 300°K in argon and 500°K in xenon) which is

$$\Gamma_n = (1/4)n_n v_n A. \tag{5.77}$$

where n_n is the neutral gas density. Let

$$v_n = \sqrt{8kT_o/\pi M_m}, \tag{5.78}$$

be the average velocity of the neutrals (M_m is the mass of the atom and k is the Boltzmann constant). Therefore,

$$T_{ni} = v_n \dot{m}_n = v_n \Gamma_n M_m = (1/4)n_n v_n^2 A M_m, \tag{5.79}$$

where $v_{n,\ ex} = v_n$ is the exhaust velocity of the neutrals. For an ionized propellant, it must first be converted into ions before being accelerated to high velocity and ejected out. Only a fraction of the stored propellant is transformed to ions and produces thrust. Mass or propellant utilization efficiency is therefore given by

$$\eta_m = \frac{\dot{m}_{\text{ions}}}{\dot{m}_f} \equiv \frac{\dot{m}_i}{\dot{m}_f} = \frac{i_b M_{\text{ions}}}{q \dot{m}_f}. \tag{5.80}$$

Consequently,

$$T = \dot{m}_i c_e = \dot{m}_f c_e \eta_m, I_{sp} = T/\dot{m}_f g = \eta_m \dot{m}_f c_e/\dot{m}_f g. \tag{5.81}$$

Generally mass utilization efficiency, $\eta_m \approx 1$.

In electrical propulsion, the energy is supplied from an external source, which means the propellant and energy necessary to create momentum are uncoupled, in contrast to chemical propulsion. The power supply must be carried by the vehicle. Contrary to a chemical rocket, the mass of the energy source of an electrical rocket does not decrease in time. It corresponds to a dry mass. As a consequence, the power-to-mass ratio of the thruster is a critical design parameter. Electric propulsion is one case when not only is the fuel burning rate $\dot{m}$ a constant but also very small. However, there is a power source constraint. In the case of electric propulsion, one may assume that the power-plant power supplied is given by $P_s = m_s h_s$, where h_s is the specific power of the power plant. The power P_s is related to the actual power converted into useful power by the jet as $P_T = \eta P_s$ where η is the overall mechanical power conversion efficiency. Assuming a uniform burn rate, it related to the propellant mass, m_e, the relative exhaust velocity, c_e, and the burn time, t_b, by the relation

$$P_T = \frac{1}{2}\left(\frac{m_e}{t_b}\right)c_e^2. \tag{5.82}$$

Hence, it follows that the mass of the structure m_s and the mass of the expended propellant m_e are related by

$$m_s = m_e \frac{c_e^2}{2\eta t_b h_s} = m_e \frac{c_e^2}{v_c^2}, v_c^2 = 2\eta t_b h_s. \tag{5.83}$$

Recall that the relative exhaust velocity is c_e, the vehicle velocity is v, the characteristic velocity of thruster is v_c, the propulsive efficiency is $\eta_p = 2\eta_v / (1 + \eta_v^2)$, vehicle to exhaust velocity ratio is $\eta_v = v/c_e$, the overall mechanical efficiency is $\eta = P_{\text{out}}/P_{\text{input}} = P_T/P_s$, the specific power of the power plant is h_s, and the specific mass ratio is $\alpha = 1/h_s = m_w/P_{\text{input}} = m_s/P_s$.

Thus, the initial mass of the system is the sum of the mass of the structure m_s, the mass of the expended propellant m_e and the payload mass m_p. Thus,

$$m_0 = m_s + m_e + m_p = m_e \left(\frac{v_c^2 + c_e^2}{v_c^2} \right) + m_p \text{ and } m_0 - m_p = m_e \left(\frac{v_c^2 + c_e^2}{v_c^2} \right). \tag{5.84}$$

Hence, we may write

$$\begin{aligned}
m_e &= (m_0 - m_p) \left(\frac{v_c^2}{v_c^2 + c_e^2} \right), \\
m_s &= (m_0 - m_p) \left(\frac{c_e^2}{v_c^2 + c_e^2} \right), \\
\frac{m_s}{m_0} &= \left(1 - \frac{m_p}{m_0} \right) \left(\frac{c_e^2}{v_c^2 + c_e^2} \right), \\
\frac{m_s}{m_0} + \frac{m_p}{m_0} &= \left(\frac{c_e^2}{v_c^2 + c_e^2} \right) + \left(\frac{m_p}{m_0} \right) \left(1 - \frac{c_e^2}{v_c^2 + c_e^2} \right).
\end{aligned} \tag{5.85}$$

Substituting in the rocket equation, and eliminating m_s, one obtains an estimate of the Δv_m that could be obtained from electric propulsion for a given specific power output, h_s; burn time, t_b; overall mechanical power conversion efficiency, η; payload to overall mass ratio and relative exhaust velocity, c_e; as

$$\Delta v_m = c_e \ln \left(\frac{m_0}{m_p + m_s} \right) = c_e \ln \left(\frac{1 + \frac{v_c^2}{c_e^2}}{1 + \frac{m_p}{m_0} \frac{v_c^2}{c_e^2}} \right). \tag{5.86}$$

An alternate form of this equation is

$$\frac{\Delta v_m}{v_c} = \frac{c_e}{v_c} \ln \left(\frac{m_0}{m_p + m_s} \right) = \frac{c_e}{v_c} \ln \left(\frac{1 + \frac{c_e^2}{v_c^2}}{\frac{m_p}{m_0} + \frac{c_e^2}{v_c^2}} \right). \tag{5.87}$$

Thus,

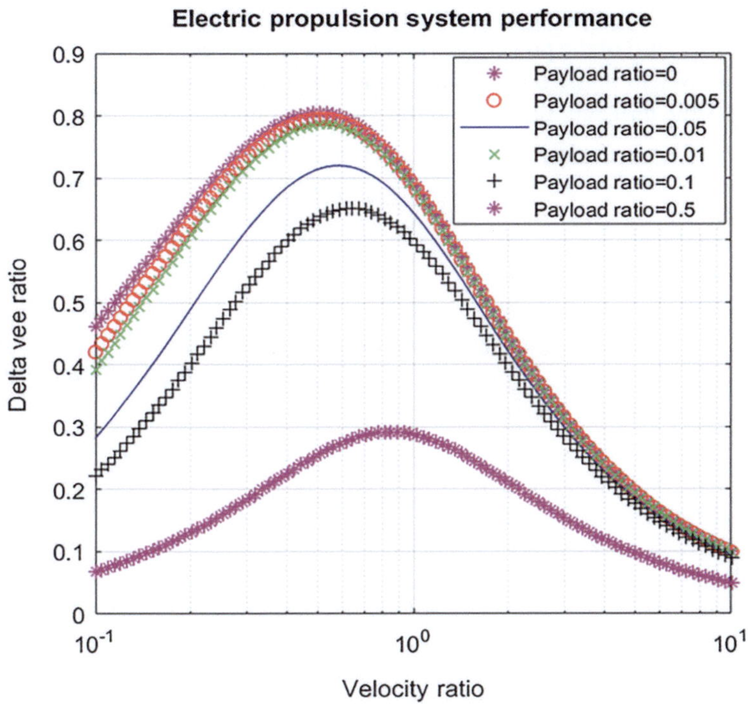

Fig. 5.9 Plot of $\Delta v_m/v_c$ versus c_e/v_c based on Eq. 5.89

$$\frac{m_p + m_s}{m_0} = \exp\left(-\frac{\Delta v_m}{c_e}\right), 1 + \frac{c_e^2}{v_c^2} = \exp\left(\frac{\Delta v_m}{c_e}\right)\left(\frac{m_p}{m_0} + \frac{c_e^2}{v_c^2}\right). \tag{5.88}$$

Rearranging and solving for the payload mass ratio is expressed as

$$\frac{m_p}{m_0} = 1 - \left(1 - \exp\left(-\frac{\Delta v_m}{c_e}\right)\right) - \frac{c_e^2}{v_c^2}\left(1 - \exp\left(-\frac{\Delta v_m}{c_e}\right)\right). \tag{5.89}$$

A plot of $\Delta v_m/v_c$ versus c_e/v_c is shown in Fig. 5.9.

There are a few underlying assumptions on which Eq. 5.89 is based. The above expression assumes that the initial mass of the system is made up of the structural mass (m_s), the mass of the expended propellant ($m_e = m_f$) and the mass of the payload (m_p) and that entire structural mass is equal to the propulsion system mass. The fuel mass is the entire expended or propellant mass. In reality, the structural mass m_s may be considered to include the mass of the fuel tanks (m_t) as well as the propulsion system mass (m_w). The fuel tanks are jettisoned after the fuel is consumed and the tanks are empty.

To modify this expression for the payload mass ratio $\mu_p = m_p/m_0$, define the specific mass ratio of the power system mass $\alpha = 1/h_s = m_w/P_{\text{input}}$ as the ratio of the propulsion system mass to the power input to the thrusters and $t_b = m_f/\dot{m} = m_e/\dot{m}$ which is as the thrusting time. But, $T/P_{\text{input}} = 2\eta/c_e$ and $T = \dot{m}c_e$. Also,

$$T = \sqrt{2\eta m_s h_s \frac{m_e}{t_b}}. \tag{5.90}$$

Thus,

$$m_w = \alpha c_e T/2\eta = \alpha c_e^2 m_e/2\eta t_b, \eta \neq 0. \tag{5.91}$$

Note (assuming $\eta = 1$ is max) that

$$P_T = P_{out} = \frac{1}{2}\dot{m}c_e^2 = \frac{1}{2}\left(\frac{m_e}{t_b}\right)c_e^2. \tag{5.92}$$

Recall that

$$m_0 = m_s + m_e + m_p. \tag{5.93}$$

Hence, one has

$$\frac{m_p}{m_0} = 1 - \frac{m_s}{m_0} - \frac{m_e}{m_0}, \frac{m_e}{m_0} = 1 - \frac{m_s}{m_0} - \frac{m_p}{m_0} = 1 - \exp\left(-\frac{\Delta v_m}{c_e}\right). \tag{5.94}$$

Since the structural mass is the sum of the propulsion system mass and the mass of the tanks, $m_s = m_t + m_w$, it follows that $m_0 = m_t + m_w + m_e + m_p$ and that

$$\frac{m_p}{m_0} = 1 - \frac{m_t}{m_0} - \frac{m_w}{m_0} - \frac{m_e}{m_0}. \tag{5.95}$$

Thus, the payload mass ratio (μ_p) is related to the tank mass ratio (μ_t), the power system mass ratio (μ_w) and expended fuel mass ratio (μ_e) by

$$\mu_p = 1 - \mu_t - \mu_w - \mu_e. \tag{5.96}$$

Assuming that

$$\begin{aligned}\mu_t &= K_t\mu_e, \mu_w = \alpha c_e^2 \mu_e/2\eta t_b, v_c^2 = 2\eta t_b/\alpha, \mu_e = 1 - \exp(-\Delta v_m/c_e), \mu_w \\ &= (c_e^2/v_c^2)\mu_e, \mu_p = 1 - \mu_e - \left(K_t + \frac{c_e^2}{v_c^2}\right)\mu_e,\end{aligned} \tag{5.97}$$

It follows that

$$\mu_p = \frac{m_p}{m_0} = \exp\left(-\frac{\Delta v_m}{c_e}\right) - \left(K_t + \frac{c_e^2}{v_c^2}\right)\left(1 - \exp\left(-\frac{\Delta v_m}{c_e}\right)\right). \tag{5.98}$$

Consequently, Eq. 5.87 for Δv_m is modified to (Edelbaum, 1961; Oleson, 1993)

$$\frac{\Delta v_m}{v_c} = \frac{c_e}{v_c} \ln\left\{\frac{1 + K_t + \frac{c_e^2}{v_c^2}}{\frac{m_p}{m_0} + K_t + \frac{c_e^2}{v_c^2}}\right\}. \tag{5.99}$$

The modification has the effect of effectively increasing the exhaust velocity c_e or the specific impulse.

5.7.1 Application to Mission Design

In designing for a specific mission, the first objective is to design a set of feasible orbit transfer and propulsion system combinations for a given mission requirement. The steps involved are as follows:

1. Compute the maximum Δv available.
2. Compute the Δv required for all the orbit transfers. This may involve a number of orbit transfer strategies consisting of Hohmann transfers, elliptical and hyperbolic transfers, Hohmann transfer segments and spiral orbit transfers.
3. Compute the propulsion system mass for a chemical (liquid or solid) and electric propulsion. For an electric propulsion system, the above simple analysis may be used. A comparable analysis for chemical propulsion is much more complex and tedious.

One may then compute the total mass of the system and would also have a candidate propulsion system for the mission.

5.8 Chemical Propulsion

A chemical propulsion system consists to varying degrees of complexity of a (1) propellant supply and feed system, (2) combustion chamber (3) and an exhaust nozzle. Broadly, chemical thrusters can be classed into two types, liquid propellant systems which could be cold gas, bipropellant or monopropellant types, or solid propellant systems. The typical I_{sp}'s from these thrusters are (1) cold gas, $I_{sp} \sim 30$ s – 130 s; (2) monopropellant, $I_{sp} \sim 200$ s; (3) bipropellant, $I_{sp} \sim 375$ s; and (4) solid, $I_{sp} \sim$ 225–275 s. Each type of thruster relies upon the use of a nozzle; the expansion of gases through a nozzle converts random species motion into directed axial flow.

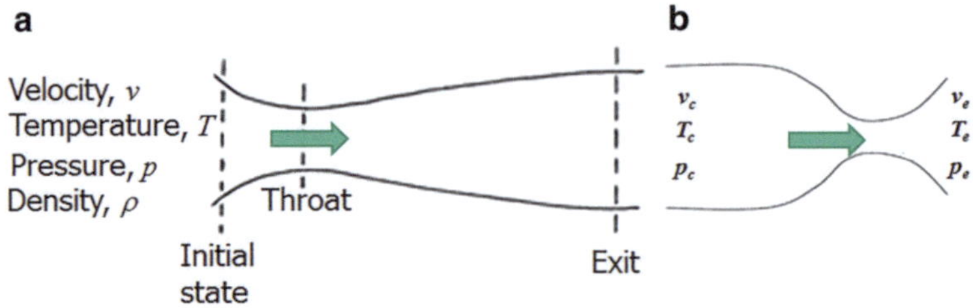

Fig. 5.10 (**a**) A typical converging-diverging nozzle and (**b**) a model to describe the conditions

The nozzle and the flow through a nozzle are common to all propulsion thrusters. In the theory of an ideal nozzle, several assumptions are made (Sutton & Biblarz, 2010):

1. The propellant in the rocket is homogeneous: the density of the fluid is uniform throughout.
2. The propellant is in a gaseous state.
3. The propellant nominally obeys the ideal gas law.
4. The flow is one-dimensional and an isentropic flow. Thus, the flow is adiabatic. This means that there is no significant change in temperature of the walls in comparison with that of the fluid.
5. Friction along the walls is negligible.
6. There are no flow discontinuities.
7. The flow is steady.

Within an isentropic and adiabatic flow, the relative entropy of the propellant is zero, thus no irreversible energy loss (Sutton & Biblarz, 2010). The perfect or universal gas law relates the pressure in the flow to its local density, and temperature in terms of the universal gas constant R_0 and the molecular weight M_w is

$$p = R\rho T = (R_0/M_w)\rho T \tag{5.100}$$

From the relation between the adiabatic pressure and density, $p\rho^{-\gamma} = constant$, where γ is the ratio of the specific heat coefficients at constant pressure and at constant volume, and the above universal gas law, the relationship between the temperature and pressure is,

$$\frac{T_2}{T_1} = \left(\frac{p_2}{p_1}\right)^{\frac{\gamma-1}{\gamma}} = \left(\frac{V_2}{V_1}\right)^{\gamma-1}, \text{with } \gamma = \frac{c_p}{c_v}, c_p = c_v + R_0, c_p = \frac{\gamma R_0}{\gamma - 1}. \tag{5.101}$$

In the case of any typical nozzle, the flow conditions at inlet throat and exit are used to characterize the nozzle. A converging-diverging nozzle, as illustrated in Fig. 5.10, has a converging section between the inlet and throat and a diffuser between the throat and the exhaust. The throat is the point where the cross-sectional area of the nozzle is the smallest. At that point, the velocity of the propellant may be sonic if there is sufficient pressure coming from the inlet (choked flow condition). The ambient pressure ranges from standard (sea level) pressure on Earth to very low ambient pressure in space. For the usage of electric propulsion, the ambient environment is assumed to be in celestial space (pressure is less than 10^{-5} Torr. 1 torr = 133.322 pascal. The torr (symbol: Torr) is a unit of pressure based on an absolute scale, now defined as exactly 1/760 of a standard atmosphere (101,325 Pa). Thus one torr is exactly 101,325/760 pascals ($\approx$ 133.32 Pa)).

Consider a typical expansion process through a nozzle. From the continuity equation for the conservation of mass,

$$\dot{m} = \rho v A. \tag{5.102}$$

From the steady flow energy balance or energy conservation (in the right units!),

$$\frac{1}{2}v^2 + c_p T = \text{constant} = \frac{1}{2}v_c^2 + c_p T_c. \tag{5.103}$$

Assume $v_c = 0$. From the temperature-pressure relation,

$$\frac{T}{T_c} = \left(\frac{\rho}{\rho_c}\right)^{\gamma-1} = \left(\frac{p}{p_c}\right)^{(\gamma-1)/\gamma}. \tag{5.104}$$

From steady flow energy balance, with $v_c = 0$,

$$\frac{v_e^2}{2} = C_p(T_c - T_e) = \frac{\gamma R T_c(1 - T_e/T_c)}{(\gamma-1)} = \frac{\gamma R_0 T_c}{(\gamma-1)M_w}\left(1 - (p_e/p_c)^{\gamma-1/\gamma}\right). \tag{5.105}$$

The isentropic equations follow.

$$\left(\frac{p_e}{p_c}\right)^{\frac{\gamma-1}{\gamma}} - 1 = \frac{\gamma-1}{2}\frac{v_e^2}{c_e^2} = \frac{\gamma-1}{2}M_e^2. \tag{5.106}$$

If the flow is isentropic along the nozzle, be it fully subsonic or supersonic from the throat, the isentropic equations apply at the throat. Thus,

$$\left(\frac{p_t}{p_c}\right)^{\frac{\gamma-1}{\gamma}} = \frac{T_t}{T_c} = 1 + \frac{\gamma-1}{2}M_e^2 \Rightarrow M_e = \sqrt{\frac{2}{\gamma-1}\left(\left(\frac{p_t}{p_c}\right)^{\frac{\gamma-1}{\gamma}} - 1\right)}. \tag{5.107}$$

Using the fact that the speed of sound is related to the temperature by $c = \sqrt{\gamma R T}$, the flow velocity at the exit (assuming supersonic) is

$$v_e = c_e M_e = \sqrt{\frac{2\gamma R T_t}{\gamma - 1}} \sqrt{\left(\left(\frac{p_c}{p_t}\right)^{\frac{\gamma-1}{\gamma}} - 1\right)} = \sqrt{\frac{2\gamma R T_t}{\gamma - 1}} \sqrt{\left(\frac{1}{1 + \frac{\gamma - 1}{2} M_e^2} - 1\right)}. \tag{5.108}$$

At the exit, it follows that

$$\dot{m}_e = \rho_e v_e A_e = \frac{\gamma p_t A_e}{\sqrt{\gamma R T_t}} M_e \left(1 + \frac{\gamma - 1}{2} M_e^2\right)^{-\frac{\gamma-2}{2}}. \tag{5.109}$$

Naming with a "*" variables at the stage where M = 1 (i.e. the sonic section, which may be a real throat within the nozzle or at some extrapolated imaginary throat downstream of a subsonic nozzle), the sonic mass flow rate and speed in the isentropic discharge through a nozzle are

$$\dot{m}_e|_{M_e=1} = \dot{m}^* = \rho^* v^* A^* = \frac{p^* A^*}{R T^*} \sqrt{\gamma R T^*} = \frac{\gamma p^* A^*}{\sqrt{\gamma R T^*}} = \frac{\gamma p_t A^*}{\sqrt{\gamma R T_t}} \left(\frac{2}{\gamma + 1}\right)^{\frac{\gamma+1}{2(\gamma-1)}}. \tag{5.110}$$

Thus, the area ratio where the speed of the flow is sonic to the exit area is

$$\frac{A^*}{A_e} = M_e \left(\frac{\frac{\gamma+1}{2}}{1 + \frac{\gamma-1}{2} M_e^2}\right)^{\frac{\gamma+1}{2(\gamma-1)}}. \tag{5.111}$$

Or the cross-sectional area ratio of the exit area, A_e, to the throat area A_t is

$$\frac{A_e}{A_t} = \frac{1}{Ma_e} \left(\frac{1 + \frac{\gamma+1}{2} Ma_e^2}{\frac{\gamma+1}{2}}\right)^{\frac{\gamma+1}{2(\gamma-1)}}. \tag{5.112}$$

The throat-inlet pressure ratio depends only on the specific heat ratio of the fluid when the flow is choked:

$$\frac{p_t}{p_i} = \left(\frac{2}{\gamma + 1}\right)^{\frac{\gamma}{\gamma-1}}. \tag{5.113}$$

Equation assumes that the inlet flow speed is zero (*Mai* = *0*). The expansion ratio at the nozzle exhaust exit plane depends on the specific heat ratio and the exit flow Mach number:

$$\frac{p_i}{p_e} = \left(1 + \frac{\gamma - 1}{2} Ma_e^2\right)^{\frac{\gamma}{\gamma - 1}}. \tag{5.114}$$

The exhaust velocity of the thruster, v_{ex}, is the velocity of the propellant as it ejects from the nozzle. This velocity depends on the ratio of the exhaust pressure to the throat pressure. The equation includes the inlet temperature, T_i, of the propellant:

$$v_{ex} = \sqrt{T_i \frac{R_0}{M_w} \frac{2\gamma}{\gamma + 1} \left(1 - \left(\frac{p_e}{p_i}\right)^{\frac{\gamma - 1}{\gamma}}\right)}, \tag{5.115}$$

Thus, it is observed that the exhaust velocity increases with

1. Increasing the temperature in the combustion chamber prior to the inlet to the nozzle, T_c (but adverse effects of higher temperatures on nozzle heat transfer and increased dissociation losses).
2. Increasing pressure ratio p_c/p_e (but limited by attendant increases in motor weight).
3. Low molecular weight M_w, since $R = R_0/M_w$.
4. To a limited extent, by reducing γ (of limited use compared to other effects).

Finally, the exhaust velocity at the exit v_{ex} is simply limited by T_c and $c_p = \gamma R_0/(\gamma - 1)$. To a large extent, the temperature within the combustion chamber T_c is determined by the adiabatic flame temperature T_A. As an example, in the case of liquid oxygen (LOX) and liquid hydrogen (LH) liquid propulsion engine, the adiabatic flame temperature is $T_A \sim 2980\ ^\circ K$. Value of $C_p = \gamma R/(\gamma - 1)$ at $2980\ ^\circ K$, with the production of H_2O, is given by $C_p > {\sim}2.4 \times 10^3 J/Kg\ ^\circ K$. When the fuel is just H_2 as in electrothermal propulsion, the corresponding value of C_p is $C_p \sim 2.0 \times 10^3 J/Kg\ ^\circ K$. If one uses unsymmetrical dimethylhydrazine (UDMH; 1,1-dimethylhydrazine), which is a chemical compound with the formula $H_2NN(CH_3)_2$, the adiabatic flame temperature is higher and equal to $T_A \sim 3436\ ^\circ K$, when UDMH is used in hypergolic rocket fuels as a bipropellant in combination with the oxidizer nitrogen tetroxide (UDMH/N_2O_4). (A hypergolic propellant combination used in a rocket engine is one whose components spontaneously ignite when they come into contact with each other.) UDMH is a derivative of hydrazine and is sometimes referred to as a hydrazine.

The ideal mass flow rate determines the most efficient thrust possible in a thruster:

$$\dot{m}_{\text{ideal}} = A_t p_i \gamma \sqrt{\frac{\left(\frac{2}{\gamma+1}\right)^{\frac{\gamma+1}{\gamma-1}} M_m}{\gamma R T_i}}. \tag{5.116}$$

Therefore, the ideal vacuum thrust in an efficient thruster (Khare & Saha, 2021) is

$$F_{t,\text{ideal}} = \dot{m}_{\text{ideal}} v_{ex}. \tag{5.117}$$

In the nonideal case,

$$F_t = \dot{m}_{\text{ideal}} v_{ex} + (p_e - p_a) A_e. \tag{5.118}$$

The direction of the exhaust velocity is not completely in the axial direction if one is using a conical nozzle, for example. The radial direction of the exhaust velocity does not contribute to the thrust. Therefore, there is loss in performance because of the divergence of the conical nozzle. To achieve the best performance, the angle of rise, α_{rise} (i.e. the cone half-angle), has to be less than 15 degrees. Knowing the angle of rise, the radii of the throat, r_t, and exit, r_e, the optimum length of the cone is (Stoute, 2013; Sutton & Biblarz, 2010)

$$L_{\text{opt\_cone}} = \left(r_t \left(\frac{r_e}{r_t} - 1 \right) + 1.5 r_t (\sec \alpha_{\text{rise}} - 1) \right) \cot \alpha_{\text{rise}}. \tag{5.119}$$

A supersonic expansion fan, known as *Prandtl-Meyer expansion*, a two-dimensional wave, is a centred expansion process that occurs when the flow is supersonic and when it turns around a convex corner. The exhaust flow of the engine forms a standing diamond pattern of shocks and expansion waves, which can be explained in part from the Prandtl-Meyer expansion equation, and gives the flow deflection angle, v_{PM}:

$$\begin{aligned} v_{PM} &= \sqrt{\frac{\gamma + 1}{\gamma - 1}} \tan^{-1} \sqrt{\frac{\gamma + 1}{\gamma - 1} (Ma^2 - 1)} - \tan^{-1} \sqrt{Ma^2 - 1}, \theta_{PM} \\ &= v_{PM}(Ma_2) - v_{PM}(Ma_1), \end{aligned} \tag{5.120}$$

where θ_{PM} is the net deflection angle of the local flow in passing through an expansion wave fan. The specific heat capacity ratio and local flow Mach number determine the value of the angle. The angles of the shocks that appear in the diamond pattern can be as seen from shock theory or numerical flow calculations.

The velocity increments and specific impulse obtained by various thrusters are briefly summarized. Thus, for a *cold gas propulsion* system,

$$I_{sp} = \left(\frac{\gamma C^*}{g}\right)\sqrt{\frac{2}{(\gamma-1)}\left(\frac{2}{\gamma+1}\right)^{\frac{\gamma+1}{\gamma-1}}\left(1-\frac{p_e}{p_c}\right)^{\frac{\gamma-1}{\gamma}}} \text{ and } v_e$$

$$= \sqrt{\frac{2\gamma T_c}{(\gamma-1)}\left(1-\frac{p_e}{p_c}\right)^{\frac{\gamma-1}{2}}}, \tag{5.121}$$

where C^* is the characteristic velocity, is the p_e/p_c exit to chamber pressure ratio, T_c is the chamber temperature and γ is the ratio of specific heat coefficients.

For a *liquid propulsion* system,

$$v_e = \sqrt{\frac{2\gamma R T_c}{(\gamma-1)}\left(1-\frac{p_e}{p_c}\right)^{\frac{\gamma-1}{2}}}, \tag{5.122}$$

where R is the universal gas constant and γ is the Poisson ratio.

For a *solid propulsion system* with a de Laval nozzle, the burn rate (r) which is dependent on a temperature coefficient (a) and a combustion index (n), and the characteristic velocity C^* are respectively given by,

$$r = a p_c^n,\ C^* = p_c A_t / \dot{m}. \tag{5.123}$$

For comparisons with other systems (Goebel & Katz, 2008), it is noted that for a *resistojet*, the thrust (τ) produced by the propellant at stagnation pressure also depends on stagnation number density of propellant ions (n_o) in m^{-3}, stagnation temperature (T_o) and the probability (χ) of a molecule exiting the expansion slot area (A_o). Specific impulse (I_{sp}) is a function of the stagnation temperature and the mass of the expended propellant (m_e):

$$\tau = A_o \chi n_o k T_o / 2,\ I_{sp} = \sqrt{\pi k T_o / 2 m_e} / g, \tag{5.124}$$

where k is the Boltzmann constant and g is the acceleration due to gravity at sea level. These equations are generally derived from considerations of energy balance, by equating the kinetic energy and either the electrostatic or the internal energy.

For a *radio frequency ion thruster*,

$$v_{e_i} = \sqrt{2 q_{\text{ion}} V_{\text{ion}} / m_{\text{ion}}},\ \tau = I_{\text{ion}-\text{beam}} \sqrt{2 m_{\text{ion}} V_{\text{ion}} / q_{\text{ion}}},\ I_{sp} = (\gamma_c \eta_m / g) \sqrt{e V_{\text{ion}} / m_{\text{ion}}}, \tag{5.125}$$

where q_{ion} is the ionic charge, V_{ion} is voltage driving the ions, m_{ion} is ionic mass, $I_{\text{ion}-\text{beam}}$ is the ionic beam current, e is the electronic charge, η_m is the thruster mass utilization efficiency and γ_c is the total thrust correction factor.

For an *electrospray thruster*,

$$\langle q_{\text{ion}}/m_{\text{ion}}\rangle = 1/(\rho Q), v_{e_i} = \sqrt{2V_{\text{ion}} I_{\text{ion -beam}} \langle q_{\text{ion}}/m_{\text{ion}}\rangle}$$
$$= \sqrt{2V_{\text{ion}} I_{\text{ion -beam}}/\rho Q}, \tau = \rho Q v_{e_i} = \sqrt{2V_{\text{ion}} I_{\text{ion -beam}} \rho Q}. \tag{5.126}$$

For *pulsed-plasma thrusters*, the Lorentz force describes the relationship between the force (thrust) produced due to charged particles moving through a self-induced magnetic field. Thrust produced depends on the charge of ions (q), sum of all collision forces per particle (propellant) over all particles ($(P_i)_k$) and particle velocity ($\widetilde{\mathbf{u}}_i$). Thus, the general expressions for the thrust and ion exit velocity are

$$\tau = m_{\text{ion}} \frac{d\widetilde{\mathbf{v}}_i}{dt} = q\left(\widetilde{\mathbf{E}} + \widetilde{\mathbf{u}}_i \times \widetilde{\mathbf{B}}\right) + \sum (P_i)_k, v_e = \frac{\tau}{\dot{m}\eta} = \frac{1}{\eta}\frac{\mu_0}{4\pi}\ln\left(\frac{R_a}{R_c}\right), \tag{5.127}$$

where μ_0 is the permeability of free space and R_a and R_c are the radii of the anode and cathode, respectively.

5.8.1 Chemical Propulsion Systems

The various cold gas, monopropellant, bipropellant and solid propellant systems are briefly discussed. Cold gas is the simplest system, typically a pressurized gas such as nitrogen, argon and propane (or even xenon, a very expensive waste!) with a control valve. A typical cold gas thruster system is illustrated in Fig. 5.11. It also consists of

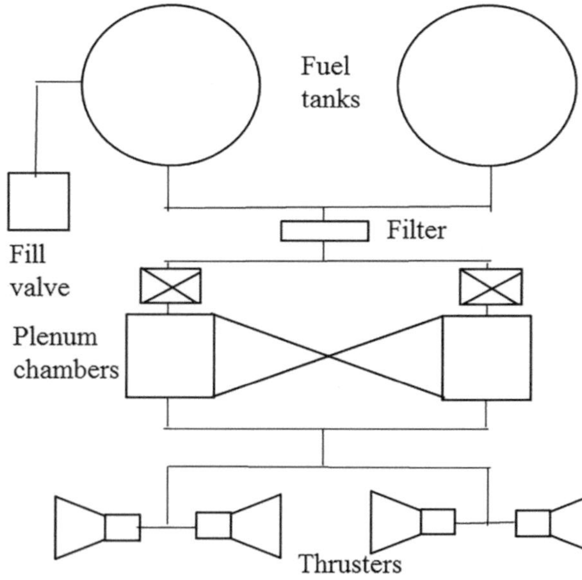

Fig. 5.11 Typical cold gas thruster system

tanks, cold gas feed lines, control valves, flow sensors, pressure transducers and solenoid valves.

Principles are the similar for both solid and liquid propellants but with some essential differences which are as follows: (i) solid fuel thrusters are comparatively simple in design (e.g. no separate propellant storage, feed system or combustion chamber); (ii) solid fuel thrusters have fixed thrust-time curves determined by the fuel grain geometry, whereas liquid rocket fuel thrusters can vary thrust and have stop/start capabilities; (iii) in general, solid propellant thrusters give lower specific impulse than liquid propellant fuel thrusters. While cold gas systems (80 ± 50 s I_{sp}) use a pressurized gas, monopropellant liquid systems (210 ± 20 s I_{sp}) use hydrazine or hydrogen peroxide and a catalyst, bipropellant liquid systems (375 ± 75 s I_{sp}) use hydrocarbons like kerosene and an oxidizer and solid propellant systems (250 ± 25 s I_{sp}) use solid fuel and an oxidizer.

Monopropellant systems are low performance propulsion thrusters characterized by simple storage (non-cryogenic), a simple feed and flow control using a high pressure tank with a "blow down" feed and also using catalytic decomposition to enhance performance. A typical monopropellant is illustrated in Fig. 5.12.

Monopropellant thrusters are generally used either with a catalyst or a heater system. However, catalysts could become poisoned due to contaminants. The thrust level one could expect from such thrusters is of the order of 10 N. The most common propellants are hydrazine (N_2H_4) which is in a liquid state in the temperature range 275–387 °K. When it is mixed with or is in contact with nitrogen, it decomposes into ammonia (NH_3) and hydrogen (H_2) via an exothermic reaction. It is sometimes used to heat the propellant which is then expanded through a nozzle. The main use of monopropellant thrusters is in attitude and orbit control. Thus, there are four thrusters located at each location with three dedicated to attitude or orientation control and one for translational velocity or orbit control and with the groups of thrusters located at four locations for complete bidirectional controllability. The main drawback of monopropellant thrusters is that the fuel could freeze either around the vanes within the storage tanks when the spacecraft becomes too cold. A typical monopropellant thruster system is illustrated in Fig. 5.13, and the corresponding locations of the thrusters around a spacecraft's body are shown in Fig. 5.14.

The thruster redundancy matrix shown in Table 5.1 determines how the thrusters are deployed for independent, positive or negative roll-pitch-yaw attitude control.

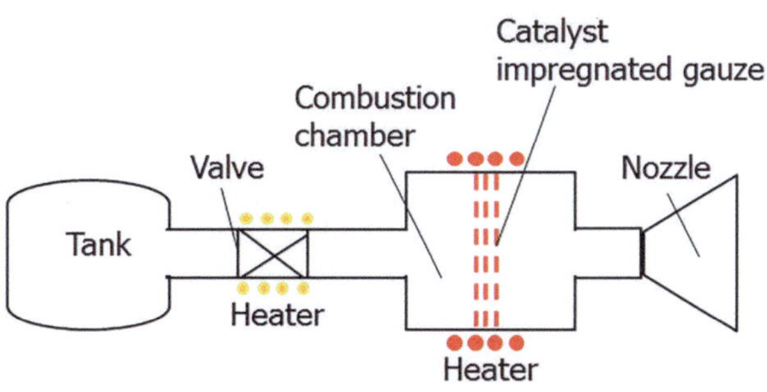

Fig. 5.12 A typical monopropellant spacecraft thruster

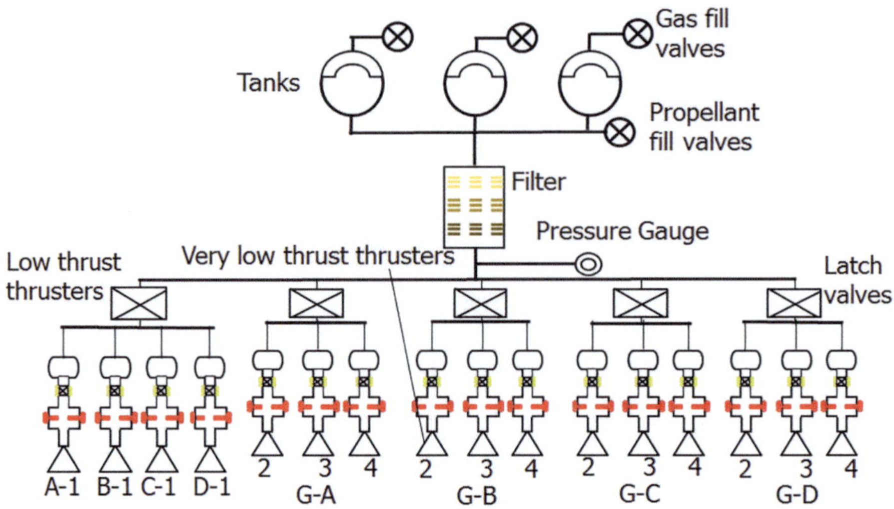

Fig. 5.13 A typical monopropellant thruster based propulsion system, using pressurized propellant storage tanks

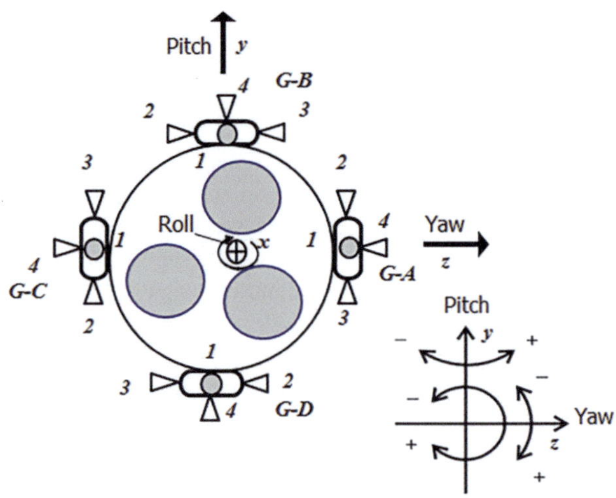

Fig. 5.14 Typical locations of thrusters and thruster groups for spacecraft attitude and orbit control; also shown are the roll, pitch and yaw axes (bottom view, looking forward towards payload)

Table 5.1 Thruster redundancy matrix

Thruster	Primary	Backup
+roll	A-2, C-2	B-2, D-2
-roll	A-3, C-3	B-3, D-3
+pitch	C-4	B-2, D-3
-pitch	A-4	B-3, D-2
+yaw	A-2, C-3	B-4

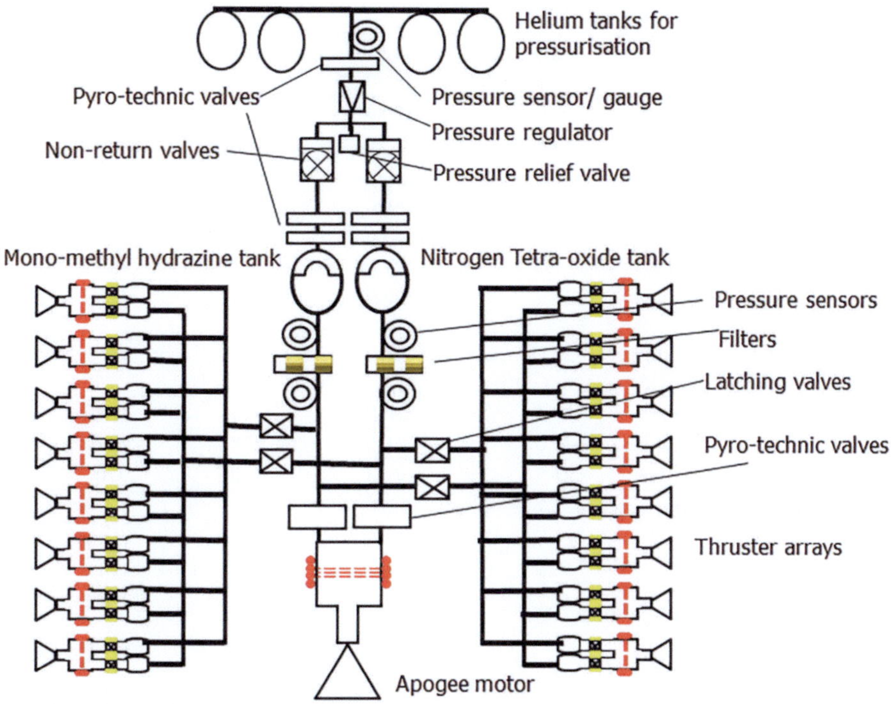

Fig. 5.15 A typical bipropellant thruster based propulsion system, using independent pressurized propellant storage

Bipropellant thrusters are known to offer the best performance of all chemical propulsion systems. However, they could be quite complex, especially when used to generate high thrust as they require pumping systems for high performance. Pumping systems could sometimes be a source of vibration and instability. The main drawbacks of bipropellant systems are that the fuel and oxidizer must be stored separately and independently. Furthermore, the liquid fuel is susceptible to "sloshing" which could be a problem for a spin-stabilized spacecraft as it induces instability. The propellants with highest performance must be stored cryogenically. A typical bipropellant system is shown in Fig. 5.15. The fuel and oxidizer are fed to the combustion chamber via an injector where they are ignited. In the case of hypergolic propellants, an igniter is not needed as the propellants ignite on contact with each other. Cryogenic propellants include LOX (stored below −253 °C) and LH_2 (−183 °C), requiring an igniter. Easily storable propellants like kerosene (RP-1), hydrazine, nitrogen tetroxide (N_2O_4) and monomethyl hydrazine (MMH, chemical formula: CH_3NHNH_2) are used with bipropellant thrusters. A liquid propellant rocket motor will supply a fuel and an oxidizer into a combustion chamber to react and produce hot gases which yield thrust by expanding and accelerating in a nozzle. High performance requires that the combustion to occur

Table 5.2 Liquid propellant summary

Fuel	Oxidizer	Molecular wt. (products)	Combustion temp. °*K*	Typical sp. impulse	Mean density (Kg/m^3)
Hydrogen	Oxygen	10.0	2980	390	280
	Fluorine	12.8	4120	410	460
Kerosene	Oxygen	23.4	3690	301	1020
	Fluorine	23.9	2920	320	1230
	Red fuming Nitric acid	25.7	3160	268	1355
	Nitrogen Tetra-oxide	26.2	3460	276	1250
	Hydrogen peroxide	22.2	3010	278	1362
Hydrazine	Oxygen	19.4	3410	313	1070
	Nitric acid	20.0	2970	278	1310
UDMH	Oxygen	21.5	3625	310	970
(Hypergolic)	Nitric acid	23.7	3220	276	1220
Monopropellants					
Hydrazine		10.3	966	200	1011
Hydrogen peroxide		22.7	1267	165	1422

at high pressure, which requires that the propellants are either pumped in or fed by a high pressure (pressure fed) in the propellant tanks.

The most common liquid propellants used are summarized in Table 5.2. All quoted values are for a combustion chamber pressure of 7 MPa with an ideal expansion to an exit pressure of 0.1 MPa. Higher combustion chamber pressures admit increases in I_{sp}. For example, at 20 MPa, LH2-LOX yields an increased specific impulse I_{sp} of ~460 s, from the quoted value of $I_{sp} \sim 390$ s.

A hypergolic propellant combination used in a rocket engine is one whose components spontaneously ignite when they come into contact with each other. Unsymmetrical dimethyl-hydrazine (UDMH; 1,1-dimethyl-hydrazine) is a chemical compound with the chemical formula $H_2NN(CH_3)_2$. UDMH is a derivative of hydrazine and is sometimes also referred to as a hydrazine. It is often used in hypergolic rocket fuels as a bipropellant in combination with the oxidizer nitrogen tetroxide. Aerozine 50 is a 50:50 mix of hydrazine and UDMH. C-Stoff is a 50:50 mix of hydrazine and methanol fuel used for the HWK 109–509 rocket motor series. UH 25 is also a fuel mixture used for rockets. It was developed for the European Ariane 2–4 launch vehicles. UH 25 was developed after a disaster during the second flight of the Ariane 1 rocket. During launch, one of the four Viking engines, with UDMH, on the first stage developed a combustion instability which led to an engine fire, subsequent explosion and destruction of the vehicle. Following this event and starting with Ariane 2, the fuel was changed from pure UDMH to the mixture UH 25. UH 25 was used in Ariane rocket versions 2 through 4 and in the Indian GSLV Mk III and the Chandrayan III missions. UH 25 is a mixture of 75% UDMH and 25%

hydrazine hydrate. It is hypergolic with dinitrogen tetroxide (N_2O_4) as oxidizer, and both can be stored as liquids at room temperature.

Considering a typical application, such as Indian Space Research Organisation's (ISRO) Chandrayan II mission to the Moon in 2023, the payload was a lunar lander, a lunar surface rover and six scientific probes (including a Langmuir plasma probe). The booster rockets were 2 × S200 strap-on solid rocket motor boosters, generating 6000 kN thrust in total, with 204 tons of fuel in each. The rocket propellant mixture in each solid rocket motor consisted of ammonium perchlorate (oxidizer, 69.6% by weight), atomized aluminium powder (fuel, 16%), iron oxide (catalyst, 0.4%), PBAN (binder that also acts as fuel, 12.04%) and an epoxy curing agent (1.96%). The stage 2 of the rocket launcher (L110) was hypergolic fuel (UH25) propelled. The stage 3 was (LOX-LH_2) cryogenic fuel propelled (118 kN thrust, 28.6 tons prop/), climbing to 219 km in 7 mins, to reach a velocity of 5.1 km/s.

Launch vehicles pump their propellants. Spacecraft use a pressure-fed system that feeds an inert pressurant gas, either maintained at a constant pressure in the propellant tank by a regulator or introduced while allowing its pressure to decrease as the propellant is expelled and the pressurant volume increases, known as the "blow-down" mode. As the propellant flows through a series of valves and filters, it may be exposed to various temperatures (possibly causing changes in its properties, such as viscosity), and dynamic effects may occur, such as slosh in the propellant tank or water hammer in the pipes. The two are compared in Fig. 5.16.

In Fig. 5.17, two different types of pump-fed systems are compared, one using a turbine pump and a gas generator and the other a conventional pump-fed system using a stored pressurant gas. Generally, primary (launch vehicles) systems use turbo pump feed while secondary thrusters use gas pressure-fed system. A gas pressure feed system has a higher tank mass. The trade-off is between the extra mass and complexity of turbo pump feed and the lighter tanks of turbo pump feed versus the simpler design of gas pressure feed system. Turbo pump based systems are used for larger launch vehicles, while the more conventional pump-fed systems using a stored pressurant gas are used in medium sized launch vehicles. Propellant tanks are designed for use either with pressure-fed systems or with pump-fed systems. The internal tank pressure depends on the feed system. In the case of pressure-fed engines, the pressure can be as high as 6 MPa. In the case of pump-fed system, lower pressures will suffice, so the pressure is less than or equal to 0.5 Mpa. The ullage volume is the volume occupied by the gas, and it allows for thermal expansion of the propellant liquids and for the accumulation of dissolved gases or of gaseous products of slow reactions within the propellant during storage.

Propellant tanks are sized after determining the required propellant load. For bipropellant systems, the oxidizer-fuel mixture ratio, *Ox*/*Fu*, is defined as the ratio of the mass flow rate of the oxidizer to the mass flow rate of the fuel, m_{ox}/m_{fuel}, the mixture ratio for the required specific impulse. The tank volumes are calculated from propellant volumes plus ullage (~5%), for design margins and residuals, plus the requirement for propellant management devices. The selection of the mixture ratio could be critical. The selection is made for overall system benefits rather than just maximizing the specific impulse. In the case of N_2O_4 /MMH designers, almost

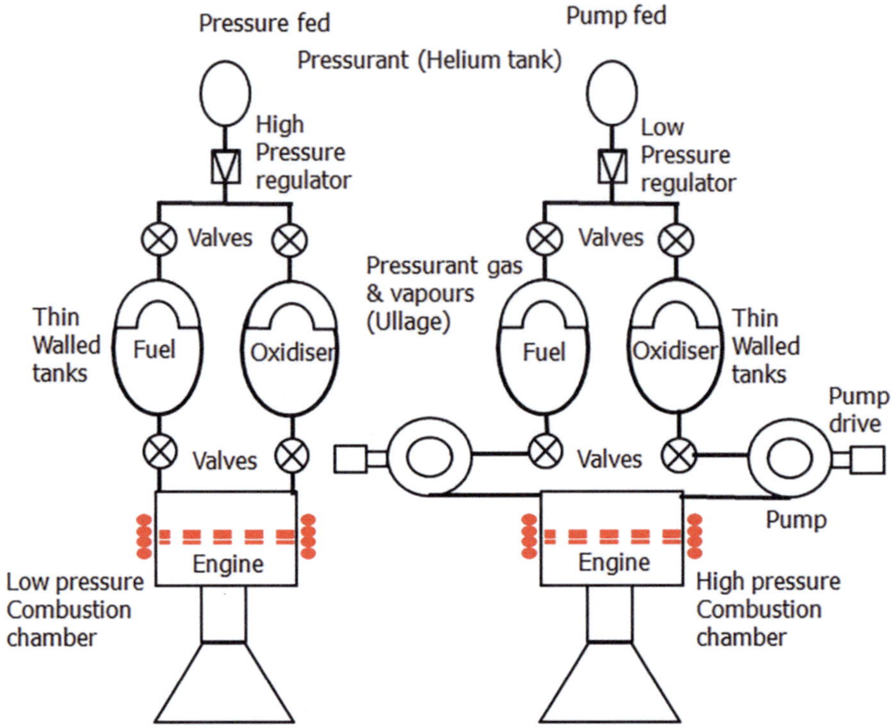

Fig. 5.16 Comparison between a pressure-fed spacecraft propulsion system and pump-fed launch vehicle propulsion system

always use an *Ox*/*Fu* ratio of 1.64 for equal tank size which simplifies tank manufacturing, configuration layout and integration. For a typical liquid propulsion system, the propellant mass is 85–93% of the total mass. The typical breakdown of the other masses is tank mass ~ 5% to 15% of propellant mass and an additional 20–30% of the overall tank mass for mounting hardware and propellant management devices.

The structural design of the tanks is based on the internal pressure the tanks are subjected to. The tank mass is calculated following the structural design to select the thickness of the pressure vessel. For spherical tanks, in terms of the radius r, wall thickness t and maximum internal pressure p, the hoop stress is given by $\sigma_{\text{hoop}} = p \times r/(2t)$. The design process involves the selection of the material, estimation of the tank size and determination of the thickness from maximum allowable stress criterion followed by the estimation of the weight. The maximum allowable stress $\sigma_{\max}$ is reduced to account for the inefficiency of the joints. For cylindrical tanks, the longitudinal stress needs to be considered as well. The longitudinal stresses in cylindrical tanks remain the same as in spherical tanks. However, the hoop stress is double that for a spherical tank of the same radius, so

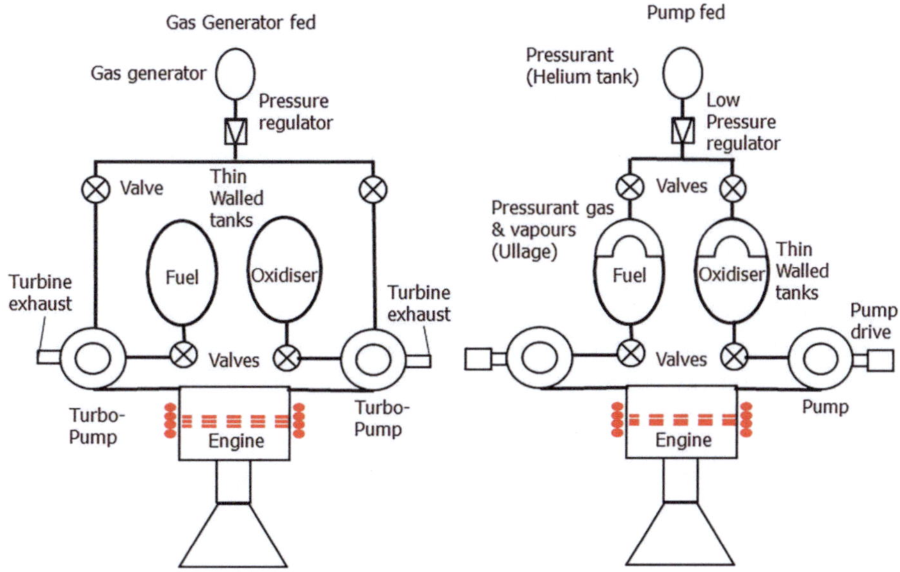

Fig. 5.17 Comparison of two different types of pump-fed systems

$\sigma_{\text{hoop}} = p \times r/t$. The design process is similar to that for a spherical tank although modifications are made for tanks with spherical end caps.

Propellant management devices are essential as the system must operate in "zero-gravity" environment. The critical "zero-gravity" problem is related to the requirement that on motor restart, the propellant must be in liquid phase at the tank outlet and with no vapour. The problem in important:

(i) For supercritical storage: Pressure is greater than critical and therefore the fuel must remain in a single phase with no vapour mixed in. The disadvantage is that there is a weight penalty due to high pressure required for storage.
(ii) For inertial systems (not suited for on-orbit propulsion): In this case one is faced with ensuring propellant "bottoming" due to the tank being accelerated by small solid rocket motor, which in turn could push the propellant up and away from the exit valve.
(iii) For positive propellant expulsion devices, which use a choice of bladder configurations, such as a diaphragm, an inflatable bladder or a piston-cylinder combination.
(iv) For capillary systems, where the vapour is kept away from the exit port by the use of baffles or screens, though restricted to tubes of small diameter on Earth. Larger tubes exhibit capillary behaviour in a zero-gravity environment in outer space. This strategy is clearly less satisfactory for low surface tension liquids (e. g. cryogenics). The pressure required to overcome surface tension is directly related to the surface tension and is $\Delta P = 2\sigma/r$, where ΔP is the pressure

required to overcome surface tension, σ is the surface tension and r is the radius of the pores.

Finally solid propellant systems are briefly discussed. Solid propellant rocket motors are characterized by some very unique features. The propellant of a solid rocket motor is stored entirely within the combustion chamber in the form of one or more shaped blocks, called grains. The main characteristic of a solid propellant rocket is its simplicity. The burning times may be from a few seconds or fractions of a second to as long as 1 or 2 min. Once the rocket has ignited, the combustion generally proceeds until all the propellant is burnt; the thrust program, i.e. is fixed. It is possible, with suitable grain designs, to obtain a thrust which increases (progressive burning), remains constant (neutral) or decreases (regressive) with burning time. In constant-geometry nozzles, thrust is more or less proportional to the chamber pressure. The combustion chamber itself is located within a skirt and insulated from it. The skirt is encased within a motor casing. The nose cap and frustum which house the drogue and other parachutes and at least four rocket motors and several booster motors along with the required propellant tanks are arranged in series to form a long cylindrical rocket structure within a series of skirts connected by attachment rings. The series is terminated by the steering system and a single nozzle extension.

Chamber is essentially a high-temperature tank containing the entire solid mass of propellant or grain. Combustion proceeds from the surface of the grain. The mass flow rate of the hot generated gases is equal to the product of the mean density of the solid propellant, the area of the burning surface and burn rate or the surface recession rate. The thrust achieved depends on grain geometry, and the variations in the geometry can provide for increasing thrust (progressive burning), constant thrust (neutral burning), decreasing thrust (regressive burning), thrust variation with dual peaks or two or more steps as well as other variations. Figure 5.18 represents a typical record of a near constant thrust motor. The time integral of the thrust gives the

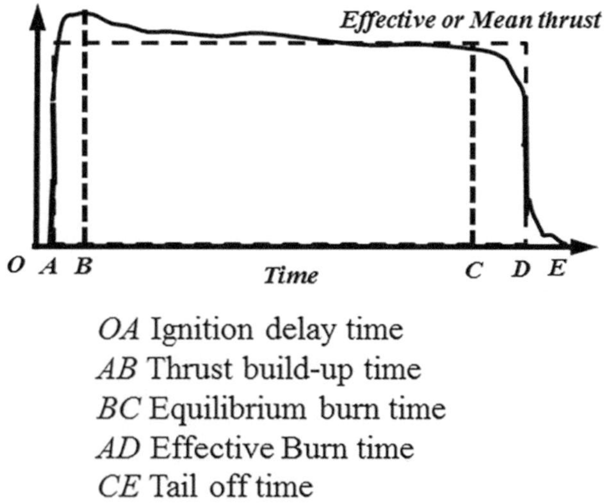

Fig. 5.18 Example of thrust variation in a constant thrust motor

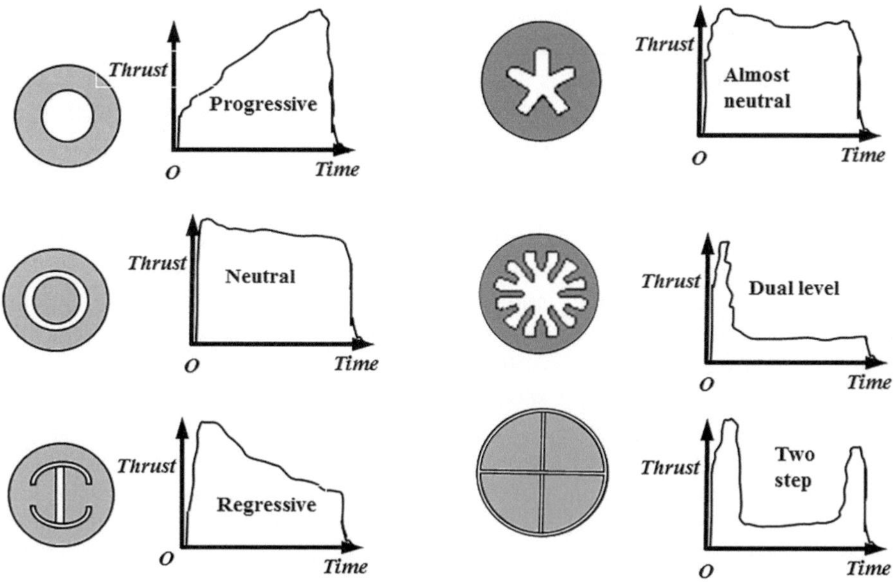

Fig. 5.19 Most common solid propellant core cross sections of grain geometries

total impulse. The effective or mean thrust is given by the ratio of the total impulse to the effective burn time. The most common solid propellant core cross-sections of grain geometries are, circular sections like, tubular (progressive thrust), rod within a tube (constant thrust), non-circular such as double anchor (regressive), star (near constant thrust), finocyl (dual thrust) and dual step (two step thrust) are shown in Fig. 5.19.

Two basic solid motor propellant classes are as follows:

1. Homogenous [single base, nitrocellulose (NC), double base (NC+ nitroglycerin, NG), triple base (double + nitroguanidine, NQ)].
2. Composite propellants (heterogeneous). Example: shuttle booster rocket propellant composition—16% atomized aluminium powder (fuel), 69.8% ammonium perchlorate (oxidizer), 0.2% iron oxide powder (catalyst), 12% polybutadiene acrylic acid acrylonitrile (binder) and 2% epoxy curing agent. Each solid rocket motor typically contains more than 450,000 kg of propellant.

Composite propellant, also known as heterogeneous (not double base propellant, DBP) propellant, consists of separate fuel and oxidizer chemicals.

DBP are typically nitrocellulose + nitroglycerin. Nitroglycerin contains enough oxygen for complete oxidation to occur, with no need for a separate oxidizer. These are chemicals manipulated so they combine in each molecule of the oxidizer and the fuel. For these DBP, there are problems of mechanical strength unless NG:NC $< 1:1$, but for high I_{sp}, one needs NG:NC $\sim 8.6:1$. Combustion produces clear gases (no smoke) with an I_{sp}: 196–250 s and with a combustion temperature of 1700–3000 K.

Hydroxyl-terminated polybutadiene (HTPB) is an oligomer of butadiene terminated at each end with a hydroxyl functional group. It reacts with isocyanates to form polyurethane polymers. HTPB is a translucent liquid with a colour similar to wax paper and a viscosity similar to corn syrup. JAXA describes the propellant as "HTPB/AP/Al = 12/68/20", which means proportioned by mass, HTPB plus curative 12% (binder and fuel), ammonium perchlorate 68% (oxidizer) and aluminium powder 20% (fuel).

For composites, the fuel and oxidizer are provided in different constituents. The fuel acts as a binder giving the desired mechanical properties. A typical composite fuel is: -polyurethane (organic) oxidizer: -NH4CLO4 (inorganic salts) and Aluminium powder is added to give an increased temperature of combustion. It is typically characterized by a higher I_{sp}, $I_{sp} \sim 230$–263 s, Tc ~ 2600–4000°K. (This was used on shuttle booster rocket, i.e. not secondary propulsion system; SBR performance: Isp ~ 260 s; Tc ~ 3480°K, Pc ~ 7 MPa, Pe ~ 0.1 MPa).

Examples and properties of propellants can be found in Sutton & Biblarz, 2010.

In summary, the key features of all solid propellant motors are simplicity, low cost and low performance but not controllability (safety problem), and it and can be fired just once so there could be a safety issue. The thrust level can be varied only by initially setting the propellant grain geometry. Moreover, the motor cannot be reignited.

Exercises

1. (a) The initial mass of a rocket m_0, is the sum of the masses of the propellant system, the fuel or propellant itself and the payload, and it is assumed that at the end of propulsion system operation, the fuel is completely expended. Using the Δv equation, $\Delta v = u_{ex} \ln (m_0/m_f)$, where m_f is the final mass of the rocket at the end of the propulsion system operation and u_{ex} is the rocket's exhaust velocity, show that,

$$\Delta t_{f0} = \frac{u_{ex}^2 \alpha}{2\eta}\left(1 + \frac{m_{\text{payload}}}{\alpha P_{\text{propulsion\_system}}}\right)\left\{\exp\left(\frac{\Delta v}{u_{ex}}\right) - 1\right\},$$

$$m_{\text{propellant}} = \alpha P_{\text{propulsion\_system}}\left(1 + \frac{m_{\text{payload}}}{\alpha P_{\text{propulsion\_system}}}\right)\left\{\exp\left(\frac{\Delta v}{u_{ex}}\right) - 1\right\}.$$

In the above relations, $\Delta t_{f0} = t_f - t_0$ is the total time of propulsion system operation; $m_{\text{propellant}}$ is the mass of the propellant required; m_{payload} is the mass of the payload; $P_{\text{propulsion \_ system}}$ is the power delivered by the propulsion system; α is the specific mass of the propulsion system, i.e. the propulsion system mass per unit of the power delivered by it; and η is the overall power conversion efficiency, i.e. the ratio of the mechanical power of the rocket to the power delivered by the propulsion system. You may assume maximum propulsive efficiency when estimating the mechanical power of the rocket.

(b) Given that the required $\Delta v = 5$ km/s for a mission, power system specific mass is 15 kg/kW, a specific impulse of 1000s, an overall power conversion efficiency of 20% and the power delivered by the propulsion system is 10 kW and that propellant mass available is 720 kg, estimate the maximum allowable mass of the payload and the total time of propulsion system operation.

(c) If the payload mass is required to be 1000 kg, determine the propellant mass required as well as total time of propulsion system operation. Hence, compare the total mass of the rocket with the corresponding mass obtained in part b).

2. (a) Describe the principal features of a liquid propulsion system which might be used on board a satellite for primary and secondary propulsion. Highlight the way in which these systems differ from liquid propulsion systems used in launch vehicles.

(b) A geostationary satellite of mass 3000 kg is injected by a launch vehicle into a transfer orbit having the properties indicated in Table 5.3. The satellite has an on-board propulsion system which is used initially to inject the satellite into geostationary orbit (GEO) using a single thruster firing—it is then used to maintain the GEO for a 5-year mission and then to finally inject the satellite into a "graveyard" orbit. The Δv per year for maintenance is 50 m/s/year, and the injection to the graveyard orbit has a Δv of 40 m/s.

Calculate the total mass of fuel required by the propulsion system if it operates with a specific impulse of 320 s.

3. (a) Briefly explain why multistaging of a launch vehicle is used in all currently available launch-to-orbit systems.

(b) The mass ratio at burnout of a rocket can be defined in terms of the variable R defined as ratio of the initial mass of the rocket to the mass at burnout. Thus, if the mass of the payload is m_p, the expended mass of the propellant and oxidizer or fuel is m_e, and the mass of the remaining structure is m_s—the ratio of the initial mass of the rocket to the mass at burnout could be defined.

(i) For a single-stage rocket, state the velocity change at burnout in terms of the initial to final mass ratio. How does this equation change for a multistage rocket?

Table 5.3 Properties of a transfer orbit

Transfer orbit semi major axis (a)	24,500 km
Apogee on transfer orbit	Geostationary altitude
Transfer orbit inclination to equator (i)	5°
Transfer orbit argument of perigee (ω)	0°
Geostationary orbit period	23 h 56 min
Geostationary orbit inclination (i)	0°

(ii) If we define a payload mass ratio p and the structural coefficient s, respectively, as $p = m_p/(m_p + m_s + m_e)$, $s = m_s/(m_s + m_e)$, show that

$$R = 1/(p + (1-p)s), p = (1-sR)/(R(1-s)), s = (1-pR)/(R(1-p)).$$

(iii) For a single-stage rocket, show that the expended fuel mass ratio and the structure mass ratio at burnout are, respectively, given as

$$m_e/\sum m = (1-s)(1-p), m_s/\sum m = s(1-p).$$

(c) A three stage launcher has each stage of the launcher with an identical payload ratio, p. At lift-off, the total mass of the launcher is 15,000 kg. The launcher is used to place a satellite in an elliptical orbit of 300 km perigee and 800 km apogee. The final stage is ignited at 300 km, when the vehicle is travelling at 6 km/s horizontally. This final stage has a I_{sp} of 350 s and has a structure to fuel mass ratio of $m_s/m_e = 0.1$.

Calculate the payload mass of the final stage.

4. (a) Explain the primary features of geosynchronous and a geostationary orbit. Explain the differences between these two orbits, with reference to the ground track and the significance of these differences.
   (b) Identify two different practical approaches that may be adopted to launch a satellite into a geostationary Earth orbit. Explain the primary difference between these two approaches.
   (c) Explain how to estimate the maximum latitude that can be viewed from a satellite in any orbit, and calculate this value for a geostationary satellite.
   (d) Identify at least two disadvantages of geostationary satellites for remote sensing applications.
   (e) It is desired to launch a Sun-synchronous satellite. Explain the key features of such a satellite.

5. (a) A satellite is approaching the planet Jupiter with a relative velocity of 20 km/s at the edge of its sphere of influence. The eccentricity of the satellite's orbit about Jupiter is 1.07. Explaining all your assumptions and calculation steps, estimate (i) the semi major axis of the orbit and the distance of closest approach to Jupiter, (ii) the excess velocity and the total energy of the spacecraft as it approaches the planet, (iii) the turning angle, (iv) the velocity increment the satellite acquires as it swings by the planet Jupiter and (v) the excess velocity of the satellite as it departs the planet Jupiter.
   (b) There is an international legal requirement for satellites operating in geostationary orbit to have sufficient fuel mass at the end of the mission to enter a "graveyard" orbit. Identify the principal reason for this and the issues driving the selection of the graveyard orbital elements.

Table 5.4 Parameters relating to the design solution and other relevant data

	Design value	Actual
End of life wet mass (including propellant)	3000 kg	Unknown
Specific impulse	240 s	210 s
Geo orbit radius	42,164 km	

(c) A satellite has been designed with sufficient fuel to inject it at the end of life from geostationary orbit (GEO) into a coplanar circular graveyard orbit having a radius of 300 km larger than its operational orbit. The parameters relating to the design solution and other relevant data are given in Table 5.4. During the mission, two faults arise: (i) fuel gauging has resulted in less residual propellant so that only 80% of the design value remains in the tank; (ii) the specific impulse of the propulsion unit has reduced to 210 s. If the legal requirement for a graveyard orbit is that it must have a circular orbit greater than 200 km above the GEO value, will the satellite meet this legal requirement? State all the assumptions you make.

6. (a) Obtain, from first principles, the relationship between the area ratio and the exit Mach number for the flow through a choked nozzle, stating all your assumptions and the basic physical principles used. How does this relationship change if the flow at the throat is already supersonic?
(b) What are the consequences of a poorly designed nozzle?
(c) Under what conditions can a flow external to a spacecraft be considered to be a free molecular flow? Explain three important features that differentiate free molecular flow from a continuum.
(d) The power input to a thruster is 2500 MW. The power output efficiency is 0.35 and the efficiency of the thruster is 0.6. The I_{sp} is given as 5000 s. Find the mass flow rate at the nozzle exit and the net thrust generated by the thruster.

References

Edelbaum, T. N. (1961). Propulsion requirements for controllable satellites. *ARS Journal, 31*, 1079–1089.

Fortescue, P. W., Swinerd, G. S., & Stark, J. P. W. (2011). *Spacecraft systems engineering* (4th ed.). Wiley.

Goebel, D. M., & Katz, I. (2008). *Fundamentals of electric propulsion: Ion and hall thrusters* (pp. 6–7, 22, 26, 148, 189, 192, 243–244, 325, 327). Jet Propulsion Laboratory (JPL).

Jensen, J., Townsend, G., Kork, J., & Kraft, D. (1962). *Design guide to orbital flight*. McGraw Hill Book Co.

Khare, S., & Saha, U. K. (2021). Rocket nozzles: 75 years of research and development. *Sādhanā, 46*. https://doi.org/10.1007/s12046-021-01584-6

Oleson, S. R. (1993) *An analytical optimization of electric propulsion orbit transfers*. NASA Contractor Report 191129.

Prussing, J. E., & Conway, B. A. (1993). *Orbital mechanics*. Oxford University Press.

Stoute, C. A. B. (2013). *Hybrid electric thruster using gas mixtures*, Ph. D. Dissertation, York University, Toronto, Ontario, Canada.

Sutton, G. P., & Biblarz, O. (2010). *Rocket propulsion elements* (8th ed.). Wiley.

Chapter 6
Trajectory Optimization and Feedback Control

6.1 Trajectory Optimization Principles: Introduction

Considering the orbit transfers dealt with so far in the preceding chapters, two classes seem to emerge. For chemical rockets, the orbit manoeuvres were assumed to be velocity impulses or velocity increments, dv. The trajectories were assembled from a series of unpowered elliptic arcs, which were connected with each other at the points where the "impulses" were applied. Control was exercised at discrete points along the orbit. For electric propulsion, however, in spite of its limited introduction so far, the orbit manoeuvres were assumed to involve continuously applied low magnitude thrust forces, operating over finite periods of time. The trajectories were continuous arcs, no impulses were used and control was applied continuously. These trajectories facilitate continuous optimization from start to finish.

In the case of discrete impulsive manoeuvres, the mass fraction with the payload (p/l) mass was defined as

$$R = m_0/m_f = \left(m_s + m_{prop} + m_{p/l}\right)/\left(m_s + m_{p/l}\right). \tag{6.1}$$

The initial mass was defined to be m_0, while the final mass was assumed to be m_f. The mass of the structure (w/o the payload and the propellant) was defined as m_s. The propellant mass was defined as m_{prop}. With these definitions, the velocity increment was given as

$$\Delta v = gI_{sp}\ln(R) = gI_{sp}\ln\left(m_0/m_f\right). \tag{6.2}$$

The Hohmann transfer represents an optimum orbit transfer with discrete impulses only based on the solutions to the two-body problem. The corresponding transfer times are obtained by the use of Lambert's theorem. However, the above relations do not apply to continuous transfers which must be considered afresh.

R. Vepa, *Space Vehicle Maneuvering, Propulsion, Dynamics and Control*,
https://doi.org/10.1007/978-3-031-65518-0_6

6.2 Calculus of Variations and Pontryagin's Minimum Principle

Principles of minimization form one of the most wide-ranging means of formulating mathematical models governing the dynamical motion of space vehicles. Moreover, many popular numerical integration schemes such as the powerful finite element method are also founded upon a minimization paradigm. The basic mathematical analysis of nonlinear minimization principles on infinite-dimensional function spaces, a subject known as the "calculus of variations", and Pontryagin's minimum principle play a key role in the definition of optimal trajectories in general and in all space systems related optimization problems. Optimization is not a matter of choice in the case of space systems but an absolute necessity, which must be considered as a matter of course. Classical solutions to optimization problems in the calculus of variations are prescribed by boundary value problems involving certain types of differential equations, known as the Euler-Lagrange equations. An introduction to the calculus of variations may be found in Forsyth (1960).

6.2.1 Functionals, Variations and Pontryagin's Minimum Principle

To develop the equations for the optimised transfer trajectories, it is important to introduce certain important mathematical concepts. The first of these is the notion of a functional. Consider a continuous function of t, $x(t)$, and the integral of its function, $f(x(t))$, defined in the closed interval $[t_0, t_f]$, and let

$$J = \int_{t_0}^{t_f} f(x(t))dt. \tag{6.3}$$

An example of J is given by

$$J = \int_{t_0}^{t_f} (x(t))^2 dt. \tag{6.4}$$

$J(x(t))$ is a functional of $x(t)$. Formally, a functional is a mapping (function) from a space of functions into the underlying field (usually the real or complex numbers). The space of functions is also called the domain of the functional and the space of admissible functions.

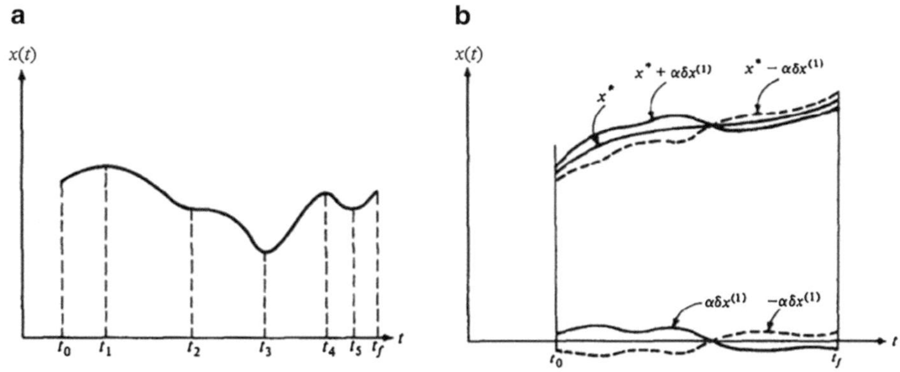

Fig. 6.1 Illustration of (**a**) a function and (**b**) a variation

The variation plays the same role in determining the extreme values of functional as the differential does in finding the maxima and minima of a function. Typical examples of a function and the variation of a function are illustrated in Fig. 6.1a, b.

The problem of optimal control is to derive the necessary conditions for the optimal control from the basics of calculus of variations. Thus, the problem is first defined in terms of the calculus of variations with differential constraint equations that must also be satisfied. The conditions for optimality are then developed. It also involves the derivation of the transversality and corner point conditions.

Consider the functional

$$J = \int_{t_0}^{t_f} L(x(t), \dot{x}(t), t) dt. \tag{6.5}$$

In Eq. 6.5, the variable $x(t)$ is required to satisfy

$$f(x(t), \dot{x}(t), t) = 0. \tag{6.6}$$

Since it is not really feasible to eliminate the variables, an extended functional is considered. Thus the functional is extended to

$$J(x, \lambda) = \int_{t_0}^{t_f} (L(x(t), \dot{x}(t), t) + \lambda f(x(t), \dot{x}(t), t)) dt. \tag{6.7}$$

In Eq. 6.7, the variables $\lambda = \lambda(t)$ are unknown and known as Lagrange multipliers. There are as many of them as there are constrain equations, one for each differential constraint. Consider a variation of the extended functional, which in accordance with Pontryagin's minimum principle must satisfy at an extremum, the equality:

$$\delta J(x, \delta x, \lambda, \delta\lambda) = \int_{t_0}^{t_f} (\widetilde{g}_x \delta x + \widetilde{g}_{\dot{x}} \delta\dot{x} + f\delta\lambda) dt = 0. \tag{6.8}$$

In Eq. 6.8,

$$\widetilde{g}(x(t), \dot{x}(t), t) = L(x(t), \dot{x}(t), t) + \lambda f(x(t), \dot{x}(t), t). \tag{6.9}$$

Integrating Eq. 6.8 by parts,

$$\delta J(x, \delta x, \lambda, \delta\lambda) = \int_{t_0}^{t_f} \left(\widetilde{g}_x \delta x - \frac{d}{dt} \widetilde{g}_{\dot{x}} \delta x + f\delta\lambda \right) dt = 0. \tag{6.10}$$

Since $f(x(t), \dot{x}(t), t) = 0$, it follows that it holds, for any λ, for the optimal solution as well. Thus, the variational condition that the extended functional must satisfy reduces to

$$\delta J(x, \delta x, \lambda, \delta\lambda) = \int_{t_0}^{t_f} \left(\widetilde{g}_x - \frac{d}{dt} \widetilde{g}_{\dot{x}} \right) \delta x dt = 0. \tag{6.11}$$

It follows that the condition an extremum of $J(x(t))$ is given by

$$\left(\widetilde{g}_x - \frac{d}{dt} \widetilde{g}_{\dot{x}} \right) = 0. \tag{6.12}$$

Substituting for $\widetilde{g}(x(t), \dot{x}(t), t)$ in Eq. 6.12, one has

$$L_x + \lambda f_x - \frac{d}{dt} (L_{\dot{x}} - \lambda f_{\dot{x}}) = 0. \tag{6.13}$$

Equation 6.12 is Euler's equation for optimality.

6.2.2 Application to the Optimum Control Problem

The optimum control problem may now be generalized, as one needs to find an admissible vector u^* that makes the following system:

$$\dot{x}(t) = f(x(t), u(t), t), \quad x(t_0) = x_0, \tag{6.14}$$

to follow a feasible state vector trajectory x^* that minimizes the functional

$$J = \Phi\big(x(t_f), t_f\big) + \int_{t_0}^{t_f} L(x(t), u(t), t)dt. \tag{6.15}$$

The functional in Eq. 6.15 is also expressed as

$$J = \Phi(x(t_0), t_0) + \int_{t_0}^{t_f} \left(L(x(t), u(t), t) + \frac{d}{dt}\Phi(x(t), t) \right) dt. \tag{6.16}$$

The functional in Eq. 6.16 is also expressed as

$$J = \Phi(x(t_0), t_0) + \int_{t_0}^{t_f} (L + \Phi_x \dot{x} + \Phi_t) dt. \tag{6.17}$$

The corresponding extended functional, to accommodate the constraint Eq. 6.14, is

$$J = \Phi(x(t_0), t_0) + \int_{t_0}^{t_f} (L + \Phi_x \dot{x} + \Phi_t + \lambda f - \lambda \dot{x}) dt. \tag{6.18}$$

Moreover, since $x(t_0) = x_0$ is fixed, the extended functional is

$$J = \int_{t_0}^{t_f} (L + \Phi_x \dot{x} + \Phi_t + \lambda f - \lambda \dot{x}) dt. \tag{6.19}$$

Assuming that the Φ terms satisfy the Euler's equation for optimality, one has

$$(\Phi_x \dot{x} + \Phi_t)_x - \frac{d}{dt}\left((\Phi_x \dot{x} + \Phi_t)_{\dot{x}}\right) = 0. \tag{6.20}$$

Equation 6.20 is identically satisfied, as long as Φ is continuously differentiable. Thus, the extended functional reduces to

$$J = \int_{t_0}^{t_f} (L + \lambda f - \lambda \dot{x}) dt. \tag{6.21}$$

Consider a variation of the extended functional, which must satisfy at an extremum the equality

$$\delta J = \int_{t_0}^{t_f} \left\{ \left(L_x + \lambda f_x + \dot{\lambda} \right) \delta x + (L_u + \lambda f_u) \delta u + (f - \dot{x}) \delta \lambda \right\} dt. \tag{6.22}$$

Hence, the conditions of optimality, subject to the boundary conditions, which must be derived, are

$$L_x + \lambda f_x + \dot{\lambda}(t) = 0, \;\; L_u + \lambda f_u = 0, \;\; \dot{x}(t) = f, \tag{6.23}$$

To derive the boundary conditions, consider the conditions on the boundary. On the boundary,

$$\begin{aligned} &\left(\Phi_x \left(x^* \left(t_f \right), t_f \right) - \lambda^* \left(t_f \right) \right) \delta x_f \\ &\quad + \left(L \left(x^* \left(t_f \right), u^* \left(t_f \right), t_f \right) + \Phi_t \left(x^* \left(t_f \right), t_f \right) + \lambda f \left(x^* \left(t_f \right), u^* \left(t_f \right), t_f \right) \right) \delta t_f = 0. \end{aligned} \tag{6.24}$$

From Eq. 6.24, one can derive the transversality conditions for the control problem. To this end, define a Hamiltonian:

$$H = L + \lambda f. \tag{6.25}$$

The optimality conditions and the boundary conditions, in terms of the Hamiltonian H, are, respectively, given by

$$H_x + \dot{\lambda}(t) = 0, \;\; H_u = 0, \;\; \dot{x}(t) = f, \tag{6.26}$$

$$\begin{aligned} &\left(\Phi_x \left(x^* \left(t_f \right), t_f \right) - \lambda^* \left(t_f \right) \right) \delta x_f \\ &\quad + \left(H \left(x^* \left(t_f \right), u^* \left(t_f \right), t_f \right) + \Phi_t \left(x^* \left(t_f \right), t_f \right) \right) \delta t_f = 0. \end{aligned} \tag{6.27}$$

The transversality conditions can be derived for each specific case of boundary conditions from the above. These are necessary (and sufficient if H is convex) conditions.

6.2.3 Boundary Conditions

Boundary conditions are particularly important, and they could be natural to the problem or imposed. If they are not natural to the problem, they may have to be imposed, provided they reflect the real situation. In optimal control, this leads to the two-point boundary value problem (TPBVP).

If in Eq. 6.27, $\delta t_f = 0$ and t_f is prescribed, the boundary condition that must be imposed is

$$\Phi_x\left(x^*\left(t_f\right), t_f\right) = \lambda^*\left(t_f\right). \tag{6.28}$$

If in Eq. 6.27, $\delta x_f = 0$ and x_f is prescribed, the boundary condition that must be imposed is

$$\Phi_t\left(x^*\left(t_f\right), t_f\right) = -H\left(x^*\left(t_f\right), u^*\left(t_f\right), t_f\right). \tag{6.29}$$

In Eq. 6.27, if both $\delta t_f = 0$ and $\delta x_f = 0$ and both t_f and x_f are prescribed, it follows that $x(t_0) = x_0$ and $x(t_f) = x_f$, and the above derived conditions given by Eqs. 6.28 and 6.29 do not apply. The applicable boundary conditions and the corresponding variations are summarized in Table 6.1.

6.2.4 Example 1

Before plunging into optimization of spacecraft orbital trajectories, a few simpler examples will be considered first. Consider the cost function:

Table 6.1 Summary of applicable boundary conditions and variations

Problem	Description	Variation	Boundary condition
t_f is fixed	$x(t_f) = x_f$ specified	$\delta x(t_f) = \delta x_f = 0$, $\delta t_f = 0$	$x^*(t_0) = x_0$, $x^*(t_f) = x_f$
t_f is fixed	x_f is unspecified	$\delta x(t_f) = \delta x_f$, $\delta t_f = 0$	$x^*(t_0) = x_0$, $\Phi_x(x^*(t_f), t_f) = \lambda^*(t_f)$
t_f is free	$x(t_f) = x_f$ specified	$\delta x_f = 0$	$x^*(t_0) = x_0$, $x^*(t_f) = x_f$, $H(x^*(t_f), u^*(t_f), t_f) + \Phi_t(x^*(t_f), t_f) = 0$
t_f is free	x_f is unspecified		$x^*(t_0) = x_0$, $\Phi_x(x^*(t_f), t_f) = \lambda^*(t_f)$, $H(x^*(t_f), u^*(t_f), t_f) + \Phi_t(x^*(t_f), t_f) = 0$
	$x(t_f)$ is on a moving point $q(t)$	$x\left(t_f\right) = \left(\frac{dq\left(t_f\right)}{dt}\right)\delta t_f$	$x^*(t_0) = x_0$, $x^*(t_f) = q(t_f)$, $H\left(x^*\left(t_f\right), u^*\left(t_f\right), t_f\right) + \Phi_t\left(x^*\left(t_f\right), t_f\right) + \left(\Phi_x\left(x^*\left(t_f\right), t_f\right) - \lambda^*\left(t_f\right)\right)\left(\frac{dq\left(t_f\right)}{dt}\right) = 0$

$$J = \int_{t_0}^{t_f} \left(u^2/2\right) dt. \tag{6.30}$$

subject to the differential constraints, in the form of the state space equations:

$$\dot{x}_1(t) = x_2(t), \quad \dot{x}_2(t) = -x_2(t) + u. \tag{6.31}$$

The Hamiltonian is defined by

$$H = \left(u^2/2\right) + \lambda_1 x_2 - \lambda_2 x_2 + \lambda_2 u. \tag{6.32}$$

The necessary/sufficient conditions for optimality of the control input are

$$H_u = u^* + \lambda_2^* = 0, \quad u^* = -\lambda_2^*. \tag{6.33}$$

Consequently, it follows from Eqs. 6.31 and 6.33 for the optimal control input that

$$\dot{x}_1^*(t) = x_2^*(t), \quad \dot{x}_2^*(t) = -x_2^*(t) - \lambda_2^*. \tag{6.34}$$

The corresponding Lagrange multiplier dynamics or co-state equations are

$$-H_{x_1} = 0 = \dot{\lambda}_1, \quad -H_{x_2} = -\lambda_1^* + \lambda_2^* = \dot{\lambda}_2^*. \tag{6.35}$$

Thus, there are four coupled differential equations to solve.

6.2.5 Transversality Conditions

One needs four boundary conditions for solving the integration constants. If the state equations are assumed to be either scalar or vector equations, there are four corresponding scalar or vector boundary conditions. These are of the form

$$x(t_0) = x_0, \quad x\left(t_f\right) = x_f. \tag{6.36}$$

If boundary points are free, one needs the transversality conditions, given by

$$\Phi_x\left(x^*\left(t_f\right), t_f\right) = \lambda^*\left(t_f\right) = 0, 0 = \lambda_1^*\left(t_f\right), \quad 0 = \lambda_2^*\left(t_f\right). \tag{6.37}$$

along with the initial conditions

$$x_1^*(t_0)=x_{10}^*,\ \ x_2^*(t_0)=x_{20}^*. \tag{6.38}$$

6.2.6 Example 2

Consider the cost function, with acceleration inputs, $a_x(t)$ and $a_y(t)$, given by

$$J=\frac{1}{2}\int_{t_0}^{t_f}\left(a_x^2+a_y^2\right)dt. \tag{6.39}$$

subject to the differential constraints, in the form of the state space equations:

$$\begin{aligned}\dot{x}_1(t)=x_3(t),\dot{x}_2(t)=x_4(t),\ \ \dot{x}_3(t)=&-\frac{\mu x_1(t)}{\left(x_1^2+x_2^2\right)^{3/2}}+a_x\\ \equiv -f_3(\mathbf{x})+a_x,\ \ \dot{x}_4(t)=&-\frac{\mu x_2(t)}{\left(x_1^2+x_2^2\right)^{3/2}}+a_y\equiv -f_4(\mathbf{x})+a_y.\end{aligned} \tag{6.40}$$

The Hamiltonian is defined by

$$H=(1/2)\left(a_x^2+a_y^2\right)+\lambda_1x_3+\lambda_2x_4+\lambda_3(a_x-f_3(\mathbf{x}))+\lambda_4\left(a_y-f_4(\mathbf{x})\right). \tag{6.41}$$

The necessary/sufficient conditions for optimality of the control input are

$$H_{a_x}=a_x+\lambda_3=0,\ \ H_{a_y}=a_y+\lambda_4=0,\ \ a_x=-\lambda_3,\ \ a_y=-\lambda_4. \tag{6.42}$$

Consequently, it follows from Eqs. 6.40 and 6.42 for the optimal control input that

$$\begin{aligned}\dot{x}_1(t)=x_3(t),\ \ \dot{x}_2(t)=x_4(t),\ \ \dot{x}_3(t)=&-\frac{\mu x_1(t)}{\left(x_1^2+x_2^2\right)^{3/2}}+a_x\\ \equiv -f_3(\mathbf{x})-\lambda_3,\ \ \dot{x}_4(t)=&-\frac{\mu x_2(t)}{\left(x_1^2+x_2^2\right)^{3/2}}+a_y\equiv -f_4(\mathbf{x})-\lambda_4.\end{aligned} \tag{6.43}$$

The corresponding Lagrange multiplier dynamics or co-state equations are

$$H_{x_1} = -\lambda_3 f_{3,x_1}(\mathbf{x}) - \lambda_4 f_{4,x_1}(\mathbf{x}) = -\dot{\lambda}_1,\ H_{x_2} = -\lambda_3 f_{3,x_2}(\mathbf{x}) - \lambda_4 f_{4,x_2}(\mathbf{x})$$
$$= -\dot{\lambda}_2,\ H_{x_3} = -\lambda_1 = \dot{\lambda}_3,\ H_{x_4} = -\lambda_2 = \dot{\lambda}_4. \tag{6.44}$$

Thus, there are four more coupled differential equations to solve which constitute a two-point boundary value problem, TPBVP. The equations are given by

$$\begin{aligned}
&\dot{x}_1(t) = x_3(t),\\
&\dot{x}_2(t) = x_4(t),\\
&\dot{x}_3(t) = -\frac{\mu x_1(t)}{\left(x_1^2 + x_2^2\right)^{3/2}} + a_x \equiv -f_3(\mathbf{x}) - \lambda_3,\\
&\dot{x}_4(t) = -\frac{\mu x_2(t)}{\left(x_1^2 + x_2^2\right)^{3/2}} + a_y \equiv -f_4(\mathbf{x}) - \lambda_4.\\
&-\lambda_1 = \dot{\lambda}_3,\quad -\lambda_2 = \dot{\lambda}_4,\\
&\lambda_3 \frac{\mu\left(2x_1^2 - x_2^2\right)}{\left(x_1^2 + x_2^2\right)^{5/2}} + \lambda_4 \frac{3\mu x_1 x_2}{\left(x_1^2 + x_2^2\right)^{5/2}} = -\dot{\lambda}_1,\\
&\lambda_4 \frac{\mu\left(2x_2^2 - x_1^2\right)}{\left(x_1^2 + x_2^2\right)^{5/2}} + \lambda_3 \frac{3\mu x_1 x_2}{\left(x_1^2 + x_2^2\right)^{5/2}} = -\dot{\lambda}_2,
\end{aligned} \tag{6.45}$$

subject to the boundary conditions

$$x_k(t_0) = x_{k0},\ \lambda_k\left(t_f\right) = \lambda_{kf},\ k = 1, \cdots, 4. \tag{6.46}$$

6.3 Optimal Tracking Trajectory Synthesis

In considering the optimization of spacecraft trajectories for interplanetary transfer, Schwenzfeger (1974) underscored the importance of using the most appropriate regularized state variables in defining the transfer problem. The spacecraft trajectory optimization and synthesis problem can be described as the synthesis of an optimized trajectory that meets certain optimality requirements and satisfies all the physically meaningful constraints that must imposed on the problem, including initial and terminal conditions. In Shirazi et al. (2018), a review of the methods of solving spacecraft trajectory optimization problems as well as the models and objective functions employed was undertaken. Consider the planar orbital motion of a satellite of mass m around a central body with control forces u_1 and u_2 and disturbance forces w_1 and w_2 in the radial and tangential directions, respectively. With the control forces and disturbance forces included, the equations of motion in the radian and tangential direction are

$$m\left(\ddot{r} - r\dot{\theta}^2\right) = u_1 - \mu\left(m/r^2\right) + w_1, \quad m\left(r\ddot{\theta} + 2\dot{r}\dot{\theta}\right) = u_2 + w_2. \tag{6.47}$$

While the nonlinear equations of motion may be directly used to solve nonlinear optimal control problem, the equations of motion will first be linearized before formulating and solving the linear optimal control problem. To linearize the equations of motion, define

$$r = \overline{r} + \delta r, \quad \theta = \omega t + \delta\theta. \tag{6.48}$$

In Eq. 6.48, $\overline{r}$ and ω are assumed to satisfy

$$m\left(\overline{r}\omega^2\right) = \mu\left(m/\overline{r}^2\right) \Rightarrow \omega = \left(1/\overline{r}\right)\sqrt{\left(\mu/\overline{r}\right)}. \tag{6.49}$$

Substituting for r, θ from Eqs. 6.48 in Eq. 6.47,

$$m\left(\delta\ddot{r} - \overline{r}\omega^2 - \delta r\omega^2 - 2\overline{r}\omega\delta\dot{\theta}\right) = u_1 - \mu\frac{m}{\overline{r}^2}\left(1 - 2\frac{\delta r}{\overline{r}}\right) + w_1. \tag{6.50}$$

Equation 6.50 reduces to

$$\begin{aligned} m\left(\delta\ddot{r} - \delta r\omega^2 - 2\overline{r}\omega\delta\dot{\theta}\right) &= u_1 + 2\mu\frac{m}{\overline{r}^3}\delta r + w_1 \\ \Rightarrow m\left(\delta\ddot{r} - 3\delta r\omega^2 - 2\overline{r}\omega\delta\dot{\theta}\right) &= u_1 + w_1. \end{aligned} \tag{6.51}$$

From the second of Eqs. 6.47,

$$m\left(\overline{r}\delta\ddot{\theta} + 2\omega\delta\dot{r}\right) = u_2 + w_2. \tag{6.52}$$

Combining Eqs. 6.51 and 6.52, in state space, one has the linearized equations:

$$\dot{\mathbf{x}} = \begin{bmatrix} 0 & 0 & 1 & 0 \\ 0 & 0 & 0 & 1 \\ 3\omega^2 & 0 & 0 & 2\overline{r}\omega \\ 0 & 0 & -2\omega/\overline{r} & 0 \end{bmatrix}\mathbf{x} + \frac{1}{m\overline{r}}\begin{bmatrix} 0 & 0 \\ 0 & 0 \\ \overline{r} & 0 \\ 0 & 1 \end{bmatrix}\left(\begin{bmatrix} u_1 \\ u_2 \end{bmatrix} + \begin{bmatrix} w_1 \\ w_2 \end{bmatrix}\right), \quad \mathbf{x} = \begin{bmatrix} \delta r \\ \delta\theta \\ \delta\dot{r} \\ \delta\dot{\theta} \end{bmatrix}. \tag{6.53}$$

The corresponding measurement equation, given the measurement noise v, is,

$$\mathbf{y} = [0 \ \ 1 \ \ 0 \ \ 0]\mathbf{x} + v. \tag{6.54}$$

When one is interested in the problem of finding the control forces,

$$\mathbf{u} \equiv [u_1 \quad u_2]^T = \mathbf{u}(t), \ \ t_0 \le t \le t_f, \tag{6.55}$$

the state vector time history is sought, such that it minimizes the cost functional:

$$J = \frac{1}{2}\int_0^{t_f} \mathbf{u}^T\mathbf{R}\mathbf{u}dt + 0.5(\mathbf{x}(t) - \mathbf{x}_d)^T\mathbf{Q}_f(\mathbf{x}(t) - \mathbf{x}_d)\big|_{t=t_f} = \frac{1}{2}\int_0^{t_f} \mathbf{u}^T\mathbf{R}\mathbf{u}dt + \Phi\{\mathbf{x}(t)\}\big|_{t=t_f}, \tag{6.56}$$

subject to Eqs. 6.53 and 6.55. Following Bryson Jr. and Ho (1969), one may construct the corresponding Hamiltonian function, which is

$$H = \frac{1}{2}\mathbf{u}^T\mathbf{R}\mathbf{u} + \boldsymbol{\lambda}_{\mathbf{x}}^T(\mathbf{A}\mathbf{x}) + \boldsymbol{\lambda}_{\mathbf{x}}^T(\mathbf{B}\mathbf{u}), \tag{6.57}$$

where $\mathbf{R}$ is a positive definite and symmetric weighting matrix. In our case, the relative costs of applying an acceleration in any direction are the same, and for this reason, we may choose $\mathbf{R} \equiv \mathbf{I}_{2\times 2}$ to be the 2×2 identity matrix. The corresponding co-state differential equations are

$$\frac{d}{dt}\boldsymbol{\lambda}_{\mathbf{x}}^T(t) = -\frac{\partial H}{\partial \mathbf{x}} = -\boldsymbol{\lambda}_{\mathbf{x}}^T\mathbf{A}. \tag{6.58}$$

It may be observed that the time history of the co-state vector $\boldsymbol{\lambda}_{\mathbf{x}}$ is independent of the control forces. The optimal control is obtained from

$$\frac{\partial H}{\partial \mathbf{u}} = \mathbf{u}^T\mathbf{R} + \boldsymbol{\lambda}_{\mathbf{x}}^T\mathbf{B} = 0 \Rightarrow \mathbf{u} = -\mathbf{R}^{-1}\mathbf{B}^T\boldsymbol{\lambda}_{\mathbf{x}}. \tag{6.59}$$

For the co-state boundary conditions, one has

$$\boldsymbol{\lambda}_{\mathbf{x}}(t_f) = \frac{\partial\Phi\{\mathbf{x}(t)\}}{\partial \mathbf{x}}\bigg|_{t=t_f} = \mathbf{Q}_f(\mathbf{x}(t_f) - \mathbf{x}_d). \tag{6.60}$$

Equations 6.53 and 6.58 are numerically integrated with the appropriate boundary conditions, and once the control is found from Eq. 6.59, Eq. 6.53 is used to define the optimal state vector.

In a boundary value problem, the goal is to find a solution to an ordinary differential equation that also satisfies certain specified boundary conditions. The boundary conditions specify a relationship between the values of the solution at two or more locations in the interval of integration. In the simplest case, the boundary conditions apply at either the initial point or at the final point of the interval, but not at both. In a more general case, the boundary conditions are specified at both the initial and final points.

In Appendix 6.1, a general numerical method of solving a two-point boundary value problem, based on the Lagrange interpolation polynomial, is presented. MATLAB®, a commercial product for solving matrix and other equations, also provides two boundary value problem solvers bvp4c.m and bvp5c.m which are designed to handle systems of ordinary differential equations of the form

$$y'(x) = f(y, x). \tag{6.61}$$

where x is the independent variable, y is the dependent variable and $y'(x)$ represents the derivative of y with respect to the independent variable x and is also written as dy/dx. An important part of the process of using the two boundary value problem solvers is providing an initial guess for the required solution. The quality of this guess can be critical for the solver performance and even for a successful computation. The solver bvp4c.m is a finite difference code that implements the three-stage Lobatto IIIa formula. The three-stage Lobatto IIIa formula belongs to the Lobatto family of collocation formulae, where a collocation polynomial provides a continuous solution that is fourth-order accurate uniformly in the interval of integration. Mesh selection and error control are based on the residual of the continuous solution. The collocation technique uses a mesh of points to divide the interval of integration into subintervals. The solver determines a numerical solution by solving a global system of algebraic equations resulting from the boundary conditions and the collocation conditions imposed on all the subintervals. The solver then estimates the error of the numerical solution on each subinterval. If the solution does not satisfy the tolerance criteria, then the solver adapts the mesh and repeats the process. The user must provide the points of the initial mesh, as well as an initial approximation of the solution at the mesh points.

The solver bvp5c.m is a finite difference code that implements the four-stage Lobatto IIIa formula. This is another collocation formula in the Lobatto family where a collocation polynomial provides a continuous solution that is fifth-order accurate uniformly over the interval of integration. The formula is implemented as an implicit Runge-Kutta formula. bvp5cm solves the algebraic equations directly, whereas bvp4cm uses analytical condensation. bvp4c.m handles unknown parameters directly, while bvp5c.m augments the system with trivial differential equations for unknown parameters. More information about these solvers may be found in the documentation associated with MATLAB-SIMULINK.

6.3.1 Application to Spacecraft Absolute Motion

Consider the dynamical equations of a multi-body planetary system. For a general n-body problem, the total force $\boldsymbol{f}_i$ acting upon the planetary body of mass m_i, due to the other $n - 1$ planetary bodies with masses m_j, is

$$\boldsymbol{f}_i = G \sum_{\substack{j=1 \\ j \neq i}}^{n} \frac{m_i m_j}{r_{ij}^3} \left(\boldsymbol{r}_j - \boldsymbol{r}_i \right). \tag{6.62}$$

The equation of motion of the ith body is (Battin, 1999)

$$m_i \frac{d^2 \boldsymbol{r}_i}{dt^2} = \boldsymbol{f}_i + \boldsymbol{F}_{p_i} = G \sum_{\substack{j=1 \\ j \neq i}}^{n} \frac{m_i m_j}{r_{ij}^3} \left(\boldsymbol{r}_j - \boldsymbol{r}_i \right) + \boldsymbol{F}_{p_i}. \tag{6.63}$$

In Eq. 6.62, the force $\boldsymbol{F}_{p_i}$ is the sum total of the perturbation and disturbance forces acting on the satellite. By convention, it is assumed $j = 1$ represents the satellite, $j = 3$ represents the planet Earth, $j = 3$ represents the target planet (assuming that the mission is an interplanetary one) or Moon and $j = 4$ represents the Sun and so on. The satellite is modelled as a point mass so all gravity gradient and perturbation forces due to the satellite's finite volume are neglected. For the satellites motion, one has

$$\frac{d^2 \boldsymbol{r}_1}{dt^2} = G \sum_{\substack{j=1 \\ j \neq 1}}^{N} \frac{m_j}{r_{1j}^3} \left(\boldsymbol{r}_j - \boldsymbol{r}_1 \right), \quad n < N. \tag{6.64}$$

The influence of the satellite on the planetary bodies may be ignored. Hence,

$$\frac{d^2 \boldsymbol{r}_i}{dt^2} = G \sum_{\substack{j=1 \\ j \neq i}}^{N} \frac{m_j}{r_{ij}^3} \left(\boldsymbol{r}_j - \boldsymbol{r}_i \right), \quad n < N, i = 2, \ldots, N. \tag{6.65}$$

The equation of motion of the satellite may be expressed in terms of the control force or in terms of the thrust magnitude T, which is the magnitude of the specific thrust or thrust per unit mass applied to the satellite and $\boldsymbol{u}$ which defines the direction of the thrust force:

$$\frac{d\boldsymbol{r}_1}{dt}=\boldsymbol{v}_1,\quad \frac{d\boldsymbol{v}_1}{dt}=G\sum_{j=2}^{N}\frac{m_j}{r_{1j}^3}\left(\boldsymbol{r}_j-\boldsymbol{r}_1\right)+\boldsymbol{F}_{c_1},\quad \boldsymbol{F}_{c_1}=T\boldsymbol{u},\quad \frac{d\Delta \boldsymbol{v}}{dt}=T. \tag{6.66}$$

The direction of the thrust force can be specified in terms of two angles, an azimuth angle α and a declination β, as

$$\boldsymbol{u}=\left[\sin\alpha\cos\beta\cos\alpha\cos\beta\sin\beta\right]. \tag{6.67}$$

The rate of change of mass of the satellite may be expressed in terms of the specific impulse and the total magnitude of the thrust as

$$dm/dt=-mT/\left(I_{sp}g_0\right). \tag{6.68}$$

The optimal control problem may now be stated. One is interested in the problem of finding the optimum steering control input, $\boldsymbol{u}(t)$, $t_0 \le t \le t_f$, the thrust direction time history which minimizes a cost functional, subject to the constraint dynamical equations of motion of the satellite (6.66, 6.67 and 6.68) defined above. The cost function is defined as

$$J=\left|\boldsymbol{r}_1(t)-\boldsymbol{r}_{1f}\right|\Big|_{t=t_f}=\Phi\{\boldsymbol{r}_1(t)\}\big|_{t=t_f}. \tag{6.69}$$

The time varying parameter vectors $\boldsymbol{r}_i$ for $i=2,\ldots,N$, in the dynamical equations of motion of the satellite (6.66, 6.67 and 6.68) satisfy Eq. 6.65.

The equations may also be expressed in terms of the relative position from the target planet. Thus, given that

$$\boldsymbol{r}_{1j}=\boldsymbol{r}_1-\boldsymbol{r}_j,\quad \boldsymbol{r}_{1j_t}=\boldsymbol{r}_1-\boldsymbol{r}_{j_t},\quad \frac{d\boldsymbol{r}_1}{dt}=\boldsymbol{v}_1,\quad \frac{d\boldsymbol{v}_1}{dt}=G\sum_{j=2}^{N}m_j\frac{\left(\boldsymbol{r}_j-\boldsymbol{r}_1\right)}{r_{1j}^3}+\boldsymbol{F}_{c_1}, \tag{6.70}$$

the equations of relative motion with respect to the target planet are

$$\frac{d\boldsymbol{r}_{1j_t}}{dt}=\boldsymbol{v}_{1j_t},\quad \frac{d\boldsymbol{v}_{1j_t}}{dt}=G\sum_{j=2}^{N}m_j\frac{\left(\boldsymbol{r}_j-\boldsymbol{r}_1\right)}{r_{1j}^3}-G\sum_{\substack{j=2\\ j\ne j_t}}^{N}m_j\frac{\boldsymbol{r}_{1j_t}}{r_{1j_t}^3}+\boldsymbol{F}_{c_1}. \tag{6.71}$$

In the vicinity of the target body ($j=3\equiv j_t$), given that $j=4$ is the central body,

$$\frac{d\boldsymbol{v}_{1j_t}}{dt} = -Gm_J\frac{\boldsymbol{r}_{1j_t}}{r_{1j_t}^3} + \boldsymbol{F}_{c_1} - G\sum_{\substack{j=2\\ j\neq 1,J}}^{N} m_j\left\{\frac{(\boldsymbol{r}_1-\boldsymbol{r}_j)}{r_{1j}^3} - \frac{(\boldsymbol{r}_{j_t}-\boldsymbol{r}_j)}{r_{j_tj}^3}\right\}, \tag{6.71}$$

In the SOI of the target,

$$\sum_{\substack{j=2\\ j\neq 1,J}}^{N} m_j\left\{\frac{(\boldsymbol{r}_1-\boldsymbol{r}_j)}{r_{1j}^3} - \frac{(\boldsymbol{r}_{j_t}-\boldsymbol{r}_j)}{r_{j_tj}^3}\right\} \cong 0, \tag{6.72}$$

In the vicinity of the target body ($j = 3 \equiv j_t$), given that $j = 4$ is the central body, one has the approximation

$$\frac{d\boldsymbol{r}_{1j_t}}{dt} = \boldsymbol{v}_{1j_t}, \quad \frac{d\boldsymbol{v}_{1j_t}}{dt} \cong -Gm_J\frac{\boldsymbol{r}_{1J}}{r_{1J}^3} + \boldsymbol{F}_{c_1}, \tag{6.73}$$

The transversality conditions ensure that the initial and final states are selected optimally within the feasible regions of the states. For the transversality conditions, one may write

$$\lambda_{r_1}(t_f) = \left.\frac{\partial\Phi\{\boldsymbol{r}_1(t)\}}{\partial r_1}\right|_{t=t_f} = sign(\boldsymbol{r}_1(t_f) - \boldsymbol{r}_{1f}), \quad \lambda_{v_1}(t_f) = \left.\frac{\partial\Phi\{\boldsymbol{r}_1(t)\}}{\partial \boldsymbol{v}_1}\right|_{t=t_f} = 0,$$
$$\lambda_m(t_f) = 0. \tag{6.74}$$

The problem of synthesizing an optimal trajectory to Mars was considered earlier by Miele and Wang (1997). As an example, a typical optimal trajectory from Earth to the planet Mars is shown in Fig. 6.2, computed with only the gravitational effects of the Sun, Earth, Mars and the satellite included. Figure 6.2 shows the same trajectory with the influence of Venus also included, showing the relatively negligible influence of Venus, as the spacecraft does not approach close to it (Fig. 6.3).

6.3.2 Application to Spacecraft Relative Motion

To begin with, the linear state space representation of the linearized HCW may be derived from Eqs. 1.101, 1.102 and 1.103 in Chap. 1, in terms of the mean rate of motion n, and may be expressed in matrix form, in terms of the traditional orbital states, as

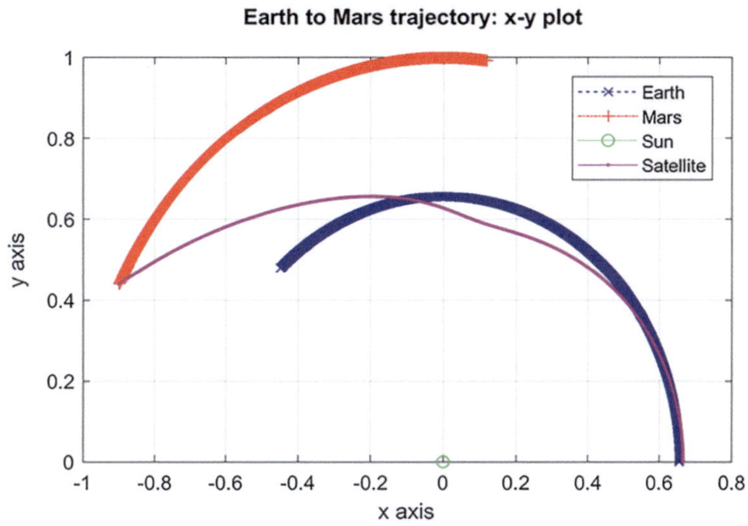

Fig. 6.2 Typical optimal transfer orbit from Earth to Mars in a heliocentric frame with only the four principal bodies

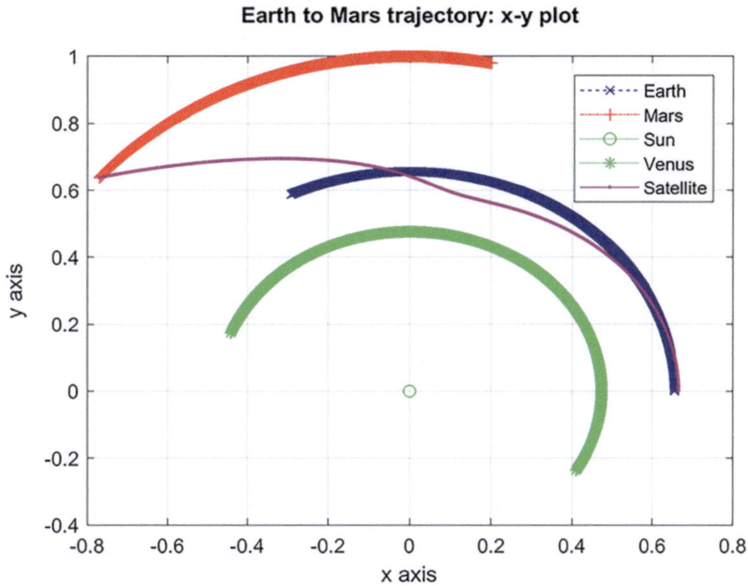

Fig. 6.3 Typical optimal transfer orbit from Earth to Mars in a heliocentric frame with five principal bodies

$$\frac{d\mathbf{q}}{dt} = \mathbf{A}\mathbf{q} + \mathbf{B}\mathbf{u}. \tag{6.75}$$

Matrices **A** and **B** may be appropriately defined as

$$\mathbf{A} = \begin{bmatrix} 0 & 0 & 0 & 1 & 0 & 0 \\ 0 & 0 & 0 & 0 & 1 & 0 \\ 0 & 0 & 0 & 0 & 0 & 1 \\ 3n^2 & 0 & 0 & 0 & 2n & 0 \\ 0 & 0 & 0 & -2n & 0 & 0 \\ 0 & 0 & -n^2 & 0 & 0 & 0 \end{bmatrix}, \quad \mathbf{B} = \begin{bmatrix} 0 & 0 & 0 \\ 0 & 0 & 0 \\ 0 & 0 & 0 \\ 1 & 0 & 0 \\ 0 & 1 & 0 \\ 0 & 0 & 1 \end{bmatrix}. \tag{6.76}$$

In Eq. 6.75, the state vector and the control accelerations are, respectively, defined as

$$\mathbf{q} = \begin{bmatrix} x & y & z & v_x & v_y & v_z \end{bmatrix}^T, \tag{6.77}$$

$$\mathbf{u} = \begin{bmatrix} a_{Tx} & a_{Ty} & a_{Tz} \end{bmatrix}^T. \tag{6.78}$$

One is interested in the problem of finding the optimal control accelerations:

$$\mathbf{u} = \mathbf{u}(t), \quad t_0 \leq t \leq t_f. \tag{6.79}$$

One could seek the state vector time history, such that it minimizes the cost functional:

$$J = \frac{1}{2}\int_0^\infty \left(\mathbf{u}^T\mathbf{R}\mathbf{u}\right)dt + 0.5(\mathbf{q}(t) - \mathbf{q}_d)^T\mathbf{Q}_f(\mathbf{q}(t) - \mathbf{q}_d)\big|_{t=t_f} = \frac{1}{2}\int_0^\infty \times \left(\mathbf{u}^T\mathbf{R}\mathbf{u}\right)dt + \Phi\{\mathbf{q}(t)\}\big|_{t=t_f}, \tag{6.80}$$

subject to Eq. 6.75. Following Bryson Jr. and Ho (1969), to choose the weighting matrix **R**, one may construct the corresponding Hamiltonian function, which is

$$H = (1/2)\mathbf{u}^T\mathbf{R}\mathbf{u} + \lambda_{\mathbf{q}}^T(\mathbf{A}\mathbf{q}) + \lambda_{\mathbf{q}}^T(\mathbf{B}\mathbf{u}), \tag{6.81}$$

where **R** is a positive definite and symmetric matrix. In our case, the relative costs of applying an acceleration in any direction are the same, and for this reason, we may choose $\mathbf{R} \equiv r\mathbf{I}_{3\times 3}$ to be the 3×3 identity matrix and r is a scalar. The corresponding co-state differential equations are

Table 6.2 Orbital elements of Vestas

	a(in AU)	e	i	Ω	ω	M
Vestas	2.36151	0.098758	6.39234°	103.85136°	151.19853°	20.86384°

$$\frac{d}{dt}\boldsymbol{\lambda}_{\mathbf{q}}^{T}(t) = -\frac{\partial H}{\partial \mathbf{q}} = -\boldsymbol{\lambda}_{\mathbf{q}}^{T}\mathbf{A}. \tag{6.82}$$

It may be observed that the time history of the co-state vector $\boldsymbol{\lambda}_{\mathbf{q}}$ is independent of the control accelerations. The optimal control is obtained from

$$\partial H/\partial \mathbf{u} = \mathbf{u}^{T}\mathbf{R} + \boldsymbol{\lambda}_{\mathbf{q}}^{T}\mathbf{B} = 0 \Rightarrow \mathbf{u} = -\mathbf{R}^{-1}\mathbf{B}^{T}\boldsymbol{\lambda}_{\mathbf{q}}. \tag{6.83}$$

For the co-state boundary conditions, one has

$$\boldsymbol{\lambda}_{\mathbf{q}}(t_f) = \left.\frac{\partial \Phi\{\mathbf{q}(t)\}}{\partial \mathbf{q}}\right|_{t=t_f} = \mathbf{Q}_f(\mathbf{q}(t_f) - \mathbf{q}_d). \tag{6.84}$$

Equations 6.59 and 6.75 are numerically integrated, and once the control is found from Eq. 6.83, Eq. 6.75 is used to define the optimal state vector.

An example demonstrates the application of the optimal trajectories is of spacecraft flying in close proximity of asteroid in a circular orbit. The asteroid Vestas which is in a near circular orbit is assumed to be in a circular orbit. The initial position and velocity of the asteroid Vestas are obtained from the initial classical orbital elements and reproduced in Table 6.2.

The control law was synthesized using the HCW equations in terms of the traditional orbital states. All the simulations were initially carried out in MATLAB using the Gauss perturbation equations of motion based on the modified equinoctial elements defined in Chap. 2, Eq. 2.19 (see, e.g. Walker et al., 1985; Kechichian, 1990), to illustrate the influence of the feedback control thrust. The control inputs over the powered segment were determined using the optimal trajectory synthesis methodology outlined earlier. The relative orbit data obtained from the optimization was used to smoothly interpolate the satellite orbit from its initial to its final position. The post-powered phase of the orbit was obtained by continuing with the integration of the four body equations beyond the powered phase, to verify that the satellite and the asteroid traverse the same orbit.

Initially, the satellite is assumed to be orbiting the Sun within the SOI of the asteroid Vestas. Following the application of an optimum "Δv" impulse, the satellite is on an in-plane transfer orbit. In Fig. 6.4 are shown the optimum orbital states in the planar transfer of the spacecraft from its original orbit to a point on the desired final orbit. This optimum transfer takes place in 13.2717 days. The optimal trajectory synthesis problem is scaled and transformed to an in-plane transfer problem. The corresponding orbital states and control inputs are shown in Fig. 6.4. The optimum position and velocity coordinates in the LVLH coordinate frame of the reference

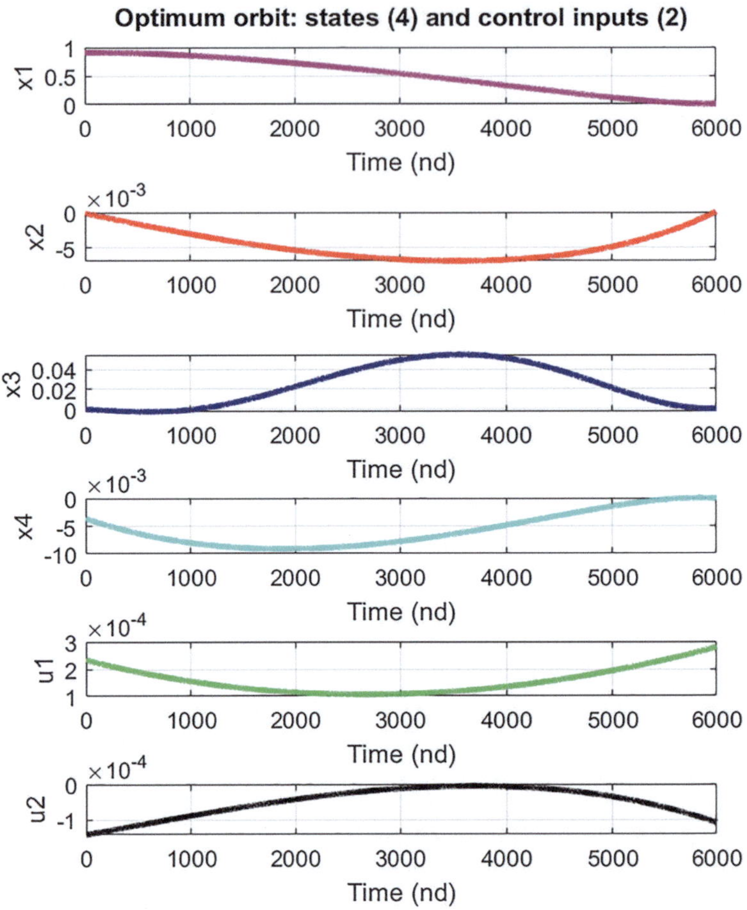

Fig. 6.4 In-plane Cartesian position and velocity orbital states and control inputs for the optimum trajectory synthesis

Vestas orbit are shown in Fig. 6.5. It is seen that all the position and velocities reach the prescribed final position and velocity as the relative position coordinates and velocity components are all zero.

In Fig. 6.6 is shown the segment of the optimum transfer orbit in three-dimensional space. At the end of the transfer, a feedback control law is used to precisely track the desired or prescribed orbit of the satellite. The methodology of feedback control to precisely track the desired or prescribed orbit of the satellite will be discussed in Sect. 6.3.5. The position and velocity coordinates are then converted to the global space fixed coordinate frame, and the orbit frame is define in terms of the modified equinoctial elements (MEE), usually labelled as p, f, g, h, k, L. The evolution of the MEE during the optimal transfer is shown in Fig. 6.7. It is seen from these orbital elements that the orbit is now as desired.

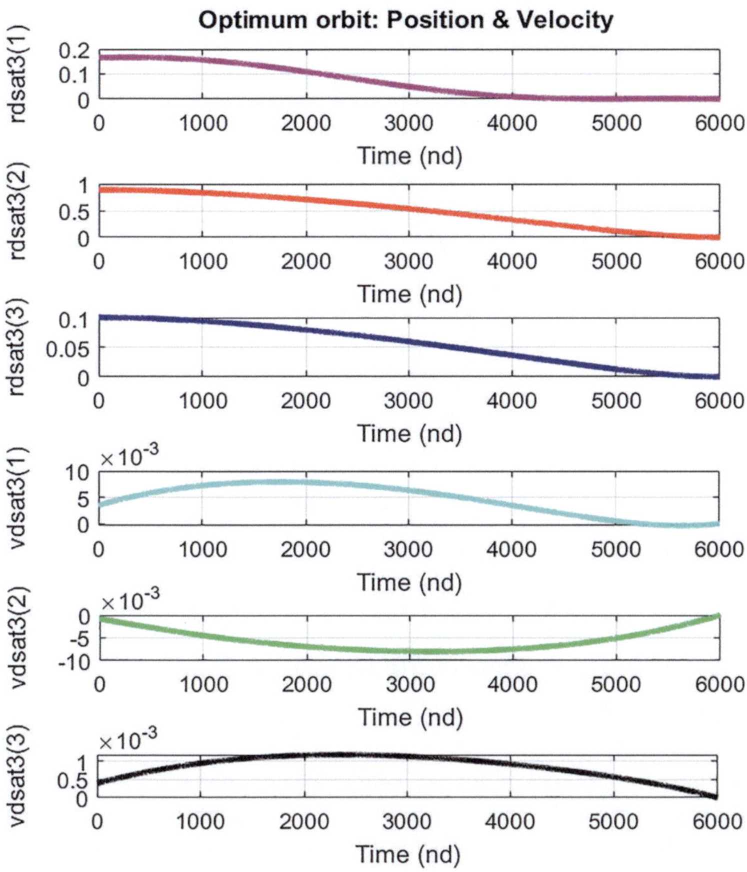

Fig. 6.5 The optimum position and velocities in LVLH coordinates of the reference Vestas orbit

Fritz and Turkoglu (2017) have also considered the synthesis of optimal trajectories for missions to multiple asteroids including the benefits of the influence gravity assist, enroute to the target asteroids. Englander et al. (2017) have adopted a multidisciplinary mission optimization technique to solve the problem of a mission to small celestial bodies, including the consideration of classical trajectory design, the choice of spacecraft power and propulsion systems and also the scientific value of the targets.

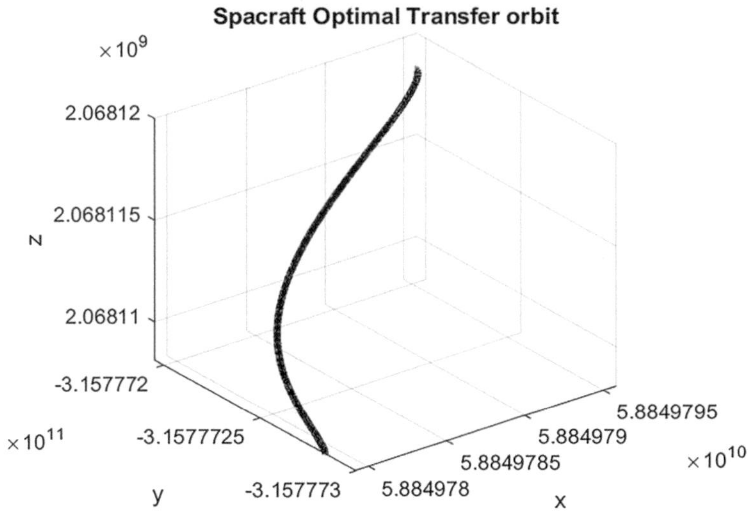

Fig. 6.6 Position and velocity coordinate time histories of the optimum transfer orbit in three-dimensional space

6.3.3 Application to Spacecraft Motion in Spherical Coordinates

In this section, the problem of constructing reference orbital trajectories in a spherical coordinate frame is discussed in some detail. The formulation is based on Vepa (2019). Thus, the translational kinematics of a space vehicle in a typical orbit is first considered. Consider a 3-1-3 sequence of Euler angles, where the angles are θ the longitude, a 90° rotation about the "1" axis and ϕ the latitude. The corresponding transformations are

$$T_3(\theta)=[\cos\,\theta \sin\theta 0-\ \sin\theta\cos\theta 0001],\ \ T_1(90\,^\circ)=\begin{bmatrix}1 & 0 & 0\\ 0 & 0 & 1\\ 0 & -1 & 0\end{bmatrix}, \qquad (6.85)$$

$$T_3(\phi)=[\cos\,\phi \sin\phi 0-\ \sin\phi\cos\phi 0001].$$

The Earth-fixed to body-fixed transformation is given by

$$T_{BE}=T_3(\theta)T_1(90\,^\circ)T_3(\phi),$$
$$T_{BE}=[\cos\,\theta\cos\phi\cos\theta\sin\phi\sin\theta-\ \sin\theta\cos\phi-\ \sin\theta\sin\phi\cos\theta\sin\phi-\ \cos\phi 0]. \qquad (6.86)$$

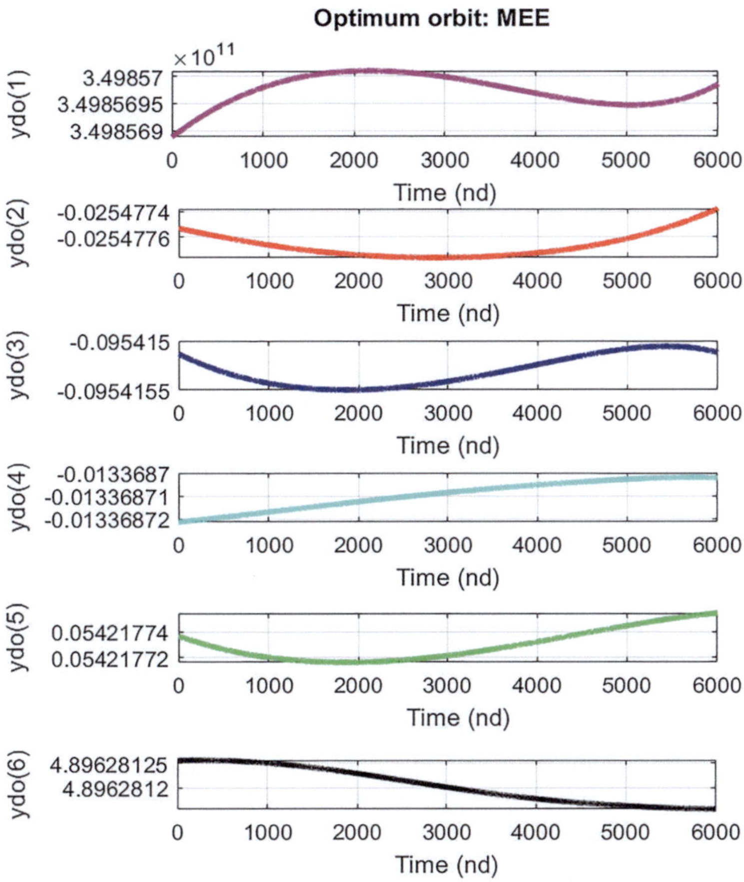

Fig. 6.7 The evolution of the MEE during the optimal transfer. The first is the semi-latus rectum and the last is the true anomaly

Combined with the radial position r, θ the longitude and ϕ the latitude, constitute a set of spherical coordinates. The angular velocity components for the above 3-1-3 sequence are

$$\boldsymbol{\omega} = \left[\dot{\theta}\sin\phi \quad \dot{\theta}\cos\phi \quad \dot{\phi}\right]^T. \tag{6.87}$$

Now let γ be the flight path angle, ψ be the heading or yaw angle and v the magnitude of the vehicle's velocity. Then the kinematic equations relating the velocity components to the rate of change of position in the spherical coordinates are

$$\dot{r} = v\sin\gamma, \quad r\dot{\phi} = v\cos\gamma\sin\psi, \quad r\cos\phi\dot{\theta} = v\cos\gamma\cos\psi. \tag{6.88}$$

Equation. 6.88 is non-singular as long as the variable ϕ satisfies $-\pi/2 < \phi < \pi/2$; i.e. the vehicle does not overfly the poles.

For completeness, the translational kinetics of the space vehicle orbiting the Earth in a low Earth orbit is also briefly presented. Consider first a nonrotating planet. The local acceleration due to gravitation, positive down, which is also the negative radial component may be defined as

$$g = \mu/r^2. \tag{6.89}$$

The kinetic equations are expressed as

$$\dot{v} = a_s - g \sin\gamma, \tag{6.90}$$

$$v \cos\gamma \, \dot{\psi} = a_w - \left(v^2/r\right) \cos^2\gamma \cos\psi \tan\phi, \tag{6.91}$$

$$v\dot{\gamma} = a_n - \left(g - \left(v^2/r\right)\right) \cos\gamma, \tag{6.92}$$

$$\dot{m} = -T/v_e, \tag{6.93}$$

Equation 6.93 represents the rate of change of mass of the vehicle due to the expulsion of mass with relative velocity v_e by the propulsion system. In Eqs. 6.90, 6.91 and 6.92, the aerodynamic forces in the tangential, normal and binormal directions are, respectively, defined by

$$\begin{aligned} ma_s &= T\cos\alpha - D, \quad ma_w = (L + T\sin\alpha)\sin\phi_r, \quad ma_n \\ &= (L + T\sin\alpha)\cos\phi_r. \end{aligned} \tag{6.94}$$

In Eq. 6.94, ϕ_r is the bank angle, T is the body fixed thrust, L is the aerodynamic lift which is linearly dependent on the elevator angle, D is the aerodynamic drag and α is the angle of attack, which is equal to the pitch plus the flight path angles, $\alpha = \theta_p + \gamma$. The Euler angles defining the attitude of the vehicle are ψ, θ_p, ϕ_r (yaw, pitch roll or 3-2-1 sequence) which can be obtained from the quaternion components. Given the Euler parameters or quaternion components,

$$\eta = \cos(\phi/2); \ \boldsymbol{\varepsilon} = \left[\varepsilon_1 \quad \varepsilon_2 \quad \varepsilon_3\right]^T = \mathbf{n}\sin(\phi/2), \tag{6.95}$$

the transformation relating an orbiting reference frame to a body fixed frame is given by

$$\mathbf{T}_{BR}(\mathbf{q}) = \{T_{ij}\}$$
$$= \begin{bmatrix} \eta^2 + \varepsilon_1^2 - \varepsilon_2^2 - \varepsilon_3^2 & 2(\varepsilon_1\varepsilon_2 + \varepsilon_3\eta) & 2(\varepsilon_1\varepsilon_3 - \varepsilon_2\eta) \\ 2(\varepsilon_1\varepsilon_2 - \varepsilon_3\eta) & \eta^2 - \varepsilon_1^2 + \varepsilon_2^2 - \varepsilon_3^2 & 2(\varepsilon_2\varepsilon_3 + \varepsilon_1\eta) \\ 2(\varepsilon_1\varepsilon_3 + \varepsilon_2\eta) & 2(\varepsilon_2\varepsilon_3 - \varepsilon_1\eta) & \eta^2 - \varepsilon_1^2 - \varepsilon_2^2 + \varepsilon_3^2 \end{bmatrix}. \tag{6.96}$$

Thus, it follows that

$$\begin{aligned} &t = Trace(T_{ij}) = 3\eta^2 - \varepsilon_1^2 - \varepsilon_2^2 - \varepsilon_3^2 = 4\eta^2 - 1; \;\; r = \sqrt{1+t} = 2\eta; \\ &\eta = r/2 = \sqrt{1+t}/2s = 1/2r = 1/(4\eta); \;\; \varepsilon_1 = (T_{23} - T_{32})s, \\ &\varepsilon_2 = (T_{31} - T_{13})s, \;\; \varepsilon_3 = (T_{12} - T_{21})s. \end{aligned} \tag{6.97}$$

Hence, the quaternion components may be related to the 3-2-1 Euler angle sequence components from the elements of the attitude transformation matrix.

In the case of a rotating planet with the atmosphere rotating with it, one has

$$\dot{v} = a_s + g_s + cf_v, \;\; \dot{m} = -T/v_e, \tag{6.98}$$

$$v\cos\gamma\,\dot{\psi} = a_w + g_w - \left(v^2/r\right)\cos^2\gamma\cos\psi\tan\phi + \left(cf_\psi + v\cos\gamma co_\psi\right), \tag{6.99}$$

$$v\dot{\gamma} = a_n + \left(g_n + \left(v^2/r\right)\right)\cos\gamma + \left(cf_\gamma + vco_\gamma\right). \tag{6.100}$$

The centrifugal (*cf*) and Coriolis (*co*) terms due to the planet rotation rate (Ω) are

$$cf_v = \Omega^2 r\cos\phi(\sin\gamma\cos\phi - \cos\gamma\sin\phi\sin\psi), \tag{6.101}$$

$$cf_\psi = -\Omega^2 r\sin\phi\cos\phi\cos\psi, \tag{6.102}$$

$$cf_\gamma = -\Omega^2 r\cos\phi(\cos\gamma\cos\phi + \sin\gamma\sin\phi\sin\psi), \tag{6.103}$$

$$co_\psi = 2\Omega(\tan\gamma\cos\phi\sin\psi - \sin\phi), \;\; co_\gamma = 2\Omega\cos\phi\cos\psi. \tag{6.104}$$

In this problem, the only constraints applied are the kinematic equations, as the velocity vector itself is assumed to be a control variable, so as to be able to define a reference trajectory for the space vehicle. Generally, when constraints need to be applied to the kinematic variables, it is often convenient to express the kinematics in a frame where the constraints reduce to equality relations applied to the variables. In deriving the optimal velocity control input, for the kinematic equations, Eq. 6.88 is used. Thus, the Hamiltonian is defined by

$$H = \lambda_r v \sin\gamma + \lambda_\theta \frac{v\cos\gamma\cos\psi}{r\cos\varphi} + \lambda_\varphi \frac{v\cos\gamma\sin\psi}{r}. \tag{6.105}$$

The differential equations that the co-states must satisfy are

$$\dot{\lambda}_r = -\frac{\partial H}{\partial r} = \lambda_\theta \frac{v\cos\gamma\cos\psi}{r^2\cos\varphi} + \lambda_\varphi \frac{v\cos\gamma\sin\psi}{r^2}, \tag{6.106}$$

$$\dot{\lambda}_\theta = -\frac{\partial H}{\partial \theta} = 0,\ \dot{\lambda}_\varphi = -\frac{\partial H}{\partial \varphi} = -\lambda_\theta \frac{v\cos\gamma\cos\psi}{r(\cos\varphi)^2}\sin\varphi. \tag{6.107}$$

The control input is expressed in terms of the fixed magnitude $v = |\mathbf{v}|_{\max}$ and the direction of the velocity vector. Thus, the optimality conditions for the control parameters are

$$\frac{\partial H}{\partial \gamma} = \lambda_r v\cos\gamma - \lambda_\theta \frac{v\sin\gamma\cos\psi}{r\cos\varphi} - \lambda_\varphi \frac{v\sin\gamma\sin\psi}{r} = 0, \tag{6.108}$$

$$\frac{\partial H}{\partial \psi} = -\lambda_\theta \frac{v\cos\gamma\sin\psi}{r\cos\varphi} + \lambda_\varphi \frac{v\cos\gamma\cos\psi}{r} = 0. \tag{6.109}$$

Hence, co-states satisfy the relation

$$v[\lambda_r \quad \lambda_\theta \quad \lambda_\varphi] = |v[\lambda_r \quad \lambda_\theta \quad \lambda_\varphi]|[\sin\ \gamma r\cos\phi\cos\gamma\cos\psi r\cos\gamma\sin\psi]. \tag{6.110}$$

Thus, the optimal control parameters defining the direction of the velocity vector are given by

$$\begin{aligned}&[\sin\ \gamma\cos\gamma\cos\psi\cos\gamma\sin\psi]^T\\ &= \begin{bmatrix} 1 & 0 & 0 & \\ 0 & r\cos & & \phi 000r \end{bmatrix}^{-1} \frac{[\lambda_r \quad \lambda_\theta \quad \lambda_\varphi]^T}{|[\lambda_r \quad \lambda_\theta \quad \lambda_\varphi]|}.\end{aligned} \tag{6.111}$$

For the co-state boundary conditions, one has

$$\begin{aligned}&\lambda_r(t_f) = (r(t) - r_d)|_{t=t_f},\ \lambda_\theta(t_f) = (\theta(t) - \theta_d)|_{t=t_f},\\ &\lambda_\phi(t_f) = (\phi(t) - \phi_d)|_{t=t_f}.\end{aligned} \tag{6.112}$$

Formulations in spherical coordinates based on the kinematics alone are used to construct reference trajectories for re-entry vehicles. This means that one does not need to precisely define the aerodynamic lift and drag forces acting on the vehicle a priori, which are generally difficult to predict. Once the reference velocity is defined, the kinetic equations may be used to define the optimal thrust required. Two examples of other applications of optimal control for defining the reference re-entry trajectory for apace vehicle may be found in Yang et al. (2018) and in Zhao et al. (2023).

6.3.4 Application to Spacecraft Motion in a CR3BP Setting

Nikolichev and Sesyukalov (2023) and Pan et al. (2022) have proposed optimal trajectory synthesis methods in the context of associated multi-body problems. There are many approaches for using the CR3BP to synthesize an interplanetary trajectory. Ikawa (1988) derived a coplanar, three-body trans-Earth-lunar and return trajectory simulation methodology. Miele and Mancuso (2001) have solved for the optimal trajectory to the Moon from the Earth and back using a three-body approach. Eapen and Sharma (2014) developed a transfer trajectory to Mars using the Lagrangian points of the spacecraft-Sun-Earth three-body system and the spacecraft-Sun-Mars three-body system. The complete trajectory consists of three phases, a trajectory from a near-Earth circular parking orbit to a halo orbit around Sun-Earth Lagrangian point L2, a transfer trajectory from Sun-Earth L2 halo orbit to Sun-Mars L1 halo orbit and a Sun-Mars L1 halo orbit to a circular orbit around Mars. The intermediate transfer arc is designed using a two-body Lambert's problem. A novel, optimal, three-body approach with ballistic capture by Mars is proposed by Mingotti et al. (2011). However, it is not possible to construct a ballistic capture trajectory in the CR3BP setting as it requires at least an elliptic restricted three-body model.

In this section, an example of the synthesis of an optimal relative motion trajectory, based on the linearized equations of motion, is considered. The linearized equations of motion of a spacecraft in the vicinity of a colinear equilibrium point were defined by Eqs. 1.160 and 1.161, in terms of the single parameter, B_L, introduced by Farquhar (1968). The equations of motion may be expressed in matrix form, in terms of the traditional orbital states, in the same form as Eq. 6.75 but with the matrices **A** and **B** defined as

$$\mathbf{A} = \begin{bmatrix} 0 & 0 & 0 & 1 & 0 & 0 \\ 0 & 0 & 0 & 0 & 1 & 0 \\ 0 & 0 & 0 & 0 & 0 & 1 \\ (2B_L + 1)n^2 & 0 & 0 & 0 & 2n & 0 \\ 0 & -(B_L - 1)n^2 & 0 & -2n & 0 & 0 \\ 0 & 0 & -B_L n^2 & 0 & 0 & 0 \end{bmatrix}, \mathbf{B} = \begin{bmatrix} 0 & 0 & 0 \\ 0 & 0 & 0 \\ 0 & 0 & 0 \\ 1 & 0 & 0 \\ 0 & 1 & 0 \\ 0 & 0 & 1 \end{bmatrix}. \tag{6.113}$$

Equations 6.59 and 6.75 may be numerically integrated, with the matrices **A** and **B** defined by Eq. 6.112, and once the control is found from Eq. 6.83, Eq. 6.75 is used to define the optimal state vector.

As an example, the case of flying in close proximity to the asteroid Vestas is revisited. The results presented in Figs. 6.4, 6.5, 6.6 and 6.7 correspond to the case with $B_L = 1$. Here, the results corresponding to $B_L = 4$ which are relevant to the spacecraft's motion in a three-body setting, relative to a colinear equilibrium point, will be presented. In Fig. 6.8 are shown the optimum orbital states in the planar transfer of the spacecraft from its original orbit to a point on the desired final orbit. This optimum transfer takes place in 57.45 days. The figures corresponding to 6.4, 6.5, 6.6 and 6.7 now with $B_L = 4$ are shown in Figs. 6.8, 6.9, 6.10 and 6.11.

6.3.5 Linear Optimal and Tracking Feedback Control

The optimal trajectory planner is based on the theory of the calculus of variations and solves a nonlinear optimization problem. The linear-quadratic regulator (LQR) method solves the optimal trajectory problem when the dynamics of a system is linear and is defined in the state space domain and in the form

$$d\mathbf{x}/dt = \mathbf{A}\mathbf{x} + \mathbf{B}\mathbf{u} \tag{6.114}$$

where **x** is the state vector, $d\mathbf{x}/dt$ is the time derivative of the state vector and **u** is the input vector with **A** and **B** as constant matrices. The system is subject to minimizing the cost functional

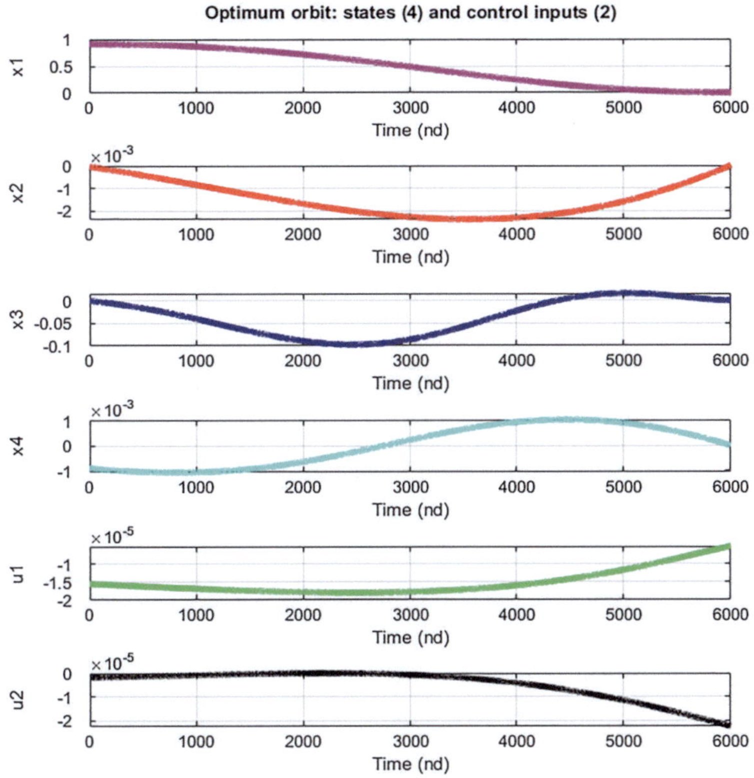

Fig. 6.8 In-plane Cartesian position and velocity orbital states and control inputs for the optimum trajectory synthesis with $B_L = 4$

$$J = \frac{1}{2}\mathbf{x}^T(t_f)\mathbf{H}_f\mathbf{x}(t_f) + \frac{1}{2}\int_{t_0}^{t_f}\left(\mathbf{x}^T(t)\mathbf{Q}\mathbf{x}(t) + \mathbf{u}^T(t)\mathbf{R}\mathbf{u}(t)\right)dt \tag{6.115}$$

where $\mathbf{H}_f$, $\mathbf{Q}$ and $\mathbf{R}$ are real symmetric and positive definite matrices. Bryson Jr. and Ho (1969) have proposed a system method for the choice of the weighting matrices. Alternate methods based on reinforcement learning and optimization methods have been proposed by Kasyap and Vepa (2022) and Ghoreishi et al. (2011), respectively. From the cost, the Hamiltonian functional is formed as

$$\mathbf{H} = \frac{1}{2}\left(\mathbf{x}^T(t)\mathbf{Q}\mathbf{x}(t) + \mathbf{u}^T(t)\mathbf{R}\mathbf{u}(t)\right) + \boldsymbol{\lambda}^T(\mathbf{A}\mathbf{x} + \mathbf{B}\mathbf{u}) \tag{6.116}$$

where $\boldsymbol{\lambda}$ is a Lagrange multiplier vector which is also known as the co-state vector. The necessary conditions for optimality are given by

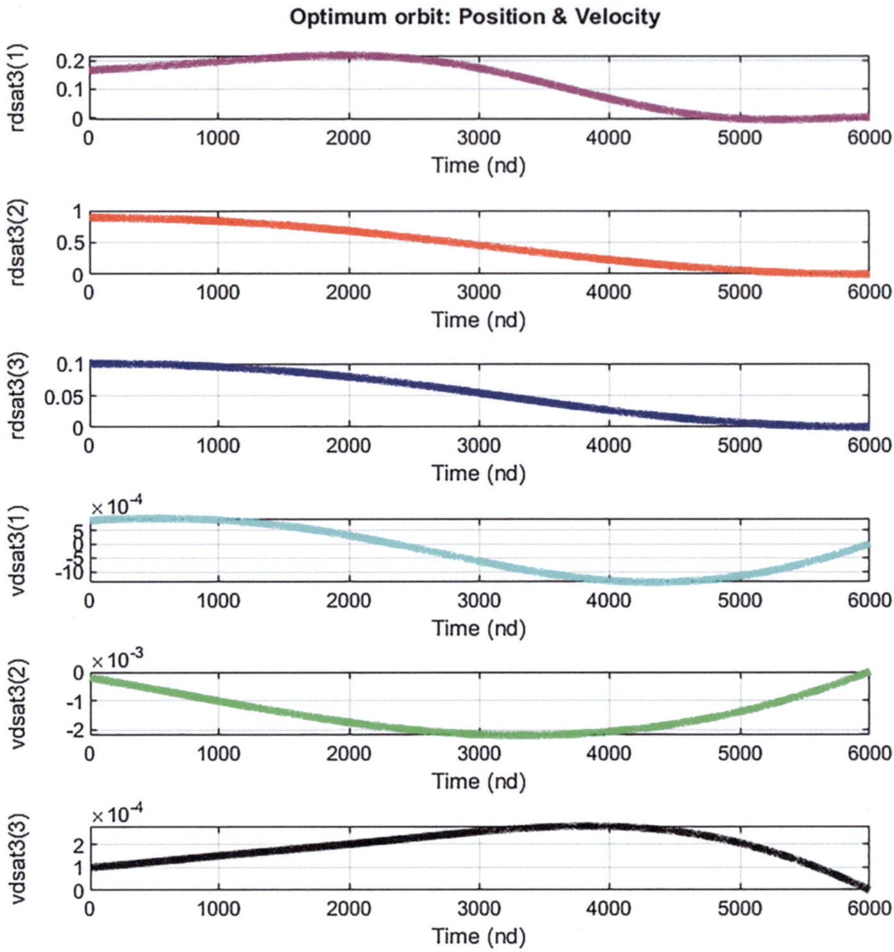

Fig. 6.9 The optimum position and velocities in LVLH coordinates of the reference Vestas orbit with $B_L = 4$

$$\begin{aligned} \dot{\mathbf{x}} &= \frac{\partial \mathbf{H}}{\partial \boldsymbol{\lambda}} = \mathbf{A}\mathbf{x} + \mathbf{B}\mathbf{u}, \ \dot{\boldsymbol{\lambda}} = -\frac{\partial \mathbf{H}}{\partial \mathbf{x}} = -\mathbf{Q}\mathbf{x} - \mathbf{A}^T\boldsymbol{\lambda}, \\ 0 &= \frac{\partial \mathbf{H}}{\partial \mathbf{u}} = \mathbf{R}\mathbf{u} + \mathbf{B}^T\boldsymbol{\lambda}. \end{aligned} \tag{6.117}$$

Finally, the optimal input $\mathbf{u}^*(t)$ for an open loop system is given by $\mathbf{u}^* = -\mathbf{R}^{-1}\mathbf{B}^T\boldsymbol{\lambda}$. The $\boldsymbol{\lambda}$ vector must be generally specified at the final time to meet the optimality requirements.

In linear optimal control theory, since the eigenvalues of $\mathbf{A}$ and $\mathbf{A}^T$ are the same, the solutions for $\boldsymbol{\lambda}$ and $\mathbf{x}$ are linearly related, and it follows that

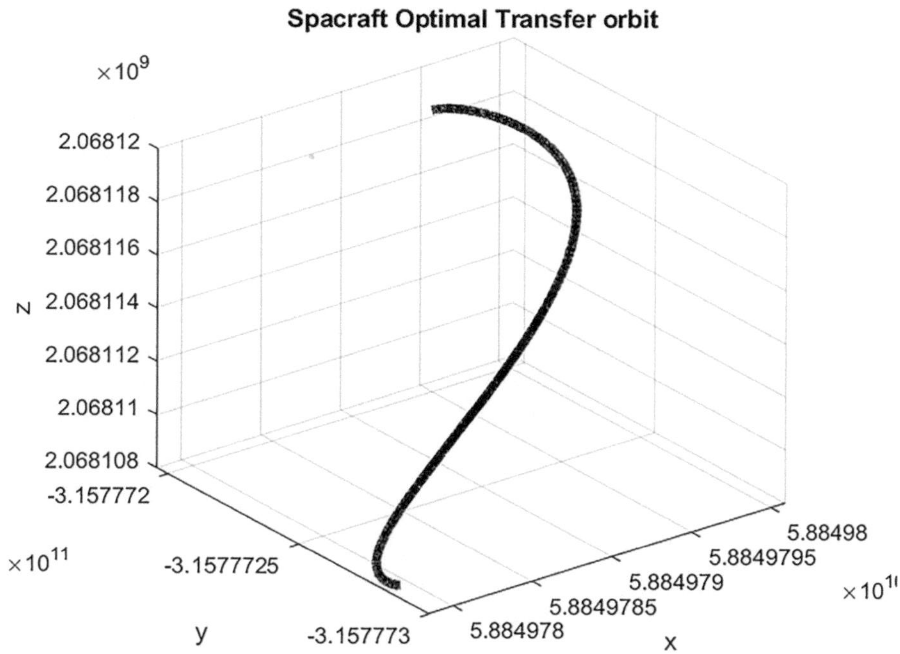

Fig. 6.10 Position and velocity coordinate time histories of the optimum transfer orbit in three-dimensional space with $B_L = 4$

$$\boldsymbol{\lambda} = \mathbf{P}\mathbf{x}, \quad \text{and} \quad \mathbf{u}^* = -\mathbf{R}^{-1}\mathbf{B}^T\boldsymbol{\lambda} = -\mathbf{R}^{-1}\mathbf{B}^T\mathbf{P}\mathbf{x}, \tag{6.118}$$

where $\mathbf{P}$ satisfies a matrix Riccati differential equation, since

$$\dot{\boldsymbol{\lambda}} = \dot{\mathbf{P}}\mathbf{x} + \mathbf{P}\dot{\mathbf{x}} = \dot{\mathbf{P}}\mathbf{x} + \mathbf{P}\mathbf{A}\mathbf{x} + \mathbf{P}\mathbf{B}\mathbf{u} = \dot{\mathbf{P}}\mathbf{x} + \mathbf{P}\mathbf{A}\mathbf{x} - \mathbf{P}\mathbf{B}\mathbf{R}^{-1}\mathbf{B}^T\mathbf{P}\mathbf{x} = -\mathbf{Q}\mathbf{x} - \mathbf{A}^T\mathbf{P}\mathbf{x}. \tag{6.119}$$

Hence, $\mathbf{P}$ satisfies the matrix Riccati differential equation:

$$\dot{\mathbf{P}} + \mathbf{P}\mathbf{A} + \mathbf{A}^T\mathbf{P} - \mathbf{P}\mathbf{B}\mathbf{R}^{-1}\mathbf{B}^T\mathbf{P} + \mathbf{Q} = \mathbf{0}, \quad \text{with} \quad \mathbf{P}(t_f) = \mathbf{P}_f. \tag{6.120}$$

The conditions at the terminal time, $t = t_f$, are chosen so as to satisfy the transversality conditions. In the case when only the steady state control laws are desired, $\dot{\mathbf{P}} = \mathbf{0}$, and $\mathbf{P}$ satisfies an algebraic matrix Riccati equation given by

$$\mathbf{P}\mathbf{A} + \mathbf{A}^T\mathbf{P} - \mathbf{P}\mathbf{B}\mathbf{R}^{-1}\mathbf{B}^T\mathbf{P} + \mathbf{Q} = \mathbf{0}. \tag{6.121}$$

Once the algebraic matrix Riccati equation is solved, the matrix Riccati differential equation may be solved either by integrating it backwards in time or by applying

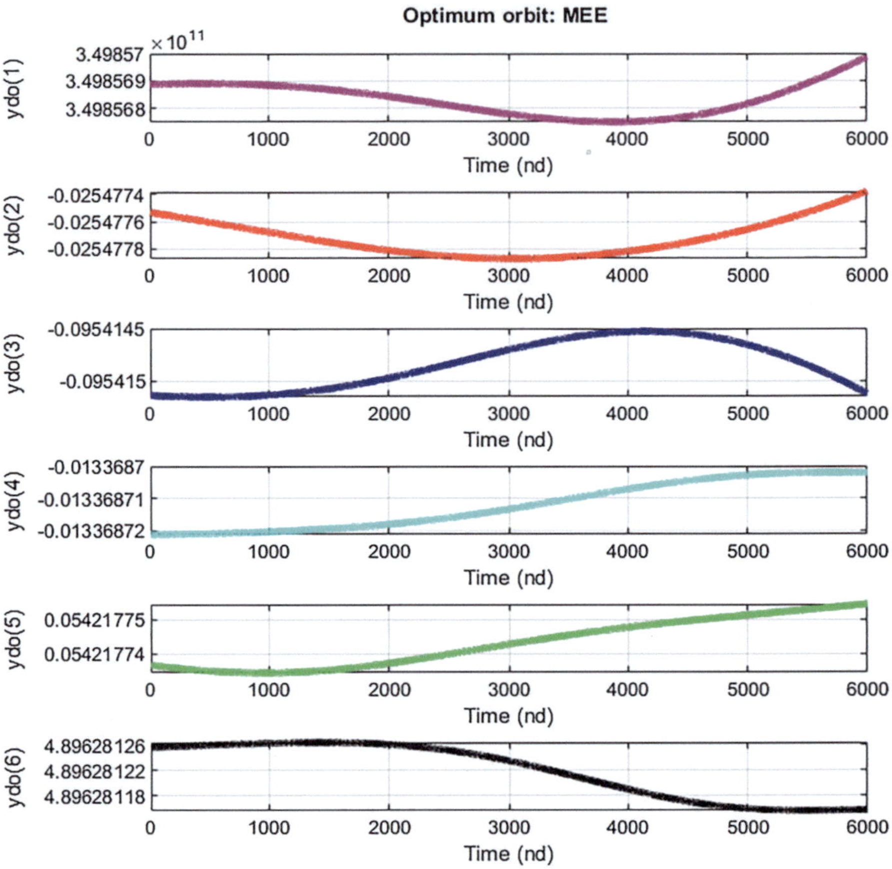

Fig. 6.11 The evolution of the MEE during the optimal transfer with $B_L = 4$. The first is the semi-latus rectum and the last is the true anomaly

the method of eigenvalue-eigenvector decomposition. Once **P** is obtained, one has the optimal feedback control law:

$$\mathbf{u}^* = -\mathbf{R}^{-1}\mathbf{B}^T\mathbf{P}\mathbf{x}. \tag{6.122}$$

A more useful solution is the optimal linear-quadratic tracking (LQT) method. Instead of finding an optimal control that drives the state to zero, the LQT finds an optimal control that drives the state to a time varying reference trajectory. In this case, the cost J is given by

$$J = \Phi(x(t_f), t_f) + \frac{1}{2} \int_{t_0}^{t_f} \left(\left(\mathbf{x}(t) - \mathbf{x}_{ref}(t) \right)^T \mathbf{Q} \left(\mathbf{x}(t) - \mathbf{x}_{ref}(t) \right) + \mathbf{u}^T(t) \mathbf{R} \mathbf{u}(t) \right) dt, \tag{6.123}$$

where

$$\Phi(\mathbf{x}(t_f), t_f) = \frac{1}{2} \left(\mathbf{x}(t_f) - \mathbf{x}_{ref}(t_f) \right)^T \mathbf{H} \left(\mathbf{x}(t_f) - \mathbf{x}_{ref}(t_f) \right), \tag{6.124}$$

and $\mathbf{x}_{ref}(t)$ is reference trajectory. The Hamiltonian is given by

$$\mathbf{H} = \frac{1}{2} \left(\left(\mathbf{x}(t) - \mathbf{x}_{ref}(t) \right)^T \mathbf{Q} \left(\mathbf{x}(t) - \mathbf{x}_{ref}(t) \right) + \mathbf{u}^T(t) \mathbf{R} \mathbf{u}(t) \right) + \boldsymbol{\lambda}^T (\mathbf{A}\mathbf{x} + \mathbf{B}\mathbf{u}). \tag{6.125}$$

The necessary conditions for optimality are given by

$$\begin{aligned} \dot{\mathbf{x}} &= \frac{\partial \mathbf{H}}{\partial \boldsymbol{\lambda}} = \mathbf{A}\mathbf{x} + \mathbf{B}\mathbf{u}, \quad \dot{\boldsymbol{\lambda}}(t) = -\frac{\partial \mathbf{H}}{\partial \mathbf{x}} = -\mathbf{Q}\left(\mathbf{x}(t) - \mathbf{x}_{ref}(t) \right) - \mathbf{A}^T \boldsymbol{\lambda}(t), \\ 0 &= \frac{\partial \mathbf{H}}{\partial \mathbf{u}} = \mathbf{R}\mathbf{u} + \mathbf{B}^T \boldsymbol{\lambda}. \end{aligned} \tag{6.126}$$

Finally, the optimal input $\mathbf{u}^*(t)$ for an open loop system is given by $\mathbf{u}^* = -\mathbf{R}^{-1}\mathbf{B}^T\boldsymbol{\lambda}$, which can be shown to reduce to

$$\mathbf{u}^* = -\mathbf{R}^{-1}\mathbf{B}^T\mathbf{P}\left(\mathbf{x}(t) - \mathbf{x}_{ref}(t) \right), \tag{6.127}$$

where $\mathbf{P}$ satisfies the matrix Riccati differential equation (6.92). Once the algebraic matrix Riccati equation is solved, the matrix Riccati differential equation may be solved either by integrating it backwards in time or by applying the method of eigenvalue-eigenvector decomposition. Once $\mathbf{P}$ is obtained, one has the optimal feedback tracking control law given by Eq. 6.98.

As an example, the case of flying in close proximity to the asteroid Vestas is revisited. Once the transfer to the desired orbit is complete, the optimal feedback control laws are applied to track the prescribed orbit. The tracking control law is derived using the HCW equations with the traditional orbital states (Eq. 6.75) but implemented with the Gauss planetary equations in terms of the MEE, after transforming the control laws to the appropriate global frame and to the corresponding state vector. The tracking controller was operational over a time frame of 1500 days. The integration time step was 6000 s. The complete orbits of the spacecraft with the optimal cost minimization tracking control laws, with two different sets of weighting matrices, **Q** and **R**, and the simulated orbit of the asteroid

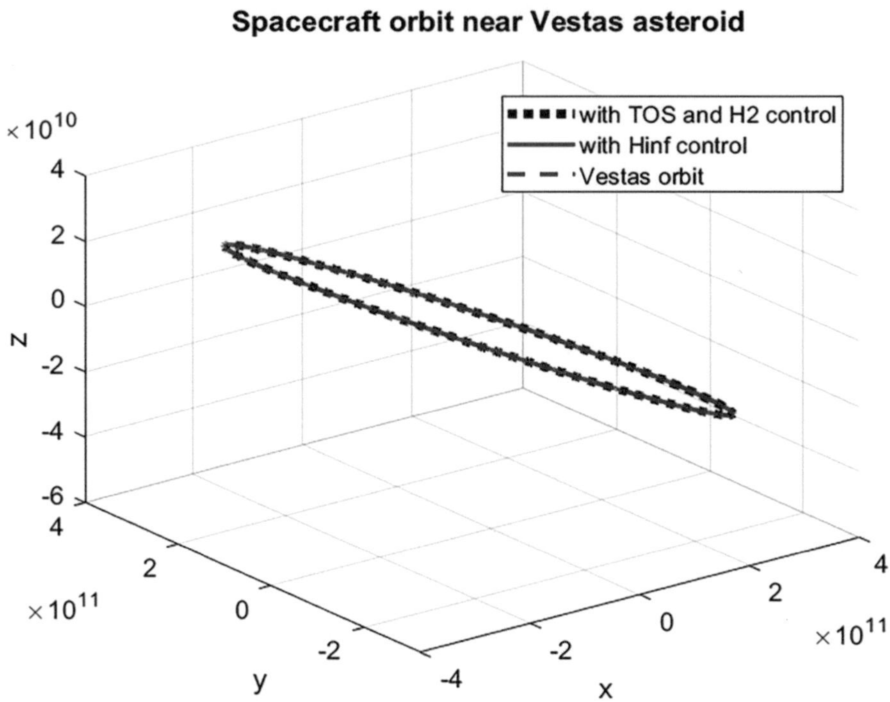

Fig. 6.12 Spacecraft closed loop orbits in the vicinity of the asteroid Vestas (The Cartesian position coordinates are in metres)

Vestas are compared in Fig. 6.12. The results show that the closed loop orbits are not very sensitive to the choice of the weighting matrices, **Q** and **R**, which were chosen carefully to avoid a collision between the spacecraft and the asteroid. A closer examination of the orbits reveals that the spacecraft never collides with the asteroid for both choices of **Q** and **R**.

6.4 Continuous Trajectory Correction Manoeuvres Using Feedback

In a recent paper, Vepa (2022) proposed a generic empirical feedback control law for spacecraft in heliocentric orbits, for trajectory correction applications, based on Newton's gravitation law, given by

$$\boldsymbol{F}_{c_1} = \frac{\boldsymbol{f}_{track,ref}}{m_1} = -KGm_{Sun}\frac{\boldsymbol{r}_1 - \boldsymbol{r}_{1,ref}}{\left\|\boldsymbol{r}_1 - \boldsymbol{r}_{1,ref}\right\|^3}. \tag{6.128}$$

In the above feedback control law, which defines a continuous trajectory correction force, K is the proportional gain which is assumed to be sufficiently small to ensure that control accelerations can be adequately generated by the electric thrusters. The choice of K, a scalar, is made to i) minimize Δv over the time frame of the flight and ii) to ensure that the spacecraft is in the desired orbit around the targeted central body or within the target body's SOI at the end of the time frame.

As an example, a typical optimal trajectory from Earth to the planet Mars shown in Fig. 6.3, computed with only the gravitational effects of the Sun, Earth, Mars, Venus and the satellite included, is reconsidered. The feedback trajectory correction control law is now applied with the reference trajectory defined as in Fig. 6.3. Inspired by the concept of ballistic capture, the control law seeks to place the satellite slightly ahead of the planet and its orbit. The spacecraft's actual trajectory with the feedback control law for trajectory correction to define the thrust force and the maximum thrust limited to 0.25 mN and with $K = 1.3$ is also shown in Fig. 6.13. The spacecraft's optimal reference and actual trajectories with the feedback control law for trajectory correction to define the thrust force but with the maximum thrust limited to a reduced level of 0.2 mN and with a reduced gain $K = 0.3$ are shown in Fig. 6.14. If one reduces the gain further, there is the distinct possibility that the spacecraft would miss crossing the trajectory of the target planet.

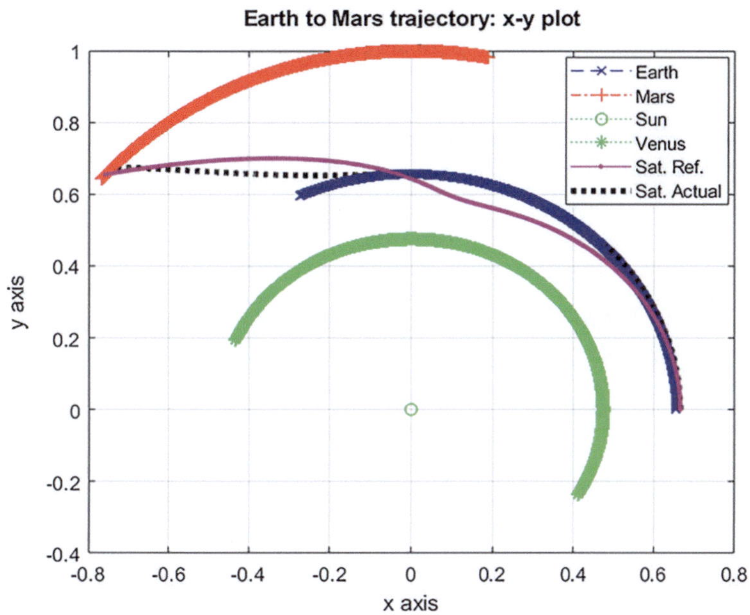

Fig. 6.13 Typical optimal transfer orbit from Earth to Mars in a heliocentric frame with five principal bodies and also showing the actual trajectory of the spacecraft with the feedback control law for trajectory correction

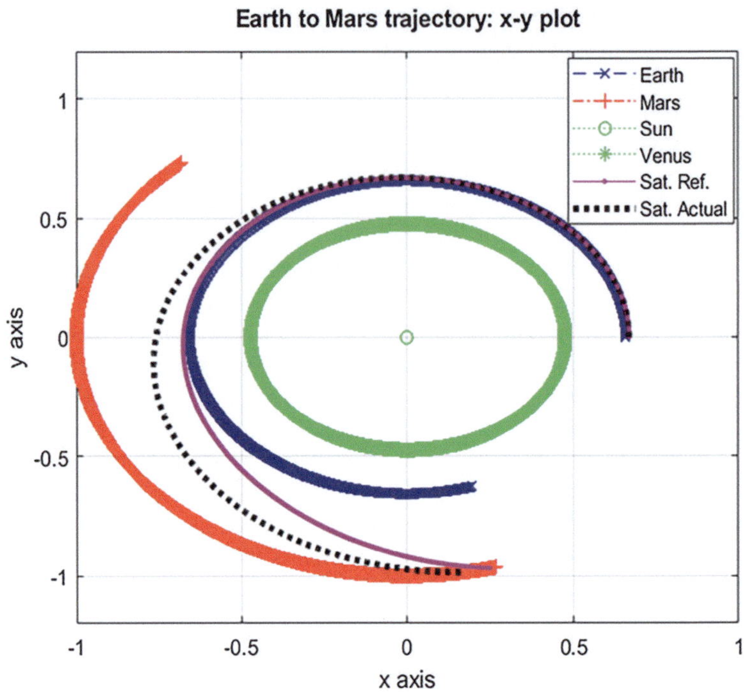

Fig. 6.14 Typical optimal transfer orbit from Earth to Mars in a heliocentric frame with five principal bodies and also showing the actual trajectory of the spacecraft with a reduced feedback control gain and limited thrust

One can also estimate the total Δv required for the mission, but these calculations are not shown here. The reader is referred to the original paper by Vepa (2022) for the details.

6.5 Control Laws Due to Edelbaum, Petropolous and Others

Edelbaum (1961), Pollard (2001) and Pollard (2000) proposed simplified feedback control laws in terms of orbital elements. Examples of these in terms of the required velocity increment (or decrement) for changes in the orbital elements a, e, i, ω, Ω are, respectively, given as

$$\Delta v = \left|\sqrt{\frac{\mu}{a_0}} - \sqrt{\frac{\mu}{a_1}}\right|, \quad \Delta v = \sqrt{\frac{\mu}{a}\frac{2}{3}}\left|\sin^{-1}e_1 - \sin^{-1}e_2\right|,$$

$$\Delta v = v\sqrt{2\left(1 - \cos\left(\frac{\pi}{2}\Delta i\right)\right)}, \quad \Delta v = \sqrt{\frac{\mu}{a}\frac{2}{3}}\frac{e}{\sqrt{1-e^2}}|\Delta\omega|, \quad \Delta v = \sqrt{\frac{\mu}{a}\frac{\pi}{2}}\sin i|\Delta\Omega|. \tag{6.129}$$

Unfortunately, such control laws do not take into account the interrelationships between the orbital elements. To deal with this problem, Petropoulos (2003) adopted a Lyapunov approach. A refined Lyapunov feedback control method developed by Petropoulos (2005) and called the Q law solves the orbit transfer in an inverse square gravity field where there is no constraint on the final true anomaly. The Q law method, as proposed and refined by Petropoulos (2003, 2005), provides means of deriving the optimal thrusting angles and the coast phases for an arbitrary transfer between two orbits around a central body. Petropoulos (2003, 2005) and Lee et al. (2005) derived analytic expressions for the optimal thrust directions and optimal orbit locations for changing each of the classical orbital elements and proposed two control law strategies that include a mechanism for tracking coasting trajectories based on the "effectivity" of the manoeuvre. "Effectivity" is a concept introduced by several authors. Thus, a direct Lyapunov feedback control method could be used during the initial stages of the mission design, avoiding high precision calculations. The Q law is based on the evaluation of a proximity quotient, Q, which represents the relations between the orbital elements by means of scaling functions that quantify the proximity of the osculating orbit to the target orbit. Essentially, it is a model predictive estimate of the time-to-go and has units of time squared. During the transfer at each instant, the Q law method chooses the thrust angles that reduce the Q value the most quickly. A coasting mechanism is also incorporated that is based on variable "effectivity" of the thrust in reducing Q at different true anomalies. The Q law is implemented in the orbital element space.

Nominally the dynamical equations for the change in the orbital elements or the Gauss planetary equations for the evolution of the classic orbital elements (COE) may be used for defining the optimal control forces. When the inclination i or eccentricity e are equal to zero, the dynamical equations for the COE will become singular. In low-thrust transfer, the COE are changed continuously, so the singularity cannot be avoided. So one uses an alternate set of orbital elements, such as the modified equinoctial elements. To formulate the Gauss planetary equations in a compact form in terms of the modified equinoctial elements, they are expressed in vector form. In vector form, the Gauss planetary equations in terms of the modified equinoctial elements can be expressed as

$$\dot{y} = \frac{dy}{dt} = \begin{bmatrix} \mathbf{A}_r(\mathbf{y}) \\ \mathbf{a}_L \end{bmatrix} \mathbf{F} + \mathbf{b}, \quad \frac{dm}{dt} = -\frac{T}{I_{sp}g_0} = -\frac{m|\mathbf{F}|}{I_{sp}g_0},$$

$$\mathbf{A}_r(\mathbf{y}) = \sqrt{\frac{p}{\mu}} \begin{bmatrix} 0 & 2p/w & 0 \\ \sin L & ((w+1)\cos L + f)/w & g(h\sin L - l\cos L)/w \\ -\cos L & ((w+1)\sin L + g)/w & f(h\sin L - l\cos L)/w \\ 0 & 0 & \dfrac{s^2\cos L}{2w} \\ 0 & 0 & \dfrac{s^2\sin L}{2w} \end{bmatrix}. \tag{6.130}$$

In Eq. 6.130, the subscripts of the modified equinoctial elements have been deleted for brevity. Moreover, the vectors, $\boldsymbol{y}$, $\mathbf{F} = |\mathbf{F}|\mathbf{u}$, $\mathbf{b}$ and $\mathbf{a}_L$, are, respectively, defined as

$$\begin{aligned} &\boldsymbol{y} = [p_{mee} \quad f_{mee} \quad g_{mee} \quad h_{mee} \quad k_{mee} \quad L_{mee}]^T, \\ &\mathbf{F} = |\mathbf{F}|\mathbf{u} = (T/m)[\sin\,\alpha\cos\beta\cos\alpha\cos\beta\sin\beta], \\ &\mathbf{b} = \sqrt{\mu p}\left(\frac{w}{p}\right)^2 [0 \quad 0 \quad 0 \quad 0 \quad 0 \quad 1]^T, \ \mathbf{a}_L = (h\sin L - k\cos L)\sqrt{\frac{p}{\mu}}[0 \quad 0 \quad 1]. \end{aligned} \tag{6.131}$$

The optimization problem may now be stated. One is interested in the problem of finding the steering control, $\mathbf{u} = \mathbf{u}(t)$, $t_0 \le t \le t_f$, the thrust direction time history which minimizes the cost functional

$$J = \frac{1}{I_{sp}g_0}\int_{t_0}^{t_f} |\mathbf{F}\mathbf{u}|dt = \frac{|\mathbf{F}|}{I_{sp}g_0}\int_{t_0}^{t_f} |\mathbf{u}|dt, \tag{6.132}$$

subject to the constraint differential Eq. (6.130) being satisfied, with the constraint

$$\left|\boldsymbol{r}(t) - \boldsymbol{r}_f\right|\Big|_{t=t_f} = \Phi\{\boldsymbol{r}(t)\}\big|_{t=t_f} = 0. \tag{6.133}$$

One may construct the corresponding Hamiltonian function, which is

$$H = \boldsymbol{\lambda}^T(t)(\mathbf{A}(\mathbf{y})|\mathbf{F}|\mathbf{u} + \mathbf{b}) + (|\mathbf{u}| - \lambda_m(t)m)\left(|\mathbf{F}|/I_{sp}g_0\right). \tag{6.134}$$

The corresponding co-state differential equations are

$$\frac{d}{dt}\boldsymbol{\lambda}(t) = -\frac{\partial H}{\partial \mathbf{y}}, \quad \frac{d}{dt}\lambda_m(t) = -\frac{\partial H}{\partial m}. \tag{6.135}$$

Substituting for the Hamiltonian, the co-state equations reduce to

$$\frac{d}{dt}\boldsymbol{\lambda}(t) = -\frac{\partial H}{\partial \mathbf{y}} = -\boldsymbol{\lambda}^T(t)\left(\frac{\partial \mathbf{A}(\mathbf{y})}{\partial \mathbf{y}}|\mathbf{F}|\mathbf{u}\right), \quad \frac{d}{dt}\lambda_m(t) = -\frac{\partial H}{\partial m} = \lambda_m(t)\frac{|\mathbf{F}|}{I_{sp}g_0}. \tag{6.135}$$

The optimal control is obtained by solving the pair or equations

$$\begin{aligned}\partial H/\partial\alpha &= \boldsymbol{\lambda}^T(t)\mathbf{A}(\mathbf{y})|\mathbf{F}|\partial\mathbf{u}/\partial\alpha = |\mathbf{F}||\partial\mathbf{u}/\partial\alpha| = |\mathbf{F}|\cos\beta,\\ \partial H/\partial\beta &= \boldsymbol{\lambda}^T(t)\mathbf{A}(\mathbf{y})|\mathbf{F}|\partial\mathbf{u}/\partial\beta = |\mathbf{F}||\partial\mathbf{u}/\partial\beta| = |\mathbf{F}|,\end{aligned} \tag{6.136}$$

where

$$\boldsymbol{u} = [\sin\,\alpha\cos\beta\cos\alpha\cos\beta\sin\beta], \quad \text{and,} \quad m(t_f) \neq 0, \;\; \lambda_m(t_f) = 0. \tag{6.137}$$

Reconsidering the Gauss planetary equations for near circular orbits, they can be written as

$$\dot{a} = \frac{da}{dt} = \frac{2a}{V}f_N, \tag{6.138}$$

$$\dot{e} = \frac{de}{dt} = \frac{f_S\sin\nu + 2f_N\cos\nu}{V}, \tag{6.139}$$

$$\dot{\omega} + \dot{\Omega} = \frac{d\omega}{dt} + \frac{d\Omega}{dt} = \frac{f_N\sin\nu - f_S\cos\nu}{eV}, \tag{6.140}$$

$$\dot{i} = \frac{di}{dt} = \frac{\cos(\nu + \omega)}{V}f_W, \quad \theta = \omega + \nu, \tag{6.141}$$

$$\dot{\Omega} = \frac{d\Omega}{dt} = \frac{\sin(\theta + \omega)}{Vi}f_W, \tag{6.142}$$

$$\dot{\nu} = \frac{d\nu}{dt} = \frac{h}{r^2} - \left(\frac{d\omega}{dt} + \frac{d\Omega}{dt}\cos i\right) \approx n. \tag{6.143}$$

In Edelbaum's method, only out-of-plane and the tangential thrust is applied, resulting in $\alpha = 0$ and thus $f_S = 0$. ω could be set equal to 0; i.e. $\omega = 0$. Edelbaum assumed ω and Ω to be constants. Therefore, the equations for these two elements can be ignored. In reality,ω and Ω will vary during a manoeuvre, but in Edelbaum's theory, only complete revolutions are considered after which it is assumed that ω and Ω will remain at their initial values at all time.

The reduced orbit equations, in terms of the tangential force and the force normal to the orbit plane, are

$$\frac{da}{dt} = \frac{2a}{V}F_t, \quad \frac{di}{dt} = \frac{\cos\theta}{V}F_n, \quad \frac{d\theta}{dt} = n. \tag{6.144}$$

Next, the angular position θ is averaged out by integrating with respect to θ. Thus, given that the yaw angle is β, one has

$$\begin{aligned} &\frac{da}{dt} = \frac{2a\cos\beta}{V}\frac{T}{m}, \quad \frac{di}{dt} = \frac{2\sin\beta}{\pi V}\frac{T}{m}, \quad \frac{dV}{dt} = -\frac{T}{m}\cos\beta \quad \text{and} \\ &\frac{di}{dV} = -\frac{2\tan\beta}{\pi V}. \end{aligned} \tag{6.145}$$

Edelbaum (1961) has shown the following, given that the velocity increment is Δv and the change in inclination is Δi:

$$\tan\beta_0 = \frac{\sin(\pi\Delta i/2)}{(v_i/v_f) - \cos(\pi\Delta i/2)}, \tag{6.146}$$

$$\Delta v = v_i(\cos\beta_0 - \sin\beta_0/(\tan(\beta_0 + \pi\Delta i/2))). \tag{6.147}$$

For, $t(k) = (k - 1)\Delta t$,

$$inc(k) = inc_i - \frac{2}{\pi}(\beta(k) - \beta_0), \tag{6.148}$$

$$\tan\beta(k) = \frac{v_i\sin(\beta_0)}{v_i\cos(\beta_0) - f\times t(k)}, \tag{6.149}$$

or

$$\tan\left(\beta(k)-\frac{\pi}{2}\right)=-\frac{v_i\cos(\beta_0)-f\times t(k)}{v_i\sin(\beta_0)}, \tag{6.150}$$

and

$$v^2(k)=v_i^2+f^2\times t^2(k)-2v_if\times t(k)\cos\beta_0. \tag{6.151}$$

The case of constant low thrust is yet another special case, for a circular orbit of a particle under the influence of retarding force, which could be modelled by Newton's law and integrated, resulting in a spiral trajectory. The associated equations are

$$v=\sqrt{\mu/r},\ \ \dot{v}=-f \tag{6.152}$$

Integrating,

$$v=v_0-f(t-t_0). \tag{6.153}$$

At $t=t_f$,

$$v_f=v_0-f(t_f-t_0)\Rightarrow f=(v_f-v_0)/(t_f-t_0). \tag{6.154}$$

Thus,

$$v^2=\left(v_0-\frac{v_f-v_0}{(t_f-t_0)}(t-t_0)\right)^2=\frac{\mu}{r}. \tag{6.155}$$

The resulting trajectory is a spiral given by

$$r=r_0v_0^2\left(v_0-\frac{v_f-v_0}{(t_f-t_0)}(t-t_0)\right)^{-2}. \tag{6.156}$$

One can now compare the low specific thrust, electrically propelled, orbit raising trajectory with the optimal trajectory, maximizing the radius, for a LEO to GEO orbit transfer. The trajectories are compared in Fig. 6.15. A three-dimensional simulation of a low specific thrust orbit raising trajectory for a LEO to GEO orbit transfer in the ecliptic plane is shown in Fig. 6.16. The distance unit is the Earth's radius.

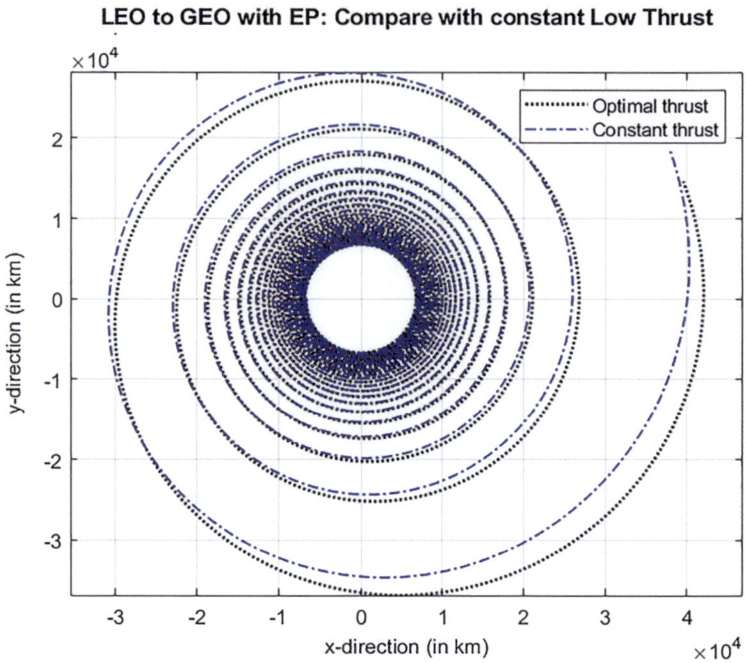

Fig. 6.15 Comparison of a low specific thrust orbit raising trajectory with the optimal trajectory, for a LEO to GEO orbit transfer, over 26 orbits in 4 days

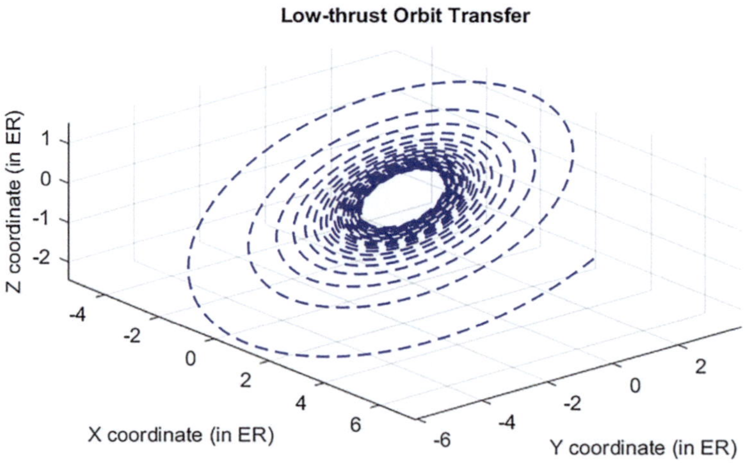

Fig. 6.16 Three-dimensional simulation of a low specific thrust orbit raising trajectory for a LEO to GEO orbit transfer in the ecliptic plane. The distance unit is the Earth's radius (ER)

6.5.1 Modifications of the Q Law

Varga and Sánchez-Pérez (2016) propose a Lyapunov feedback control method, based on the original Q law. They propose modifications to the Q law algorithms such as formulating the problem in equinoctial elements to avoid the singularities occurring for circular or zero-inclination orbits. The modifications to the algorithm accommodate singularities in the original equations and include the most relevant perturbations.

First a weighted Q function is defined, in terms of certain weights and a performance cost functional. Also defined are a set of perturbation states. These definitions are given as

$$Q=(1+W_P P)\mathbf{W}_y\mathbf{S}_y\mathbf{J},\quad \delta y=y-y_{\text{ref}}. \tag{6.157}$$

In Eq. 6.157, $\mathbf{W}_y$ and W_P are weighting factors, P is a penalty function that is yet to be specified and $\mathbf{S}_y$ are scaling factors. Next, the performance cost functional $\mathbf{J}$ is defined as

$$\mathbf{J}=\frac{1}{2}\left[\left(\frac{\delta y}{\max_\nu(\dot{y})}\right)^2\right]=\frac{1}{2}\left[\frac{\delta p^2}{\delta p_{\max}^2}\quad\frac{\delta e_x^2}{\delta e_{\max}^2}\quad\frac{\delta e_y^2}{\delta e_{\max}^2}\quad\frac{\delta h_x^2}{\delta \chi_{\max}^2}\quad\frac{\delta h_y^2}{\delta \chi_{\max}^2}\quad 0\right]. \tag{6.158}$$

Given that the MEE satisfy the first of Eq. 6.130,

$$\dot{y}=\frac{dy}{dt}=\begin{bmatrix}\mathbf{A}_r(y)\\ \mathbf{a}_L\end{bmatrix}\mathbf{F}+\mathbf{b}, \tag{6.159}$$

then with $\mathbf{R}$ defined as a symmetric and positive definite matrix, the control law may be expressed as

$$\mathbf{F}=-\mathbf{R}^{-1}\mathbf{A}^T(y)\left(\frac{\delta Q}{\delta y}\right)^T,\quad \mathbf{F}=T\boldsymbol{u}\quad\text{and}\quad\frac{\delta Q}{\delta y}$$
$$=(1+W_P P)\left(\mathbf{W}_y\mathbf{S}_y\left[\frac{\delta y}{(\max_\nu(\dot{y}))^2}\right]\right)^T. \tag{6.160}$$

In deriving Eq. 6.160, it was tacitly assumed that the penalty function P is not a function of the perturbation states. In Eq. 6.160,

$$T=|\mathbf{F}|\quad\text{and}\quad \mathbf{F}/|\mathbf{F}|=\boldsymbol{u}=[\sin\alpha\cos\beta\cos\alpha\cos\beta\sin\beta]. \tag{6.161}$$

Furthermore, to obtain the control law, one must evaluate the derivatives:

$$\frac{\delta \mathbf{W}_y \mathbf{S}_y \mathbf{J}}{\delta \boldsymbol{y}} = \left[W_p S_p \frac{\delta p}{\delta p_{\max}^2} \quad W_e S_e \frac{\delta e_x}{\delta e_{\max}^2} \quad W_e S_e \frac{\delta e_y}{\delta e_{\max}^2} \quad W_h S_h \frac{\delta h_x}{\delta \chi_{\max}^2} \quad W_h S_h \frac{\delta h_y}{\delta \chi_{\max}^2} \quad 0 \right]. \tag{6.162}$$

The control law is explicitly given by

$$\mathbf{F} = -\mathbf{R}^{-1} \mathbf{A}^T(\boldsymbol{y}) \left(\frac{\delta Q}{\delta \boldsymbol{y}} \right)^T = -\mathbf{R}^{-1} \mathbf{A}^T(\boldsymbol{y})$$

$$\left[W_p S_p \frac{\delta p}{\delta p_{\max}^2} \quad W_e S_e \frac{\delta e_x}{\delta e_{\max}^2} \quad W_e S_e \frac{\delta e_y}{\delta e_{\max}^2} \quad W_h S_h \frac{\delta h_x}{\delta \chi_{\max}^2} \quad W_h S_h \frac{\delta h_y}{\delta \chi_{\max}^2} \quad 0 \right]^T. \tag{6.163}$$

In the above, $\mathbf{R}^{-1}$ could be independently defined as a symmetric positive definite matrix. One could, in principle, also derive the control law defined by Eq. 6.163 based on optimal control approach.

The modified Q function for deriving the feedback control law is

$$Q_m = \mathbf{W}_y \mathbf{S}_y \mathbf{J} + (1/2) \mathbf{F}^T \mathbf{R} \mathbf{F}. \tag{6.164}$$

Then following the methodology of optimal control, one obtains Eq. 6.163.

One could also establish an alternate Q function formulation of the feedback tracking control law. In this case, the Q function is replaced by

$$Q_m = -\frac{\mu_J}{\boldsymbol{r}_{1J}} + \frac{1}{2} \boldsymbol{v}_{1J}^2 + V_{pot}. \tag{6.165}$$

In Eq. 6.165, V_{pot} is a potential function that is yet to be chosen. It follows that

$$\dot{Q}_m = \mu_J \frac{\boldsymbol{r}_{1J}}{r_{1J}^3} \boldsymbol{v}_{1J} + \dot{\boldsymbol{v}}_{1J} \boldsymbol{v}_{1J} + \frac{\partial V_{pot}}{\partial \boldsymbol{r}_{1J}} \boldsymbol{v}_{1J} = \left(\widehat{\boldsymbol{F}}_{c_1} + \frac{\partial V_{pot}}{\partial \boldsymbol{r}_{1J}} \right) \boldsymbol{v}_{1J} = -\frac{K_v}{2} \boldsymbol{v}_{1J}^2. \tag{6.166}$$

After making an appropriate choice of the potential function V_{pot}, the control law is given by

$$\widehat{\boldsymbol{F}}_{c_1} = -\frac{\partial V_{pot}}{\partial \boldsymbol{r}_{1J}} - \frac{K_v}{2}\boldsymbol{v}_{1J}, \quad \frac{\partial V_{pot}}{\partial \boldsymbol{r}_{1J}} = K_p\mu_J\left(\frac{\boldsymbol{r}_1}{r_{1J}^3} - \frac{\boldsymbol{r}_{1,ref}}{r_{1J,ref}^3}\right),$$
$$\widehat{\boldsymbol{F}}_{c_1} = \frac{\boldsymbol{f}_c}{m_1} = -K_p\mu_J\left(\frac{\boldsymbol{r}_1}{r_{1J}^3} - \frac{\boldsymbol{r}_{1,ref}}{r_{1J,ref}^3}\right) - \frac{K_v}{2}\boldsymbol{v}_{1J}. \tag{6.167}$$

Let $K_v = \varepsilon$, where ε is a very small positive number and $\varepsilon \to 0$ the control introduced earlier is expressed as

$$\widehat{\boldsymbol{F}}_{c_1} = \frac{\boldsymbol{f}_c}{m_1} = -K_p\mu_J\left(\frac{\boldsymbol{r}_1}{r_{1J}^3} - \frac{\boldsymbol{r}_{1,ref}}{r_{1J,ref}^3}\right) - \frac{\varepsilon}{2}\boldsymbol{v}_{1J}, \quad \varepsilon \to 0. \tag{6.168}$$

6.5.2 Other Methods

Another very important approach for synthesizing the control laws is based on the homotopy approach proposed by Bertrand and Epenoy (2002). The homotopy method is presented as a smoothing technique by them. Recently, a fast homotopy method which facilitates achieving a global fuel-optimal solution while decreasing the computational time was proposed by Saghamanesh and Baoyin (2018). The homotopy method is particularly useful in designing on-off controls for trajectory tracking.

To apply the homotopy method to the optimization problem seeking to minimize, Eq. (6.132), the cost function is modified to,

$$J = \left(1/I_{sp}g_0\right)\int_{t_0}^{t_f} |\mathbf{F}|\left((1-\varepsilon)\mathbf{u} + \varepsilon\mathbf{u}^2\right)dt. \tag{6.169}$$

In Eq. 6.138, the control $\boldsymbol{u}$ is an on-off control input. The corresponding Hamiltonian function is

$$H = \boldsymbol{\lambda}^T(t)(\mathbf{A}(\mathbf{y})|\mathbf{F}|\mathbf{u} + \mathbf{b}) + \left((1-\varepsilon)\mathbf{u} + \varepsilon\mathbf{u}^2 - \lambda_m(t)m\mathbf{u}\right)\left(|\mathbf{F}|/I_{sp}g_0\right). \tag{6.170}$$

The Hamiltonian is also expressed as

$$H = \left(\boldsymbol{\lambda}^T(t)\mathbf{A}(\mathbf{y})I_{sp}g_0 - \lambda_m(t)m + 1 - \varepsilon + \varepsilon\mathbf{u}\right)\left(|\mathbf{F}|/I_{sp}g_0\right)\mathbf{u} + \boldsymbol{\lambda}^T(t)\mathbf{b},$$
$$or\ as, \quad H = (\rho - \varepsilon(1-\mathbf{u}))\left(|\mathbf{F}|/I_{sp}g_0\right)\mathbf{u} + \boldsymbol{\lambda}^T(t)\mathbf{b}. \tag{6.171}$$

The switching function is defined as

$$\rho = 1 + \boldsymbol{\lambda}^T(t)\mathbf{A}(\mathbf{y})I_{sp}g_0 - \lambda_m(t)m. \tag{6.172}$$

Since the switching function (ρ) takes the value of zero only at finite isolated points, **u** is either zero or one on any interval. The thrust is either zero or maximum along the optimal solution, leading to a "bang-bang" control profile. The optimal control is

$$\mathbf{u} = 0, \rho > \varepsilon, \quad \mathbf{u} = 1, \rho < -\varepsilon, \quad \mathbf{u} = (1/2)(1 - \rho/\varepsilon), \|\rho\| \le \varepsilon. \tag{6.173}$$

The optimal trajectory is derived from the controls and the state and co-state equations. Further details may be found in Bertrand and Epenoy (2002) and in Pan and Pan (2020).

Exercises

1. Consider the optimal transfer of a spacecraft from a low Earth orbit (LEO) to a low Mars orbit (LMO). The transfer problem may be formulated as a restricted multi-body, continuous thrust, optimal trajectory synthesis problem in that the spacecraft is considered subject to the gravitational fields of Earth, Mars and Sun (and possibly Venus and Jupiter, for any gravity assist) along the entire trajectory. This is done to achieve increased accuracy compared with the methods of patched conics. Following a similar formulation to that of Miele and Wang (1997) by assuming also the presence of Venus to benefit from possible gravity assist, reformulate the optimal trajectory problem to the planet Mars and solve the same using a MATLAB environment. You may assume that the Sun and planetary trajectories are not affected by the presence of the spacecraft, while the spacecraft itself is influenced by the presence of these celestial bodies. You may assume that initially the spacecraft is in a circular Earth orbit at an altitude of 600 km and it is desired that the final orbit is a circular Mars orbit with an altitude of 400 km.
2. Reconsider exercise 1 and the problem of synthesizing an optimum trajectory to the planet Jupiter via Mars to take advantage of the gravity assist provided by the latter planet, and synthesize the continuous thrust, optimal multi-body trajectory in a MATLAB environment, using the same approach as in exercise 1. Rather than transferring to a circular orbit around Mars, the spacecraft now continues its journey to Jupiter. You may assume that the final orbit is a circular orbit around Jupiter with an altitude of 1000 km.
3. Reconsider exercises 1 and 2 and the problem of synthesizing an optimum trajectory to the planet Mercury via Venus to take advantage of the gravity assist provided by the latter planet, and synthesize the continuous thrust, optimal multi-body trajectory in a MATLAB environment, using the same approach as in

exercise 1. You may assume that the final orbit is a circular orbit around Mercury with an altitude of 400 km.

4. Consider the problem of synthesizing a continuous thrust, optimum trajectory to the asteroids like Hylonome, Hindalgo and Ceres, via Jupiter, to take advantage of the gravity assist provided by the latter planet, and synthesize the optimal multi-body trajectory in a MATLAB environment, using the results of Fritz and Turkoglu (2017) to initialize the optimal trajectory synthesis problem and the methods presented in this chapter.
5. Miele and Mancuso (2001) have solved for the low energy or minimum fuel-optimal trajectory to the Moon from the Earth using a three-body approach, similar to theirs. There results for the time of flight exceed the time of flight for Apollo missions to the Moon. Repeat their solution, trying to reduce the time of flight in the process, but without using a minimum time approach and by using a range of other or alternate optimizing cost functions or by altering the weights dynamically, over different phases of the flight.
6. The governing equations for the lunar lander are described as follows:

$$dr/dt = v,\ \ d\phi/dt = u/r,\ \ du/dt = -vu/r + (T/m)\cos\beta,$$
$$dv/dt = u^2/r - \mu/r + (T/m)\sin\beta,\ \ dm/dt = -T/\left(gI_{sp}\right),$$

where r and ϕ are the radial distance and the angular position of the lunar lander from the origin of the moon, u and v are the transverse and the radial velocity, m is the mass of the lander and μ is the gravitational parameter of the moon. The angular position is defined from the initial position of a de-orbit burn phase. The mass flow rate *dm*/*dt* is defined in terms of the specific impulse I_{sp} and the acceleration due to gravity on the surface of the Earth. The thrust vector is parameterized in terms of the thrust magnitude T, and thrust vector angle β, which is chosen as the control input variable, during the powered descent phase. Minimizing the total fuel consumed over the flight is equivalent to maximizing the final landing mass problem. For this reason, the optimal landing trajectory is required to maximize the mass of the lunar lander on landing on the surface of the moon. The cost function is thus chosen as $J = -m_f$, where $m_f = m(t)$ at $t = t_f$. The initial conditions $r_0 = r_{\text{moon}} + h_0$, $\phi_0 = 180°$, $u_0 =$ unspecified, $v_0 = 0$, $m_0 =$ specified, where r_{moon} is the mean radius of the moon. The terminal conditions are $r_f = r_{\text{moon}}$, $\phi_f =$ unspecified, $u_f = r_{\text{moon}}\omega_{\text{moon}}$, $v_f = 0$ and $m_f =$ unspecified. A suitable value may be assumed for the preliminary orbital height h_0 of the spacecraft.

It is required to define a Hamiltonian and the co-state equations based on optimal control theory so the optimality conditions can be applied to define the equation for the optimal control. The boundary conditions the co-states must satisfy can also be obtained from optimal control theory.

Solve the associated two-point boundary value problem numerically and obtain the optimal landing trajectory.

Appendix 6.1: Solving the Two-Point Boundary Value Problem for the Optimal Trajectory

To apply the pseudo-spectral method for solving the optimal trajectory synthesis problem, in the first instance, the time domain given by $[t_0, t_f]$ is transformed to the domain $[-1, 1]$.

The linear transformation of the variable t to τ is

$$t=(1/2)\left(t_f-t_0\right)\tau+(1/2)\left(t_f+t_0\right). \tag{6.174}$$

Thus,

$$dt=(1/2)\left(t_f-t_0\right)d\tau. \tag{6.175}$$

Thus, the time derivative is transformed to

$$d/dt=\left(2/\left(t_f-t_0\right)\right)d/d\tau \text{and} d/d\tau=(1/2)\left(t_f-t_0\right)d/dt. \tag{6.176}$$

The solutions to the state variables are approximated by the Lagrange interpolation polynomials that match the states at a set of N interpolation points and are given by

$$f(\tau)=\sum_{i=0}^{N} l_i(\tau)x_i, l_i(\tau)=\frac{\prod_{j\neq i,j=0}^{N}\left(\tau-\tau_j\right)}{\prod_{j\neq i,j=0}^{N}\left(\tau_i-\tau_j\right)}, x_i=x(\tau_i). \tag{6.177}$$

The polynomials $l_i(\tau)$ have the property that $l_i(\tau_j) = \delta_{ij}$.

A popular choice of the interpolation points used is the Chebyshev-Gauss-Lobatto points, which are the extreme points of the N-order Chebyshev polynomials:

$$T_N(\tau)=\cos\left(N\cos^{-1}\tau\right). \tag{6.178}$$

The Chebyshev-Gauss-Lobatto points are defined as

$$\tau_j=\cos(\pi(N-j)/N), j=0, 1, \cdots, N. \tag{6.179}$$

Another simpler and equally popular choice is to use small equal segments defined by

$$\tau_j = (2j - N)/N, j = 0, 1, \cdots, N. \tag{6.180}$$

The scalar differentiation functions $D_{ji}(\tau_j, \tau_i)$ are used to define the temporal derivatives of these approximated functions in the transformed domain. Thus,

$$\frac{d}{d\tau} f(\tau)\bigg|_{\tau=\tau_j} \equiv \dot{f}(\tau_j) = \sum_{i=0}^{N} D_{ji}(\tau_j, \tau_i) x_i. \tag{6.181}$$

Since

$$\frac{d}{d\tau} l_i(\tau) = \frac{\prod\limits_{j\neq i, j\neq 0, j=0}^{N} (\tau - \tau_j) + \prod\limits_{j\neq i, j\neq 1, j=0}^{N} (\tau - \tau_j) + \prod\limits_{j\neq i, j\neq 2, j=0}^{N} (\tau - \tau_j) \cdots + \prod\limits_{j\neq i, j\neq N, j=0}^{N} (\tau - \tau_j)}{\prod\limits_{j\neq i, j=0}^{N} (\tau_i - \tau_j)}, \tag{6.182}$$

it follows that

$$D_{ji}(\tau_j, \tau_i) = \frac{\sum\limits_{\substack{m=0 \\ m\neq i}}^{N} \left\{ \prod\limits_{\substack{k=0 \\ k\neq i, m}}^{N} (\tau_j - \tau_k) \right\}}{\prod\limits_{k\neq i, k=0}^{N} (\tau_i - \tau_k)}, i = 0, 1, \cdots, N, j = 1, 2, \cdots, N. \tag{6.183}$$

When there are $2n$ state and co-state variables, one equation each for a state or co-state variable together with the N differentiation equations constitutes a set of $N + 1$ equations. Thus, there are a total of $2(N + 1)n$ equations for as many variables. With the initial and final conditions, the equations are solvable after deleting the trivial equations, for the $2Nn$ unknowns.

The state vector is an n vector, and it is assumed to satisfy

$$\frac{d\mathbf{x}}{dt} = \mathbf{A}(t)\mathbf{x} + \mathbf{B}(t)\mathbf{u} + \mathbf{H}(t)\mathbf{d}. \tag{6.184}$$

It follows that

$$\begin{aligned}\frac{d\mathbf{x}(\tau_j)}{d\tau} &= \sum_{i=0}^{N} D_{ji}(\tau_j, \tau_i)\mathbf{I}_{n\times n}\mathbf{x}(\tau_i) \\ &= \frac{t_f - t_0}{2}\{\mathbf{A}(\tau_j)\mathbf{x}(\tau_j) + \mathbf{B}(\tau_j)\mathbf{u}(\tau_j) + \mathbf{H}(\tau_j)\mathbf{d}(\tau_j)\}.\end{aligned} \tag{6.185}$$

Let

$$\begin{aligned}&\mathbf{X} = \begin{bmatrix}\mathbf{x}^T(\tau_1) & \mathbf{x}^T(\tau_2) & \cdots & \mathbf{x}^T(\tau_N)\end{bmatrix}^T, \mathbf{U} = \begin{bmatrix}\mathbf{u}^T(\tau_1) & \mathbf{u}^T(\tau_2) & \cdots & \mathbf{u}^T(\tau_N)\end{bmatrix}^T, \\ &\widehat{\mathbf{d}} = \begin{bmatrix}\mathbf{d}^T(\tau_1) & \mathbf{d}^T(\tau_0) & \cdots & \mathbf{d}^T(\tau_N)\end{bmatrix}^T \text{and}\, \Delta t = t_f - t_0\end{aligned} \tag{6.186}$$

Then,

$$\frac{2}{\Delta t}\dot{\mathbf{X}} \equiv \mathbf{D}_{1,N}\mathbf{X} + \mathbf{D}_0\mathbf{x}(\tau_0), \tag{6.187}$$

and

$$\begin{aligned}\mathbf{D}_{1,N}\mathbf{X} + \mathbf{D}_0\mathbf{x}(\tau_0) = &\begin{bmatrix}\mathbf{A}_1 & \mathbf{0} & \cdots & \mathbf{0} \\ \mathbf{0} & \mathbf{A}_2 & \cdots & \mathbf{0} \\ \cdots & \cdots & \cdots & \cdots \\ \mathbf{0} & \mathbf{0} & \cdots & \mathbf{A}_N\end{bmatrix}\mathbf{X} + \begin{bmatrix}\mathbf{B}_1 & \mathbf{0} & \cdots & \mathbf{0} \\ \mathbf{0} & \mathbf{B}_2 & \cdots & \mathbf{0} \\ \cdots & \cdots & \cdots & \cdots \\ \mathbf{0} & \mathbf{0} & \cdots & \mathbf{B}_N\end{bmatrix}\mathbf{U} \\ &+ \begin{bmatrix}\mathbf{H}_1 & \mathbf{0} & \cdots & \mathbf{0} \\ \mathbf{0} & \mathbf{H}_2 & \cdots & \mathbf{0} \\ \cdots & \cdots & \cdots & \cdots \\ \mathbf{0} & \mathbf{0} & \cdots & \mathbf{H}_N\end{bmatrix}\widehat{\mathbf{d}}.\end{aligned} \tag{6.188}$$

where

$$\begin{aligned}&\mathbf{D}_{1,N} = \frac{2}{\Delta t}\begin{bmatrix}D_{11}\mathbf{I}_{n\times n} & D_{12}\mathbf{I}_{n\times n} & \cdots & D_{1N}\mathbf{I}_{n\times n} \\ D_{21}\mathbf{I}_{n\times n} & D_{22}\mathbf{I}_{n\times n} & \cdots & D_{2N}\mathbf{I}_{n\times n} \\ \cdots & \cdots & \cdots & \cdots \\ D_{N1}\mathbf{I}_{n\times n} & D_{N2}\mathbf{I}_{n\times n} & \cdots & D_{NN}\mathbf{I}_{n\times n}\end{bmatrix}, \mathbf{D}_0 = \frac{2}{\Delta t}\begin{bmatrix}D_{10}\mathbf{I}_{n\times n} \\ D_{20}\mathbf{I}_{n\times n} \\ \cdots \\ D_{N0}\mathbf{I}_{n\times n}\end{bmatrix}, \\ &\mathbf{A}_j = \mathbf{A}(\tau_j), \mathbf{B}_j = \mathbf{B}(\tau_j), \quad \mathbf{H}_j = \mathbf{H}(\tau_j).\end{aligned} \tag{6.189}$$

Thus,

$$\left\{\mathbf{D}_{1,N} - \begin{bmatrix} \mathbf{A}_1 & \mathbf{0} & \cdots & \mathbf{0} \\ \mathbf{0} & \mathbf{A}_2 & \cdots & \mathbf{0} \\ \cdots & \cdots & \cdots & \cdots \\ \mathbf{0} & \mathbf{0} & \cdots & \mathbf{A}_N \end{bmatrix}\right\}\mathbf{X} = \begin{bmatrix} \mathbf{B}_1 & \mathbf{0} & \cdots & \mathbf{0} \\ \mathbf{0} & \mathbf{B}_2 & \cdots & \mathbf{0} \\ \cdots & \cdots & \cdots & \cdots \\ \mathbf{0} & \mathbf{0} & \cdots & \mathbf{B}_N \end{bmatrix}\mathbf{U}$$

$$+ \begin{bmatrix} \mathbf{H}_1 & \mathbf{0} & \cdots & \mathbf{0} \\ \mathbf{0} & \mathbf{H}_2 & \cdots & \mathbf{0} \\ \cdots & \cdots & \cdots & \cdots \\ \mathbf{0} & \mathbf{0} & \cdots & \mathbf{H}_N \end{bmatrix}\widehat{\mathbf{d}} - \mathbf{D}_0\mathbf{x}(\tau_0). \tag{6.190}$$

Similarly, in general, the co-state vector is assumed to satisfy

$$\frac{d}{dt}\boldsymbol{\lambda}(t) = -\mathbf{A}(t)^T\boldsymbol{\lambda}(t) + \mathbf{Q}(t)\mathbf{x}(t). \tag{6.191}$$

It follows that

$$\frac{d\boldsymbol{\lambda}(\tau_j)}{d\tau} = \sum_{i=0}^{N} D_{ji}(\tau_j, \tau_i)\mathbf{I}_{n\times n}\boldsymbol{\lambda}(\tau_i) = -\frac{t_f - t_0}{2}\left\{\mathbf{A}^T(\tau_j)\boldsymbol{\lambda}(\tau_j) - \mathbf{Q}(\tau_j)\mathbf{x}(\tau_j)\right\}. \tag{6.192}$$

Thus,

$$\frac{2}{\Delta t}\dot{\boldsymbol{\Lambda}}$$

$$\equiv \mathbf{D}_{0,N-1}\boldsymbol{\Lambda} + \mathbf{D}_N\boldsymbol{\lambda}(\tau_N) = -\begin{bmatrix} \mathbf{A}_0^T & \mathbf{0} & \cdots & \mathbf{0} \\ \mathbf{0} & \mathbf{A}_1^T & \cdots & \mathbf{0} \\ \cdots & \cdots & \cdots & \cdots \\ \mathbf{0} & \mathbf{0} & \cdots & \mathbf{A}_{N-1}^T \end{bmatrix}\boldsymbol{\Lambda}$$

$$+ \begin{bmatrix} \mathbf{Q}_0 & \mathbf{0} & \cdots & \mathbf{0} \\ \mathbf{0} & \mathbf{Q}_1 & \cdots & \mathbf{0} \\ \cdots & \cdots & \cdots & \cdots \\ \mathbf{0} & \mathbf{0} & \cdots & \mathbf{Q}_{N-1} \end{bmatrix}\mathbf{X}, \tag{6.193}$$

with

$$\mathbf{D}_0 \equiv \frac{2}{\Delta t}\begin{bmatrix} D_{0N}\mathbf{I}_{n\times n} \\ D_{1N}\mathbf{I}_{n\times n} \\ \cdots \\ D_{N-1,N}\mathbf{I}_{n\times n} \end{bmatrix}, \boldsymbol{\Lambda} = \begin{bmatrix} \boldsymbol{\lambda}^T(\tau_0) & \boldsymbol{\lambda}^T(\tau_1) & \cdots & \boldsymbol{\lambda}^T(\tau_{N-1}) \end{bmatrix}^T. \tag{6.194}$$

Thus,

$$\left\{\mathbf{D}_{0,N-1}+\begin{bmatrix}\mathbf{A}_0^T & \mathbf{0} & \cdots & \mathbf{0}\\ \mathbf{0} & \mathbf{A}_1^T & \cdots & \mathbf{0}\\ \cdots & \cdots & \cdots & \cdots\\ \mathbf{0} & \mathbf{0} & \cdots & \mathbf{A}_{N-1}^T\end{bmatrix}\right.\mathbf{\Lambda}=\begin{bmatrix}\mathbf{Q}_0 & \mathbf{0} & \cdots & \mathbf{0}\\ \mathbf{0} & \mathbf{Q}_1 & \cdots & \mathbf{0}\\ \cdots & \cdots & \cdots & \cdots\\ \mathbf{0} & \mathbf{0} & \cdots & \mathbf{Q}_{N-1}\end{bmatrix}\mathbf{X}-\mathbf{D}_N\boldsymbol{\lambda}(\tau_N). \tag{6.195}$$

Finally, the two sets of equations for the state and co-state vectors are solved simultaneously with the initial conditions applied to the state vector, the final conditions applied to the co-state vector and the control given by

$$\mathbf{u}(t)=-\mathbf{R}^{-1}\mathbf{B}^T(t)\boldsymbol{\lambda}(t). \tag{6.196}$$

Thus, if

$$\widehat{\mathbf{A}}=\begin{bmatrix}\mathbf{A}_1 & \mathbf{0} & \cdots & \mathbf{0}\\ \mathbf{0} & \mathbf{A}_2 & \cdots & \mathbf{0}\\ \cdots & \cdots & \cdots & \cdots\\ \mathbf{0} & \mathbf{0} & \cdots & \mathbf{A}_N\end{bmatrix},\widehat{\mathbf{H}}\equiv\begin{bmatrix}\mathbf{H}_1 & \mathbf{0} & \cdots & \mathbf{0}\\ \mathbf{0} & \mathbf{H}_2 & \cdots & \mathbf{0}\\ \cdots & \cdots & \cdots & \cdots\\ \mathbf{0} & \mathbf{0} & \cdots & \mathbf{H}_N\end{bmatrix}, \tag{6.197}$$

the pair of coupled linear equations to solve for $\mathbf{X}$ and $\mathbf{\Lambda}$ are

$$\begin{aligned}&\left\{\widehat{\mathbf{A}}-\mathbf{D}_{1,N}\right\}\mathbf{X}-\begin{bmatrix}\mathbf{0} & \mathbf{0} & \cdots & \mathbf{0}\\ \mathbf{0} & \mathbf{B}_1\mathbf{R}^{-1}\mathbf{B}_1^T & \cdots & \mathbf{0}\\ \cdots & \cdots & \cdots & \cdots\\ \mathbf{0} & \mathbf{0} & \cdots & \mathbf{B}_{N-1}\mathbf{R}^{-1}\mathbf{B}_{N-1}^T\end{bmatrix}\mathbf{\Lambda}\\ &=\begin{bmatrix}\mathbf{0}\\ \mathbf{0}\\ \cdots\\ \mathbf{B}_N\mathbf{R}^{-1}\mathbf{B}_N^T\end{bmatrix}\boldsymbol{\lambda}(\tau_f)-\widehat{\mathbf{H}}\widehat{\mathbf{d}}+\mathbf{D}_0\mathbf{x}(\tau_0),\\ &\begin{bmatrix}\mathbf{Q}_0 & \mathbf{0} & \cdots & \mathbf{0}\\ \mathbf{0} & \mathbf{Q}_1 & \cdots & \mathbf{0}\\ \cdots & \cdots & \cdots & \cdots\\ \mathbf{0} & \mathbf{0} & \cdots & \mathbf{Q}_{N-1}\end{bmatrix}\mathbf{X}+\left\{-\begin{bmatrix}\mathbf{A}_0^T & \mathbf{0} & \cdots & \mathbf{0}\\ \mathbf{0} & \mathbf{A}_1^T & \cdots & \mathbf{0}\\ \cdots & \cdots & \cdots & \cdots\\ \mathbf{0} & \mathbf{0} & \cdots & \mathbf{A}_{N-1}^T\end{bmatrix}-\mathbf{D}_{0,N-1}\right\}\mathbf{\Lambda}\\ &=\mathbf{D}_N\boldsymbol{\lambda}(\tau_f).\end{aligned} \tag{6.198}$$

These equations may also be derived from the Hamiltonian equations for the state and co-state vectors, for the optimal closed loop. Thus, with $\widehat{\mathbf{d}} = \mathbf{0}$, the Hamiltonian equations for the closed loop are

$$\frac{d}{dt}\begin{bmatrix} \mathbf{x}(t) \\ \boldsymbol{\lambda}(t) \end{bmatrix} = \begin{bmatrix} \mathbf{A} & -\mathbf{BRB}^T \\ \mathbf{Q} & -\mathbf{A}^T \end{bmatrix}\begin{bmatrix} \mathbf{x}(t) \\ \boldsymbol{\lambda}(t) \end{bmatrix}. \tag{6.199}$$

Thus, the solution is found by solving the discretized Hamiltonian equations.

When the conditions specified on the states and co-states are both at the final time $\tau = \tau_f$, the pair of coupled linear equations to solve for $\mathbf{X}$ and $\boldsymbol{\Lambda}$ are

$$\begin{gathered}
\left\{\begin{bmatrix} \mathbf{A}_0 & \mathbf{0} & \cdots & \mathbf{0} \\ \mathbf{0} & \mathbf{A}_1 & \cdots & \mathbf{0} \\ \cdots & \cdots & \cdots & \cdots \\ \mathbf{0} & \mathbf{0} & \cdots & \mathbf{A}_{N-1} \end{bmatrix} - \mathbf{D}_{0,N-1}\right\} \\
\mathbf{X} - \begin{bmatrix} \mathbf{B}_0\mathbf{R}^{-1}\mathbf{B}_0^T & \mathbf{0} & \cdots & \mathbf{0} \\ \mathbf{0} & \mathbf{B}_1\mathbf{R}^{-1}\mathbf{B}_1^T & \cdots & \mathbf{0} \\ \cdots & \cdots & \cdots & \cdots \\ \mathbf{0} & \mathbf{0} & \cdots & \mathbf{B}_{N-1}\mathbf{R}^{-1}\mathbf{B}_{N-1}^T \end{bmatrix}\boldsymbol{\Lambda} \\
= \begin{bmatrix} 0 \\ \mathbf{0} \\ \cdots \\ \mathbf{B}_N\mathbf{R}^{-1}\mathbf{B}_N^T \end{bmatrix}\boldsymbol{\lambda}(\tau_f) - \widehat{\mathbf{H}}\widehat{\mathbf{d}} + \mathbf{D}_N\mathbf{x}(\tau_f), \\
\begin{bmatrix} \mathbf{Q}_0 & \mathbf{0} & \cdots & \mathbf{0} \\ \mathbf{0} & \mathbf{Q}_1 & \cdots & \mathbf{0} \\ \cdots & \cdots & \cdots & \cdots \\ \mathbf{0} & \mathbf{0} & \cdots & \mathbf{Q}_{N-1} \end{bmatrix}\mathbf{X} + \left\{-\begin{bmatrix} \mathbf{A}_0^T & \mathbf{0} & \cdots & \mathbf{0} \\ \mathbf{0} & \mathbf{A}_1^T & \cdots & \mathbf{0} \\ \cdots & \cdots & \cdots & \cdots \\ \mathbf{0} & \mathbf{0} & \cdots & \mathbf{A}_{N-1}^T \end{bmatrix} - \mathbf{D}_{0,N-1}\right\}\boldsymbol{\Lambda} \\
= \mathbf{D}_N\boldsymbol{\lambda}(\tau_f).
\end{gathered} \tag{6.200}$$

References

Battin, R. (1999). *An introduction to the mathematics and methods of astrodynamics, Revised Edition* (AIAA education series, 1801 Alexander Bell Drive). Revised Edition.

Bertrand, R., & Epenoy, R. (2002). New smoothing techniques for solving bang–bang optimal control problems—Numerical results and statistical interpretation. *Optimal Control Applications and Methods, 23*(4), 171–197. https://doi.org/10.1002/oca.709

Bryson, A. E., Jr., & Ho, Y. C. (1969). *Applied optimal control.* Ginn and Company.

Eapen, R. T., & Sharma, R. K. (2014). Mars interplanetary trajectory design via Lagrangian points. *Astrophysics and Space Science, 353*, 65–71. https://doi.org/10.1007/s10509-014-2012-x

Edelbaum, T. N. (1961). Propulsion requirements for controllable satellites. *ARS Journal, 31*, 1079–1089.

Englander, J. A., Vavrina, M. A., Lim, L. F., McFadden, L. A., Rhoden, A. R., & Noll, K. S. (2017). Trajectory optimization for missions to small bodies with a focus on scientific merit. *Computing in Science & Engineering, 19*(4), 18–28. https://doi.org/10.1109/MCSE.2017.3151246

Farquhar, R. W. (1968). Chapter 3: Linear feedback control. In *The control and use of Libration-point satellites* (SUDAAR NO. 350, Department of Aeronautics and Astronautics) (pp. 73–91). Stanford University.

Forsyth, A. R. (1960). *Calculus of variations.* Dover Publ.

Fritz, S., & Turkoglu, K. (2017). Optimal trajectory determination and mission design for Asteroid/deep-space exploration via multibody gravity assist maneuvers. *International Journal of Aerospace Engineering (Hindawi), 2017*, 6801023(1–12). https://doi.org/10.1155/2017/6801023

Ghoreishi, S. A., Nekoui, M. A., & Basiri, S. O. (2011). Optimal design of LQR weighting matrices based on intelligent optimization methods. *International Journal of Intelligent Information Processing, 2*(1), 63.

Ikawa, H. (1988). Coplanar three-body trans-Earth-lunar trajectory simulation methodology. In *AIAA 26th aerospace science meeting, Reno, NV.*

Kasyap, V., & Vepa, R. (2022). *Reinforcement learning based linear quadratic regulator for the control of a quadcopter.* Paper presented to AIAA 2022 region VII student conference, 28–-29 November 2022, University of Adelaide, Adelaide, Australia.

Kechichian, J. A. (1990). *Equinoctial orbit elements: Application to optimal transfer problems.* In *AIAA 90-2976, AIAA/AAS astrodynamics conference, Portland, OR, 20–22 August 1990.*

Lee, S., Petropoulos, A., & von Allmen, P. (2005). Low-thrust orbit transfer optimization with refined Q-law and multi-objective genetic algorithm. In *AAS 05-393, AAS/AIAA astrodynamics specialists conference Lake Tahoe, CA, USA, August 7–11, 2005.*

Miele, A., & Mancuso, S. (2001). Optimal trajectories for earth–moon–earth flight. *Acta Astronautica, 49*(2), 59–71.

Miele, A., & Wang, T. (1997). Optimal trajectories for earth-to-Mars flight. *Journal of Optimization Theory and Applications, 95*, 467–499.

Mingotti, G., Topputo, F., & Bernelli-Zazzera, F. (2011). Earth-Mars transfers with ballistic escape and lowthrust capture. In *Celestial mechanics and dynamical astronomy* (pp. 169–188). https://doi.org/10.1007/s10569-011-9343-5

Nikolichev, I., & Sesyukalov, V. (2023). Design of a low-Energy Earth-Moon Flight Trajectory Using a planar auxiliary problem. *Applied Sciences, 13*, 1967. https://doi.org/10.3390/app13031967

Pan, X., & Pan, B. (2020). Practical homotopy methods for finding the best minimum-fuel transfer in the circular restricted three-body problem. *IEEE Access, 8*, 47845–47862.

Pan, S., Topputo, F., Hou, X., & Wang, Y. (2022). On optimal three-impulse Earth-Moon transfers in a four-body model. In *73rd International Astronautical Congress (IAC), Paris, France, 18–-22 September 2022, IAC-22,C1,IP,22,x73449* (pp. 1–9).

Petropoulos, A. E. (2003). Simple control laws for low-thrust orbit transfers. In *AAS/AIAA astrodynamics specialists conference.*

Petropoulos, A. E. (2005). Refinements to the Q-law for low-thrust orbit transfers. *Advances in the Astronautical Sciences, 120*, 963–982.

Pollard, J. E. (2000). *Simplified analysis of low-thrust orbital manoeuvers*. Aerospace Report No. TR-2000(8565)-10.

Pollard, J. E. (2001). Simplified approach for assessment of low-thrust elliptical orbit transfers. In *27th international electric propulsion conference*.

Saghamanesh, M., & Baoyin, H. (2018). A robust homotopic approach for continuous variable low-thrust trajectory optimization. *Advances in Space Research (ASR), 62*, 3095–3113.

Schwenzfeger, K. J. (1974, April). *Low thrust space vehicle trajectory optimization using regularized variables*, NASA TR-426.

Shirazi, A., Ceberio, J., & Lozano, J. A. (2018). Spacecraft trajectory optimization: A review of models, objectives, approaches and solutions. *Progress in Aerospace Sciences, 102*, 76–98. https://doi.org/10.1016/j.paerosci.2018.07.007

Varga, G., & Sánchez-Pérez, J. M. (2016). Many-revolution low-thrust orbit transfer computation using equinoctial Q-law including J2 and eclipse effects. In *6th international conference on astrodynamics tools and techniques, ICATT-111-21-97, ESA, March 2016* (pp. 2463–2481).

Vepa, R. (2019). *Optimal manoeuver trajectory synthesis for autonomous space and aerial vehicles and robots*. Presented at: TAROS 2019, 20th towards autonomous robotic systems conference, 3–5th July 2019, London, UK.

Vepa, R. (2022). Feedback tracking control of optimal reference trajectories for spacecraft relative motion. *Advances in Space Research, 69*(9), 3478–3489.

Walker, M. J. H., Ireland, B., & Owens, J. (1985). A set of modified equinoctial orbital elements. *Celestial Mechanics, 36*, 409–419.

Yang, L., Liu, X., Chen, W., & Zhou, H. (2018). Autonomous entry guidance using linear Pseudospectral model predictive control. *Aerospace Science and Technology, 80*, 38–55.

Zhao, P., Meng, J., Chen, Z., Zhang, H., & Yu, Y. (2023). Analytical trajectory prediction for skip re-entry of lifting vehicle. *Proceedings of the IMechE Part G: Journal of Aerospace Engineering, Institution of Mechanical Engineers, 237*(9), 2139–2157.

Chapter 7
Patched Three-Body Approximation: Application to Interplanetary Trajectory Planning

7.1 Three-Body Problem

Although the three-body problem and its circular restricted solution were introduced in Chap. 1, it deserves a much more extensive and detailed coverage. For this reason it will be discussed to a much greater depth in this chapter.

The three-body problem is one of the oldest problems in classical dynamics that continues to throw up surprises. It has challenged scientists from Newton's time to the present. It arose in an attempt to understand the Sun's effect on the motion of the Moon around the Earth. This was of much practical importance in marine navigation, where lunar tables were necessary to accurately determine longitude at sea.

The study of the three-body problem led to the discovery of the planet Neptune and it explains the location and stability of the Trojan asteroids and has furthered our understanding of the stability of the solar system. The three-body problem admits many "regular" solutions such as the collinear and equilateral periodic solutions of Euler and Lagrange as well as the more recently discovered figure-∞ solution. On the other hand, it can also display chaos as originally discovered by Poincaré (1993). Although a general solution in closed form is still unknown, Sundman (1912) discovered a representation of the solutions in fractional powers of time which is an exceptionally slowly converging series.

In this chapter, the problem of three bodies is considered, modelled as point masses (m_j with position vectors r_j for $j = 1, 2, 3$) moving under their mutual gravitational attraction. This system has 9 degrees of freedom, whose dynamics is determined by 9 coupled second-order nonlinear ODEs:

$$m_j \frac{d^2 \mathbf{r}_j}{dt^2} = \sum_{i \neq j} G m_j m_i \frac{\mathbf{r}_i - \mathbf{r}_j}{\left|\mathbf{r}_i - \mathbf{r}_j\right|^3}. \tag{7.1}$$

The three components of momentum, three components of angular momentum and the energy are respectively given by

R. Vepa, *Space Vehicle Maneuvering, Propulsion, Dynamics and Control*,
https://doi.org/10.1007/978-3-031-65518-0_7

$$\mathbf{P}=\sum_j m_j\dot{\mathbf{r}}_j,\ \ \mathbf{L}=\sum_j \mathbf{r}_j\times m_j\dot{\mathbf{r}}_j,\ \ T=\frac{1}{2}\sum_j m_j\dot{\mathbf{r}}_j^2,\ \ E=\frac{1}{2}\sum_j m_j\dot{\mathbf{r}}_j^2-\sum_{i\neq j}\frac{Gm_jm_i}{|\mathbf{r}_i-\mathbf{r}_j|}=T+V. \tag{7.2}$$

The restricted three-body problem is a simplified version of the three-body problem where the mass of the third body m_3 is assumed much smaller than the masses of the primaries m_1 and m_2. Thus the bodies with masses m_1 and m_2 move in Keplerian orbits which are not affected by the mass of the third body, m_3. In the planar circular restricted three-body problem (CR3BP), the primaries move in fixed circular orbits around their common centre of mass with the angular speed $\Omega=\sqrt{G(m_1+m_2)/d^3}$ given by Kepler's third law and m_3 moves in the same plane as m_1 and m_2. Here, d is the separation between the primaries. This system has 2 degrees of freedom associated to the planar motion of m_3, and therefore, a four-dimensional phase space just like the planar Kepler problem for the reduced mass. However, unlike the two-body Kepler problem which has three conserved quantities (energy, z-component of angular momentum and direction of the Laplace-Runge-Lenz (LRL) vector, $\mathbf{A}_{\mathrm{LRL}}=\mathbf{P}\times\mathbf{L}-\mu m(\mathbf{r}_2-\mathbf{r}_1)/|\mathbf{r}_2-\mathbf{r}_1|$, which is constant both in direction and magnitude and describes the shape and orientation of the orbit of a body around another) and is exactly solvable, the planar restricted three-body problem has only one known conserved quantity, the "Jacobi integral", which is the energy of m_3 in the co-rotating (non-inertial) frame of the primaries.

The restricted three-body problem is most easily understood and analysed in a coordinate system rotating with the two primary bodies. Since the solution of the two-body problem is that of the two primary masses rotating at constant angular velocity Ω about their centre of mass, in a coordinate system rotating with angular velocity Ω, these two bodies are stationary. The third body moves in their gravitational fields and experiences the additional acceleration terms associated with the rotation of the coordinate system.

The circular restricted three-body problem (CR3BP) requires that the two bodies have their masses significantly larger than the third body ($m_1>>m_3$ and $m_2>>m_3$) and that the motions of m_1 and m_2 are limited to circular orbits around their centre of mass. In the literature in celestial mechanics devoted to the CR3BP, m_1 and m_2 are typically referred to as the primaries (Szebehely 1967), but in stellar dynamics m_1 and m_2 are called the primary and secondary, respectively, with the assumption that $m_1\geq m_2$. In this chapter, following most authors, the former nomenclature is used as a clear distinction between m_1 and m_2 is not necessary; however, the second nomenclature is used when a distinction between m_1 and m_2 is necessary.

When the third mass moves in three dimensions, this case is called the 3D CR3BP. However, if it moves in the same plane as the primaries, this case is the planar CR3BP. Since the gravitational influence of the third body on the primaries is negligible, the orbits of the primaries could be described by the two-body problem, whose solutions are well-known. Once the solutions for the primaries are known, they can be used to determine the motion of the third body resulting from the gravitational field of the primaries. From a mathematical point of view, the problem becomes simpler than the general three-body problem. Nevertheless, for some sets of initial conditions the resulting motions of the third body remain unpredictable.

While trajectory design has traditionally been initiated with a baseline mission concept associated with the two-body problem and the resulting conics, the patched three-body method as proposed by Koon et al. (2000) and further explained in Koon et al. (2007) can offer a better solution for most interplanetary trajectories. In this chapter, the systematic construction of the trajectories in the case of a multi-body problem using invariant manifolds or trajectory patches from multiple three-body systems and using the patched three-body approximation in each situation is now described. The objective is to document some of the key aspects of the patched three-body method and its application to interplanetary trajectory synthesis. The objective of the patched three-body method is to take maximum advantage of the three-body gravitational fields to reduce fuel consumption, which is especially useful in the design of interplanetary trajectories designed to visit one or more celestial bodies. These may include a low energy transfer mission to and from the Earth to the Moon or trajectories for a low energy mission to orbit another planet, such as Mars or Jupiter. Two examples of applications are presented. The advantages of the patched three-body approximation method over the traditional patched conic approximation using trajectory patches of Keplerian orbits are discussed in the concluding paragraph.

The key cost of a mission is the amount of fuel required to complete the whole mission successfully. To this end the primary metric used to compare various trajectory options is the propulsive velocity increment $|\Delta \vec{v}|$ required to escape the departure planet and reach the desired trajectory insertion point, cruise along the trajectory till one reaches a region close to the destination planet and the finally the entry velocity at the destination planet, before successfully returning back, if such a return is envisaged.

7.2 Key Features of the Circular Restricted Three-Body Problem

To fully comprehend the patched three-body approximation method, and its applications, which rely on being able to rapidly transform from one set of synodic coordinates to another, it is important to review the key features of the three-body problem. Consider a two-body system of mass m_1 and m_2 in an inertial references frame. A small mass m_3 which does not affect the large masses is also included. The origin of a non-rotating coordinate system (Fig. 7.1) is chosen to be at the centre of mass of the primaries, so one can write $m_1\mathbf{r}_1 = m_2\mathbf{r}_2$ where m_1 and m_2 are the masses of the primaries with $\mathbf{r}_1$ and $\mathbf{r}_2$ denoting their position vectors with respect to the origin of the coordinate system. To describe the problem mathematically, the approach of Szebehely (1967), Roy (2005), Danby (1988), Musielak and Quarles (2014) and Eberle et al. (2008) is followed. Further details may be found in Musielak and Quarles (2014).

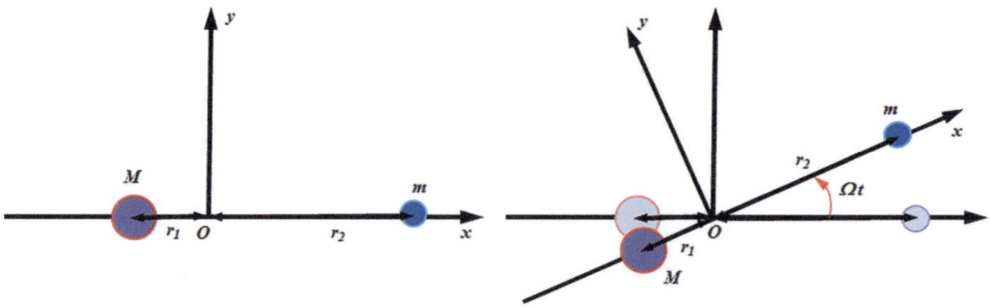

Fig. 7.1 Non-rotating and rotating coordinate systems used to describe the CR3BP. (Re-drawn from Eberle et al. (2008). Copyright 2008 Astronomy & Astrophysics)

The components of the position vectors are as follows: $\mathbf{r}_1 = [x_1 \quad y_1 \quad z_1]^T$, $\mathbf{r}_2 = [x_2 \ y_2 \ z_2]^T$ and $\mathbf{r}_3 = [x_3 \ y_3 \ z_3]^T$. Furthermore, $\mathbf{r}_{ij} = \mathbf{r}_i - \mathbf{r}_j$. Thus

$$\frac{d^2\mathbf{r}_j}{dt^2} = \sum_{i \neq j} Gm_i \frac{\mathbf{r}_{ij}}{\left|\mathbf{r}_{ij}\right|^3}. \tag{7.3}$$

Since it is only the motion of the satellite that is of interest

$$\frac{d^2\mathbf{r}_3}{dt^2} = \sum_{i \neq j} Gm_i \frac{\mathbf{r}_{i3}}{\left|\mathbf{r}_{i3}\right|^3}. \tag{7.4}$$

Let $m = m_1 + m_2$ be the total mass of the primaries, $\mu = m_2/m$ the mass ratio of the secondary, so $m_2 = \mu m$ and $m_1 = (1 - \mu)m$, $d = r_{12}$, $r_{13} = |\mathbf{r}_{13}|$ and $r_{23} = |\mathbf{r}_{23}|$, and equating the gravitational force to the centrifugal force:

$$\frac{m_3 v_j^2}{r_{j3}} = \frac{Gm_3 m_{3-j}}{d^2} = \frac{m_3}{r_{j3}} r_{j3}^2 \Omega^2 = \frac{Gm_3 m r_{j3}}{d^3}, \tag{7.5}$$

where $r_{j3}^2 = \left\{ \left(x_j - x_3\right)^2 + \left(y_j - y_3\right)^2 + \left(z_j - z_3\right)^2 \right\}$. The bodies are illustrated in Fig. 7.2. This results in Kepler's third law

$$\Omega^2 = Gm/d^3, \quad Gm = G(m_1 + m_2) = \mu_1 + \mu_2 = d^3\Omega^2; \quad T = 2\pi\sqrt{d^3/(\mu_1 + \mu_2)}. \tag{7.6}$$

Now, since the two large masses are in circular orbits, one may write: $x_{13} \equiv x_1 = r_{13} \cos \Omega t$, $y_{13} \equiv y_1 = r_{13} \sin \Omega t$, $x_{23} \equiv x_2 = -r_{23} \cos \Omega t$, $y_{23} \equiv y_2 = -r_{23} \sin \Omega t$, $z_1 = z_2 = 0$ if the orbits are in the same plane. Introducing $x_3 = d\tilde{x}$, $y_3 = d\tilde{y}$, $z_3 = d\tilde{z}$, $\tau = \Omega t$, the satellite satisfies the following equations in three dimensions:

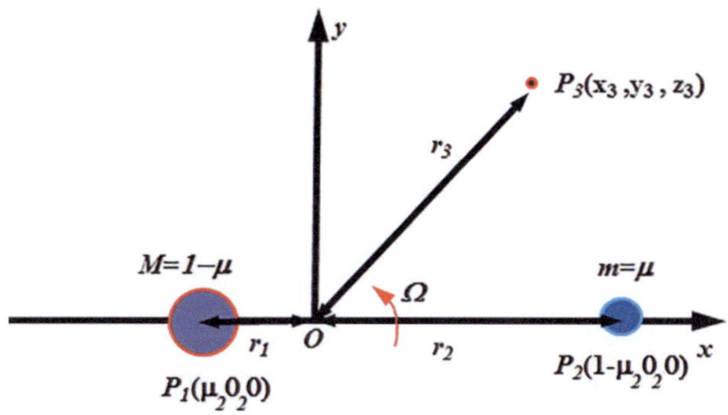

Fig. 7.2 Coordinate system and non dimensionalization based on Fig. 7.1

$$\frac{d^2}{d\tau^2}\begin{bmatrix}\tilde{x}\\ \tilde{y}\\ \tilde{z}\end{bmatrix} = -\frac{(1-\mu)}{\tilde{r}_{13}^3}\begin{bmatrix}\tilde{x}-\mu\cos\tau\\ \tilde{y}-\mu\sin\tau\\ \tilde{z}\end{bmatrix} - \frac{\mu}{\tilde{r}_{23}^3}\begin{bmatrix}\tilde{x}+(1-\mu)\cos\tau\\ \tilde{y}+(1-\mu)\sin\tau\\ \tilde{z}\end{bmatrix}, \tag{7.7}$$

with $\tilde{r}_{13}^2 = \left\{(\tilde{x}-\mu\cos\tau)^2+(\tilde{y}-\mu\sin\tau)^2+\tilde{z}^2\right\}, \tilde{r}_{23}^2 = \left\{(\tilde{x}+(1-\mu)\cos\tau)^2+(\tilde{y}+(1-\mu)\sin\tau)^2+\tilde{z}^2\right\}$. This set of equations describes the CR3BP in the inertial coordinate system.

One may write the above set of equations in the rotating (synodic) coordinate system (Fig. 7.1) by using the following relationships between the coordinates $\tilde{x}$, $\tilde{y}$, $\tilde{z}$ and their derivatives in the inertial system, and the coordinates x, y and z, in the synodic system:

$$\begin{bmatrix}\tilde{x}\\ \tilde{y}\\ \tilde{z}\end{bmatrix} = \begin{bmatrix}\cos\varpi\tau & -\sin\varpi\tau & 0\\ \sin\varpi\tau & \cos\varpi\tau & 0\\ 0 & 0 & 1\end{bmatrix}\begin{bmatrix}x\\ y\\ z\end{bmatrix}, \begin{bmatrix}\dot{\tilde{x}}\\ \dot{\tilde{y}}\\ \dot{\tilde{z}}\end{bmatrix}$$

$$= -\varpi\begin{bmatrix}\sin\varpi\tau & \cos\varpi\tau & 0\\ -\cos\varpi\tau & \sin\varpi\tau & 0\\ 0 & 0 & 0\end{bmatrix}\begin{bmatrix}x\\ y\\ z\end{bmatrix} + \begin{bmatrix}\cos\varpi\tau & -\sin\varpi\tau & 0\\ \sin\varpi\tau & \cos\varpi\tau & 0\\ 0 & 0 & 1\end{bmatrix}\begin{bmatrix}\dot{x}\\ \dot{y}\\ \dot{z}\end{bmatrix}.$$

$$\begin{bmatrix}\dot{\tilde{x}}\\ \dot{\tilde{y}}\\ \dot{\tilde{z}}\end{bmatrix} = \begin{bmatrix}\cos\varpi\tau & -\sin\varpi\tau & 0\\ \sin\varpi\tau & \cos\varpi\tau & 0\\ 0 & 0 & 1\end{bmatrix}\left\{\begin{bmatrix}\dot{x}\\ \dot{y}\\ \dot{z}\end{bmatrix} - \varpi\begin{bmatrix}0 & 1 & 0\\ -1 & 0 & 0\\ 0 & 0 & 0\end{bmatrix}\begin{bmatrix}x\\ y\\ z\end{bmatrix}\right\}$$

$$= \begin{bmatrix}\cos\varpi\tau & -\sin\varpi\tau & 0\\ \sin\varpi\tau & \cos\varpi\tau & 0\\ 0 & 0 & 1\end{bmatrix}\begin{bmatrix}\dot{x}-\varpi y\\ \dot{y}+\varpi x\\ \dot{z}\end{bmatrix},$$

$$\begin{bmatrix}\ddot{\tilde{x}}\\ \ddot{\tilde{y}}\\ \ddot{\tilde{z}}\end{bmatrix} = \begin{bmatrix}\cos\varpi\tau & -\sin\varpi\tau & 0\\ \sin\varpi\tau & \cos\varpi\tau & 0\\ 0 & 0 & 1\end{bmatrix}\begin{bmatrix}\ddot{x}-2\varpi\dot{y}-\varpi^2 x\\ \ddot{y}+2\varpi\dot{x}-\varpi^2 y\\ \ddot{z}\end{bmatrix}. \tag{7.8}$$

Therefore, the set of equations describing the CR3BP in the synodic coordinate system with $\varpi = 1$ can be written as

$$\frac{d^2}{d\tau^2}\begin{bmatrix} x \\ y \\ z \end{bmatrix} - 2\frac{d}{d\tau}\begin{bmatrix} y \\ -x \\ 0 \end{bmatrix} = \begin{bmatrix} x \\ y \\ 0 \end{bmatrix} - \frac{(1-\mu)}{r_1^3}\begin{bmatrix} x-\mu \\ y \\ z \end{bmatrix} - \frac{\mu}{r_2^3}\begin{bmatrix} x+(1-\mu) \\ y \\ z \end{bmatrix}. \tag{7.9}$$

In the above equations, $r_1^2 = \left\{(x-\mu)^2 + y^2 + z^2\right\}, r_2^2 = \left\{(x+1-\mu)^2 + y^2 + z^2\right\}$. In state space form

$$\dot{x}_1 = \dot{x} = x_3, \quad \dot{x}_2 = \dot{y} = x_4, \dot{x}_5 = \dot{z} = x_6, \tag{7.10}$$

$$\begin{aligned} \dot{x}_3 &= 2\dot{x}_2 + x_1 - \frac{\mu}{r_2^3}(x_1 + 1 - \mu) - \frac{1-\mu}{r_1^3}(x_1 - \mu), \\ \dot{x}_4 &= -2\dot{x}_1 + x_2\left(1 - \frac{\mu}{r_2^3} - \frac{1-\mu}{r_1^3}\right), \dot{x}_6 = -x_5\left(\frac{\mu}{r_2^3} + \frac{1-\mu}{r_1^3}\right). \end{aligned} \tag{7.11}$$

7.2.1 *Altering the Coordinate System*

In what follows it is intended to change the coordinate system to the more commonly used frame of reference shown in Fig. 7.3.

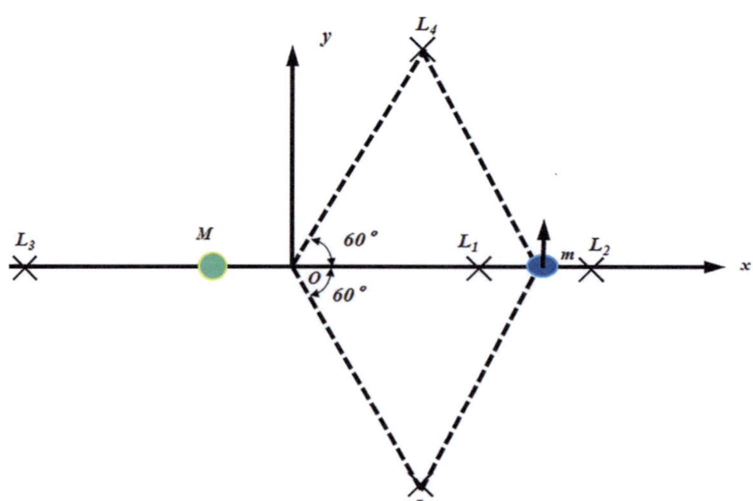

Fig. 7.3 Rotating frame of reference used by most authors, showing also the relative location of the equilibrium points

The transformation to the new frame of reference is expressed as

$$\begin{bmatrix} x \\ y \\ z \end{bmatrix}_{new} = -\begin{bmatrix} x \\ y \\ z \end{bmatrix}_{old}. \tag{7.12}$$

Hence equations of motion reduce to

$$\frac{d^2}{d\tau^2}\begin{bmatrix} x \\ y \\ z \end{bmatrix} - 2\frac{d}{d\tau}\begin{bmatrix} y \\ -x \\ 0 \end{bmatrix} = \begin{bmatrix} x \\ y \\ 0 \end{bmatrix} - \frac{(1-\mu)}{r_1^3}\begin{bmatrix} x+\mu \\ y \\ z \end{bmatrix} - \frac{\mu}{r_2^3}\begin{bmatrix} x-(1-\mu) \\ y \\ z \end{bmatrix}, \tag{7.13}$$

with $r_1^2 = \left\{(x+\mu)^2 + y^2 + z^2\right\}, r_2^2 = \left\{(x-1+\mu)^2 + y^2 + z^2\right\}$.

In state space form

$$\dot{x}_1 = \dot{x} = x_3,\ \ \dot{x}_2 = \dot{y} = x_4,\ \ \dot{x}_5 = \dot{z} = x_6, \tag{7.14}$$

$$\dot{x}_3 = 2\dot{x}_2 + x_1 - \frac{\mu}{r_2^3}(x_1 - 1 + \mu) - \frac{1-\mu}{r_1^3}(x_1 + \mu), \tag{7.15}$$

$$\dot{x}_4 = -2\dot{x}_1 + x_2\left(1 - \frac{\mu}{r_2^3} - \frac{1-\mu}{r_1^3}\right), \dot{x}_6 = -x_5\left(\frac{\mu}{r_2^3} + \frac{1-\mu}{r_1^3}\right). \tag{7.16}$$

with $r_1^2 = \left\{(x+\mu)^2 + y^2 + z^2\right\}, r_2^2 = \left\{(x-1+\mu)^2 + y^2 + z^2\right\}$.

7.2.2 *The Equilibrium or Lagrange Points*

The equilibrium points are obtained by setting

$$\frac{d^2}{d\tau^2}\begin{bmatrix} x \\ y \\ z \end{bmatrix} = \frac{d}{d\tau}\begin{bmatrix} x \\ y \\ z \end{bmatrix} = \begin{bmatrix} 0 \\ 0 \\ 0 \end{bmatrix}. \tag{7.17}$$

Thus, one has

$$\begin{bmatrix} x \\ y \\ 0 \end{bmatrix} - \frac{(1-\mu)}{r_1^3}\begin{bmatrix} x+\mu \\ y \\ z \end{bmatrix} - \frac{\mu}{r_2^3}\begin{bmatrix} x-(1-\mu) \\ y \\ z \end{bmatrix} = \begin{bmatrix} 0 \\ 0 \\ 0 \end{bmatrix}. \tag{7.18}$$

If one chooses $z \equiv 0$

$$\begin{bmatrix} x \\ y \end{bmatrix} - \frac{(1-\mu)}{r_1^3}\begin{bmatrix} x+\mu \\ y \end{bmatrix} - \frac{\mu}{r_2^3}\begin{bmatrix} x-(1-\mu) \\ y \end{bmatrix} = \begin{bmatrix} 0 \\ 0 \end{bmatrix}. \tag{7.19}$$

From the first of the above two equilibrium equations for x and y

$$x - \frac{(1-\mu)}{r_1^3}(\mu+x) - \frac{\mu}{r_2^3}(\mu+x-1)=0. \tag{7.20}$$

In the above, $y = 0$, and

$$r_1 = |\mu+x|, r_2 = |\mu+x-1|, x - \frac{(1-\mu)}{|\mu+x|^3}(\mu+x) - \frac{\mu}{|\mu+x-1|^3}(\mu+x-1)=0. \tag{7.21}$$

The absolute values are considered by replacing them by the original expressions multiplied by a factor which is either +1 or −1, depending on the sign of the expression. For instance, $|x|$ would be replaced by x/s, and s can afterwards be assigned the value −1 if x is negative, or +1 if x is positive or zero. Such substitutions lead to an equivalent equation which is of fifth order in x and fourth order in μ, after deleting the common denominator (which is only zero at the positions of either of the two heavy bodies, and those positions are not considered anyway on physical grounds):

$$x - \frac{s_1(1-\mu)}{(\mu+x)^2} - \frac{s_2\mu}{(1-\mu-x)^2} = 0, \tag{7.22}$$

where s_1 is $(x+\mu)/|x+\mu|$ and s_2 is $(x-1+\mu)/|x-1+\mu|$. Thus one has the quintic polynomial

$$x(\mu+x-1)^2(\mu+x)^2 + s_1(1-\mu)(\mu+x-1)^2 + s_2\mu(\mu+x)^2 = 0. \tag{7.23}$$

To start the solution by iteration, one uses the solutions obtained by perturbation analysis:

$$\begin{aligned} &L_1 = 1 - \frac{1}{3}z - \frac{1}{9}z^3 + \frac{58}{81}z^4,\ L_2 = 1 - \frac{1}{3}z - \frac{1}{9}z^3 + \frac{50}{81}z^4, \\ &z = \left(\frac{\mu}{3}\right)^{1/3},\ L_3 = 1 + \frac{5}{12}\mu - \frac{1127}{20736}\mu^3 - \frac{7889}{248832}\mu^4. \end{aligned} \tag{7.24}$$

In the above $z = (\mu/3)^{1/3}$ is the Hill radius and signifies the region of influence of the second body in the context of the three-body problem.

Alternately

$$x \mp \frac{(1-\mu)}{(\mu+x)^2} \mp \frac{\mu}{(1-\mu-x)^2} = 0. \tag{7.25}$$

The above can be reduced to the following equations for γ_i, where $\gamma_1 = -(x-1+\mu)$, $\gamma_2 = -(x - 1 + \mu)$ and $\gamma_3 = -(x + \mu)$ to define the relative location of the first three equilibrium points

$$\gamma_i^5 \mp \gamma_i^4(3-\mu) + \gamma_i^3(3-2\mu) \mp \mu\gamma_i^2 \pm 2\mu\gamma_i - \mu = 0, \text{ for } i = 1, 2, \tag{7.26}$$

and

$$\gamma_i^5 + \gamma_i^4(3+(\mu-1)) + \gamma_i^3(3+2(\mu-1)) - (\mu-1)\gamma_i^2 + 2(\mu-1)\gamma_i + (\mu-1) = 0, \\ \text{for } i = 3. \tag{7.27}$$

From the second of the two equilibrium equations

$$y\left(1 - \frac{(1-\mu)}{r_1^3} - \frac{\mu}{r_2^3}\right) = 0. \tag{7.28}$$

If one chooses $y \neq 0$,

$$\left(1 - \frac{1}{r_1^3} + \frac{\mu}{r_1^3} - \frac{\mu}{r_2^3}\right) = 0. \tag{7.29}$$

The above equation admits the solution

$$r_1 = r_2 = 1, \text{so } x = \mu - 1/2 \text{ and } r_1^2 = r_2^2 = \left(1/4 + y^2\right) = 1. \Rightarrow y = \pm\sqrt{3/4}. \tag{7.30}$$

Thus overall there are five equilibrium points. The location of the equilibrium points are shown in Fig. 7.3. The locations of the two primaries are only shown.

7.2.3 The Hill Sphere and the Sphere of Influence

The region around a planet within which the body's gravitational force, acting on a third body, exceeds that of the central body it orbits is called the Hill sphere, whose radius is called the Hill radius. In more precise terms, the Hill sphere approximates the gravitational sphere of influence of a smaller body, on a third body in its vicinity and also orbiting the smaller body, in the face of perturbations from a more massive body. In contrast, the sphere of influence (SOI), which is also known as the Laplace

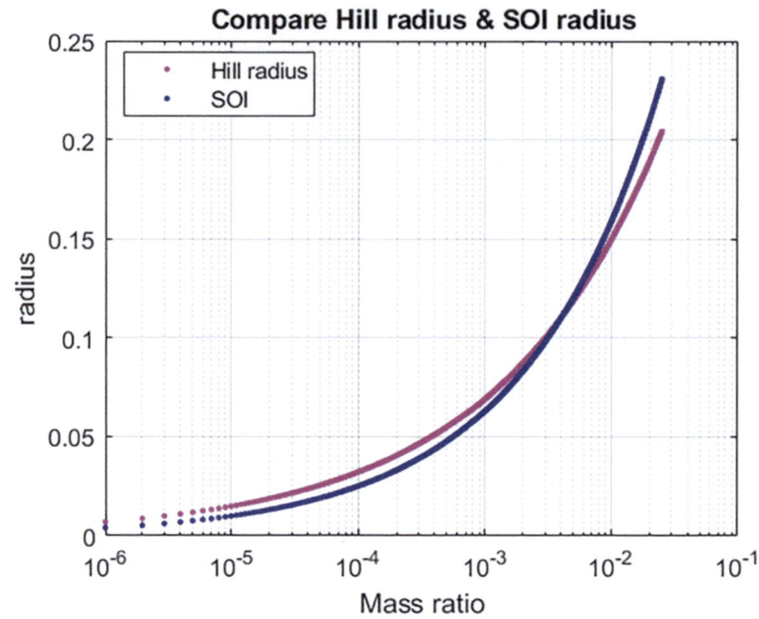

Fig. 7.4 Comparison of the Hill and SOI radii for low values of μ (the mass ratio) on the horizontal axis

SOI, is the oblate-spheroid-shaped region around a celestial body where the primary gravitational influence on an orbiting object is that body, in the presence of a third larger body. For values of $\mu < 0.0041$, the minimum SOI radius ($(\mu/(1-\mu))^{0.4}$) is smaller than the Hill sphere radius ($z = (\mu/3)^{1/3}$) and is enclosed by it as shown in Fig. 7.4 for different values of μ on the horizontal axis. For higher values of μ, the minimum SOI radius is greater than the Hill radius. It is interesting to observe that in the case of the class of three-body problems associated with the Sun, a planet and the satellite, the value of μ is always less than 0.0041. Only one of the five equilibrium points is within the Hill sphere of the smaller of the two primaries. It is of primary importance and is denoted as L1. The second is just outside the Hill sphere and collinear with the line joining the two primaries and L1. The third is also collinear with the line joining the two primaries as well as with L1 and L2.

7.2.4 *Jacobi's Integral and Constant*

To transform the equations of motion into a canonical form, introduce a potential function ϕ defined as

$$\phi = \frac{1}{2}\left(x^2 + y^2\right) + \frac{1-\mu}{r_1} + \frac{\mu}{r_2},\ r_1^2 = \left\{(x+\mu)^2 + y^2 + z^2\right\}, r_2^2 = \left\{(x-1+\mu)^2 + y^2 + z^2\right\}. \tag{7.31}$$

Therefore, the set of equations describing the CR3BP in the synodic coordinate system can be written as

$$\frac{d^2}{d\tau^2}\begin{bmatrix} x \\ y \\ z \end{bmatrix} - 2\frac{d}{d\tau}\begin{bmatrix} y \\ -x \\ 0 \end{bmatrix} = \begin{bmatrix} \partial/\partial x \\ \partial/\partial y \\ \partial/\partial z \end{bmatrix}\phi. \tag{7.32}$$

Thus

$$\begin{aligned}\frac{d}{d\tau}[x \quad y \quad z]\left\{\frac{d^2}{d\tau^2}\begin{bmatrix} x \\ y \\ z \end{bmatrix} - 2\frac{d}{d\tau}\begin{bmatrix} y \\ -x \\ 0 \end{bmatrix}\right\} &= \frac{1}{2}\frac{d}{d\tau}\left(\frac{d}{d\tau}[x \quad y \quad z]\frac{d}{d\tau}\begin{bmatrix} x \\ y \\ z \end{bmatrix}\right) \\ &= \frac{d}{d\tau}[x \quad y \quad z]\begin{bmatrix} \partial/\partial x \\ \partial/\partial y \\ \partial/\partial z \end{bmatrix}\phi = \frac{d\phi}{d\tau}.\end{aligned} \tag{7.33}$$

Integrating, one obtains an energy relation given by

$$\frac{1}{2}\left(\frac{d}{d\tau}[x \quad y \quad z]\frac{d}{d\tau}\begin{bmatrix} x \\ y \\ z \end{bmatrix}\right) = \phi + C \Rightarrow \frac{1}{2}\mathbf{v}^2 = \phi + C \Rightarrow \frac{1}{2}\mathbf{v}^2 - \phi = C. \tag{7.34}$$

The Jacobi integral is thus defined by

$$J = \frac{\dot{x}^2 + \dot{y}^2 + \dot{z}^2}{2} - \frac{x^2 + y^2}{2} - \frac{(1-\mu)}{r_1} - \frac{\mu}{r_2} = C. \tag{7.35}$$

Hence, it follows that

$$\frac{\dot{x}^2 + \dot{y}^2 + \dot{z}^2}{2} = J + \frac{x^2 + y^2}{2} + \frac{(1-\mu)}{r_1} + \frac{\mu}{r_2} = J + \phi = \phi + C. \tag{7.36}$$

The energy in the synodic coordinates is

$$E = \frac{(\dot{x} - y)^2 + (x + \dot{y})^2 + \dot{z}^2}{2} - \frac{(1-\mu)}{r_1} - \frac{\mu}{r_2}. \tag{7.37}$$

The moment of momentum is

$$H = x^2 + y^2 + x\dot{y} - y\dot{x}. \tag{7.38}$$

It follows that

$$E - H = \frac{\dot{x}^2 + \dot{y}^2 + \dot{z}^2 - x^2 - y^2 - 2\dot{x}y + 2x\dot{y} + \dot{z}^2}{2} - \frac{(1-\mu)}{r_1} - \frac{\mu}{r_2} - x\dot{y} + y\dot{x} = J. \tag{7.39}$$

The constant C may be evaluated by evaluating the preceding equation at the initial conditions. Furthermore the *zero velocity curves* follow from

$$\frac{1}{2}\mathbf{v}^2 = \phi + C = 0; \cdot \frac{\dot{x}^2 + \dot{y}^2 + \dot{z}^2}{2} = \frac{x^2 + y^2}{2} + \frac{(1-\mu)}{r_1} + \frac{\mu}{r_2} + C = 0. \tag{7.40}$$

Once C is determined by the initial conditions, the ZVC encloses a region in space where $\mathbf{v}^2 = 2\phi + 2C \geq 0$ and where the third body must be found. The third body cannot be outside of this region because its velocity would be imaginary since $\mathbf{v}^2 = 2\phi + 2C < 0$.

Having obtained the Jacobi integral in the rotating coordinate system, it may be also written explicitly in a non-rotating coordinate system as

$$\dot{\tilde{x}}^2 + \dot{\tilde{y}}^2 + \dot{\tilde{z}}^2 - 2\left(\tilde{x}\dot{\tilde{y}} - \tilde{y}\dot{\tilde{x}}\right) = \frac{2(1-\mu)}{\tilde{r}_{13}} + \frac{2\mu}{\tilde{r}_{23}} + 2\tilde{C} = 0. \tag{7.41}$$

7.2.5 HALO and Periodic Orbits: Stability and Linearization

Lissajous trajectories are fully three-dimensional quasi-periodic solutions in the multi-dimensional phase space associated with the three-body problem that can be generated by a careful selection of in-plane and out-of-plane initial conditions. Under certain conditions, it is possible to generate a subset of such Lissajous trajectories that are not only precisely periodic solutions but are also multi-dimensional that have been named as Halo orbits. Halo orbits as originally defined in the circular restricted three-body problem, as precisely periodic orbits, do not exist in a general multi-body solution including perturbations such as solar radiation pressure and non-periodic primary motions. Recent missions to the Lagrange or libration points and orbits around them are the Sun-Earth connections missions included the L_1 libration point mission SOHO, the Triana mission, an Earth observation mission, to L_1, (which was renamed by NASA as the Deep Space Climate Observatory, DSCOVR), the L_1/L_2 WIND mission to study radio waves and plasma that occur in the solar wind and in the Earth's magnetosphere and the James Webb

Space Telescope and the Terrestrial Planet Finder missions to the L_2 libration point. As conventional tools do not currently incorporate any theoretical understanding of the multi-body problem and do not fully exploit dynamical relationships, the nonlinear dynamical systems theory-based approaches, including the use of the Lissajous and periodic orbits, use of the stable and unstable manifolds and the use of selected patch points are fast becoming popular as the primary tools for trajectory design. Barden et al. (1996) have discussed the features of the orbits around the equilibrium points, as well as transfers from Earth to Halo orbits, Halo-to-Earth orbits and Halo-to-Halo orbits. An in-depth discussion of identified types of motion near the Halo orbit in the centre manifold was given by Barden and Howell (1998). Orbits may be constructed by considering small perturbations about the equilibrium points. Often it is essential to establish their stability, prior to calculating the trajectories. The stability of planar periodic orbits (Lyapunov orbits) with respect to perturbations perpendicular to the plane was studied by Hénon (1973). The equations of motion, in state space form

$$\dot{x}_1=\dot{x}=x_3,\ \ \dot{x}_2=\dot{y}=x_4,\ \ \dot{x}_5=\dot{z}=x_6,\ \ \dot{x}_3=2\dot{x}_2+x_1-\frac{\mu}{r_2^3}(x_1-1+\mu)-\frac{1-\mu}{r_1^3}(x_1+\mu), \tag{7.42}$$

or

$$\dot{x}_3=2x_4+x_1\left(1-\frac{\mu}{r_2^3}-\frac{1-\mu}{r_1^3}\right)+(1-\mu)\mu\left(\frac{1}{r_2^3}-\frac{1}{r_1^3}\right), \tag{7.43}$$

$$\dot{x}_4=-2x_3+x_2\left(1-\frac{\mu}{r_2^3}-\frac{1-\mu}{r_1^3}\right),\ \ \dot{x}_6=-x_5\left(\frac{\mu}{r_2^3}+\frac{1-\mu}{r_1^3}\right), \tag{7.44}$$

$$r_1^2=\left\{(x+\mu)^2+y^2+z^2\right\},r_2^2=\left\{(x-1+\mu)^2+y^2+z^2\right\}. \tag{7.45}$$

The following derivatives may be used to linearize the equations of motion:

$$\frac{d}{dx}\left(\frac{1}{r_i^3}\right)=-\frac{3}{2}\frac{1}{r_i^5}\frac{dr_i^2}{dx},\ \ \frac{d}{dx}r_1^2=2(x+\mu),\ \frac{d}{dx}r_2^2=2(x-1+\mu),\ \ \frac{d}{dy}\left(\frac{1}{r_i^3}\right)=-3\frac{y}{r_i^5},$$
$$\frac{d}{dz}\left(\frac{1}{r_i^3}\right)=-3\frac{z}{r_i^5},\ \ \frac{d}{dx}\left(\frac{1}{r_1^3}\right)=-3\frac{(x+\mu)}{r_1^5},\ \ \frac{d}{dx}\left(\frac{1}{r_2^3}\right)=-3\frac{(x-1+\mu)}{r_2^5}. \tag{7.46}$$

The linearized equations about the equilibrium points $(x_e\ \ y_e\ \ z_e)$ are given by

$$\dot{x}_1=\dot{x}=x_3,\ \ \dot{x}_2=\dot{y}=x_4,\ \ \dot{x}_5=\dot{z}=x_6,$$

$$
\begin{aligned}
\dot{x}_3 &= 2x_4 + x_1\left(1 - \frac{\mu}{r_{2e}^3} - \frac{1-\mu}{r_{1e}^3}\right) + 3x_e\left(\frac{\mu(x_e+\mu)}{r_{2e}^5} + \frac{(1-\mu)(x_e-1+\mu)}{r_{1e}^5}\right)x_1 \\
&+3x_e\left(\frac{\mu y_e}{r_{2e}^5} + \frac{(1-\mu)y_e}{r_{1e}^5}\right)x_2 \\
&+3x_e\left(\frac{\mu z_e}{r_{2e}^5} + \frac{(1-\mu)z_e}{r_{1e}^5}\right)x_5 - 3(1-\mu)\mu\left(\left(\frac{x_e-1+\mu}{r_{2e}^5} - \frac{x_e+\mu}{r_{1e}^5}\right)x_1\right. \\
&\left.+\left(\frac{y_e}{r_{2e}^5} - \frac{y_e}{r_{1e}^5}\right)x_2 + \left(\frac{z_e}{r_{2e}^5} - \frac{z_e}{r_{1e}^5}\right)x_5\right) \\
\dot{x}_4 &= -2x_3 + x_2\left(1 - \frac{\mu}{r_{2e}^3} - \frac{1-\mu}{r_{1e}^3}\right) + 3y_e\left(\frac{\mu(x_e+\mu)}{r_{2e}^5} + \frac{(1-\mu)(x_e-1+\mu)}{r_{1e}^5}\right)x_1 \\
&+3y_e\left(\frac{\mu y_e}{r_{2e}^5} + \frac{(1-\mu)y_e}{r_{1e}^5}\right)x_2 + 3y_e\left(\frac{\mu z_e}{r_{2e}^5} + \frac{(1-\mu)z_e}{r_{1e}^5}\right)x_5, \\
\dot{x}_6 &= -x_5\left(\frac{\mu}{r_{2e}^3} + \frac{1-\mu}{r_{1e}^3}\right) + 3z_e\left(\frac{\mu(x_e+\mu)}{r_{2e}^5} + \frac{(1-\mu)(x_e-1+\mu)}{r_{1e}^5}\right)x_1 \\
&+3z_e\left(\frac{\mu y_e}{r_{2e}^5} + \frac{(1-\mu)y_e}{r_{1e}^5}\right)x_2 + 3z_e\left(\frac{\mu z_e}{r_{2e}^5} + \frac{(1-\mu)z_e}{r_{1e}^5}\right)x_5.
\end{aligned} \tag{7.47}
$$

Hence when $z_e = 0$

$$
\begin{aligned}
&\dot{x}_1 = \dot{x} = x_3, \quad \dot{x}_2 = \dot{y} = x_4, \quad \dot{x}_5 = \dot{z} = x_6, \\
\dot{x}_3 &= 2x_4 + x_1\left(1 - \frac{\mu}{r_{2e}^3} - \frac{1-\mu}{r_{1e}^3}\right) + 3x_e\left(\frac{\mu(x_e+\mu)}{r_{2e}^5} + \frac{(1-\mu)(x_e-1+\mu)}{r_{1e}^5}\right)x_1 \\
&+3x_e\left(\frac{\mu y_e}{r_{2e}^5} + \frac{(1-\mu)y_e}{r_{1e}^5}\right)x_2 \\
&-3(1-\mu)\mu\left(\frac{x_e-1+\mu}{r_{2e}^5} - \frac{x_e+\mu}{r_{1e}^5}\right)x_1 - 3(1-\mu)\mu\left(\frac{y_e}{r_{2e}^5} - \frac{y_e}{r_{1e}^5}\right)x_2, \\
&\dot{x}_6 = -x_5\left(\frac{\mu}{r_{2e}^3} + \frac{1-\mu}{r_{1e}^3}\right).
\end{aligned} \tag{7.48}
$$

The solutions to the linearized equations of motion allow the generation of entire families of orbits. The stability of the solutions to the linearized equations are then easily assessed. Unstable trajectories from an equilibrium point are useful when one wishes to depart rapidly from the vicinity of a planet.

The first three Lagrangian points are stable only in the plane perpendicular to the line between the two bodies. This can be seen most easily by considering the L_1 point. If a body located at the L_1 point drifted closer to one of the masses, the gravitational attraction it felt from that mass would be greater, and it would be pulled closer. However, a body displaced perpendicularly from the central line would feel a force pulling it back towards the equilibrium point. This is because the lateral components of the two masses' gravity would add to produce this force, whereas the components along the axis between them would balance out. Although the L_1, L_2 and L_3 points are nominally unstable, it turns out that, at least in the restricted three-body problem, it is possible to find stable periodic orbits, the Halo orbits, around these points in the plane perpendicular to the line joining the primaries.

The last two Lagrange points (or Trojan points), L_4 and L_5, are stable equilibrium points, provided the ratio of the primary masses is larger than 24.9605. This is the case for the Sun-Earth, Sun-Jupiter, Earth-Moon and all Sun-planet systems.

7.2.6 Linearization About Co-linear Libration Points

The scalar Eq. (7.13) may be expressed as

$$\begin{aligned}\ddot{x}-2\dot{y}-x &= -(1-\mu)\left(\frac{x+\mu}{\|\mathbf{r}_1\|^3}\right)-\mu\left(\frac{x-1+\mu}{\|\mathbf{r}_2\|^3}\right)+u_x,\cdot\\ \ddot{y}+2\dot{x}-y &= -(1-\mu)\left(\frac{y}{\|\mathbf{r}_1\|^3}\right)-\mu\left(\frac{y}{\|\mathbf{r}_2\|^3}\right)+u_y,\\ \ddot{z} &= -(1-\mu)\left(\frac{z}{\|\mathbf{r}_1\|^3}\right)-\mu\left(\frac{z}{\|\mathbf{r}_2\|^3}\right)+u_z,\end{aligned} \tag{7.49}$$

$$\mathbf{r}_1^2=(x+\mu)^2+y^2+z^2,\ \ \mathbf{r}_2^2=(x+\mu-1)^2+y^2+z^2. \tag{7.50}$$

The equations may be linearized about an equilibrium point. Consider a co-linear equilibrium point, along the line joining the two primary masses, and let x_0 be the distance of a collinear equilibrium point from the barycentre. Then, linearized equilibrium equations can be shown to be

$$\ddot{x}-2\dot{y}-(2B_L+1)x=u_x,\ \ \ddot{y}+2\dot{x}+(B_L-1)y=u_y,\ \ \ddot{z}+B_Lz=u_z, \tag{7.51}$$

or

$$\ddot{x}-2\dot{y}-(3-2\Delta)x=u_x,\ \ \ddot{y}+2\dot{x}-\Delta y=u_y,\ \ \ddot{z}+(1-\Delta)z=u_z, \tag{7.52}$$

$$\begin{aligned}B_L=\left(\frac{(1-\mu)}{(x_0+\mu)^3}+\frac{\mu}{(x_0+\mu-1)^3}\right)&=1-\frac{\mu(1-\mu)}{x_0}\\ \times\left(\frac{1}{(x_0+\mu)^3}-\frac{1}{(x_0+\mu-1)^3}\right)&=1-\Delta.\end{aligned} \tag{7.53}$$

If one lets $x_0 = 1 + \gamma - \mu$, i.e. γ is the distance of the co-linear equilibrium point from centre of mass of the second body, it follows that

$$B_L = \left(\frac{(1-\mu)}{(1+\gamma)^3} + \frac{\mu}{|\gamma|^3} \right). \tag{7.54}$$

In the above the overhead "dot" $(\cdot)$ represents differentiation with respect to non-dimensional time. The stability of the system can now be characterized in terms of the parameter, B_L. For no instability in the out of plane motion, B_L must be real and positive. The equilibrium points are unstable for all real and positive values of B_L.

7.3 Transformation of Synodic Reference Frames

The most important step in patched three-body method is the transformation from one set of synodic coordinates to another. Transformations from one set of synodic coordinates to another involve both re-scaling of both the distance and time. From Eq. (7.8),

$$\begin{bmatrix} \tilde{x} \\ \tilde{y} \\ \tilde{z} \end{bmatrix} = \begin{bmatrix} \cos\Omega_1 t & -\sin\Omega_1 t & 0 \\ \sin\Omega_1 t & \cos\Omega_1 t & 0 \\ 0 & 0 & 1 \end{bmatrix} \frac{1}{d_1} \begin{bmatrix} x \\ y \\ z \end{bmatrix}; \quad \begin{bmatrix} x \\ y \\ z \end{bmatrix} = \begin{bmatrix} \cos\Omega_1 t & \sin\Omega_1 t & 0 \\ -\sin\Omega_1 t & \cos\Omega_1 t & 0 \\ 0 & 0 & 1 \end{bmatrix} d_1 \begin{bmatrix} \tilde{x} \\ \tilde{y} \\ \tilde{z} \end{bmatrix}. \tag{7.55}$$

For another set of synodic coordinates, assuming inferior conjunction of both planets

$$\begin{bmatrix} \tilde{\tilde{x}} \\ \tilde{\tilde{y}} \\ \tilde{\tilde{z}} \end{bmatrix} = \begin{bmatrix} \cos\Omega_2 t & -\sin\Omega_2 t & 0 \\ \sin\Omega_2 t & \cos\Omega_2 t & 0 \\ 0 & 0 & 1 \end{bmatrix} \frac{1}{d_2} \begin{bmatrix} x \\ y \\ z \end{bmatrix} = \begin{bmatrix} \cos\Omega_2 t & -\sin\Omega_2 t & 0 \\ \sin\Omega_2 t & \cos\Omega_2 t & 0 \\ 0 & 0 & 1 \end{bmatrix} \times \begin{bmatrix} \cos\Omega_1 t & \sin\Omega_1 t & 0 \\ -\sin\Omega_1 t & \cos\Omega_1 t & 0 \\ 0 & 0 & 1 \end{bmatrix} \frac{d_1}{d_2} \begin{bmatrix} \tilde{x} \\ \tilde{y} \\ \tilde{z} \end{bmatrix}. \tag{7.56}$$

$$\begin{bmatrix} \tilde{\tilde{x}} \\ \tilde{\tilde{y}} \\ \tilde{\tilde{z}} \end{bmatrix} = \begin{bmatrix} \cos\Omega_2 t & -\sin\Omega_2 t & 0 \\ \sin\Omega_2 t & \cos\Omega_2 t & 0 \\ 0 & 0 & 1 \end{bmatrix} \frac{1}{d_2} \begin{bmatrix} x \\ y \\ z \end{bmatrix}$$
$$= \begin{bmatrix} \cos(\Omega_1-\Omega_2)t & \sin(\Omega_1-\Omega_2)t & 0 \\ -\sin(\Omega_1-\Omega_2)t & \cos(\Omega_1-\Omega_2)t & 0 \\ 0 & 0 & 1 \end{bmatrix} \frac{d_1}{d_2} \begin{bmatrix} \tilde{x} \\ \tilde{y} \\ \tilde{z} \end{bmatrix}. \tag{7.57}$$

One defines $\Omega_1^2 = G(m_s+m_1)/d_1^3$, $\Omega_2^2 = G(m_s+m_2)/d_2^3$. $\Omega_{12}' = (\Omega_1-\Omega_2)/\Omega_2$. The new time scale is τ. If $\tau=\tau_2=\Omega_2 t$, $\Omega_{12}' = (\Omega_1-\Omega_2)/\Omega_2$ $\tau_1 = \Omega_1 t = (\Omega_1/\Omega_2)\tau$. Hence, $\tau = (\Omega_2/\Omega_1)\tau_1$. Transforming the times, $\tau = (\Omega_2/\Omega_1)\tau_1$, the position vector is transformed as

$$\begin{bmatrix} \tilde{\tilde{x}} \\ \tilde{\tilde{y}} \\ \tilde{\tilde{z}} \end{bmatrix} = \begin{bmatrix} \cos\Omega_{12}'\tau & \sin\Omega_{12}'\tau & 0 \\ -\sin\Omega_{12}'\tau & \cos\Omega_{12}'\tau & 0 \\ 0 & 0 & 1 \end{bmatrix} \frac{d_1}{d_2} \begin{bmatrix} \tilde{x} \\ \tilde{y} \\ \tilde{z} \end{bmatrix}. \tag{7.58}$$

For $\tau = 0$,

$$\begin{bmatrix} \tilde{\tilde{x}} \\ \tilde{\tilde{y}} \\ \tilde{\tilde{z}} \end{bmatrix} = \frac{d_1}{d_2} \begin{bmatrix} \tilde{x} \\ \tilde{y} \\ \tilde{z} \end{bmatrix}. \tag{7.59}$$

The time derivative is

$$\frac{d}{d\tau}\begin{bmatrix} \tilde{\tilde{x}} \\ \tilde{\tilde{y}} \\ \tilde{\tilde{z}} \end{bmatrix} = -\Omega_{12}' \begin{bmatrix} \sin\Omega_{12}'\tau & -\cos\Omega_{12}'\tau & 0 \\ \cos\Omega_{12}'\tau & \sin\Omega_{12}'\tau & 0 \\ 0 & 0 & 0 \end{bmatrix} \frac{d_1}{d_2} \begin{bmatrix} \tilde{x} \\ \tilde{y} \\ \tilde{z} \end{bmatrix}$$
$$+ \begin{bmatrix} \cos\Omega_{12}'\tau & \sin\Omega_{12}'\tau & 0 \\ -\sin\Omega_{12}'\tau & \cos\Omega_{12}'\tau & 0 \\ 0 & 0 & 1 \end{bmatrix} \frac{d_1}{d_2}\frac{d}{d\tau} \begin{bmatrix} \tilde{x} \\ \tilde{y} \\ \tilde{z} \end{bmatrix}, \tag{7.60}$$

which could be expressed as

$$\frac{d}{d\tau}\begin{bmatrix} \tilde{\tilde{x}} \\ \tilde{\tilde{y}} \\ \tilde{\tilde{z}} \end{bmatrix} = \begin{bmatrix} \cos\Omega_{12}'\tau & \sin\Omega_{12}'\tau & 0 \\ -\sin\Omega_{12}'\tau & \cos\Omega_{12}'\tau & 0 \\ 0 & 0 & 1 \end{bmatrix} \frac{d_1}{d_2} \left\{ \frac{d}{d\tau}\begin{bmatrix} \tilde{x} \\ \tilde{y} \\ \tilde{z} \end{bmatrix} - \Omega_{12}' \begin{bmatrix} 0 & -1 & 0 \\ 1 & 0 & 0 \\ 0 & 0 & 0 \end{bmatrix} \begin{bmatrix} \tilde{x} \\ \tilde{y} \\ \tilde{z} \end{bmatrix} \right\}. \tag{7.61}$$

Consequently

$$\frac{d}{d\tau}\begin{bmatrix}\tilde{\tilde{x}}\\ \tilde{\tilde{y}}\\ \tilde{\tilde{z}}\end{bmatrix}=\begin{bmatrix}\cos\Omega'_{12}\tau & \sin\Omega'_{12}\tau & 0\\ -\sin\Omega'_{12}\tau & \cos\Omega'_{12}\tau & 0\\ 0 & 0 & 1\end{bmatrix}\frac{d_1}{d_2}\begin{bmatrix}d\tilde{x}/d\tau+\Omega'_{12}\tilde{y}\\ d\tilde{y}/d\tau-\Omega'_{12}\tilde{x}\\ d\tilde{z}/d\tau\end{bmatrix}. \tag{7.62}$$

But

$$\frac{d}{d\tau}=\frac{\Omega_1}{\Omega_2}\frac{d}{d\tau_1}=\left(1+\Omega'_{12}\right)\frac{d}{d\tau_1}. \tag{7.63}$$

Hence

$$\frac{d}{d\tau}\begin{bmatrix}\tilde{\tilde{x}}\\ \tilde{\tilde{y}}\\ \tilde{\tilde{z}}\end{bmatrix}=\begin{bmatrix}\cos\Omega'_{12}\tau & \sin\Omega'_{12}\tau & 0\\ -\sin\Omega'_{12}\tau & \cos\Omega'_{12}\tau & 0\\ 0 & 0 & 1\end{bmatrix}\frac{d_1}{d_2}\begin{bmatrix}\left(1+\Omega'_{12}\right)d\tilde{x}/d\tau_1+\Omega'_{12}\tilde{y}\\ \left(1+\Omega'_{12}\right)d\tilde{y}/d\tau_1-\Omega'_{12}\tilde{x}\\ \left(1+\Omega'_{12}\right)d\tilde{z}/d\tau_1\end{bmatrix}. \tag{7.64}$$

For $\tau=0$,

$$\frac{d}{d\tau}\begin{bmatrix}\tilde{\tilde{x}}\\ \tilde{\tilde{y}}\\ \tilde{\tilde{z}}\end{bmatrix}=\frac{d_1}{d_2}\begin{bmatrix}\left(1+\Omega'_{12}\right)d\tilde{x}/d\tau_1+\Omega'_{12}\tilde{y}\\ \left(1+\Omega'_{12}\right)d\tilde{y}/d\tau_1-\Omega'_{12}\tilde{x}\\ \left(1+\Omega'_{12}\right)d\tilde{z}/d\tau_1\end{bmatrix}. \tag{7.65}$$

7.3.1 Example 1: Earth-Moon ($\mu = 0.01215$), Sun-Earth ($\mu = 3.036\mathrm{e}-06$) and Satellite Systems

The orbit period of the Moon is 27.32166 days, the orbital rate of the Moon: $\Omega_1 =$ 2.661699624635926e-06, the orbital rate of the Earth in rads/sec is $\Omega_2 =$ 19.91063865436695e-08 and the ratio $\Omega_1/\Omega_2 =$ 13.368228266510888 and $\Omega_2/\Omega_1 =$ 0.074804228358752. The differential ratio $\Omega'_{12} = (\Omega_1 - \Omega_2)/\Omega_2 =$ 12.368228266510888 and $\tau = (\Omega_2/\Omega_1)\tau_1$, $\Omega_2/\Omega_1 =$ 0.074804228358752.

The Earth-Moon distance is $d_1 =$ 384,400 km while the Earth-Sun distance is $d_2 =$ 149,597,870 km. Thus the ratio $d_2/d_1 =$ 389.1723985431842 and its inverse is $d_1/d_2 =$ 0.002569555301823.

7.3.2 Example 2: Sun-Earth ($\mu = 3.036\text{e} - 06$), Sun-Jupiter ($\mu = 9.5364\text{e} - 04$) and Satellite Systems

In this example, the Sun-Earth distance is $d_1 = 1AU$, Sun-Jupiter distance is $d_2 = 5.202887AU$, the length of the Earth year is 365.2421875 days, the time taken by Jupiter to orbit the Sun is 11.8618 years, the orbital rate of the Earth in rads/sec is 19.91063865436695e-08, the orbital rate of Jupiter is 1.678551202546573e-08, and the difference between the two is $\Omega_{12} = \Omega_1 - \Omega_2 = 18.232087451820377\text{e-}08$. The ratio $d_1/d_2 = 0.192200983799956$. The new time scale τ is defined as $\tau = \tau_2 = \Omega_2 t$, so $\Omega'_{12} = (\Omega_1 - \Omega_2)/\Omega_2 = 10.861800000000006$, $\tau_1 = \Omega_1 t = (\Omega_1/\Omega_2)\tau$, the ratio $\Omega_1/\Omega_2 = 11.8618$, $\Omega_2/\Omega_1 = 0.084304237130958$, and $\tau = (\Omega_2/\Omega_1)\tau_1$.

Transforming the times, $\tau = (\Omega_2/\Omega_1)\tau_1$. The transformation in the new time scale is

$$\begin{bmatrix} \tilde{\tilde{x}} \\ \tilde{\tilde{y}} \\ \tilde{\tilde{z}} \end{bmatrix} = \begin{bmatrix} \cos \Omega'_{12}\tau & \sin \Omega'_{12}\tau & 0 \\ -\sin \Omega'_{12}\tau & \cos \Omega'_{12}\tau & 0 \\ 0 & 0 & 1 \end{bmatrix} \frac{d_1}{d_2} \begin{bmatrix} \tilde{x} \\ \tilde{y} \\ \tilde{z} \end{bmatrix}. \tag{7.66}$$

For $\tau = 0$

$$\begin{bmatrix} \tilde{\tilde{x}} \\ \tilde{\tilde{y}} \\ \tilde{\tilde{z}} \end{bmatrix} = \frac{d_1}{d_2} \begin{bmatrix} \tilde{x} \\ \tilde{y} \\ \tilde{z} \end{bmatrix}, \quad \frac{d}{d\tau} \begin{bmatrix} \tilde{\tilde{x}} \\ \tilde{\tilde{y}} \\ \tilde{\tilde{z}} \end{bmatrix} = \frac{d_1}{d_2} \begin{bmatrix} (1 + \Omega'_{12}) d\tilde{x}/d\tau_1 + \Omega'_{12}\tilde{y} \\ (1 + \Omega'_{12}) d\tilde{y}/d\tau_1 - \Omega'_{12}\tilde{x} \\ (1 + \Omega'_{12}) d\tilde{z}/d\tau_1 \end{bmatrix}. \tag{7.67}$$

7.4 Methods for Generating Orbit Families

There are indeed several methods for generating families of feasible trajectories (Singh et al. (2021), Rogov 2016), Haapala (2014), Geisel (2013), Escribano (2010), Gordon (2008), Topputo et al. (2005), Jarabek (2004), Farquhar and Kamel (1973)). Kim and Hall (2002) have discussed the methods of generating periodic orbits about the Lagrange points. All of these involve the integration of the equations of motion, (forward/backward) starting at the following:

(i) Points close to the Lagrange or equilibrium points without and with velocity perturbations
(ii) Points close to the Lagrange points in directions of eigenvectors of perturbed linear dynamics or along directions on the manifold surfaces

(iii) Lyapunov orbits or planar periodic orbits in one or more of the orbit planes (by the methods of differential continuation or collocation)
(iv) Halo orbits (non-planar periodic orbits)
(v) Points close to the minor or major planets after regularization without and with velocity perturbations.

7.4.1 Plotting the Manifolds and Trajectories

In order to generate a manifold for the CR3BP, the two-dimensional nonlinear equations of motion were numerically integrated for a range of initial conditions. The state space representation of the equations of motion is programmed in MATLAB. The numerical integration requires an initial position and velocity so that it can generate a trajectory over a specified time interval. The stable manifolds are obtained by propagating the equations of motion by introducing a small perturbation to the initial state in the stable eigenvector direction, backwards in time. The unstable manifolds are obtained by propagating the equations of motion forwards in time from a state perturbed along the unstable eigenvector direction. The methodology for the computation of invariant manifold has been outlined by Barden et al. (1996).

An initial velocity of zero is required to find the stable manifold to travel to a Lagrange point. The stable manifold can be propagated forward and backward in time so that a spacecraft is able to travel to and from a Lagrange point on the manifold. The forward time propagation is done using the equation of motion. In order to perform backward time propagation, the equations of motion are modified by changing the independent variable to –*t*.

7.5 The Bi-circular Restricted Four-Body Problem

Observe that equations of the CR3BP may be expressed as

$$\frac{d^2}{d\tau^2}\begin{bmatrix} x \\ y \\ z \end{bmatrix} - 2\frac{d}{d\tau}\begin{bmatrix} y \\ -x \\ 0 \end{bmatrix} = \begin{bmatrix} x \\ y \\ 0 \end{bmatrix} - \frac{(1-\mu)}{r_1^3}\begin{bmatrix} x+\mu \\ y \\ z \end{bmatrix} - \frac{\mu}{r_2^3}\begin{bmatrix} x-(1-\mu) \\ y \\ z \end{bmatrix}. \tag{7.68}$$

The equations are expressed in matrix form as

$$\frac{d^2}{d\tau^2}\mathbf{X} - \mathbf{A}\dot{\mathbf{X}} - \mathbf{B}\mathbf{X} = -(1-\mu)\frac{\boldsymbol{\rho}+\mu\mathbf{r}}{\|\boldsymbol{\rho}+\mu\mathbf{r}\|^3} - \mu\frac{\boldsymbol{\rho}-(1-\mu)\mathbf{r}}{\|\boldsymbol{\rho}-(1-\mu)\mathbf{r}\|^3}, \tag{7.69}$$

with $\boldsymbol{\rho} = [x \quad y \quad z]^T$ and $\mathbf{r} = [1 \quad 0 \quad 0]^T$.

In the case of the bi-circular restricted four-body problem in the Earth (with mass m_2) and the Moon (with mass equal to m_3) synodic frame, and with Sun as the fourth body with mass m_1, if $\mu = \frac{m_3}{m_2+m_3}$ and $\mu_1 = \frac{m_1}{m_2+m_3}$, equations of motion of the satellite (Singh et al. (2021)) may be expressed as

$$\begin{aligned} \frac{d^2}{d\tau^2}\mathbf{X} - \mathbf{A}\dot{\mathbf{X}} - \mathbf{B}\mathbf{X} = & -(1-\mu)\frac{\boldsymbol{\rho} + \mu\mathbf{r}}{\|\boldsymbol{\rho} + \mu\mathbf{r}\|^3} - \mu\frac{\boldsymbol{\rho} - (1-\mu)\mathbf{r}}{\|\mathbf{p} - (1-\mu)\mathbf{r}\|^3} \\ & +\mu_1\left((1-\mu)\frac{\mathbf{R} - \mu\mathbf{r}}{\|\mathbf{R} - \mu\mathbf{r}\|^3} + \mu\frac{\mathbf{R} + (1-\mu)\mathbf{r}}{\|\mathbf{R} + (1-\mu)\mathbf{r}\|} - \frac{\mathbf{R} + \boldsymbol{\rho}}{\|\mathbf{R} + \boldsymbol{\rho}\|^3}\right), \end{aligned} \tag{7.70}$$

where the position vector of the Sun, from the Earth-Moon barycentre in the Earth-Moon synodic frame, is $\mathbf{R} = [R_x \quad R_y \quad 0]^T$.

The most significant effect of the bi-circular restricted four-body problem is the fact that the equilibrium points are no longer stationary, due to the motion of the Sun, although the motion of the stationary points is small. Assuming that the Sun is moving slowly, the equilibrium points satisfy

$$\begin{aligned} \mathbf{B}\mathbf{X} = & (1-\mu)\frac{\boldsymbol{\rho} + \mu\mathbf{r}}{\|\boldsymbol{\rho} + \mu\mathbf{r}\|^3} + \mu\frac{\boldsymbol{\rho} - (1-\mu)\mathbf{r}}{\|\boldsymbol{\rho} - (1-\mu)\mathbf{r}\|^3} \\ & -\mu_1\left((1-\mu)\frac{\mathbf{R} - \mu\mathbf{r}}{\|\mathbf{R} - \mu\mathbf{r}\|^3} + \mu\frac{\mathbf{R} + (1-\mu)\mathbf{r}}{\|\mathbf{R} + (1-\mu)\mathbf{r}\|} - \frac{\mathbf{R} + \boldsymbol{\rho}}{\|\mathbf{R} + \boldsymbol{\rho}\|^3}\right). \end{aligned} \tag{7.71}$$

For the circular restricted case, if one lets the satellite position vector in the orbit plane to be $\boldsymbol{\rho} = [x \quad y]^T$, the position vector of the Sun to be $\mathbf{R} = [R_x \quad R_y]^T$ and $\mathbf{r} = [1 \quad 0]^T$

$$\begin{aligned} \boldsymbol{\rho} = & (1-\mu)\frac{\boldsymbol{\rho} + \mu\mathbf{r}}{\|\boldsymbol{\rho} + \mu\mathbf{r}\|^3} + \mu\frac{\boldsymbol{\rho} - (1-\mu)\mathbf{r}}{\|\boldsymbol{\rho} - (1-\mu)\mathbf{r}\|^3} + \mu_1\left(\frac{\mathbf{R} + \boldsymbol{\rho}}{\|\mathbf{R} + \boldsymbol{\rho}\|^3} - (1-\mu)\frac{\mathbf{R} - \mu\mathbf{r}}{\|\mathbf{R} - \mu\mathbf{r}\|^3}\right. \\ & \left. -\mu\frac{\mathbf{R} + (1-\mu)\mathbf{r}}{\|\mathbf{R} + (1-\mu)\mathbf{r}\|}\right). \end{aligned} \tag{7.72}$$

The above equation is solved numerically using the Newton-Raphson method for increasing values of μ_1. Observe that

$$\frac{\partial \boldsymbol{\rho}}{\partial x} = [1 \quad 0]^T \quad \text{and} \quad \frac{\partial}{\partial x}\left(\frac{1}{\|\mathbf{R} + \boldsymbol{\rho}\|^3}\right) = -\frac{3}{2}\frac{1}{\|\mathbf{R} + \boldsymbol{\rho}\|^5}\frac{d\|\mathbf{R} + \boldsymbol{\rho}\|^2}{dx} = -3\frac{(R_x + x)}{\|\mathbf{R} + \boldsymbol{\rho}\|^5}. \tag{7.73}$$

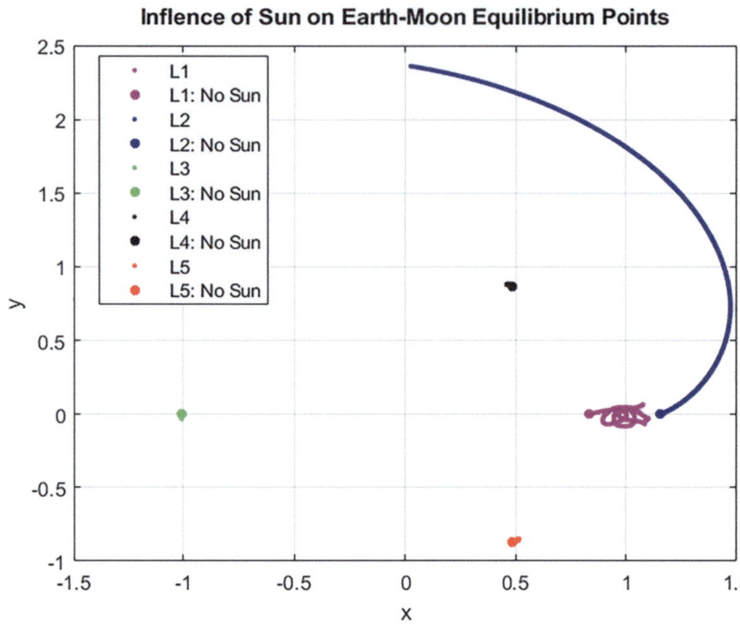

Fig. 7.5a Influence of the motion of the Sun on the orbit of a satellite located at one of the Earth-Moon synodic frame equilibrium points

Hence

$$\mu_1 \frac{\partial}{\partial x}\left(\frac{\mathbf{R}+\boldsymbol{\rho}}{\|\mathbf{R}+\boldsymbol{\rho}\|^3}\right) = \mu_1\left(\frac{[1 \quad 0]^T}{\|\mathbf{R}+\boldsymbol{\rho}\|^3}\right) - 3\mu_1 \frac{(\mathbf{R}+\boldsymbol{\rho})(R_x + x)}{\|\mathbf{R}+\boldsymbol{\rho}\|^5}. \tag{7.74}$$

By numerically integrating the position of satellite at the equilibrium point L_1, it is observed that it can move up to 10–15% of the distance from the larger primary. Thus the equilibrium could be significantly disturbed. The most unstable of the equilibrium points is L_2 while the other three are relatively stable. The effect of the Sun's movement is illustrated in Fig. 7.5a. The trajectories were obtained by integrating the equations of motion of the bi-circular restricted four-body problem.

7.5.1 Motion of the Spacecraft in a Multi-body Environment

Considering the motion of a spacecraft in an inertial frame of reference, in terms of its position vector $\mathbf{r}$ and the position vectors of the planet $\mathbf{r}_p$ and that of the Sun $\mathbf{r}_s$, in the vicinity of other celestial bodies at position vectors given by $\mathbf{r}_m$, with the gravitational constants of the planet, Sun and other celestial bodies given, respectively, by μ_p, μ_s and μ_m, the general multi-body equations are given by

$$\frac{d^2\mathbf{r}}{dt^2} + \frac{\mu_p}{|\mathbf{r}|^3}\mathbf{r} + \mu_s\left(\frac{\mathbf{r}_s}{|\mathbf{r}_s|^3} + \frac{\mathbf{r} - \mathbf{r}_s}{|\mathbf{r} - \mathbf{r}_s|^3}\right) = -\sum_{m=1}^{M}\mu_m\left(\frac{\mathbf{r}_m}{|\mathbf{r}_m|^3} + \frac{\mathbf{r} - \mathbf{r}_m}{|\mathbf{r} - \mathbf{r}_m|^3}\right). \quad (7.75)$$

The motion of a spacecraft is now described in an Earth-centred frame $Oxyz$ with Cartesian coordinates: the origin O is placed at the centre of the Earth, the x-axis points to the vernal point, the z-axis coincides with the Earth's rotation axis and the y-axis is placed consequently to obtain a right-hand-oriented frame. The motion can be obtained by considering an n-body problem with $n = 4$, since the gravity actions of the Moon, the Earth and the Sun are included. The dynamics of the spacecraft is then expressed by the following equations relative to the Earth-centred frame:

$$\frac{d}{dt}\begin{bmatrix} x \\ y \\ z \end{bmatrix} = \begin{bmatrix} v_x \\ v_y \\ v_z \end{bmatrix}, \quad \frac{d}{dt}\begin{bmatrix} v_x \\ v_y \\ v_z \end{bmatrix} = -\frac{\mu_s}{r_{\mathrm{ssat}}^3}\begin{bmatrix} x - x_s \\ y - y_s \\ z - z_s \end{bmatrix} - \frac{\mu_m}{r_{\mathrm{msat}}^3}\begin{bmatrix} x - x_m \\ y - y_m \\ z - z_m \end{bmatrix} - \frac{\mu_e}{r_{\mathrm{esat}}^3}\begin{bmatrix} x \\ y \\ z \end{bmatrix} + \begin{bmatrix} a_x \\ a_y \\ a_z \end{bmatrix}. \quad (7.76)$$

In the above equations, $[x \ \ y \ \ z]^T$ is the satellite's position vector; $[v_x \ \ v_y \ \ v_z]^T$ is the satellite' velocity vector; $[x_s \ \ y_s \ \ z_s]^T$ is the position vector of the Sun; $[x_m \ \ y_m \ \ z_m]^T$ is the position vector of the Moon; $[a_x \ \ a_y \ \ a_z]^T$ is the control acceleration vector; μ_s, μ_m and μ_e are the gravitational constants of the Sun, the Moon and Earth, respectively; r_{esat} is the distance of the satellite with respect to the centre of the Earth; r_{ssat} and r_{msat} are the distances between the Sun and the satellite and the Moon and the satellite. All the quantities in the equation are continuous functions of time. The quantities $[\mu_s \ \ \mu_e \ \ \mu_m]$ are the gravitational parameters. The sum $\mu_m + \mu_e$ may be expressed as $\mu_m + \mu_e = d^3\Omega^2$ and let $\mu = \mu_m/(\mu_m + \mu_e)$. Thus, $\mu_m = \mu d^3\Omega^2$ and $\mu_e = (1 - \mu)d^3\Omega^2$. Changing the time variable to $\tau = \Omega t$, choosing the unit of distance to be d

$$\frac{d}{d\tau}\begin{bmatrix} x \\ y \\ z \end{bmatrix} = \begin{bmatrix} v_x \\ v_y \\ v_z \end{bmatrix}, \quad \frac{d}{d\tau}\begin{bmatrix} v_x \\ v_y \\ v_z \end{bmatrix} = -\frac{(m_s/(m_e + m_m))d^3}{r_{\mathrm{ssat}}^3}\begin{bmatrix} x - x_s \\ y - y_s \\ z - z_s \end{bmatrix} - \frac{\mu d^3}{r_{\mathrm{msat}}^3}\begin{bmatrix} x - x_m \\ y - y_m \\ z - z_m \end{bmatrix}$$
$$- \frac{(1 - \mu)d^3}{r_{\mathrm{esat}}^3}\begin{bmatrix} x \\ y \\ z \end{bmatrix} + \frac{1}{\Omega^2}\begin{bmatrix} a_x \\ a_y \\ a_z \end{bmatrix}. \quad (7.77)$$

Transforming all quantities, i.e. $[x \ \ y \ \ z]^T$ the satellite's position vector, $[v_x \ \ v_y \ \ v_z]^T$ the satellite's velocity vector, $[x_s \ \ y_s \ \ z_s]^T$ the position vector of the Sun, $[x_m \ \ y_m \ \ z_m]^T$ the position vector of the Moon, and $[a_x \ \ a_y \ \ a_z]^T$ the control accelerations to a rotating frame of reference, rotating about the z-axis (the notation is not altered for convenience), Eq. (7.77) should be

$$\frac{d}{d\tau}\begin{bmatrix} x \\ y \\ z \end{bmatrix} = \begin{bmatrix} v_x \\ v_y \\ v_z \end{bmatrix}, \ \frac{d}{d\tau}\begin{bmatrix} v_x \\ v_y \\ v_z \end{bmatrix} - 2\begin{bmatrix} v_y \\ -v_x \\ 0 \end{bmatrix} - \begin{bmatrix} x \\ y \\ 0 \end{bmatrix}$$

$$= -\frac{(m_s/(m_e+m_m))d^3}{r_{\text{ssat}}^3}\begin{bmatrix} x-x_s \\ y-y_s \\ z-z_s \end{bmatrix} - \frac{\mu d^3}{r_{\text{msat}}^3}\begin{bmatrix} x-x_m \\ y-y_m \\ z-z_m \end{bmatrix} - \frac{(1-\mu)d^3}{r_{\text{esat}}^3}\begin{bmatrix} x \\ y \\ z \end{bmatrix} + \frac{1}{\Omega^2}\begin{bmatrix} a_x \\ a_y \\ a_z \end{bmatrix}. \tag{7.78}$$

Moving the origin of the rotating frame to a point given by $(\mu \quad 0 \quad 0)$ results in

$$\frac{d}{d\tau}\begin{bmatrix} x \\ y \\ z \end{bmatrix} = \begin{bmatrix} v_x \\ v_y \\ v_z \end{bmatrix}, \ \frac{d}{d\tau}\begin{bmatrix} v_x \\ v_y \\ v_z \end{bmatrix} - 2\begin{bmatrix} v_y \\ -v_x \\ 0 \end{bmatrix} - \begin{bmatrix} x \\ y \\ 0 \end{bmatrix} = -\frac{\mu_1 d^3}{r_{\text{ssat}}^3}\begin{bmatrix} x-x_s \\ y-y_s \\ z-z_s \end{bmatrix}$$
$$- \frac{\mu d^3}{r_{\text{msat}}^3}\begin{bmatrix} x-x_m \\ y-y_m \\ z-z_m \end{bmatrix} - \frac{(1-\mu)d^3}{r_{\text{esat}}^3}\begin{bmatrix} x+\mu \\ y \\ z \end{bmatrix} + \frac{1}{\Omega^2}\begin{bmatrix} a_x \\ a_y \\ a_z \end{bmatrix}, \tag{7.79}$$

with $\mu_1 = m_s/(m_e + m_m)$. Since d is the unit of distance, the state space equations are

$$\frac{d}{d\tau}\begin{bmatrix} x \\ y \\ z \end{bmatrix} = \begin{bmatrix} v_x \\ v_y \\ v_z \end{bmatrix}, \ \frac{d}{d\tau}\begin{bmatrix} v_x \\ v_y \\ v_z \end{bmatrix} - 2\begin{bmatrix} v_y \\ -v_x \\ 0 \end{bmatrix} - \begin{bmatrix} x \\ y \\ 0 \end{bmatrix} = -\frac{\mu_1}{r_{\text{ssat}}^3}\begin{bmatrix} x-x_s \\ y-y_s \\ z-z_s \end{bmatrix}$$
$$- \frac{\mu}{r_{\text{msat}}^3}\begin{bmatrix} x-x_m \\ y-y_m \\ z-z_m \end{bmatrix} - \frac{(1-\mu)}{r_{\text{esat}}^3}\begin{bmatrix} x+\mu \\ y \\ z \end{bmatrix} + \frac{1}{\Omega^2}\begin{bmatrix} a_x \\ a_y \\ a_z \end{bmatrix}. \tag{7.80}$$

In this rotating frame of reference if the moon is in a circular orbit, then it follows that

$$[x_m \quad y_m \quad z_m]^T = [1-\mu \quad 0 \quad 0]^T. \tag{7.81}$$

Assuming further that the Sun is infinitely far away from the Earth and its Moon, the equations reduce to the ones given earlier for Earth-Moon system, in the synodic frame. It must be emphasized that, in the above equations, $[x \quad y \quad z]^T$ is the satellite's position vector, $[v_x \quad v_y \quad v_z]^T$ the satellite' velocity vector, $[x_s \quad y_s \quad z_s]^T$, the position vector of the Sun, $[x_m \quad y_m \quad z_m]^T$ the position vector

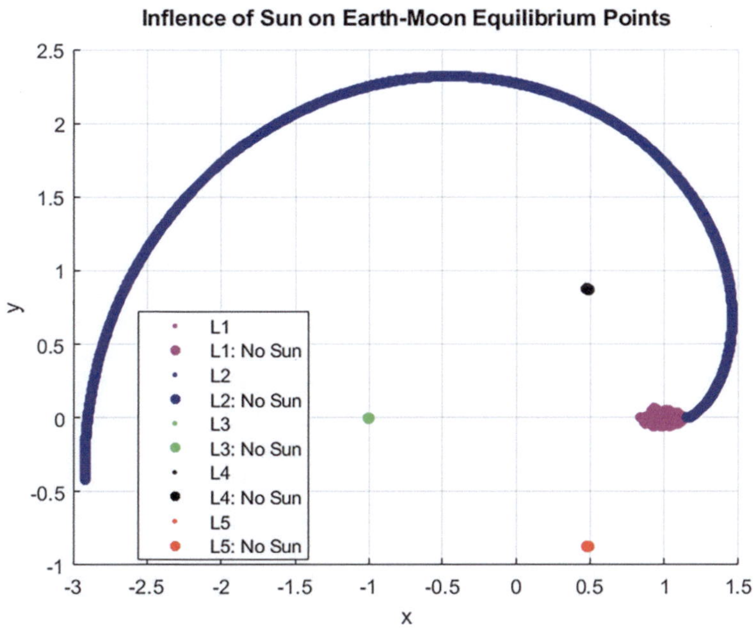

Fig. 7.5b Influence of the motion of the Sun on the orbit of a satellite located at one of the Earth-Moon synodic frame equilibrium points, assessed by using the multi-body approach

of the Moon and $[a_x \quad a_y \quad a_z]^T$ the control acceleration vector are all in the Earth-centred rotating frame of reference and not in the Earth-centred inertial frame.

The most unstable of the equilibrium points continues to be L_2 with the multi-body approach while the other three are relatively stable. In Fig. 7.5a, the Sun is stationary. The effect of the Sun's movement using the multi-body approach is illustrated in Fig. 7.5b. The value of μ is the ratio of mean values of the Earth-Moon barycentre and the radius of the Moon's orbit. In the Figs. 7.5a and 7.5b, the Lagrange points are defined based on the mean distance of the Moon from the Earth which is normalized to unity. The actual positions of both the Earth and the Moon can vary as the Moon progresses in its orbit. The trajectories were obtained by integrating the equations of motion of the multi-body problem. It is observed that in this case the equilibrium point L_2 is unstable even in the absence of any perturbation due to the Sun's motion, a feature that cannot be observed by considering Fig. 7.5a alone. To construct the relative motion equations in an Earth-centred Earth-fixed frame, one could use the equations of motion of a satellite and that of the Earth in an Earth-centred inertial frame and construct the relevant equations of motion by subtracting the later from the former.

7.6 The Patched Three-Body Approximation

The construction of the trajectories in the case of a four-body problem using invariant manifolds from multiple three-body systems using the patched three-body approximation is now described. The objective is to take maximum advantage of an N-body gravitational field to reduce fuel consumption, which is especially useful in the design of interplanetary trajectories designed to visit one or more celestial bodies. These may include a low energy transfer mission to and from the Earth to the Moon or trajectories such as a low energy mission to orbit another planet, such as Mars or Jupiter. In Koon et al. (2007), using the phase space tubes of two three-body systems, they were able to construct a transfer trajectory from the Earth which executes an un-propelled (i.e. ballistic) capture at the Moon. To design, a spacecraft trajectory from the Earth to the Moon which also takes advantage of the Sun's gravity, Koon et al. (2007) model the trajectory as two pieces: the first piece being a solution of the Sun-Earth-spacecraft three-body system (where the Moon's gravitational influence is unimportant) and the second piece being a solution of the Earth-Moon-spacecraft three-body system (where the Sun's influence is ignored). The two pieces are connected at a patch point between two three-body solution trajectories. The patch point on each of the two trajectories is at the same location or position in space but has differing velocities. The velocity discontinuity, of the norm $\left|\Delta \vec{v}\right|$, corresponds to the impulsive rocket-specific thrust which will be necessary to effect the transition between the two three-body solutions. With only a slight modification of a small mid-course $\left|\Delta \vec{v}\right|$ at the patch point, this procedure produces a realistic solution which can then be validated using the bi-circular restricted four-body problem.

7.6.1 *Example 1: Earth-Moon Transfer*

In this example at the departure stage, the appropriate three-body problem is the Sun-Earth-satellite system. As the satellite arrives at its destination, the Moon, the appropriate three-body problem is the Earth-Moon-satellite system. The equilibrium points of the Sun-Earth-satellite system are shown in Fig. 7.6.

The zero velocity curves of the Sun-Earth-satellite system are shown in Fig. 7.7 with the zero velocity curves of the Earth-Moon-satellite system overlaid to scale, assuming a 180° phase angle between the Sun-Earth and Earth-Moon axes.

After plotting the zero velocity curves for both these problems, points close to a Lagrange or equilibrium point, L_2 or L_1, are considered without velocity perturbations. Trajectories passing through these points are constructed by integrating both forwards and backwards in time. The arrival trajectories of the Earth-Moon-satellite system to and from L_1 are transformed to the Sun-Earth-satellite synodic frame. Typical examples of such trajectories overlaid on the zero velocity curves are shown in Fig. 7.8. Also shown are the Hill spheres of the Earth and the Moon.

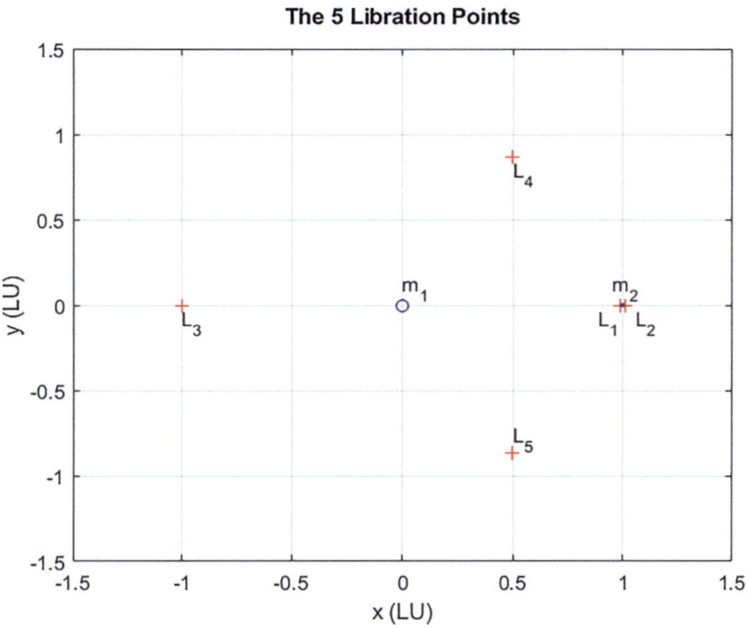

Fig. 7.6 Equilibrium points of the Sun-Earth-satellite system

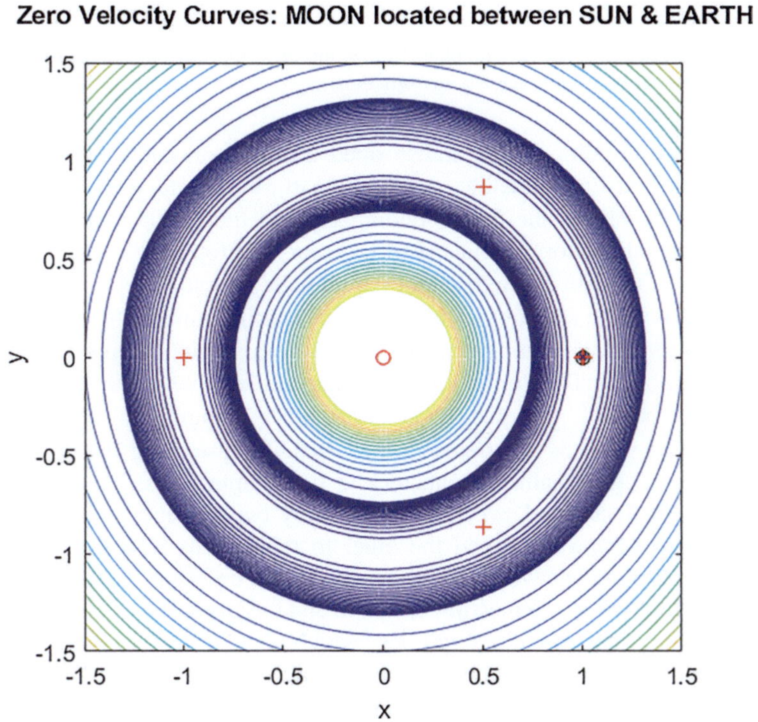

Fig. 7.7 Zero velocity curves of the Sun-Earth-satellite system with the zero velocity curves of the Earth-Moon-satellite system overlaid to scale

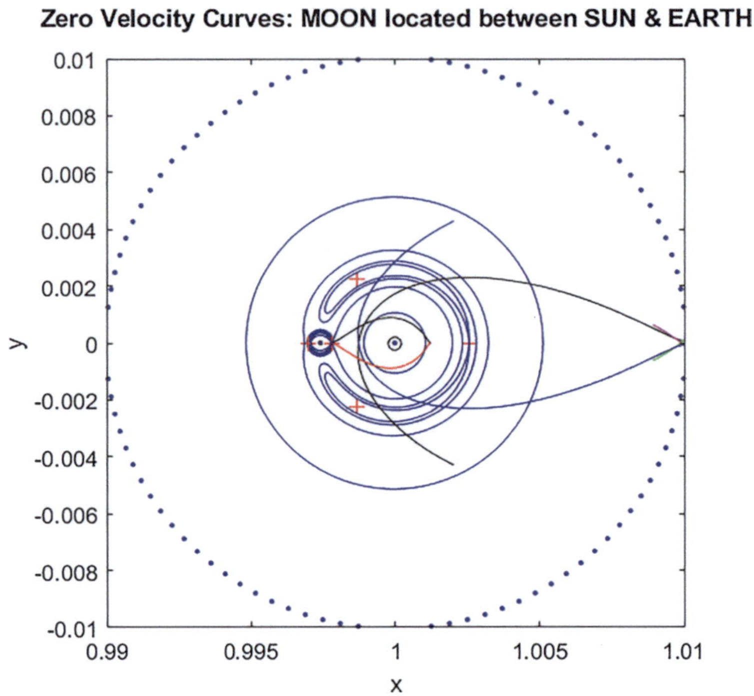

Fig. 7.8 Typical trajectories in the orbit plane from points close to a Lagrange or equilibrium point, L_2 or L_1, are considered without velocity perturbations

The intersection points of the transformed Earth-Moon-satellite system arrival trajectory and Sun-Earth-satellite system departure trajectory to and from L_2 (Fig. 7.9) are found and the corresponding $\Delta \vec{v}$ to transfer from the Sun-Earth-satellite departure trajectory to the arrival trajectory are found.

The intersection point at the top right requires the lowest $|\Delta \vec{v}|$ to transfer to the Lunar trajectory and is selected as the patch point. It is assumed that the satellite is transferred to the Earth orbiting trajectory prior to this point, from a circular orbit around the Earth. Small perturbations to swing the satellite within the lunar Hill sphere are introduced, and the trajectory with the minimum total $|\Delta \vec{v}|$ is constructed. This phase is illustrated in Fig. 7.10. A close-up of the trajectories is shown in Fig. 7.11. A ballistic trajectory is the simplest possible interplanetary trajectory. For ballistic rendezvous missions, only two impulsive manoeuvres are required: one to escape Earth and the other to rendezvous with the target body. Without any other manoeuvre in between them, the trajectory is simply a single heliocentric two-body arc connecting the Earth and the target body. In other words, a ballistic trajectory consists of a single leg, which also contains only one segment. Although the trajectory generated in this example is not a ballistic trajectory in a strict sense, there is just one leg from the parking orbit around the Earth to within the Hill sphere of the Moon. The trajectory approaching closest to the Moon requires the

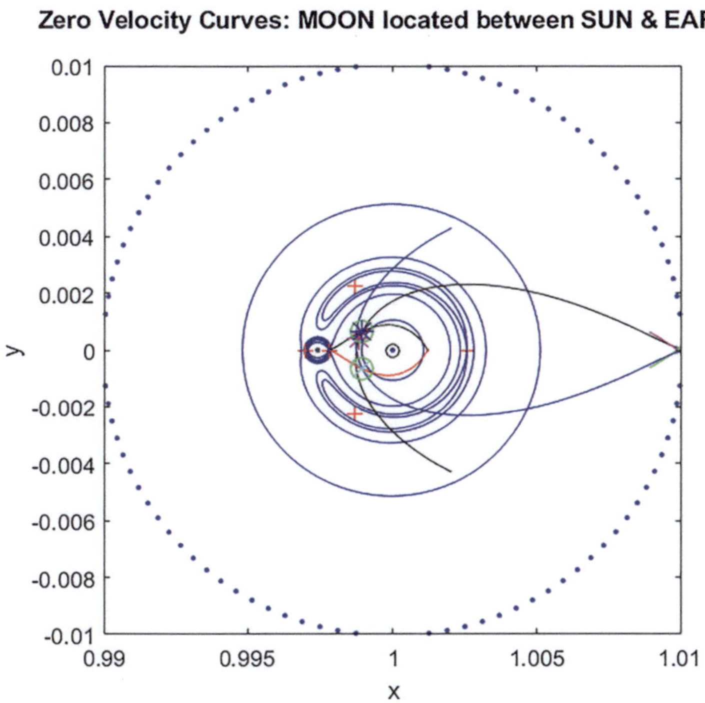

Fig. 7.9 The intersection points of the transformed Earth-Moon-satellite system arrival trajectory and Sun-Earth-satellite system departure trajectory. Also shown in the point of closest approach to the Earth

minimum $|\Delta \vec{v}|$ and is selected for the final phase. This is also within the zero-velocity loop around the Moon, which serves as the separatrix. To descend to the lunar surface, the spacecraft is set into a circular orbit around the Moon which is followed by orbit lowering. Thus the complete transfer trajectory from the Earth to the Moon may be constructed.

7.6.2 Example 2: Earth-Jupiter Transfer

In this case the entire trajectory can be divided into three phases, the departure phase, the transfer phase and the arrival phase.

The stable manifolds are used in transferring the spacecraft into the Halo orbit and the unstable manifolds help in escaping from the Halo orbit. A combination of the stable and unstable manifolds is utilized to transfer the spacecraft from the Sun-Earth system to the Sun-Jupiter system under the framework of the CRTBP. The departure phase, which refers to the spacecraft's departure from the Earth's Hill sphere, is solved by the Sun-Earth-spacecraft three-body problem. The next phase is the

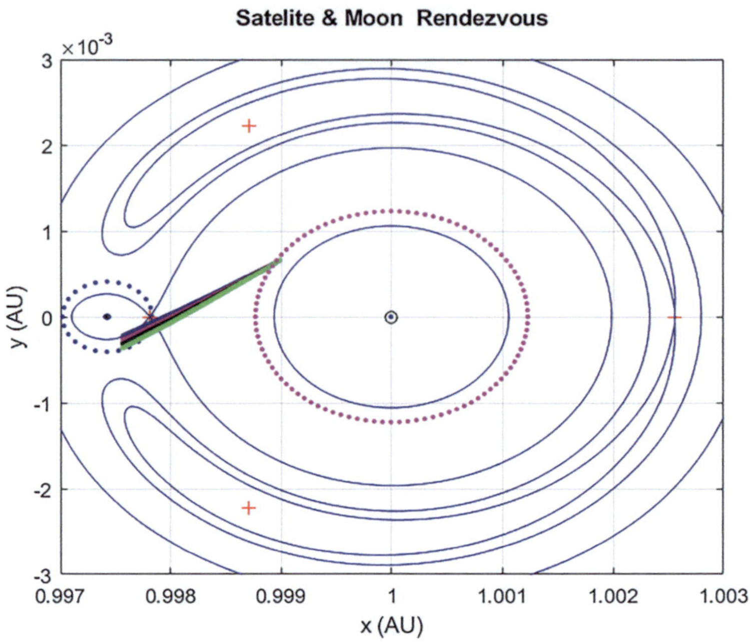

Fig. 7.10 The final phase of the orbit plane trajectory to the Moon's Hill sphere

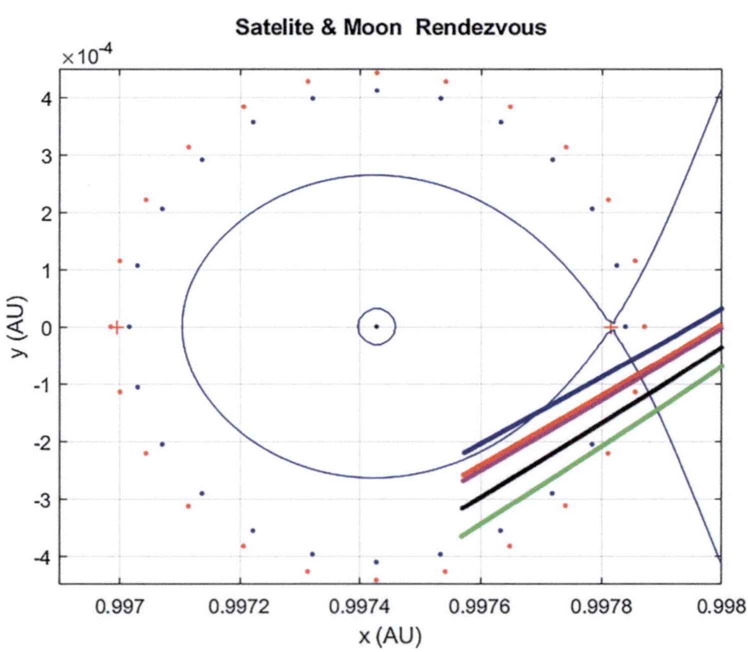

Fig. 7.11 A close-up of the final phase of the trajectory within the Moon's Hill sphere and SOI (inner and outer dotted circles, respectively)

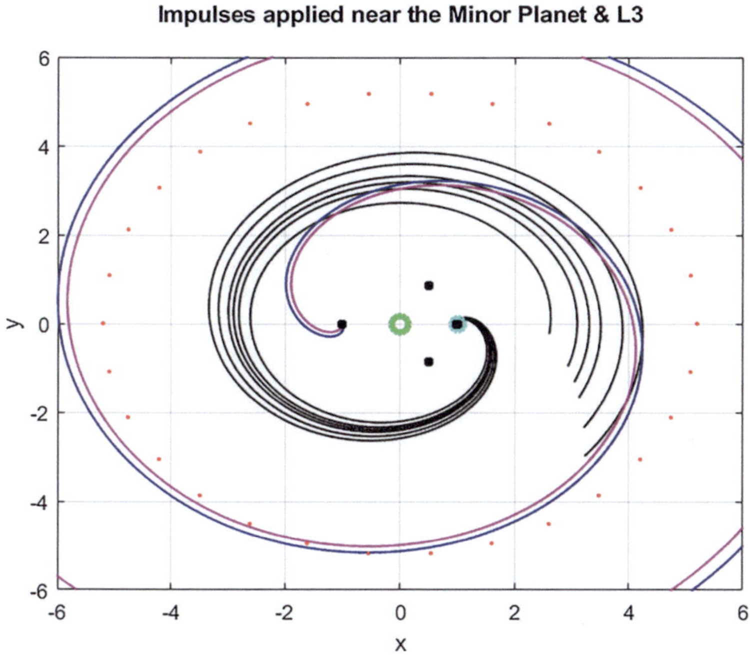

Fig. 7.12 Trajectories for leaving the Earth en-route to Jupiter (orbit shown as a dotted circle) via the orbit from the Sun-Earth-spacecraft Lagrange point L_3 (also shown are the Lagrange points and the Sun at the origin)

transfer phase which is also solved by the Sun-Earth-spacecraft three-body problem, with objective of heading out towards Jupiter. The final phase is arriving within Jupiter's Hill sphere, and this involves solving the Sun-Jupiter-spacecraft three-body problem. The first two phases are illustrated in the Sun-Earth-spacecraft synodic coordinates in Fig. 7.12. In Fig. 7.12, the departure trajectories are shown in black, leaving the Earth's Hill sphere. Initially, the departure trajectory cuts across Earth's Hill sphere, so one can transfer to it from a circular orbit around the Earth, from within the zero-velocity loop, which serves as the separatrix.

Also shown are two trajectories which start at the Sun-Earth-spacecraft Lagrange point L_3. These trajectories are suitable for heading out to Jupiter. Transfer from the Earth departure trajectory to one of these requires a small change in the $\Delta \vec{v}$ at the crossover point. Jupiter's orbit is also shown in 12° intervals. The trajectories are transformed to Sun-Jupiter-spacecraft synodic coordinates.

The trajectories in the Sun-Earth-spacecraft synodic coordinates are embedded within Fig. 7.13, which illustrates the trajectories in the Sun-Jupiter-spacecraft synodic coordinates. Also shown are the Sun-Earth-spacecraft Lagrange points and the Sun-Jupiter-spacecraft Lagrange points. The phase angle is chosen, based on phase angle of the spacecraft's direction relative to the horizontal when its distance from the Sun is equal to that of Jupiter, to ensure the transfer trajectories

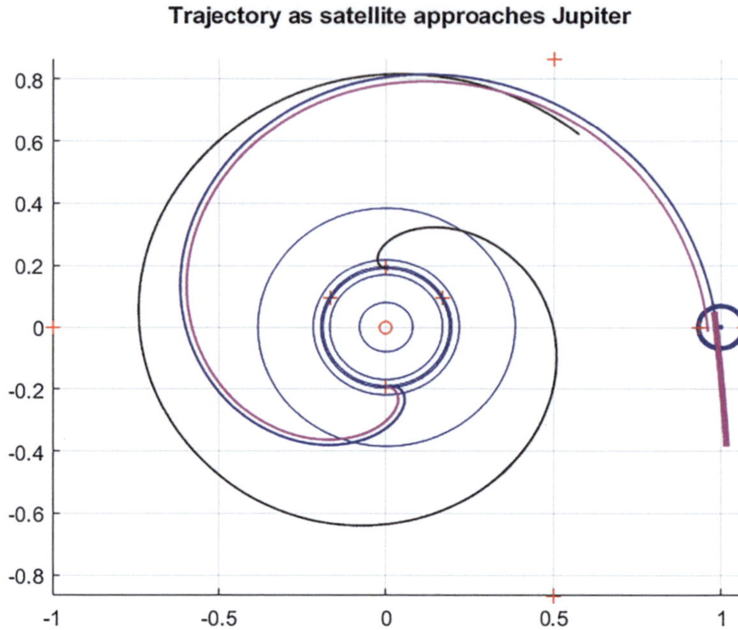

Fig. 7.13 Trajectories as spacecraft approaches Jupiter, in the Sun-Jupiter-spacecraft synodic coordinates

enter Jupiter's Hill sphere. One could also use slower departure and transfer trajectories so as to take a longer time frame to arrive at Jupiter, so as to use lesser amounts of fuel, if this is required. Only a restricted set of transfer trajectory options are shown in both Figs. 7.12 and 7.13.

One of the transfer trajectories is continued in the Sun-Jupiter-spacecraft synodic coordinates, by integrating the Sun-Jupiter-spacecraft equations of motion. A close-up of the trajectory, within Jupiter's Hill sphere and the SOI which is within the planet's Hill sphere, is shown in Fig. 7.14.

At the point of closest approach to Jupiter, another $\Delta \vec{v}$ is required, to slow the spacecraft down so it is a circular orbit around Jupiter. Alternately, one could continue en route to an outer planet including Saturn or one its moons such as Titan. The continuation of the trajectory towards Saturn would include the increment in the velocity due to the presence of Jupiter.

In this section, the application of the solution of the CR3BP in synodic coordinates and corresponding orbits to the problem of constructing interplanetary orbits departing from one planet and arriving at another has been successfully demonstrated. The patched three-body approximation has been completely documented. Trajectories of a spacecraft in a minimum of two different CR3BP settings are solved, once at the departure planet and the second at the arrival planet. Both these sets of trajectories can be transformed from one set of synodic coordinates to

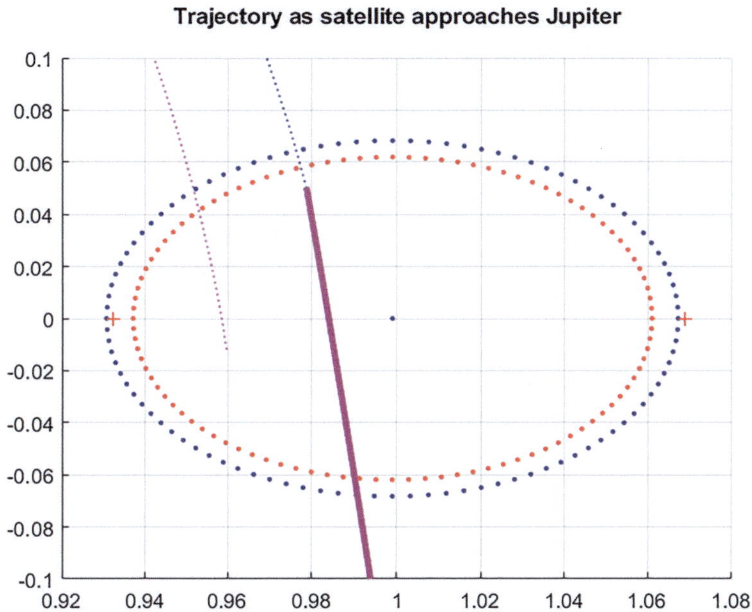

Fig. 7.14 Trajectories of the spacecraft within Jupiter's Hill sphere and SOI (the outer and inner dotted circles, respectively)

another. Moreover there are a number of trajectory prediction methods that can be used. Intersection points in a common reference frame are used to crossover from one trajectory to another. An impulsive application of a $\Delta \vec{v}$ is required at each such crossover point, in addition to the initial $\Delta \vec{v}$ applied to initiate the first transfer from circular parking orbit to the departure orbit. Generally a minimum of two impulses are required to enter the arrival planet's Hill sphere or its SOI. Here an additional impulse is required to enter a parking orbit or none, to carry on to another destination. The gravity-assisted manoeuvre at Jupiter is computed as a matter of course. If desired, one could continue with prediction of the trajectory and consider the arrival of the spacecraft at an outer planet such as Saturn and then proceed to any of its moons. All the manoeuvres required have been discussed in the earlier chapters. The complete transfer is less restrictive than a Hohmann transfer. In fact it can be shown that there are several choices for the trajectories not only in the final leg, depending on desired terminate state of the spacecraft but also over the entire mission. For this reason the total $\left|\Delta \vec{v}\right|$ required is not quoted, although it was estimated. Moreover the focus of this section is on constructing feasible trajectories. From the families of trajectories, one may then choose the most suitable trajectory for a particular mission based on the fuel requirements for the whole mission.

7.7 The Circular Restricted Three-body Problem with Applied Control Forces

There exist five equilibrium points, namely, the Lagrange points, in the synodic frame of a CRTBP (Szebehely (1967), Bate, Mueller and White (2005)). Three of these are collinear points generally denoted by L_1, L_2 and L_3 that are on the axis joining the centre of the primary and secondary bodies. It may be recalled that the collinear Lagrange points (CLPs) can be shown to be always unstable (in every system), for mass ratios of any CRTBP, such as the Earth-Moon, the Sun-Earth or Sun-Jupiter systems. The instability of CLPs also guarantees the existence of a few hyperbolic manifolds locally that can be used to generate the natural escaping/approaching motions from/to them. The possibility of harnessing the nature of the dynamics in the vicinity of the Lagrange points for interplanetary transfer applications could be explored. In particular, unstable dynamics associated with the L_1 and L_2 points and the slow dynamics and mild instability at L_3 could be fully exploited. Hence, a mechanism to reach or leave the proximity of CLPs by a suitable transfer method could be established and a transfer trajectory designed, which requires very little fuel consumption. Based on similar concepts, very efficient methods to find "zero cost" and "low cost" trajectories between Lagrange point orbits have been proposed by Farquhar et al. (1980); Gomez and Masdemont (2000); Zazzera et al. (2004); and Koon et al. (2007). Typical examples are of transfers among different Lagrange point orbits by Zeng and Zhang (2018) transfers between the Earth-Moon and the Sun-Earth systems by Howell and Kakoi (2006); low energy transfers to the Moon by Mingotti et al. (2009) and by Ozimek and Howell (2010); and minimum time transfers by Zhang et al. (2015).

Application of optimal and feedback control theories to low-thrust transfer has been extensive. Kluever (1993) successfully formulated and solved the problem of optimal low-thrust Earth-Moon transfers. Ilgen (1993) applied the Lyapunov-based controller feedback design method to Earth orbital transfers. Based on the Lyapunov approach, Petropoulos (2005) proposed the Q-law to evaluate the propellant mass and flight time for low-thrust transfer while Yuan et al. (2007) implemented the Q-law in terms of the modified equinoctial elements. Han and Yin (2011) and Li (2013) introduced the relative orbital elements, which were successfully employed by Bai et al. (2021) to develop a linearized transfer guidance approach based on the Lyapunov method. Daoud (2011) considered the application of optimal control to the CRTBP with particular emphasis on minimum time and minimum fuel control. Caillau et al. (2012) considered the minimum fuel optimal controller synthesis for the CRTBP.

Another feature that must be mentioned is the use and application of homoclinic and heteroclinic connections between the equilibrium points. Here the pioneering work of Howell et al. (1998) on using a dynamical systems approach to trajectory design to Lagrange point and of Koon et al. (2000) on the heteroclinic connections between periodic orbits and resonance transitions in celestial mechanics and the applications to trajectory design defines a whole new class of methods. Barrabes et al. (2009) and Barrabes et al. (2013) have considered numerical computation and continuation of families of homoclinic and heteroclinic connections between periodic orbits about CLPs. By constructing families of periodic orbits around the CLPs, it is possible to construct tubes in the phase space which serve as invariant manifolds, which provide physical insight into the behaviour of the trajectories in regions around the CLPs. Efficient use of Poincare sections of the phase space can provide useful visualizations of the behaviour of the orbital trajectories in the vicinity of the CLPs. Dynamical structures have been used for the low-thrust transfer trajectory design, including periodic orbits and manifolds, which provide a cost-efficient route to accomplish transfer trajectory design by Fantino et al. (2010). In Henninger and Biggs (2017) an almost minimal time low-thrust transfer from the near-Earth orbit to the Earth-Moon L_1 point was realized by the patching method and was used to connect averaged and non-averaged trajectories. In Ozimek and Howell (2010) the stable invariant manifolds were used to generate a feasible low-thrust transfer trajectory to the periodic orbits near the L_1 and L_2 points in the Earth-Moon system.

In the following sections, the objective is to establish a generic methodology for constructing complete feasible families of reasonably low-cost orbits from the low orbital regions of a departing planet to the low orbital regions of an arrival planet, using two or less patches of solutions to the CRTBP. Typical examples are transfers from the Earth to the moon, to the near-Earth planets, like Venus, Mars, Jupiter and Saturn, and back. This type of orbit transfers from one celestial body to another, involving either discrete impulses or continuous but finite time interval thrusting schedules. The challenge is to synthesize such schedules, generically and in the context of the three-body problem. There are many advantages to such as an approach, as a gravity swing by or assist which is a technique used to speed up, slow down or change the direction of a spacecraft as it approaches a planet can be incorporated in the trajectory as a matter of course. Once a feasible trajectory is synthesized, it could be further optimized to obtain an optimal trajectory by adopting techniques in optimal control such as the ones proposed by Betts and Erb (2003). Azimov (2018) has proposed a few analytical solutions to interplanetary trajectories that could also be used to find a feasible optimal solution.

In order to construct a feasible trajectory, several different representations of the circular restricted three- and four-body problems, with applied external forces, were considered, and the typical responses to classes of inputs were systematically analysed and their behavioural trends recorded. One approach is to use a component of radial thrust in addition to the tangential thrust. The use of radial thrust is not new. It was first proposed by Tsien (1953). The method has been generally discussed by Battin (1987). Boltz (1991) obtained an analytical solution for such orbits. Mengali and Quarta (2009) considered the escape from an elliptic orbit using radial thrust. Recently Bassetto et al. (2021) shown that spiral trajectories are induced by radial thrust. This and other experiential knowledge gained by numerical simulation of orbits with both tangential and radial thrust components, allowed for the right initial conditions to set for departure from a planet or Lagrange point, both with and without continuously applied low thrust, to another celestial body. Similarly, the right terminal conditions were set at the arrival body for back-propagation of the orbit towards the departure planet, both with and without continuously applied low thrust. The back-propagation of the orbits is a vital step, borrowed from optimal control theory, as it provides the understanding of the specific orbits to connect in constructing the entire trajectory. Orbits from and to a Lagrange point provide transit orbits or low-cost pathways which can provide guidelines over intervals of time, if necessary, to reduce the cost. It was then possible to pick the most appropriate thrust schedules, for the synthesis of a near-optimal trajectory from the vicinity of the departing planet to the vicinity of the arrival celestial body. Two examples are given and one discussed in some detail.

The circular restricted three-body problem (CR3PB) assumes that primary and secondary bodies are moving in a circular orbit about their centre of mass. The term "restricted" also refers to the condition that two masses are extremely heavier than the third. Among two masses, the heavier mass is considered as a primary body and the other as a secondary body. Sometimes, an additional assumption requires motion only in the orbital plane, hence the z component of equations vanishes.

Further the equation of motion are obtained for the case of CRTBP with applied control forces. To re-capitulate, a right-handed synodic (rotating) coordinate frame whose origin is located at the centre of mass of two-body system (i.e. lies between m_1 and m_2), as shown in Fig. 7.15, is considered. The masses m_1 and m_2 follow a circular orbit around their centre of mass due to the mutual gravitation force. The frame rotates about the z-axis, and consequently the masses m_1 and m_2 appear to be stationary in the rotating x-y plane.

The angular velocity of the rotating frame is constant in case of CRTBP, and given by $\mathbf{\Omega} = \Omega\hat{\mathbf{k}}$, where $\Omega = 2\pi/T$, $T = 2\pi\sqrt{r_{12}^3/Gm}$, which is given by Kepler's third law where the total mass is $m = m_1 + m_2$. In the synodic rotating coordinates,

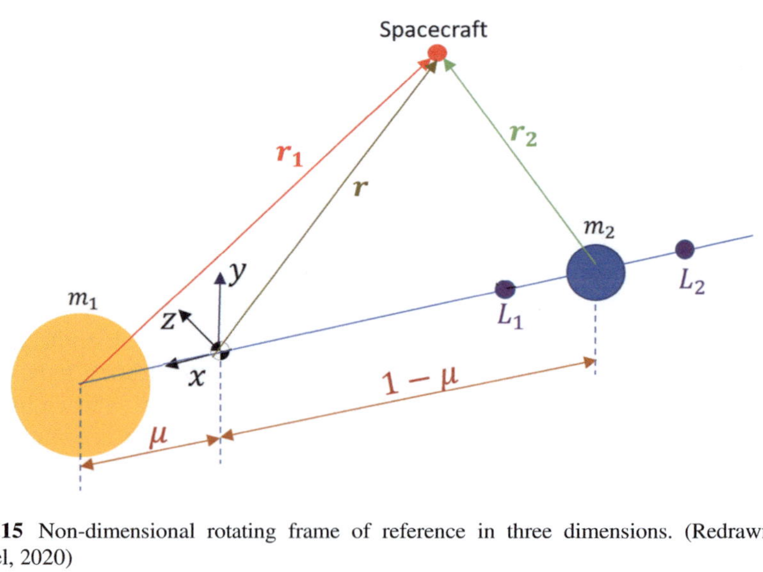

Fig. 7.15 Non-dimensional rotating frame of reference in three dimensions. (Redrawn from H. Patel, 2020)

the relative locations of m_1 and m_2 are defined relative to the centre of mass of the two bodies, given by constraint $m_1x_1 + m_2x_2 = 0$. If

$$\mu = m_2/(m_1 + m_2), \tag{7.82}$$

it follows that

$$x_1 = \mu r_{12} \text{ and } x_2 = -(1-\mu)r_{12},\ x_1 - x_2 = \mu r_{12} + (1-\mu)r_{12} = r_{12}. \tag{7.83}$$

The bodies, with masses m_1 and m_2 remain stationary in the synodic frame. When the spacecraft's position in the synodic coordinates is given by $\mathbf{r} = x\hat{\mathbf{i}} + y\hat{\mathbf{j}} + z\hat{\mathbf{k}}$, its positions relative to the bodies with masses m_1 and m_2 are respectively given by

$$\mathbf{r}_1 = (x + x_1)\hat{\mathbf{i}} + y\hat{\mathbf{j}} + z\hat{\mathbf{k}} \text{ and } \mathbf{r}_2 = (x + x_2)\hat{\mathbf{i}} + y\hat{\mathbf{j}} + z\hat{\mathbf{k}}. \tag{7.84}$$

The inertial velocity and acceleration of the spacecraft are respectively given by

$$\dot{\mathbf{r}} = \mathbf{v}_{so} + \mathbf{\Omega}\times\mathbf{r} + \dot{x}\hat{\mathbf{i}} + \dot{y}\hat{\mathbf{j}} + \dot{z}\hat{\mathbf{k}} = \mathbf{v}_{so} + \mathbf{\Omega}\times\mathbf{r} + \mathbf{v}_r = \mathbf{v}_{so} + (\dot{x} - \Omega y)\hat{\mathbf{i}} + (\dot{y} + \Omega x)\hat{\mathbf{j}} + \dot{z}\hat{\mathbf{k}}, \tag{7.85}$$

$$\ddot{\mathbf{r}} = \mathbf{a}_{so} + \mathbf{\Omega}\times(\mathbf{\Omega}\times\mathbf{r}) + 2\mathbf{\Omega}\times\mathbf{v}_r + \ddot{x}\hat{\mathbf{i}} + \ddot{y}\hat{\mathbf{j}} + \ddot{z}\hat{\mathbf{k}} = \mathbf{a}_{so} + \mathbf{\Omega}\times(\mathbf{\Omega}\times\mathbf{r}) + 2\mathbf{\Omega}\times\mathbf{v}_r + \mathbf{a}_r. \tag{7.86}$$

In the above equations, $\mathbf{v}_{so}$ and $\mathbf{a}_{so}$ are the velocity and acceleration of the origin of the synodic coordinates. When $\mathbf{v}_{so} \equiv 0$, $\mathbf{a}_{so} = 0$, and

$$\dot{\mathbf{r}} = (\dot{x} - \Omega y)\hat{\mathbf{i}} + (\dot{y} + \Omega x)\hat{\mathbf{j}} + \dot{z}\hat{\mathbf{k}}, \tag{7.87}$$

$$\ddot{\mathbf{r}} = \left(\ddot{x} - 2\Omega\dot{y} - \Omega^2 x\right)\hat{\mathbf{i}} + \left(\ddot{y} + 2\Omega\dot{x} - \Omega^2 y\right)\hat{\mathbf{j}} + \ddot{z}\hat{\mathbf{k}}. \tag{7.88}$$

A unit vector in the direction of $\mathbf{r}$ is

$$\hat{\mathbf{r}} = \left(x\hat{\mathbf{i}} + y\hat{\mathbf{j}} + z\hat{\mathbf{k}}\right)/|\mathbf{r}|. \tag{7.89}$$

A unit vector in the direction of $\dot{\mathbf{r}}$ is

$$\widehat{\dot{\mathbf{r}}} = \left((\dot{x} - \Omega y)\hat{\mathbf{i}} + (\dot{y} + \Omega x)\hat{\mathbf{j}} + \dot{z}\hat{\mathbf{k}}\right)/|\dot{\mathbf{r}}|,\ |\dot{\mathbf{r}}| = \sqrt{(\dot{x} - \Omega y)^2 + (\dot{y} + \Omega x)^2 + \dot{z}^2}. \tag{7.90}$$

From Newton's law, including the specific external force vector $\mathbf{f}_e$ or the applied acceleration not including gravitational forces

$$\ddot{\mathbf{r}} = \mathbf{f}_e - G\left(m_1\frac{\mathbf{r}_1}{|\mathbf{r}_1|^3} + m_2\frac{\mathbf{r}_2}{|\mathbf{r}_2|^3}\right) = \left(\ddot{x} - 2\Omega\dot{y} - \Omega^2 x\right)\hat{\mathbf{i}} + \left(\ddot{y} + 2\Omega\dot{x} - \Omega^2 y\right)\hat{\mathbf{j}} + \ddot{z}\hat{\mathbf{k}}. \tag{7.91}$$

The equations of motion reduce to

$$\begin{aligned}\frac{d^2}{dt^2}\begin{bmatrix}x\\y\\z\end{bmatrix} - 2\Omega\frac{d}{dt}\begin{bmatrix}y\\-x\\0\end{bmatrix} &= \Omega^2\begin{bmatrix}x\\y\\0\end{bmatrix} - \Omega^2 r_{12}^3\frac{(1-\mu)}{r_1^3}\begin{bmatrix}x+\mu r_{12}\\y\\z\end{bmatrix}\\ &\quad - \Omega^2 r_{12}^3\frac{\mu}{r_2^3}\begin{bmatrix}x-(1-\mu)r_{12}\\y\\z\end{bmatrix} + \begin{bmatrix}f_{ex}\\f_{ey}\\f_{ez}\end{bmatrix}.\end{aligned} \tag{7.92}$$

Furthermore, the external specific force vector in synodic coordinates ($[x \quad y \quad z]$) is related to the same in inertial coordinates ($[\tilde{x} \quad \tilde{y} \quad \tilde{z}]$) by the relation

$$\begin{bmatrix}f_{ex}\\f_{ey}\\f_{ez}\end{bmatrix} = \begin{bmatrix}\cos\Omega t & -\sin\Omega t & 0\\ \sin\Omega t & \cos\Omega t & 0\\ 0 & 0 & 1\end{bmatrix}\begin{bmatrix}f_{e\tilde{x}}\\f_{e\tilde{y}}\\f_{e\tilde{z}}\end{bmatrix}, \text{ where, } \begin{bmatrix}\tilde{x}\\\tilde{y}\\\tilde{z}\end{bmatrix} = \begin{bmatrix}\cos\Omega t & -\sin\Omega t & 0\\ \sin\Omega t & \cos\Omega t & 0\\ 0 & 0 & 1\end{bmatrix}\begin{bmatrix}x\\y\\z\end{bmatrix}. \tag{7.93}$$

Assuming that the unit of distance is such that $r_{12} = 1$, and the unit of time is $\tau = \Omega t$

$$\frac{d^2}{d\tau^2}\begin{bmatrix}x\\y\\z\end{bmatrix} - 2\frac{d}{d\tau}\begin{bmatrix}y\\-x\\0\end{bmatrix} = \begin{bmatrix}x\\y\\0\end{bmatrix} - \frac{(1-\mu)}{r_1^3}\begin{bmatrix}x+\mu\\y\\z\end{bmatrix} - \frac{\mu}{r_2^3}\begin{bmatrix}x-(1-\mu)\\y\\z\end{bmatrix} + \frac{1}{\Omega^2}\begin{bmatrix}f_{ex}\\f_{ey}\\f_{ez}\end{bmatrix}. \tag{7.94}$$

Re-scaling the coordinates, the spacecraft's inertial position and velocity and the inverse relationships are respectively given by

$$\begin{gathered}\mathbf{r} = r_{12}\left(x\hat{\mathbf{i}} + y\hat{\mathbf{j}} + z\hat{\mathbf{k}}\right) = r_{12}\mathbf{p}, \quad \mathbf{r}' = r_{12}\left(x'\hat{\mathbf{i}} + y'\hat{\mathbf{j}} + z'\hat{\mathbf{k}}\right) = r_{12}\mathbf{p}' = r_{12}\mathbf{v}\\ \dot{\mathbf{r}} = \Omega r_{12}\left((x'-y)\hat{\mathbf{i}} + (y'+x)\hat{\mathbf{j}} + z'\hat{\mathbf{k}}\right) = \Omega r_{12}\left(\mathbf{v} - \begin{bmatrix}0 & 1 & 0\\ -1 & 0 & 0\\ 0 & 0 & 0\end{bmatrix}\mathbf{p}\right).\end{gathered} \tag{7.95}$$

$$\mathbf{p}=\mathbf{r}/r_{12}\equiv\mathbf{r}/d,\ \ \mathbf{v}=\dot{\mathbf{r}}/(\Omega d)+\begin{bmatrix}0&1&0\\-1&0&0\\0&0&0\end{bmatrix}\mathbf{p}. \tag{7.96}$$

In the first instance, it is assumed that the thrust has three vector components, the first in the direction of the velocity vector $\dot{\bar{\mathbf{r}}}=\dot{\mathbf{r}}/\Omega d$, the second in the direction normal to it but in the orbit plane and the third normal to the orbit plane. Thus, with $|\dot{\bar{\mathbf{r}}}_p|=\sqrt{(x'-y)^2+(y'+x)^2}$

$$\frac{d^2}{d\tau^2}\begin{bmatrix}x\\y\\z\end{bmatrix}-2\frac{d}{d\tau}\begin{bmatrix}y\\-x\\0\end{bmatrix}=\begin{bmatrix}x\\y\\0\end{bmatrix}-\frac{(1-\mu)}{r_1^3}\begin{bmatrix}x+\mu\\y\\z\end{bmatrix}-\frac{\mu}{r_2^3}\begin{bmatrix}x-(1-\mu)\\y\\z\end{bmatrix}$$
$$+\frac{f_v}{r_{12}\Omega^2|\dot{\bar{\mathbf{r}}}_p|}\begin{bmatrix}x'-y\\y'+x\\0\end{bmatrix}+\frac{f_{v\perp}}{r_{12}\Omega^2|\dot{\bar{\mathbf{r}}}_p|}\begin{bmatrix}y'+x\\y-x'\\0\end{bmatrix}+\frac{f_z}{r_{12}\Omega^2}\begin{bmatrix}0\\0\\1\end{bmatrix}, \tag{7.97}$$

and f_v, $f_{v\perp}$ and f_z are scalar thrust accelerations. The quantities $f_v/r_{12}\Omega^2$, $f_{v\perp}/r_{12}\Omega^2$ and $f_z/r_{12}\Omega^2$ are the non-dimensional scalar thrust accelerations. So the equations of motion may be expressed as

$$\frac{d^2}{d\tau^2}\begin{bmatrix}x\\y\\z\end{bmatrix}-2\frac{d}{d\tau}\begin{bmatrix}y\\-x\\0\end{bmatrix}=\begin{bmatrix}x\\y\\0\end{bmatrix}-\frac{(1-\mu)}{r_1^3}\begin{bmatrix}x+\mu\\y\\z\end{bmatrix}-\frac{\mu}{r_2^3}\begin{bmatrix}x-(1-\mu)\\y\\z\end{bmatrix}$$
$$+\frac{\bar{f}_v}{|\dot{\bar{\mathbf{r}}}_p|}\begin{bmatrix}x'-y\\y'+x\\0\end{bmatrix}+\frac{\bar{f}_{v\perp}}{|\dot{\bar{\mathbf{r}}}_p|}\begin{bmatrix}y'+x\\y-x'\\0\end{bmatrix}+\bar{f}_z\begin{bmatrix}0\\0\\1\end{bmatrix}, \tag{7.98}$$

where $\bar{f}_v$, $\bar{f}_{v\perp}$ and $\bar{f}_z$ are the non-dimensional scalar thrust accelerations. The influence of low-thrust acceleration on the orbit of a spacecraft may now be considered. It may be observed that only the thrust acceleration component normal to the orbit plane is conservative. Hence the non-conservative components of the thrust acceleration in the directions of the velocity vector $\dot{\bar{\mathbf{r}}}_p$ will be considered first. The conservative component is only required to initiate a plane change, and for this reason, only the former case will be considered extensively. The equations of motion reduce to

$$\frac{d^2}{d\tau^2}\begin{bmatrix} x \\ y \\ z \end{bmatrix} - 2\frac{d}{d\tau}\begin{bmatrix} y \\ -x \\ 0 \end{bmatrix} = \begin{bmatrix} x \\ y \\ 0 \end{bmatrix} - \frac{(1-\mu)}{r_1^3}\begin{bmatrix} x+\mu \\ y \\ z \end{bmatrix} - \frac{\mu}{r_2^3}\begin{bmatrix} x-(1-\mu) \\ y \\ z \end{bmatrix} + \frac{\bar{f}_v}{\sqrt{(x'-y)^2 + (y'+x)^2}}\begin{bmatrix} x'-y \\ y'+x \\ 0 \end{bmatrix}. \tag{7.99}$$

An alternate representation of the equations of motion, with the thrust vector parameterized differently, is given by

$$\frac{d^2}{d\tau^2}\begin{bmatrix} x \\ y \\ z \end{bmatrix} - 2\frac{d}{d\tau}\begin{bmatrix} y \\ -x \\ 0 \end{bmatrix} = \begin{bmatrix} x \\ y \\ 0 \end{bmatrix} - \frac{(1-\mu)}{r_1^3}\begin{bmatrix} x+\mu \\ y \\ z \end{bmatrix} - \frac{\mu}{r_2^3}\begin{bmatrix} x-(1-\mu) \\ y \\ z \end{bmatrix} + \frac{\bar{f}_{r_p}}{|\mathbf{r}_p|}\begin{bmatrix} x \\ y \\ 0 \end{bmatrix} + \bar{f}_{np}\begin{bmatrix} 0 \\ 0 \\ 1 \end{bmatrix} + \frac{\bar{f}_\theta}{|\mathbf{r}_p|}\begin{bmatrix} y \\ -x \\ 0 \end{bmatrix}, \tag{7.100}$$

where $|\mathbf{r}_p| = \sqrt{x^2 + y^2}$.

Given that the satellite's motion is being described by one set of synodic position and velocity coordinates $\mathbf{p} = \mathbf{p}_2$, $\mathbf{v} = \mathbf{v}_2$, one would like to be able to relate them directly to another set of synodic coordinates, $\mathbf{p} = \mathbf{p}_1$, $\mathbf{v} = \mathbf{v}_1$, for the same set of inertial position and velocity coordinates. Thus given that

$$\mathbf{p}_2 = \frac{\mathbf{r}}{d_2}, \quad \mathbf{v}_2 = \frac{\dot{\mathbf{r}}}{\Omega_2 d_2} + \begin{bmatrix} 0 & 1 & 0 \\ -1 & 0 & 0 \\ 0 & 0 & 0 \end{bmatrix}\frac{\mathbf{r}}{d_2}, \tag{7.101}$$

one finds that

$$\mathbf{p}_2 = \frac{d_1}{d_2}\mathbf{p}_1, \quad \mathbf{v}_2 = \frac{\Omega_1 d_1}{\Omega_2 d_2}\mathbf{v}_1 - \frac{(\Omega_1 - \Omega_2)d_1}{\Omega_2 d_2}\begin{bmatrix} 0 & 1 & 0 \\ -1 & 0 & 0 \\ 0 & 0 & 0 \end{bmatrix}\mathbf{p}_1. \tag{7.102}$$

The inverse relationship may be easily found.

7.7.1 Planar Circular Restricted Three-Body Problem

A common approach when modelling a transfer in a three-body environment such as an Earth-Moon transfer, from an initial Earth-centred orbit to a lunar or selenocentric orbit using low-thrust propulsion, is to consider the trajectory in several phases. The orbit is assumed to be two-dimensional with the Moon's orbit about the Earth being circular. The initial phase consists of transferring from the initial parking orbit to another position inside of the Earth's sphere of influence (SOI) or the Hill sphere. There may be one or more intermediate transfers from this point on to a point within the Moon's SOI or Hill sphere. In the final phase, when the spacecraft is within the Moon's SOI or Hill sphere, it will thrust until it reaches the desired selenocentric orbit parking orbit. In the standard description of the planar CRTBP, the dynamical system will be modelled considering only the influence of gravity of the Earth and that of the Moon. Perturbations such as the *J*2 effect, solar radiation or atmospheric effects are not considered. The thrust acting on the spacecraft is considered to be the control force that is used to modify the spacecraft's trajectory.

The initial phase will consist of transferring the spacecraft from an initial parking orbit to another position within the Earth's SOI. The dynamical system is defined in an Earth-centred rotating reference frame, with the rotation rate equal to the Earth-Moon angular velocity. The equations of relative motion of the spacecraft in the primary body-centred polar coordinates are defined as follows (Kluever, 1993, pp. 26–28):

$$\dot{v}_r = -\frac{GM_E - rv_\theta^2}{r^2} - GM_M\frac{r + r_M\cos\theta}{D^3} + \frac{GM_M\cos\theta}{r_M^2} + 2\Omega v_\theta + \Omega^2 r + \frac{T}{m}\sin\alpha, \tag{7.103}$$

$$\dot{v}_\theta = -\frac{v_r v_\theta}{r} + GM_M\frac{r_M\sin\theta}{D^3} - \frac{GM_M\sin\theta}{r_M^2} - 2\Omega v_r + \frac{T}{m}\cos\alpha, \tag{7.104}$$

$$\dot{r} = v_r, \quad \dot{\theta} = v_\theta/r. \tag{7.105}$$

where v_r is the spacecraft's radial velocity, v_θ is its angular velocity, r is the radius of its orbit and θ is its angular position. The gravitational parameter of the Earth or the primary is defined as GM_E, while the corresponding lunar or secondary parameter is defined as GM_M. The angular velocity of the Moon with respect to the Earth (assuming Earth is the focus in the reference system) is defined as $\Omega = \sqrt{G(M_E + M_M)/r_M^3}$, while the radius of the Moon's orbit is represented by r_M. Furthermore, the mass of the spacecraft is defined as m, while the thrust produced by the engine is defined as T. The thrust direction, defined as the angle from the local horizon, is represented by α. The distance from the spacecraft to the Moon is derived from the cosine rule and is defined mathematically as folows:

$$D^2 = r^2 + r_M^2 - 2rr_M\cos\theta. \tag{7.106}$$

For the final phase, where the spacecraft is in a selenocentric orbit, the reference frame is rotational and lunar centred with the rotation rate equal to the Earth-Moon angular velocity. The corresponding relative equations of motion of the spacecraft in the lunar -entred polar coordinates are defined as

$$\dot{v}_{r_L} = -\frac{GM_M - r_L v_{\theta_L}^2}{r_L^2} - GM_E \frac{r_L + r_M \cos\theta_L}{D_E^3} + GM_E \frac{\cos\theta_L}{r_M^2} + 2\Omega v_{\theta_L} + \Omega^2 r_L + \frac{T}{m}\sin\alpha, \tag{7.107}$$

$$\dot{v}_{\theta_L} = -\frac{v_{r_L} v_{\theta_L}}{r_L} + GM_E \frac{r_M \sin\theta_L}{D_E^3} - GM_E \frac{\sin\theta_L}{r_M^2} - 2\Omega v_{r_L} + \frac{T}{m}\cos\alpha, \tag{7.108}$$

$$\dot{r}_L = v_{r_L}, \quad \dot{\theta}_L = v_{\theta_L}/r_L. \tag{7.109}$$

where v_{rL} is the spacecraft's radial velocity, $v_{\theta L}$ is its angular velocity, r_L is its radial position and θ_L is its angular position in the selenocentric reference frame. The distance from the spacecraft to the Earth is represented by D_E and obtained from the cosine rule.

Changing the time variable to $\tau = \Omega t$, and in terms of the mass ratio of the secondary planet, the Moon, to the total mass of the primary and secondary, $\mu = M_M/(M_E + M_M)$, the equations of relative motion of the spacecraft in the primary body-centred polar coordinates are

$$\frac{\tilde{v}_r'}{r_M} = \left(\frac{\tilde{v}_\theta}{r_M}\frac{r_M}{r} + 2\right)\frac{\tilde{v}_\theta}{r_M} - \frac{r_M^2(1-\mu)}{r^2} - \left(\mu\frac{r_M^3}{D^3} - 1\right)\frac{r}{r_M} - \left(\frac{r_M^3}{D^3} - 1\right)\mu\cos\theta + \frac{T\sin\alpha}{mr_M\Omega^2}, \tag{7.110}$$

$$\frac{\tilde{v}_\theta'}{r_M} = -\left(\frac{\tilde{v}_\theta}{r_M}\frac{r_M}{r} + 2\right)\frac{\tilde{v}_r}{r_M} + \left(\frac{r_M^3}{D^3} - 1\right)\mu\sin\theta + \frac{T\cos\alpha}{mr_M\Omega^2}, \tag{7.111}$$

$$r'/r_M = v_r/r_M\Omega = \tilde{v}_r/r_M, \quad \theta' = v_\theta/\Omega r = \tilde{v}_\theta/r. \tag{7.112}$$

For the lunar- or secondary body-centred phase, one has the corresponding equations of relative motion of the spacecraft given by

$$\frac{\tilde{v}_{r_L}'}{r_M} = \left(\frac{\tilde{v}_{\theta_L}}{r_M}\frac{r_M}{r_L} + 2\right)\frac{\tilde{v}_{\theta_L}}{r_M} - \mu\frac{r_M^2}{r_L^2} - \left((1-\mu)\frac{r_M^3}{D_E^3} - 1\right)\frac{r_L}{r_M} - \left(\frac{r_M^3}{D_E^3} - 1\right)(1-\mu)\cos\theta_L + \frac{T\sin\alpha}{mr_M\Omega^2}, \tag{7.113}$$

$$\frac{\tilde{v}_{\theta_L}'}{r_M} = -\left(\frac{\tilde{v}_{\theta_L}}{r_M}\frac{r_M}{r_L} + 2\right)\frac{\tilde{v}_{r_L}}{r_M} + \left(\frac{r_M^3}{D_E^3} - 1\right)(1-\mu)\sin\theta_L + \frac{T\cos\alpha}{mr_M\Omega^2}. \tag{7.114}$$

$$r_L'/r_M = v_{rL}/r_M\Omega = \tilde{v}_{rL}/r_M, \quad \theta_L' = v_{\theta_L}/\Omega r_L = (r_M/r_L)(\tilde{v}_{\theta_L}/r_M). \tag{7.115}$$

The above Eqs. (7.110)–(7.115) of the spacecraft, which could be expressed in non-dimensional form, may be used to study the minimum fuel, planar transfer first to a low Earth orbit and then from that orbit to a low lunar orbit by solving a set of optimal trajectory synthesis problems, as considered by Kluever (1993). The solution to the optimal low-thrust transfer trajectory from the primary to the secondary using Eqs. (7.110)–(7.115) is presented by Kluever (1993, pp. 26–36).

If the non-dimensional vector positions and position derivatives are expressed as $\mathbf{r} = r_M\mathbf{p}$, $\mathbf{r}' = r_M\mathbf{p}' = r_M\mathbf{v}$ and $r_M = d$, the vector Eq. (7.102), for transforming from one synodic coordinates to another, may now be used to transform the non-dimensional vector position and velocity derivative Eqs. (7.110)–(7.112) and Eqs. (7.113)–(7.115) from one set of synodic coordinates to another.

7.7.2 Multi-body Problems with Applied Forces and Continuous Thrust

Observe that equations of the CRTBP with externally applied forces, may be expressed as

$$\frac{d^2}{d\tau^2}\begin{bmatrix} x \\ y \\ z \end{bmatrix} - 2\begin{bmatrix} 0 & 1 & 0 \\ -1 & 0 & 0 \\ 0 & 0 & 0 \end{bmatrix}\frac{d}{d\tau}\begin{bmatrix} x \\ y \\ z \end{bmatrix} = \begin{bmatrix} x \\ y \\ 0 \end{bmatrix} - \frac{(1-\mu)}{r_1^3}\begin{bmatrix} x+\mu \\ y \\ z \end{bmatrix} - \frac{\mu}{r_2^3}\begin{bmatrix} x-(1-\mu) \\ y \\ z \end{bmatrix}$$
$$+\frac{\bar{f}_v}{\sqrt{(x'-y)^2+(y'+x)^2+z'^2}}\left\{\begin{bmatrix} x' \\ y' \\ z' \end{bmatrix} - \begin{bmatrix} 0 & 1 & 0 \\ -1 & 0 & 0 \\ 0 & 0 & 0 \end{bmatrix}\begin{bmatrix} x \\ y \\ z \end{bmatrix}\right\} \tag{7.116}$$

The equations are expressed in matrix form as

$$\frac{d^2}{d\tau^2}\mathbf{X} - 2\mathbf{A}\mathbf{X}' - \mathbf{B}\mathbf{X} = -(1-\mu)\frac{\boldsymbol{\rho}+\mu\mathbf{r}}{\|\boldsymbol{\rho}+\mu\mathbf{r}\|^3} - \mu\frac{\boldsymbol{\rho}-(1-\mu)\mathbf{r}}{\|\boldsymbol{\rho}-(1-\mu)\mathbf{r}\|^3} + \frac{\bar{f}_v}{|\mathbf{r}'|}(\mathbf{X}'-\mathbf{A}\mathbf{X}), \tag{7.117}$$

with $\boldsymbol{\rho} = [x \quad y \quad z]^T$ and $\mathbf{r} = [1 \quad 0 \quad 0]^T$.

In the case of the bi-circular restricted four-body problem in the Earth (with mass m_2) and the Moon (with mass equal to m_3) synodic frame, and with Sun as the fourth body with mass m_1, if $\mu = m_3/(m_3 + m_2)$ and $\mu_1 = m_1/(m_2 + m_3)$, equations of motion of the satellite with externally applied thrust may be expressed as

$$\frac{d^2}{d\tau^2}\mathbf{X} - 2\mathbf{A}\mathbf{X}' - \mathbf{B}\mathbf{X} = -(1-\mu)\frac{\boldsymbol{\rho}+\mu\mathbf{r}}{\|\boldsymbol{\rho}+\mu\mathbf{r}\|^3} - \mu\frac{\boldsymbol{\rho}-(1-\mu)\mathbf{r}}{\|\boldsymbol{\rho}-(1-\mu)\mathbf{r}\|^3}$$

$$+\mu_1\left((1-\mu)\frac{\mathbf{R}-\mu\mathbf{r}}{\|\mathbf{R}-\mu\mathbf{r}\|^3}+\mu\frac{\mathbf{R}+(1-\mu)\mathbf{r}}{\|\mathbf{R}+(1-\mu)\mathbf{r}\|}-\frac{\mathbf{R}+\boldsymbol{\rho}}{\|\mathbf{R}+\boldsymbol{\rho}\|^3}\right)+\frac{\bar{f}_v}{|\mathbf{r}'|}(\mathbf{X}'-\mathbf{A}\mathbf{X})$$
$$+\frac{\bar{f}_r}{|\mathbf{r}|}\mathbf{X}+\frac{\bar{f}_{\mathbf{r}\times\mathbf{r}'}}{|\mathbf{r}\times\mathbf{r}'|}\mathbf{X}\times\mathbf{X}', \tag{7.118}$$

where the position vector of the Sun, from the Earth-Moon barycentre in the Earth-Moon synodic frame, is $\mathbf{R}=\begin{bmatrix}R_x & R_y & 0\end{bmatrix}^T$. In the absence of externally allied forces, Eq. (7.118) reduces to the equations stated by Singh et al. (2021). In case when there is only a thrust component parallel to the velocity vector, one has

$$\frac{d^2}{d\tau^2}\mathbf{X}-2\mathbf{A}\mathbf{X}'-\mathbf{B}\mathbf{X}=-(1-\mu)\frac{\boldsymbol{\rho}+\mu\mathbf{r}}{\|\boldsymbol{\rho}+\mu\mathbf{r}\|^3}-\mu\frac{\boldsymbol{\rho}-(1-\mu)\mathbf{r}}{\|\boldsymbol{\rho}-(1-\mu)\mathbf{r}\|^3}$$
$$+\mu_1\left((1-\mu)\frac{\mathbf{R}-\mu\mathbf{r}}{\|\mathbf{R}-\mu\mathbf{r}\|^3}+\mu\frac{\mathbf{R}+(1-\mu)\mathbf{r}}{\|\mathbf{R}+(1-\mu)\mathbf{r}\|}-\frac{\mathbf{R}+\boldsymbol{\rho}}{\|\mathbf{R}+\boldsymbol{\rho}\|^3}\right)+\frac{\bar{f}_v}{|\mathbf{r}'|}(\mathbf{X}'-\mathbf{A}\mathbf{X}), \tag{7.119}$$

The motion of a spacecraft may also be described in an Earth-centred inertial frame $Oxyz$ with Cartesian coordinates: the origin O is placed at the centre of the Earth, the x-axis points to the vernal equinox, the z-axis coincides with the Earth's rotation axis, and the y-axis is placed consequently to obtain a right-hand-oriented frame. The motion can be obtained by considering an n-body problem with $n=4$, since the gravity actions of the Moon, the Earth and the Sun are included. The dynamics of the spacecraft is then expressed by the following equations relative to the Earth-centred inertial frame:

$$\frac{d}{dt}\begin{bmatrix}x\\y\\z\end{bmatrix}=\begin{bmatrix}v_x\\v_y\\v_z\end{bmatrix},\ \frac{d}{dt}\begin{bmatrix}v_x\\v_y\\v_z\end{bmatrix}=-\frac{\mu_s}{r_{ssat}^3}\begin{bmatrix}x-x_s\\y-y_s\\z-z_s\end{bmatrix}-\frac{\mu_m}{r_{msat}^3}\begin{bmatrix}x-x_m\\y-y_m\\z-z_m\end{bmatrix}-\frac{\mu_e}{r_{esat}^3}\begin{bmatrix}x\\y\\z\end{bmatrix}+\begin{bmatrix}a_x\\a_y\\a_z\end{bmatrix}. \tag{7.120}$$

In the above equations, $\begin{bmatrix}x & y & z\end{bmatrix}^T$ is the satellite's position vector; $\begin{bmatrix}v_x & v_y & v_z\end{bmatrix}^T$ the satellite' velocity vector; $\begin{bmatrix}x_s & y_s & z_s\end{bmatrix}^T$ the position vector of the Sun; $\begin{bmatrix}x_m & y_m & z_m\end{bmatrix}^T$ the position vector of the Moon; $\begin{bmatrix}a_x & a_y & a_z\end{bmatrix}^T$ the control acceleration vector; μ_s, μ_m and μ_e the gravitational constants of the Sun, the Moon and Earth, respectively; r_{esat} is the distance of the satellite with respect to the centre of the Earth; and r_{ssat} and r_{msat} the distances from the Sun and and the Moon to the satellite, respectively. All the quantities in the equation are continuous functions of time. The quantities $\begin{bmatrix}\mu_s & \mu_e & \mu_m\end{bmatrix}$ are the gravitational parameters. The sum $\mu_m+\mu_e$ may be expressed as $\mu_m+\mu_e=d^3\Omega^2$ and let $\mu=\mu_m/(\mu_m+\mu_e)$. Thus, $\mu_m=\mu d^3\Omega^2$ and

$\mu_e = (1 - \mu)d^3\Omega^2$. Changing the time variable to $\tau = \Omega t$, choosing the unit of distance to be d

$$\frac{d}{d\tau}\begin{bmatrix} x \\ y \\ z \end{bmatrix} = \begin{bmatrix} v_x \\ v_y \\ v_z \end{bmatrix}, \quad \frac{d}{d\tau}\begin{bmatrix} v_x \\ v_y \\ v_z \end{bmatrix} = -\frac{(m_s/(m_e+m_m))d^3}{r_{ssat}^3}\begin{bmatrix} x-x_s \\ y-y_s \\ z-z_s \end{bmatrix} - \frac{\mu d^3}{r_{msat}^3}\begin{bmatrix} x-x_m \\ y-y_m \\ z-z_m \end{bmatrix}$$
$$-\frac{(1-\mu)d^3}{r_{esat}^3}\begin{bmatrix} x \\ y \\ z \end{bmatrix} + \frac{1}{\Omega^2}\begin{bmatrix} a_x \\ a_y \\ a_z \end{bmatrix}. \tag{7.121}$$

Transforming all quantities, i.e. $[x \quad y \quad z]^T$ the satellite's position vector, $[v_x \quad v_y \quad v_z]^T$ the satellite' velocity vector, $[x_s \quad y_s \quad z_s]^T$ the position vector of the Sun, $[x_m \quad y_m \quad z_m]^T$ the position vector of the Moon, $[a_x \quad a_y \quad a_z]^T$ the control accelerations to a rotating frame of reference, rotating about the z-axis (the notation is not altered for convenience)

$$\frac{d}{d\tau}[x \quad y \quad z]^T = [v_x \quad v_y \quad v_z]^T,$$
$$\frac{d}{d\tau}\begin{bmatrix} v_x \\ v_y \\ v_z \end{bmatrix} - 2\begin{bmatrix} v_y \\ -v_x \\ 0 \end{bmatrix} - \begin{bmatrix} x \\ y \\ 0 \end{bmatrix} = -\frac{(m_s/(m_e+m_m))d^3}{r_{ssat}^3}\begin{bmatrix} x-x_s \\ y-y_s \\ z-z_s \end{bmatrix} - \frac{\mu d^3}{r_{msat}^3}\begin{bmatrix} x-x_m \\ y-y_m \\ z-z_m \end{bmatrix}$$
$$-\frac{(1-\mu)d^3}{r_{esat}^3}\begin{bmatrix} x \\ y \\ z \end{bmatrix} + \frac{1}{\Omega^2}\begin{bmatrix} a_x \\ a_y \\ a_z \end{bmatrix}. \tag{7.122}$$

Moving the origin of the rotating frame to a point given by $(\mu \quad 0 \quad 0)$ results in

$$\frac{d}{d\tau}\begin{bmatrix} x \\ y \\ z \end{bmatrix} = \begin{bmatrix} v_x \\ v_y \\ v_z \end{bmatrix}, \quad \frac{d}{d\tau}\begin{bmatrix} v_x \\ v_y \\ v_z \end{bmatrix} - 2\begin{bmatrix} v_y \\ -v_x \\ 0 \end{bmatrix} - \begin{bmatrix} x \\ y \\ 0 \end{bmatrix} = -\frac{\mu_1 d^3}{r_{ssat}^3}\begin{bmatrix} x-x_s \\ y-y_s \\ z-z_s \end{bmatrix}$$
$$-\frac{\mu d^3}{r_{msat}^3}\begin{bmatrix} x-x_m \\ y-y_m \\ z-z_m \end{bmatrix} - \frac{(1-\mu)d^3}{r_{esat}^3}\begin{bmatrix} x+\mu \\ y \\ z \end{bmatrix} + \frac{1}{\Omega^2}\begin{bmatrix} a_x \\ a_y \\ a_z \end{bmatrix}, \tag{7.123}$$

with $\mu_1 = m_s/(m_e + m_m)$. Since d is the unit of distance, the state space equations are

$$\frac{d}{d\tau}\begin{bmatrix} x & y & z \end{bmatrix}^T = \begin{bmatrix} v_x & v_y & v_z \end{bmatrix}^T,$$

$$\frac{d}{d\tau}\begin{bmatrix} v_x \\ v_y \\ v_z \end{bmatrix} - 2\begin{bmatrix} v_y \\ -v_x \\ 0 \end{bmatrix} - \begin{bmatrix} x \\ y \\ 0 \end{bmatrix} = -\frac{\mu_1}{r_{ssat}^3}\begin{bmatrix} x - x_s \\ y - y_s \\ z - z_s \end{bmatrix} - \frac{\mu}{r_{msat}^3}\begin{bmatrix} x - x_m \\ y - y_m \\ z - z_m \end{bmatrix} - \frac{(1-\mu)}{r_{esat}^3}\begin{bmatrix} x + \mu \\ y \\ z \end{bmatrix} + \frac{1}{\Omega^2}\begin{bmatrix} a_x \\ a_y \\ a_z \end{bmatrix}. \tag{7.124}$$

In this rotating frame of reference if the moon is in a circular orbit, then it follows that

$$\begin{bmatrix} x_m & y_m & z_m \end{bmatrix}^T = \begin{bmatrix} 1-\mu & 0 & 0 \end{bmatrix}^T. \tag{7.125}$$

Assuming further that the Sun is infinitely far away from the Earth and its Moon, the equations reduce to the ones given earlier for Earth-Moon system, in the synodic frame. It must be emphasized that, in the above equations, $\begin{bmatrix} x & y & z \end{bmatrix}^T$ the satellite's position vector, $\begin{bmatrix} v_x & v_y & v_z \end{bmatrix}^T$ the satellite' velocity vector, $\begin{bmatrix} x_s & y_s & z_s \end{bmatrix}^T$, the position vector of the Sun, $\begin{bmatrix} x_m & y_m & z_m \end{bmatrix}^T$ the position vector of the Moon, and $\begin{bmatrix} a_x & a_y & a_z \end{bmatrix}^T$ the control acceleration vector are all in the Earth-centred rotating frame of reference and not in the Earth-centred inertial frame. The control accelerations may be conveniently split into three vector components as was done earlier in Eqs. (7.98) and (7.100). Eq. (7.124) are a generalization of Eq. (7.94) in the presence of an additional body. The multi-body approach may be used to validate the designed trajectory by introducing appropriate changes to the initial conditions at the patch points.

7.8 Optimal Steering Control Law for Orbit Raising

Equations (7.97), for the third body or the spacecraft, may be expressed in matrix form. If the velocity and position vectors are respectively defined as $\boldsymbol{v} = \begin{bmatrix} x' & y' & z' \end{bmatrix}^T$, $\boldsymbol{r} = \begin{bmatrix} x & y & z \end{bmatrix}^T$ and the external specific force vector acting on the spacecraft as $\boldsymbol{F}_{c1} = \begin{bmatrix} a_x & a_y & a_z \end{bmatrix}^T$, it will be assumed that it can be expressed as $\boldsymbol{F}_{c1} = T\boldsymbol{u}$, where T is the magnitude of the specific thrust or thrust per unit mass and $\boldsymbol{u}$ defines the direction of the thrust force. It is the steering control input to the spacecraft. Since Eq. (7.87) are a special case of Eq. (7.124), the equations of motion given by Eq. (7.124), in matrix form, are

$$d\boldsymbol{r}/d\tau = \boldsymbol{v}, \tag{7.126}$$

where

$$\frac{d\boldsymbol{v}}{d\tau}=2\begin{bmatrix}0&1&0\\-1&0&0\\0&0&0\end{bmatrix}\boldsymbol{v}+\begin{bmatrix}1&0&0\\0&1&0\\0&0&0\end{bmatrix}\boldsymbol{r}-\frac{\mu_1}{r_{ssat}^3}(\boldsymbol{r}-\boldsymbol{r}_s)-\frac{(1-\mu)}{r_1^3}\left\{\boldsymbol{r}+\begin{bmatrix}\mu\\0\\0\end{bmatrix}\right\}$$
$$-\frac{\mu}{r_2^3}\left\{\boldsymbol{r}-\begin{bmatrix}(1-\mu)\\0\\0\end{bmatrix}\right\}+T\boldsymbol{u}\equiv\boldsymbol{V}(\boldsymbol{v},\boldsymbol{r})+T\boldsymbol{u},\tag{7.127}$$

with $r_{ssat}=|\boldsymbol{r}-\boldsymbol{r}_s|$, $\tau=\Omega t$ and

$$\boldsymbol{u}=[\,\sin\alpha\cos\beta\quad\cos\alpha\cos\beta\quad\sin\beta\,]\tag{7.128}$$

where α and β are the steering angles defining the direction of the thrust vector, which need not be explicitly computed at this stage. For simplicity, it is assumed that $\Omega=1$.

One can relate the $\Delta\boldsymbol{v}$ required to the magnitude the specific thrust by

$$d\Delta v/dt=T.\tag{7.129}$$

Normally the general cost function with a terminal cost, fixed end time and no terminal constraint is expressed as

$$J=\Phi(\boldsymbol{r}(t_f),t_f)+\int_{t_0}^{t_f}\mathcal{L}(\boldsymbol{r}(t),\boldsymbol{u}(t),t)dt\tag{7.130}$$

The terminal cost component is $\Phi(\boldsymbol{r}(t_f),t_f)$ while the running cost is $\mathcal{L}(\boldsymbol{r}(t),\boldsymbol{u}(t),t)$. When the running cost is set to zero, i.e. $\mathcal{L}(\boldsymbol{r}(t),\boldsymbol{u}(t),t)\equiv 0$, the cost functional reduces to $J=\Phi(\boldsymbol{r}(t_f),t_f)$.

When one is interested in the problem of finding the steering control

$$\boldsymbol{u}=\boldsymbol{u}(t),\quad t_0\le t\le t_f\tag{7.131}$$

the thrust direction time history is sought, such that it minimizes the cost functional

$$J=\left|\boldsymbol{r}(t)-\boldsymbol{r}_f\right|\Big|_{t=t_f}=\Phi\{\boldsymbol{r}(t)\}|_{t=t_f},\tag{7.132}$$

subject to Eqs. (7.126), (7.127) and (7.129). Introducing the single state vector, $\mathbf{x}=[\boldsymbol{r}\quad\boldsymbol{v}\quad\Delta\boldsymbol{v}]^T$, so the Eqs. (7.126), (7.128) and (7.130) are expressed as

$$\frac{d}{dt}\mathbf{x}^T = \frac{d}{dt}[\boldsymbol{r} \quad \boldsymbol{v} \quad \Delta\boldsymbol{v}] = \boldsymbol{f}^T. \tag{7.133}$$

To solve the optimization problem, seven Lagrangian multipliers or co-states are introduced given by the two 3-vectors $\boldsymbol{\lambda}_r(t)$, $\boldsymbol{\lambda}_v(t)$ and the scalar $\lambda_{\Delta v}$, denoted by a single column vector, $\boldsymbol{\lambda}(t)$. Following Bryson and Ho (1975), a Hamiltonian function is defined as

$$H = \boldsymbol{\lambda}^T(t)\boldsymbol{f}(t). \tag{7.134}$$

Hence

$$H = \boldsymbol{\lambda}_r^T \boldsymbol{v} + \boldsymbol{\lambda}_v^T \boldsymbol{V}(\boldsymbol{v}, \boldsymbol{r}) + \boldsymbol{\lambda}_v^T T\boldsymbol{u} + \lambda_{\Delta v} T. \tag{7.135}$$

The necessary conditions (Bryson & Ho, 1975) for the first variation of J to be zero include the co-state differential equations

$$d\boldsymbol{\lambda}/dt = -\partial H/\partial \boldsymbol{x}. \tag{7.136}$$

Explicitly the co-state equations are

$$\dot{\boldsymbol{\lambda}}_r(t) = -\frac{\partial \boldsymbol{V}(\boldsymbol{v}, \boldsymbol{r})}{\partial \boldsymbol{r}} \boldsymbol{\lambda}_v(t), \tag{7.137}$$

$$\dot{\boldsymbol{\lambda}}_v(t) = -\boldsymbol{\lambda}_r(t) - 2\begin{bmatrix} 0 & 1 & 0 \\ -1 & 0 & 0 \\ 0 & 0 & 0 \end{bmatrix} \boldsymbol{\lambda}_v, \tag{7.138}$$

$$\dot{\lambda}_{\Delta v} = 0, \tag{7.139}$$

where

$$\frac{\partial \boldsymbol{V}(\boldsymbol{v}, \boldsymbol{r})}{\partial \boldsymbol{r}} = \begin{bmatrix} 1 & 0 & 0 \\ 0 & 1 & 0 \\ 0 & 0 & 0 \end{bmatrix} - \left(\frac{\mu_1}{r_{\text{ssat}}^3} + \frac{(1-\mu)}{r_1^3} + \frac{\mu}{r_2^3}\right) \begin{bmatrix} 1 & 0 & 0 \\ 0 & 1 & 0 \\ 0 & 0 & 1 \end{bmatrix} - \frac{\mu_1}{r_{\text{ssat}}^5} (\boldsymbol{r} - \boldsymbol{r}_s)(\boldsymbol{r} - \boldsymbol{r}_s)^T$$

$$- \frac{(1-\mu)}{r_1^5} \left\{ \boldsymbol{r} + \begin{bmatrix} \mu \\ 0 \\ 0 \end{bmatrix} \right\} \left\{ \boldsymbol{r}^T + \begin{bmatrix} \mu \\ 0 \\ 0 \end{bmatrix}^T \right\} - \frac{\mu}{r_2^5} \left\{ \boldsymbol{r} - \begin{bmatrix} (1-\mu) \\ 0 \\ 0 \end{bmatrix} \right\} \left\{ \boldsymbol{r}^T - \begin{bmatrix} (1-\mu) \\ 0 \\ 0 \end{bmatrix}^T \right\}. \tag{7.140}$$

The optimality conditions are

$$\frac{\partial H}{\partial \alpha} = \boldsymbol{\lambda}_v \cdot T[\cos\alpha\cos\beta \quad -\sin\alpha\cos\beta \quad 0] = 0, \tag{7.141}$$

and

$$\frac{\partial H}{\partial \beta} = \boldsymbol{\lambda}_v \cdot T[-\sin\alpha\sin\beta \quad -\cos\alpha\sin\beta \quad \cos\beta] = 0. \tag{7.142}$$

Hence

$$\boldsymbol{\lambda}_v = |\boldsymbol{\lambda}_v|[\sin\alpha\cos\beta \quad \cos\alpha\cos\beta \quad \sin\beta] = \pm|\boldsymbol{\lambda}_v|\boldsymbol{u}. \tag{7.143}$$

when $\beta \equiv 0$, from Eq. (7.141), $\lambda_{v_x}\cos\alpha = \lambda_{v_x}\sin\alpha$, which can then be used to obtain the control vector $\boldsymbol{u}$, from Eq. (7.128).

In the general case, the two-parameter control vector $\boldsymbol{u}$ can be expressed as

$$\boldsymbol{u} = \pm\boldsymbol{\lambda}_v/|\boldsymbol{\lambda}_v|. \tag{7.144}$$

The choice of the sign in Eq. (7.144) will depend on the direction of the desired thrust, forward or reverse thrust. Thus the closed-loop equations of motion are

$$\frac{d}{dt}\mathbf{x} = \boldsymbol{f} = \left[\boldsymbol{v}_1 \quad \boldsymbol{V}(\boldsymbol{v},\boldsymbol{r}) \pm T\frac{\boldsymbol{\lambda}_v}{|\boldsymbol{\lambda}_v|} \quad T\right]^T, \tag{7.145}$$

with

$$\boldsymbol{V}(\boldsymbol{v},\boldsymbol{r}) = 2\begin{bmatrix} 0 & 1 & 0 \\ -1 & 0 & 0 \\ 0 & 0 & 0 \end{bmatrix}\boldsymbol{v} + \begin{bmatrix} 1 & 0 & 0 \\ 0 & 1 & 0 \\ 0 & 0 & 0 \end{bmatrix}\boldsymbol{r} - \frac{\mu_1}{r_{ssat}^3}(\boldsymbol{r} - \boldsymbol{r}_s) - \frac{(1-\mu)}{r_1^3}\left\{\boldsymbol{r} + \begin{bmatrix} \mu \\ 0 \\ 0 \end{bmatrix}\right\} - \frac{\mu}{r_2^3}\left\{\boldsymbol{r} - \begin{bmatrix} (1-\mu) \\ 0 \\ 0 \end{bmatrix}\right\}. \tag{7.146}$$

To complete the definition of the optimal solution, the boundary conditions at $t = t_f$ for the co-state system are found by applying the transversality conditions. The transversality conditions ensure that the initial and final states are selected optimally within the feasible regions of the states. For the transversality conditions, one may write

$$\lambda_r(t_f) = \partial\Phi\{\boldsymbol{r}(t)\}/\partial\boldsymbol{r}|_{t=t_f} = sign(\boldsymbol{r}(t_f) - \boldsymbol{r}_f), \tag{7.147}$$

$$\lambda_v(t_f) = \partial\Phi\{\boldsymbol{r}(t)\}/\partial\boldsymbol{v}|_{t=t_f} = 0. \tag{7.148}$$

Generally, the final mass of the spacecraft is not specified and so, in this case, the corresponding transversality condition for maintaining the final mass as a free variable is

$$\dot{\lambda}_{\Delta v} = 0. \tag{7.149}$$

Equation (7.144) is a steering control law for defining the optimal trajectory although it is not a feedback controller. The time history of the direction of the controller is given by Eq. (7.144).

While the formulation of the optimal control problem is quite straight forward, the solution of the resulting two-point boundary value problem is neither apparent nor easy, especially when the control input is limited by an upper bound. It was not possible to obtain a reasonably valid solution using, for example, MATLAB's built-in function *bvp4c.m*. Several other methods of solution also failed in providing a reasonably optimal solution. Optimal low-thrust transfers essentially rely on optimal control theory as outlined above but the key to the successful solution of the problem is the development of an appropriate initial guess for the numerical optimization. So other methods of choosing a near-optimal initial trajectory were considered and these are discussed in the next section. However, since these methods could be in any coordinates, the equations of relative motion in polar coordinates are also required. Yet only the equations of motion about the secondary body are used. In the Sun-Earth domain, Earth is the secondary body, while in the Earth-Moon domain the Moon is the secondary body and all three planetary bodies, and the spacecraft are assumed to be in the same orbit plane. Furthermore, only the most essential requirements and constraints are imposed to minimize the fuel requirements. Thus the non-dimensional relative equations of motion in polar coordinates about the secondary body are obtained from Eqs. (4.11), (4.12), and (4.13) and expressed as

$$r' = \tilde{v}_r, \quad \theta' = \tilde{v}_\theta / r. \tag{7.150}$$

$$\begin{aligned}\tilde{v}_r' = &\left(\frac{\tilde{v}_\theta}{r} + 2\right)\tilde{v}_\theta - \frac{\mu}{r^2} + \left(1 - \frac{(1-\mu)}{D_1^3}\right)r + \left(1 - \frac{1}{D_1^3}\right)(1-\mu)\cos\theta + \tilde{T}\sin\alpha \\ = &f_1(r, \theta, \tilde{v}_r, \tilde{v}_\theta) + \tilde{T}\sin\alpha,\end{aligned} \tag{7.151}$$

$$\tilde{v}_\theta' = -\left(\frac{\tilde{v}_\theta}{r} + 2\right)\tilde{v}_r - \left(1 - \frac{1}{D_1^3}\right)(1-\mu)\sin\theta + \tilde{T}\cos\alpha = f_2(r, \theta, \tilde{v}_r, \tilde{v}_\theta) + \tilde{T}\cos\alpha, \tag{7.152}$$

where $D_1 = D_1(r, \theta) = \sqrt{1 + r^2\cos^2\theta + 2r\cos\theta}$ is the distance from the spacecraft to the primary body.

Introducing a single state vector, $\mathbf{x}=\left[r \quad \theta \quad \tilde{v}_r \quad \tilde{v}_\theta \quad \Delta\tilde{v}\right]^T$, so Eqs. (7.150), (7.151) and (7.152) are expressed as

$$\frac{d}{dt}\mathbf{x}^T=\frac{d}{dt}\left[r \quad \theta \quad \tilde{v}_r \quad \tilde{v}_\theta \quad \Delta\tilde{v}\right]=\boldsymbol{f}^T=\left[\tilde{v}_r \quad \tilde{v}_\theta/r \quad f_1+\tilde{T}\sin\alpha \quad f_2+\tilde{T}\cos\alpha \quad \tilde{T}\right]. \tag{7.153}$$

$$H=\boldsymbol{\lambda}^T(t)\boldsymbol{f}(t). \tag{7.154}$$

and

$$H=\lambda_r\tilde{v}_r+\lambda_\theta(\tilde{v}_\theta/r)+\lambda_{\tilde{v}_r}f_1+\lambda_{\tilde{v}_\theta}f_2+\lambda_{\tilde{v}_r}T\sin\alpha+\lambda_{\tilde{v}_\theta}T\cos\alpha+\lambda_{\Delta v}T. \tag{7.155}$$

If α is not prescribed, the control is expressed as

$$\boldsymbol{u}=\left[\sin\alpha \quad \cos\alpha \quad 0\right].$$

In the case when α is prescribed, the co-states reduce to the adjoint states and the problem reduces to an optimization problem.

Equations (7.136) to (7.140) reduce to

$$d\boldsymbol{\lambda}/dt=-\partial H/\partial\boldsymbol{x}. \tag{7.156}$$

Explicitly the co-state equations are

$$\dot{\lambda}_r(t)=\lambda_\theta\left(\tilde{v}_\theta/r^2\right)-\lambda_{\tilde{v}_r}\partial f_1/\partial r-\lambda_{\tilde{v}_\theta}\partial f_2/\partial r,$$

$$\dot{\lambda}_\theta(t)=-\lambda_{\tilde{v}_r}\partial f_1/\partial\theta-\lambda_{\tilde{v}_\theta}\partial f_2/\partial\theta,\ \ \dot{\lambda}_{\tilde{v}_r}(t)=-\lambda_r+\lambda_{\tilde{v}_\theta}(2+\tilde{v}_\theta/r),$$

$$\dot{\lambda}_{\tilde{v}_\theta}(t)=-\lambda_\theta/r-2\lambda_{\tilde{v}_r}(1+\tilde{v}_\theta/r)+\lambda_{\tilde{v}_\theta}\tilde{v}_r/r, \tag{7.157}$$

$$\dot{\lambda}_{\Delta v}=0. \tag{7.158}$$

The optimality condition if α is not prescribed is

$$\partial H/\partial\alpha=\lambda_{\tilde{v}_r}T\cos\alpha-\lambda_{\tilde{v}_\theta}T\sin\alpha=0. \tag{7.159}$$

Thus, if α is not prescribed

$$\tan\alpha=\lambda_{\tilde{v}_r}/\lambda_{\tilde{v}_\theta}. \tag{7.160}$$

Equations similar to (7.147), (7.148), and (7.149) are used to complete the solution, with r_f prescribed and $\theta_f=0$. As the cost is only a function of the terminal conditions, the optimal solution is then easily found as there are now, several solutions available.

7.8.1 Optimal Transfer with Constrained Orbit Raising

The problem considered in the preceding section involves no constraints on the magnitude of the thrust available for applying to the spacecraft. When the thrust available is severely limited, relative to the requirements, it would be necessary to apply the maximum thrust to ensure that the spacecraft traverses as far as possible from the departure planet and as close as possible to the target planet, in a finite number of revolutions. This would not only require the application of the maximum thrust but also an optimum thrust direction angle to maximize the radial travel of the spacecraft. Through several simulations of the equations of motion of the planar circular-restricted three bodies, it was found that the optimum thrust angle was always 45°. While this result may seem surprising at first, a little reflection will show that with the thrust angle of 45°, one could expect the right balance of the distance travelled and velocity both radially and tangentially. The tangential velocity contributes to the centrifugal force which supplements the radial acceleration. Naturally, both the acceleration components together are responsible in the radial positioning of the spacecraft. Moreover there is an optimum thrust angle, at which the maximum radial position is reached.

It is natural to expect that in order to optimally raise the orbit of spacecraft, the maximal thrust has to be applied, at the optimal thrust angle, so the spacecraft completes several full revolutions before approaching the vicinity of the target planet. Only in the vicinity of the target planet would be necessary to apply the optimum steering control laws derived in the preceding section. On the other hand, if on the completion of several full revolutions about the departure planet, the spacecraft is within the sphere of influence or Hill domain of the target planet, the spacecraft could start the descent or deceleration so as to orbit the target planet.

There are other approaches to solving the problem. A related problem, which must be mentioned, is Lambert's problem where one is interested in estimating the minimum required single-impulse Δv to transfer a spacecraft to a selected orbital position in a fixed time of flight. In the case of the two-body problem, the low-thrust case is a classic problem dealt with by many authors. (See, e.g. Roa et al. (2016) and the references cited there.) The problem can also be formulated and solved in the case when the spacecraft makes multiple revolutions. Shen and Tsiotras (2003) showed that Battin's (1986) cubic equation for the time of flight can be extended to the case of multiple revolutions by adding $N\pi$ to the Eccentric anomaly, E. Bombardelli et al. (2018) have also obtained the solution of Lambert's targeting problem where the minimum required single-impulse Δv is obtained to modify the trajectory of a spacecraft in order to transfer it to a selected orbital position in a fixed time of flight for the multiple revolutions case. An alternate approach is based on solving the spiral Lambert's problem using the generalized logarithmic spirals which are generated by the application of continuous thrust to a model of the orbit, which is equivalent to a two-body problem with continuous thrust, as proposed by Roa et al. (2016). Leeghim et al. (2013) have formulated the problem as a constrained optimization problem in terms of the classical Lagrangian f and g functions. In this paper, one is interested when the thrust is applied continuously and has a low magnitude. Furthermore, in our case, it is a multi-body situation, although the two-body solution could be used as an approximation to start with. Moreover it is possible in principle to generalize the formulation of Leeghim et al. (2013) to the multi-body case with multiple revolutions and continuous thrust along the lines outlined in the preceding section. However it was decided not to adopt such an approach and instead a numerical method was adopted using Eqs. (7.150), (7.151), and (7.152), to model the orbit, including the third body and in synodic coordinates, rather than the two-body model of the orbital motion in fixed coordinates as in Roa et al. (2016). The results of the application of such an approach are presented in the next section.

7.9 Application Examples: Orbit Raising with Continuous Thrust

To start with the simulation of a circular orbit with zero thrust about the secondary body using the equations of motion of the CRTBP is considered and the spacecraft's orbit is shown in Fig. 7.16. Figure 7.17 shows the evolution of the circular orbit with thrust. Figure 7.17a is considered a low circular orbit. A uniform low thrust is applied to the spacecraft in a direction at a fixed angle of 0° to the velocity vector. The outer circular orbit radius is 2.48 times the inner circular orbit radius of

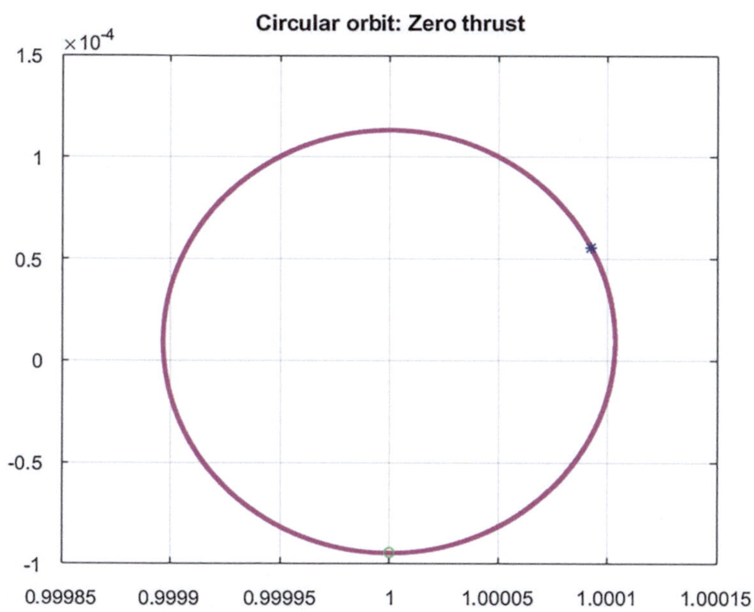

Fig. 7.16 Simulation of a circular orbit with zero thrust about the secondary body using the equations of motion of the CRTBP

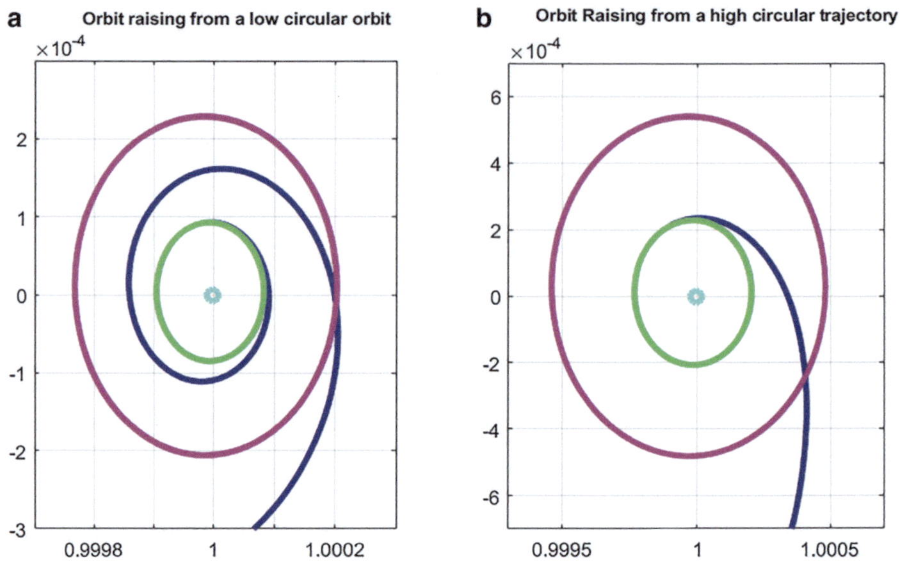

Fig. 7.17 Example of low-thrust orbit raising simulation about the secondary body using the equations of motion of the circular restricted three-body problem. (**a**) The initial orbit radius is 14142 km. (**b**) Example with initial orbit radius equal to 2.48 times 14142 km

14,142 km. Figure 7.17b is considered a relatively higher orbit equal to 2.48 times 14,142 km. The applied thrust is the same magnitude as in Fig. 7.17a. In the case of the lower orbit, the rate of increase of the orbital radius is much smaller than in the case of the higher orbit. Figure 7.18 shows the effect of increasing the thrust direction from an initial negative value to zero. Figure 7.19 shows the effect of increasing the thrust direction from zero to a final positive value. It is clear from Figs. 7.18 and 7.19 that the critical thrust direction angle for orbit raising is zero. The effect of changing the thrust direction angle on the spacecraft's direction of travel is considered in Fig. 7.20. The optimum thrust direction angle for maximum orbit raising is 45°.

Furthermore, by altering the thrust direction angle, it is possible to steer the spacecraft towards the target body although not precisely directed towards it. In Fig. 7.20a, c, the thrust direction angle is 45° while in Fig. 7.20b it is reduced to 9° and increased to 54° in Fig. 7.20d. The spacecraft is furthest radially, in a fixed time frame with the angle at 45°. While the simulations shown in Figs. 7.16, 7.17, 7.18, and 7.19 were done by using the equations of motion of the CRTBP given by Eq. (7.100), the simulations in Fig. 7.20 were performed using Eqs. (7.113)–(7.115), the planar circular restricted three-body equations of motion.

The experience gained in simulating the trajectories of spacecraft allowed for choosing a set of initial conditions, a thrust profile and a time of flight to synthesize a hybrid trajectory from the Earth to the Moon, shown in Fig. 7.21, using Eqs. (7.113)–(7.115). The spacecraft was initially in circular Earth orbit when a continuous low thrust corresponding to a total velocity change of 0.388 km/s was

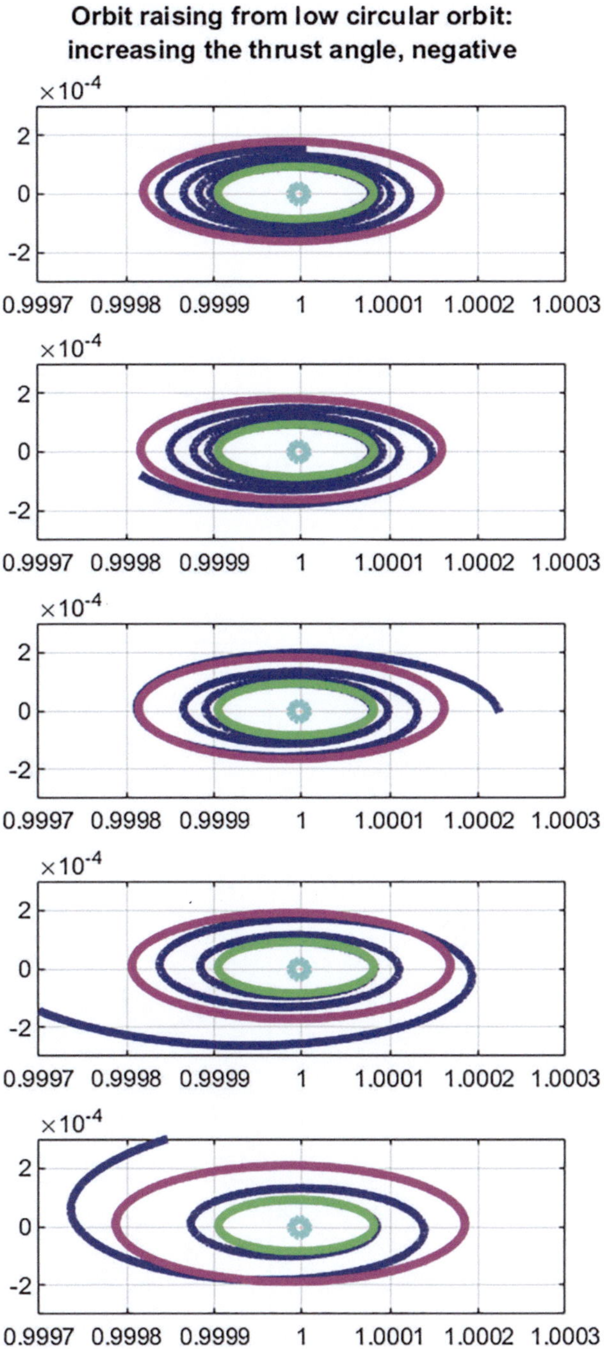

Fig. 7.18 Effect of increasing the thrust angle, with the thrust angle being negative

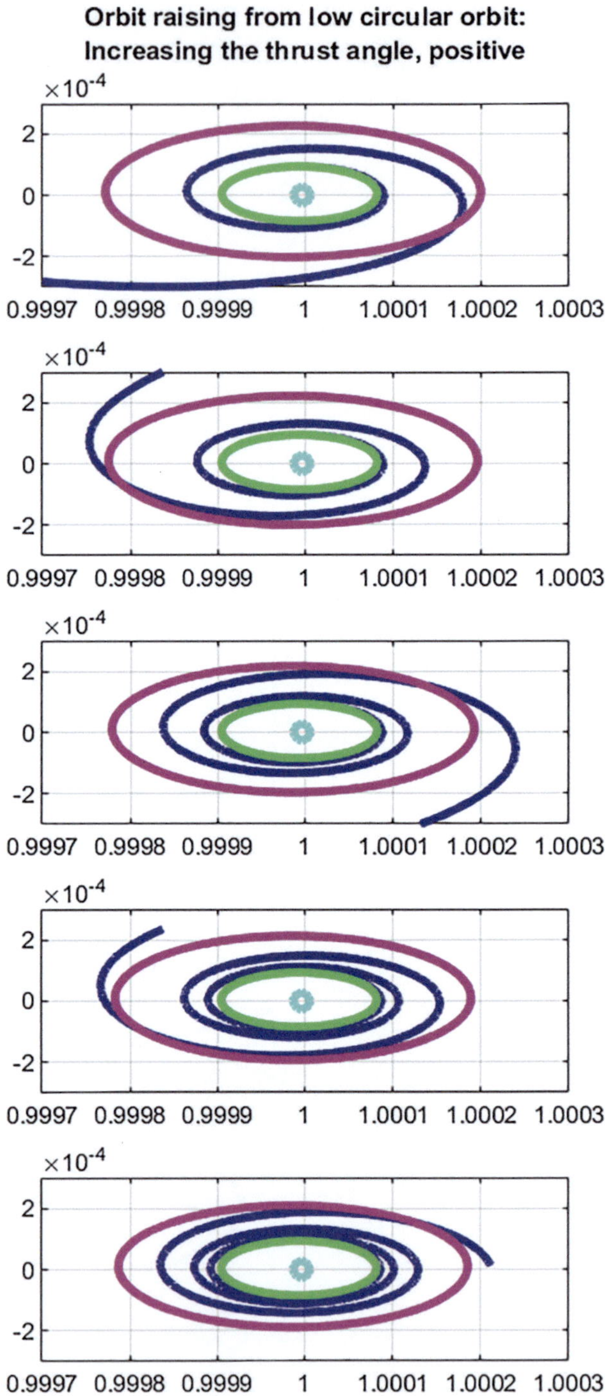

Fig. 7.19 Effect of increasing the thrust angle, with the thrust angle being positive

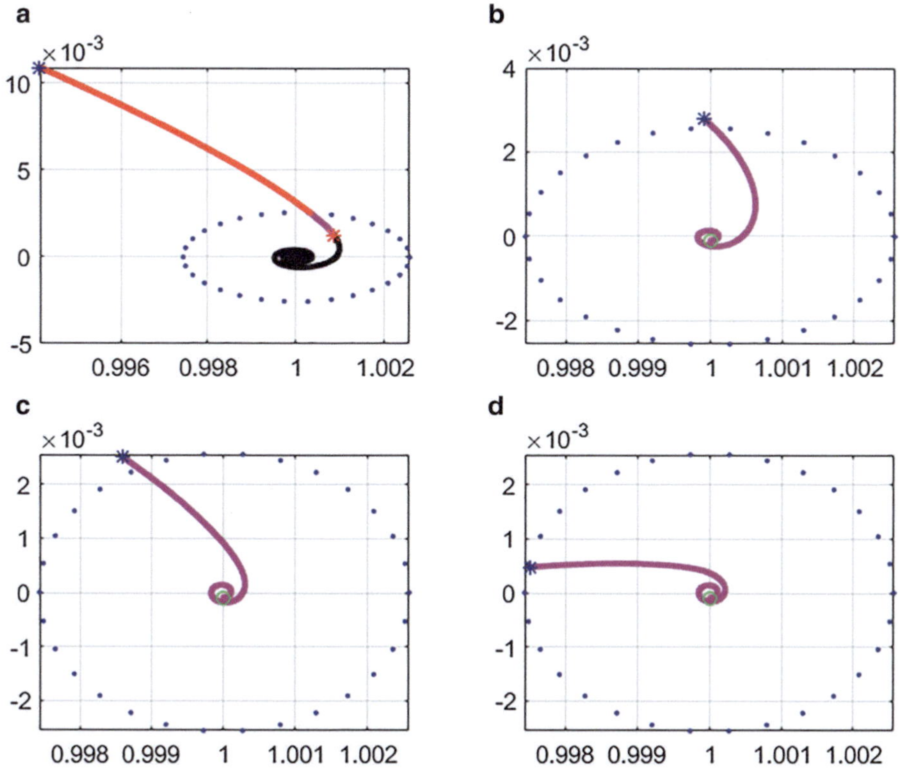

Fig. 7.20 Effect of the variation of the thrust angle: Thrust angle is 45° initially in (**a**) and increased from 9° (**b**) to 45° (**c**) and 54°(**d**) to steer the satellite towards the target body

applied to it. When it was sufficiently close to the Moon, an impulsive velocity change of 0.6873 km/s was necessary to place it into a Lunar orbit. Trajectory plots in space-fixed coordinates, in polar coordinates and in space-time coordinates of the solution shown in Fig. 7.21 meet the terminal cost requirement, given by Eq. (7.132) with $\alpha = \pi/4$, which represents an upper bound. Thus the thrust is assumed to be bounded both in magnitude and in direction. A systematic search for a better solution, with α prescribed and set to a fixed value, did not yield any results. Moreover in the steady state, the optimal steering angle $\alpha \rightarrow \pi/4$. This completely justifies the choice of $\alpha = \pi/4$. Two alternative trajectories are shown in Figs. 7.22 and 7.23.

For the trajectories shown in Figs. 7.22 and 7.23, the corresponding continuous low-thrust velocity change and the impulsive velocity change were 0.2119 and 2.1095 km/s and 0.2456 and 4.9127 km/s, respectively. In Fig. 7.21, the orbit raising segment of the trajectory, shown in Fig. 7.24, required a velocity increment of 0.1265 km/s. This can be reduced even further by considerations of staging which could be successfully exploited for orbit raising as in Jia-Richards and Lozano (2021).

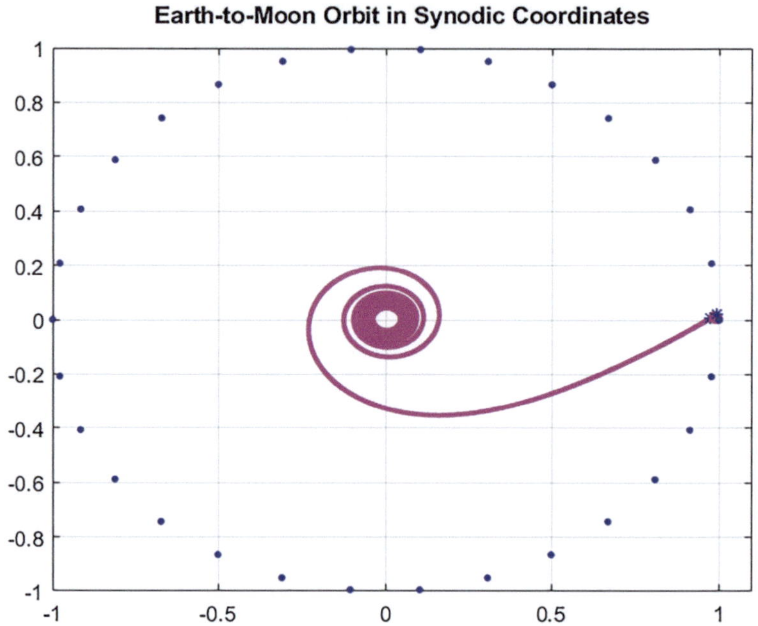

Fig. 7.21 Trajectory from the Earth to the Moon

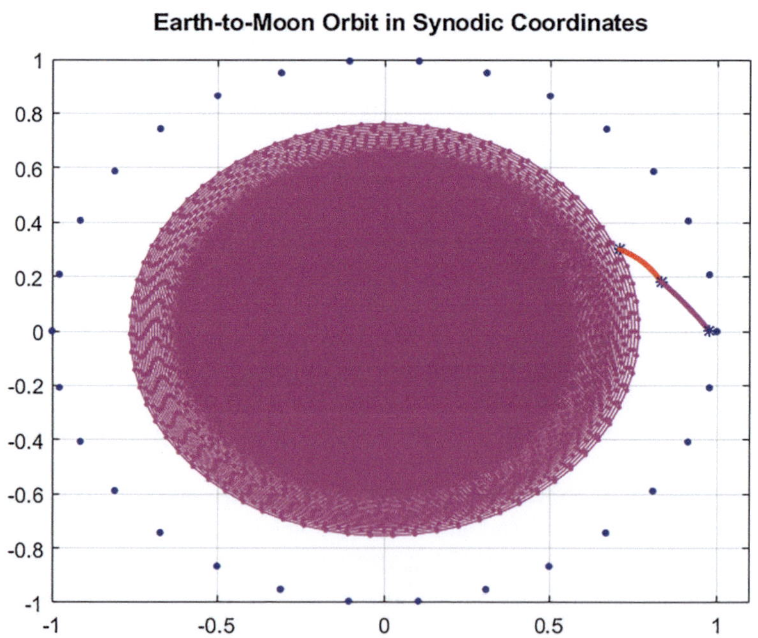

Fig. 7.22 Alternate trajectory to the Moon

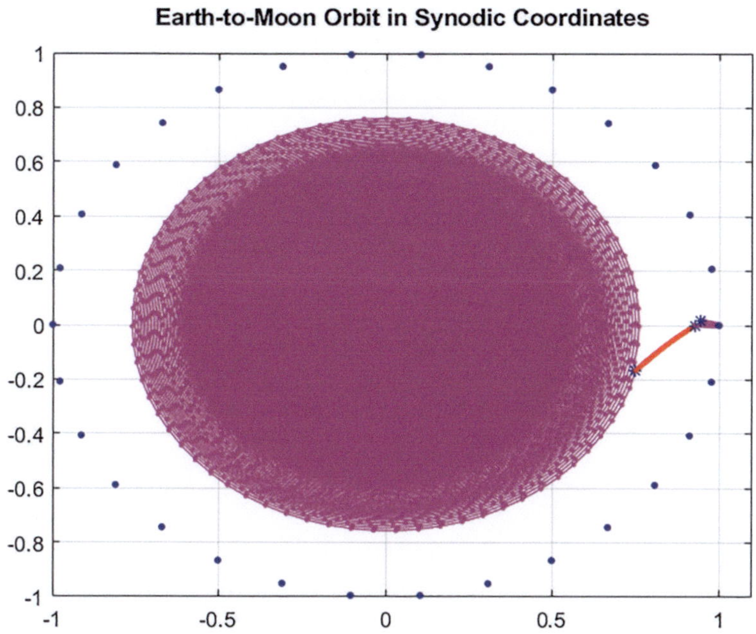

Fig. 7.23 Alternate trajectory to the Moon II

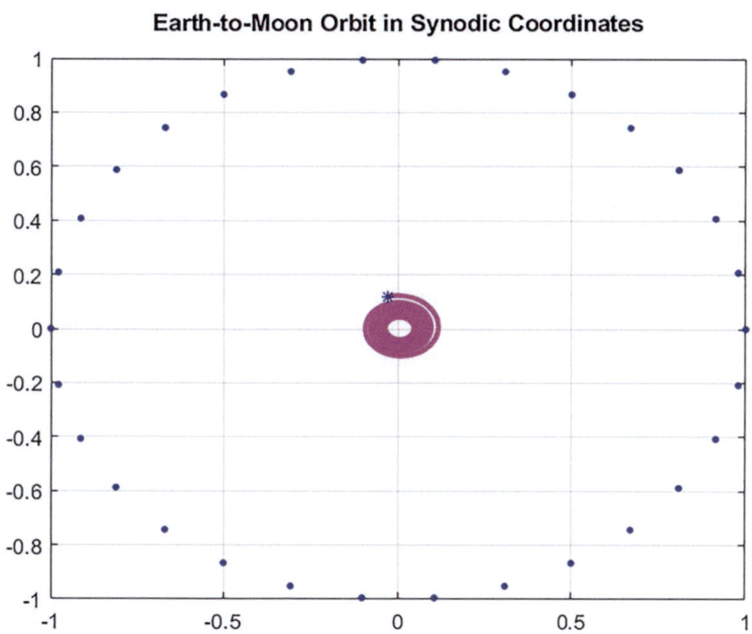

Fig. 7.24 Orbit raising by itself

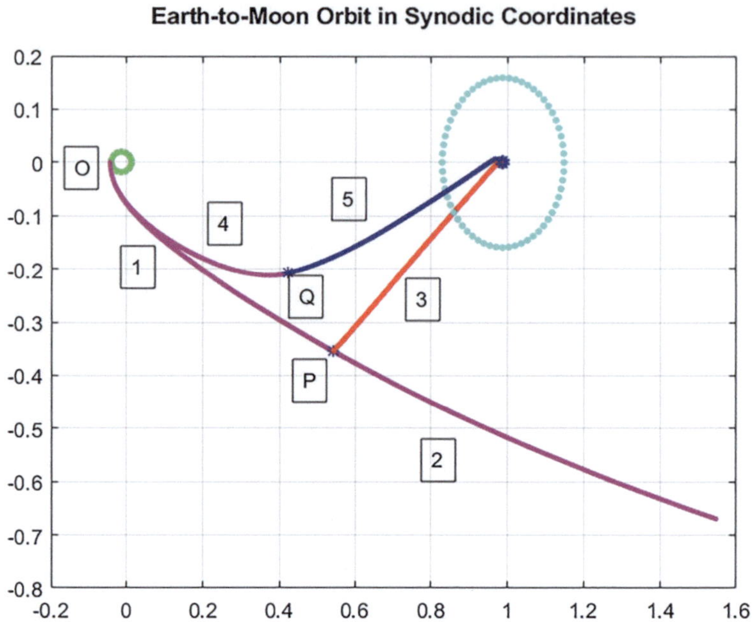

Fig. 7.25 Comparison of discrete impulse and low-thrust trajectories

Figure 7.25 provides a direct comparison of the velocity increment required for a spacecraft with continuous thrusting and one with an impulsive increment in the velocity. A spacecraft departing from Earth is injected into an orbit and with no velocity increments and would proceed along the trajectory segments 1 and 2. However if a discrete impulsive velocity change of 2.6028 km/s were applied at the patch point P, it would proceed to the vicinity of the Moon, along the coasting trajectory 3. If on the other hand, a continuous low-thrust velocity increment of 1.7799 km/s was used over segment 4, starting from the departure point O to the patch point Q, the spacecraft would proceed to the vicinity of the Moon, along the coasting trajectory 5. Also shown in Fig. 7.25 is the Hill sphere around the Moon. The trajectories shown in Fig. 7.25 were simulated using the multi-body equations, given by Eqs. (7.124). In this model, the Earth and the Moon are the primary and secondary bodies with Sun, the fourth body, orbiting relative to the Earth.

In all the examples considered so far and the corresponding trajectories shown in Figs. 7.21, 7.22, 7.23, 7.24, and 7.25, the use of the discrete impulse has not been dispensed with. Impulsive changes to the velocity conditions are often necessary to make changes to the initial conditions either at the start of a trajectory or at a patch point as well as to make mid-course corrections if and when necessary. For this reason it may not be really possible to use continuous low thrust as stand-alone thrusting strategy for most interplanetary applications. Thus the thrusting strategies adopted can at best be described as compound versions of both continuous low thrust and discrete impulsive thrust applications.

To fully comprehend the rationale behind the methodology, it is important to consider that at the departure stage the Sun-Earth-satellite CRTBP is considered. In the vicinity of the arrival region, the Earth-Moon-satellite CRTBP is considered. The reason for this is that the corresponding Lagrange points and the small regions in their vicinity provide low velocity unstable trajectories which can easily be traversed by the addition of minimal additional thrust. By patching together such low velocity unstable trajectories as well as low-thrust orbit raising and orbit lowering strategies, it is always possible to construct a feasible initial trajectory. It is for this reason, the CRTBP is central to this technique.

7.9.1 Earth-Moon Transfer with Continuous Thrust

Consider, for example, the problem of constructing a feasible trajectory from around the Earth region to around the Lunar region. Figure 7.26 shows two sets of low-thrust trajectories, one set originating from near the L_2 point of the Sun-Earth-satellite CRTBP and the other from around the L_1 point of the Earth-Moon-satellite CRTBP. The trajectory is shown in the re-scaled Earth-Moon-satellite CRTBP synodic coordinates. The intersection point of the two trajectories and its mirror image, which could serve as patch points, as beyond these points, the Lunar influence could be perceived although only marginally, are shown in the figure.

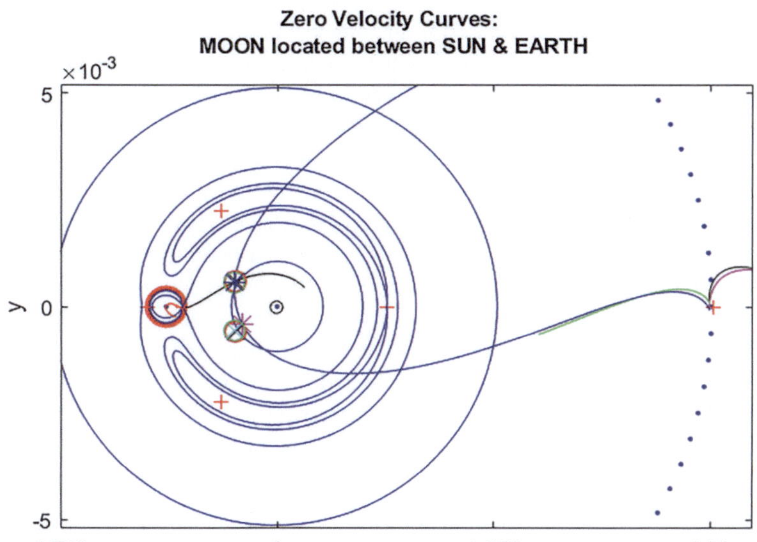

Fig. 7.26 Trajectories originating from near the L_2 point of the Sun-Earth-satellite CRTBP and from around the L_1 point of the Earth-Moon-satellite CRTBP

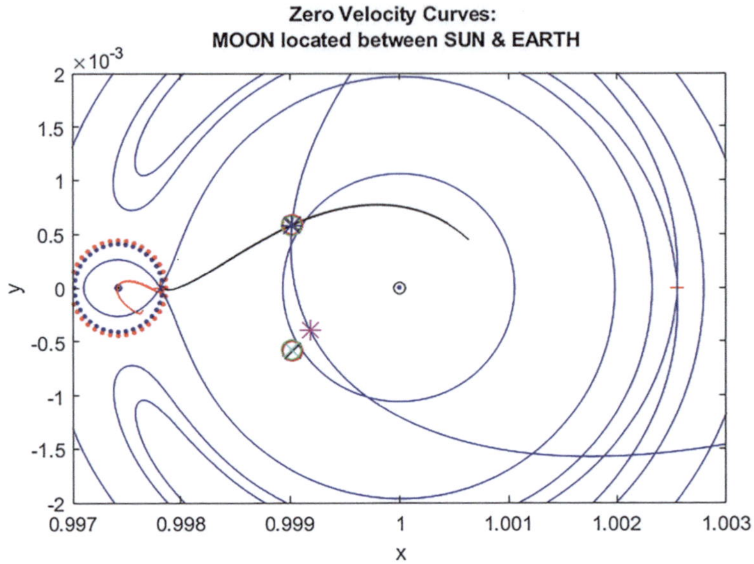

Fig. 7.27 Close-up of Earth-Moon trajectories shown in Fig. 7.26

Also shown is the nearest approach point (to Earth) of the trajectory originating from near the L_2 point of the Sun-Earth-satellite CRTBP. A close-up of Fig. 7.26 is shown in Fig. 7.27.

Also shown in Figs. 7.26 and 7.27 are the Hill circle, the inner dotted circle around the Moon and the Moon's SOI, the outer dotted circle. The magnitude of the low thrust is 15 mN. Apart from the fact that the trajectories are no longer symmetric about the *x*-axis, examining the trajectories in Fig. 7.27 closely, it is apparent that only the lunar trajectories are adequate to traverse from a point in the vicinity of Earth to a point close to the Moon. There is however a small disconnect between the two lunar trajectories in the vicinity of the Lagrange point L_1, of the Earth-Moon-satellite CRTBP, which may not be visible to the naked eye. The upper trajectory intersection point shown in Fig. 7.27 is still used as a patch point to make suitable changes to the velocity initial conditions. It also now becomes obvious that by making very small changes to the direction of the initial velocity of the final segment to the Moon, at the patch point, it is possible to ensure the spacecraft ends up within the separatrix region enclosed in Hill circle and very close to the Moon.

This is illustrated in Fig. 7.28 for five distinct but small changes to the velocity initial conditions at the patch point. The simulations of the last segment of the flight to Moon were done in a four-body environment, using the equations of motion defined by the bi-circular restricted four-body problem dynamics given by Eq. (7.119). The complete feasible trajectory may be constructed by adding to the initial phase, a low-thrust orbit raising trajectory. It then follows that almost the whole of the trajectory could be replaced one or at most two continuous segments linked by at most one patch point. This is exactly how the trajectory in Fig. 7.21 was obtained, but without the need for a discrete input at the patch point.

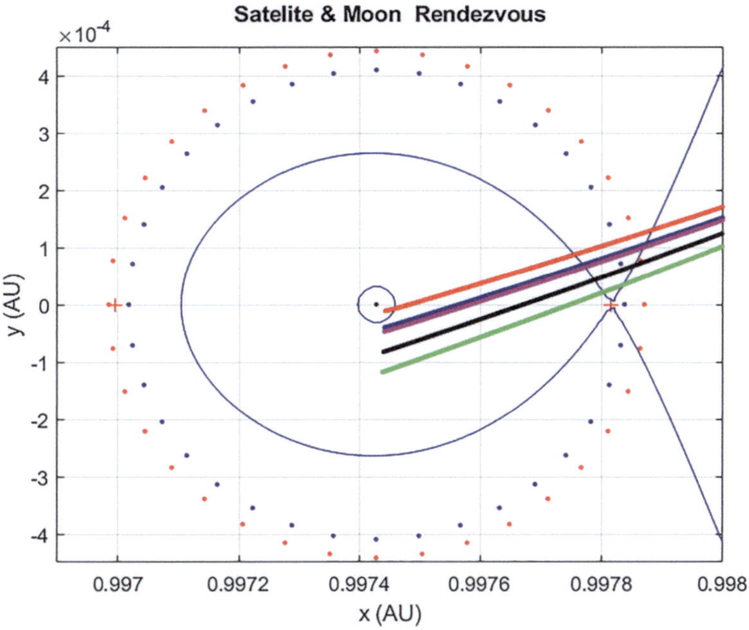

Fig. 7.28 Final phase of the Earth-Moon transfer

7.9.2 Interplanetary Transfer with Continuous Thrust

The final example considered although not presented in so much detail is the Earth-to-Jupiter low-thrust transfer. In this case, one starts with the Sun-Earth-satellite CRTBP, and it involves orbit raising well and beyond Earth's SOI. The most suitable orbits are the maximally diverging spiral orbits. Low-thrust trajectories are constructed departing from the Earth as well as from the Lagrange point L_3 and propagated up to and slightly beyond Jupiter's circular orbit as shown in Fig. 7.29 in the Sun-Earth-satellite CRTBP synodic coordinates. Jupiter's orbit is shown as a dotted line.

As one approaches Jupiter, the Sun-Jupiter-satellite CRTBP is considered. The earlier computed trajectories are transformed to the Sun-Jupiter-satellite CRTBP synodic coordinates. These are illustrated in Fig. 7.30. Also shown in Fig. 7.30 are the Hill sphere and the SOI of the planet, Jupiter.

From Fig. 7.30, it is apparent that if one is interested to going up to Jupiter, the trajectory departing from Earth and with low thrust is quite adequate. It would take the spacecraft right into Jupiter's SOI and Hill sphere, which are also shown. In the figure, Jupiter's SOI is the inner circle while the Hill sphere is the outer circle. If one wishes to travel beyond Jupiter, it is certainly advantageous to switch to one of the trajectories originating from the Lagrange point L_3, at the point of closest approach to Jupiter, after taking advantage of Jupiter's gravity.

Thus, in both the examples considered, a single trajectory was constructed from the vicinity of the departure planet to the vicinity of arrival celestial body.

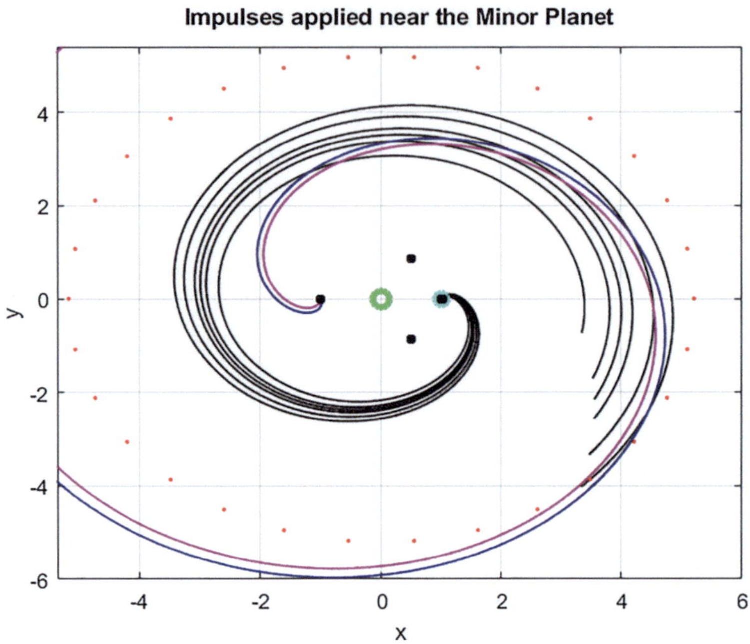

Fig. 7.29 Trajectories the L_3 Lagrange point and Earth en route to Jupiter

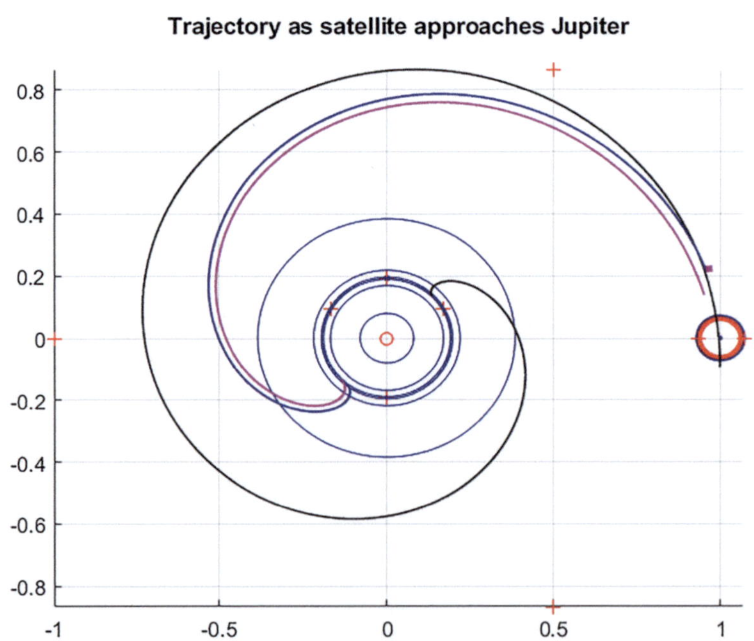

Fig. 7.30 Trajectories in the Sun-Jupiter-satellite CRTBP synodic coordinates

7.9.3 Discussion of the Application Example Results

In Sects. 7.9.1 and 7.9.2, a generic methodology was established, for constructing complete feasible families of reasonably low-cost orbits from the low orbital regions of a departing planet to the low orbital regions of an arrival planet, using two or less patches of solutions to the circular restricted three-body problem (CRTBP). Once a feasible trajectory is synthesized, it could be further optimized to obtain an optimal and acceptable trajectory by adopting well-known techniques in optimal control. Two examples are presented to illustrate the design methods.

In the first example considered, a direct synthesis of a trajectory departing from the Earth and arriving in the vicinity of the Moon was constructed using equations of motion of the Earth-Moon (and spacecraft) three-body problem in polar coordinates. The synthesis of a similar trajectory in Cartesian coordinates was also demonstrated. The trajectories originating from near the L_2 point of the Sun-Earth-satellite CRTBP and from around the L_1 point of the Earth-Moon-satellite CRTBP were first synthesized. A point of intersection of these two trajectories was used as patch point. It is assumed that the spacecraft transferred to the trajectory to the L_1 point of the Earth-Moon-satellite CRTBP prior to the patch point. At the patch point, a small Δv is delivered to the spacecraft to redirect the spacecraft towards the Moon rather than the L_1 point of the Earth-Moon-satellite CRTBP. Thus, with just one correction manoeuvre, the spacecraft is able to transfer from within the Earth's low orbital region to the Moon's low orbital region.

In the second example, the transfer of an Earth orbiting spacecraft to a low orbiting region of the planet Jupiter was demonstrated. In this case an Earth departing trajectory was found that would take the spacecraft directly with the SOI of Jupiter. If it is desired to continue on to an outer planet, a further Δv could be provided to the spacecraft when it is at its closest point of approach to the planet Jupiter. Alternately, to orbit Jupiter, and appropriate impulse is provided to place it on a circular orbit.

In this chapter, examples have been presented, to demonstrate the application of the patched three-body methodology for constructing feasible trajectories for interplanetary space travel. The key to the successful application of the methodology and the seamless generation of trajectories is the transformation from one synodic reference frame to another, in both Cartesian and polar coordinates.

Exercises

1. (a) Describe what a Lagrange point is in relation to two co-orbiting gravitational objects such as the Sun and Earth.
   (b) Draw a sketch showing approximate locations of the 5 Lagrange points in such a two-body system. Also sketch an example of a satellite orbit around a Lagrangian point. You may assume that the Earth is in circular orbit about the Sun, having a radius of 150×10^6 km.

(c) Discuss briefly the location and key features of the 5 Lagrange points of the two-body system shown in part (b).

2. (a) Consider a spacecraft orbiting the Sun in the ecliptic plane, in a heliocentric orbit. Assume that the Earth is executing a circular orbit around the Sun. The spacecraft's position vector may be expressed in a reference frame moving with the Sun-Earth line which is assumed to be rotating with an angular velocity, Ω. The moving reference frame has one axis along the Sun-Earth line and another normal to it in the orbit plane. The third axis is mutually perpendicular to the other two. The origin of the moving reference frame is located on the Sun-Earth axis at the centre of mass of the Sun-Earth two-body system.

   Obtain the transformation relating the spacecraft's position vector in the space-fixed reference frame to the spacecraft's position vector in the moving reference frame with its origin also at the centre of mass of the Sun-Earth two-body system and with two of its axes in the ecliptic plane. You may assume that the spacecraft always remains in the ecliptic plane and consider only its Cartesian coordinates in its orbit plane.

   If the same transformation is applied to the spacecraft's velocity and acceleration vectors (in the space-fixed reference frame), express the transformed vectors in terms of the spacecraft's position vector and its time derivatives, in the moving reference frame.

   (b) Draw a neat figure of the orbit plane, indicating the relative positions of the Sun, the Earth and the spacecraft in part (a).

   Given that the distance between the Sun and the Earth is d, locate the origin of the reference frames relative to the Sun and the Earth.

   Hence obtain expressions for the spacecraft's position in the orbit plane, relative to the Sun and the Earth, respectively, in terms of the Cartesian position coordinates of the spacecraft in the moving reference frame and other relevant quantities.

   (c) The total mass of the Sun and Earth is known to be m, the spacecraft's mass is given as m_s, and the ratio of the mass to the Earth to the total mass of the Sun and the Earth is given as β. The universal gravitational constant is given as G. Hence obtain an equation for the total gravitational force vector acting on the spacecraft, due to both the Sun and the Earth, in terms of the Cartesian position vector of the spacecraft in the moving reference frame.

3. Reconsider exercise 1 in the exercises of the preceding chapter, and the problem of synthesizing an optimum trajectory to the planet Jupiter via Mars to take advantage of the gravity assist provided by the latter planet, and synthesize the continuous thrust, optimal multi-body trajectory in a MATLAB environment, using the patched three-body conic approximation method. You may assume that the final orbit is a circular orbit around Jupiter with an altitude of 1000 km.

4. Reconsider exercises 1 and 2, in the exercises of the preceding chapter, and the problem of synthesizing an optimum trajectory to the planet Mercury via Venus to take advantage of the gravity assist provided by the latter planet, and synthesize the continuous thrust, optimal multi-body trajectory in a MATLAB environment, using the patched three-body conic approximation method. You may assume that the final orbit is a circular orbit around Mercury with an altitude of 400 km.

5. Consider the problem of synthesizing a continuous thrust, optimum trajectory to the asteroids like Hylonome, Hindalgo and Ceres, via Jupiter to take advantage of the gravity assist provided by the latter planet and synthesize the optimal multi-body trajectory in a MATLAB environment, using the results of Fritz and Turkoglu (2017) and Englander et al. (2017) to initialize the optimal trajectory synthesis problem and the patched three-body conic approximation method.
6. Reconsider exercise 5 in the exercises of the preceding chapter, and the problem of synthesizing an optimum, low energy or minimum fuel, trajectory to the Moon. Using the patched three-body conic approach, presented in this chapter, repeat the solution for the optimal trajectory to the Moon from the Earth, trying to reduce the time of flight in the process.
7. Consider two radiating oblate spheroids with masses m_1 and m_2 ($m_2 < m_1$), respectively, moving in a plane about their common centre of mass in a Keplerian elliptical orbit having a constant eccentricity e. The infinitesimal mass is moving in the plane of motion of the two oblate spheroid primaries. Assume that A_1 and A_2 denote the oblateness coefficients of the two primaries, respectively, where $0 < A_i < < 1, i = 1, 2$. Furthermore, assume that radiation factors q_1 and q_2 of the primaries effectively and proportionately reduce the gravitational forces, respectively.

   (a) Obtain expressions for the kinetic and potential energies and show that the for the elliptic restricted three-body problem, the equations of motion of a particle of infinitesimal mass in a non-dimensional rotating-pulsating Cartesian coordinate system $[x \quad y \quad z]$ with its origin at the barycentre of the two masses are given by

$$\frac{d^2x}{d\nu^2} - 2\frac{dy}{d\nu} - \frac{1}{1+e\cos\nu}\frac{\partial\Omega}{\partial x} = 0, \quad \frac{d^2y}{d\nu^2} + 2\frac{dx}{d\nu} - \frac{1}{1+e\cos\nu}\frac{\partial\Omega}{\partial y} = 0,$$

$$\frac{d^2z}{d\nu^2} + z - \frac{1}{1+e\cos\nu}\frac{\partial\Omega}{\partial z} = 0,$$

   where ν is the true anomaly and the functional Ω is defined in terms of two distances, r_1 and r_2, a mass ratio μ and the normalized mean motion n as

$$\Omega = \frac{x^2+y^2+z^2}{2} + \frac{1}{n^2}\left(\frac{(1-\mu)q_1}{r_1} + \frac{\mu q_2}{r_2} + \frac{(1-\mu)q_1A_1}{2r_1^3} + \frac{\mu q_2A_2}{2r_2^3}\right) + \frac{1}{2}\mu(1-\mu).$$

   The two distances r_1 and r_2 and the mass ratio μ are, respectively, defined as

$$r_1^2 = (x+\mu)^2 + y^2 + z^2, \quad r_2^2 = (x-1+\mu)^2 + y^2 + z^2, \quad \mu = m_2/(m_2+m_1).$$

   The normalized mean motion n satisfies the relation $n^2 = 1 + 3A_1/2 + 3A_2/2$. State all the assumptions you make.

(b) Assume that the three masses, the two primaries and the infinitesimal mass, are successively the Sun, the Earth and satellite and the Sun, Mars and the satellite and design a single trajectory with two discrete velocity increments, one initially and one prior to the satellite's entry into Sun-Mars, domain so the satellite enters the Hill sphere of Mars, using numerical integrations of the equations of motion. State any additional assumptions you make.

(c) Gradually reduce the second velocity increment so as to ensure the ballistic capture of the satellite by the planet Mars, when the satellite enters the Hill sphere of Mars. You may use the method of Hyeraci and Topputo (2010).

By considering the equations for the zero velocity curves for the elliptic restricted three-body problem, obtain a condition when the satellite can never enter the Hill zone around second body and hence can never be captured. Does your simulation meet these criteria?

References

Azimov, D. M. (2018). *Analytical solutions for extremal space trajectories*. Butterworth-Heinemann.

Bai, S., Han, C., & Sun, X. (2021). Linear transfer guidance based on Lyapunov method. *Journal of Physics Conference Series, 1786*, 012055.

Barden, B. T., & Howell, K. C. (1998). Fundamental motions near collinear libration points and their transitions. *The Journal of the Astronautical Sciences, 46*(4), 361–378.

Barden, B. T., Howell, K. C., & Lo, M. W. (1996). Application of dynamical systems theory to trajectory design for a libration point mission. In *AIAA/AAS astrodynamics conference, San Diego, CA, July 29–31, 1996, collection of technical papers (A96-34712 09-12)* (pp. 268–281). American Institute of Aeronautics and Astronautics.

Barrabes, E., Mondelo, J. M., & Olle, M. (2009). Numerical continuation of families of homoclinic connections of periodic orbits in the RTBP. *Nonlinearity, 22*, 2901–2918.

Barrabes, E., Mondelo, J. M., & Olle, M. (2013). Numerical continuation of families of heteroclinic connections between periodic orbits in a Hamiltonian system. *Nonlinearity, 26*, 2747–2765. https://doi.org/10.1088/0951-7715/26/10/2747

Bate, R. R., Mueller, D. D., & White, J. E. (2005). *Fundamentals of astrodynamics*. Dover Publications, Inc.

Battin, R. H. (1986). An elegant Lambert algorithm. *Journal of Guidance, Control, and Dynamics, 7*(6), 662–670.

Battin, R. H. (1987). *An introduction to the mathematics and methods of astrodynamics* (pp. 408–418). AIAA Educational Series, AIAA.

Bassetto, M., Quarta, A. A., Mengali, G., & Cipolla, V. (2021). Spiral trajectories induced by radial thrust with applications to generalized sails. *Astrodynamics, 5*(2), 121–137. https://doi.org/10.1007/s42064-020-0093-6

Betts, J. T., & Erb, S. O. (2003). Optimal low thrust trajectories to the moon. *SIAM Journal on Applied Dynamical Systems, 2*. https://doi.org/10.1137/S1111111102409080

Boltz, F. W. (1991). Orbital motion under continuous radial thrust. *Journal of Guidance, Control, and Dynamics, 14*(3), 667–670. https://doi.org/10.2514/3.20690

Bombardelli, C., Gonzalo, J. L., & Roa, J. (2018). Approximate analytical solution of the multiple revolution Lambert's targeting problem. *Journal of Guidance, Control, and Dynamics, 41*(2), 1–10. https://doi.org/10.2514/1.G002887

Bryson, A. E., & Ho, Y. C. (1975). *Applied optimal control: Optimization, estimation and control*. CRC Press.

Caillau, J.-B., Daoud, B., & Gergaud, J. (2012). Minimum fuel control of the planar circular restricted three-body problem. *Celestial Mechanics and Dynamical Astronomy, 114*, 137–150. https://doi.org/10.1007/s10569-012-9443-x

Danby, J. M. A. (1988). *Fundamentals of celestial mechanics*. Willmann-Bell.

Daoud, B. (2011) *On the optimal control of the circular restricted three body problem*. Doctoral Dissertation, Université de Bourgogne.

Eberle, J., Cuntz, M., & Musielak, Z. E. (2008). The instability transition for the restricted 3-body problem. I. Theoretical approach. *Astronomy and Astrophysics, 489*, 1329–1335.

Englander, J. A., Vavrina, M. A., Lim, L. F., McFadden, L. A., Rhoden, A. R., & Noll, K. S., (2017) Trajectory Optimization for Missions to Small Bodies with a Focus on Scientific Merit, Comput Sci Eng., 19(4), pp-18–28, Published online 2017 Aug 17. https://doi.org/10.1109/MCSE.2017.3151246.

Escribano, T. M. V. (2010). *Poincaré sections and resonant orbits in the restricted three-body problem*. Master of Science (Aeronautics and Astronautics) thesis, Purdue University.

Fantino, F., Gomez, G., Masdemont, J. J., & Ren, Y. (2010). A note on libration point orbits, temporary capture and low-energy transfers. *Acta Astronautica, 67*, 1038–1052. https://doi.org/10.1016/j.actaastro.2010.06.037

Farquhar, R. W., & Kamel, A. A. (1973). Quasi-periodic orbits about the translunar libration point. *Celestial Mechanics, 7*, 458–473.

Farquhar, R. W., Muhonen, D. P., Newman, C. R., & Heubergerg, H. S. (1980). Trajectories and orbital maneuvers for the first libration-point satellite. *Journal of Guidance, Control, and Dynamics, 3*(6), 549–554.

Fritz, S., & Turkoglu, K. (2017). Optimal trajectory determination and mission design for Asteroid/deep-space exploration via multibody gravity assist maneuvers. *International Journal of Aerospace Engineering (Hindawi), 2017,* 6801023(1–12). https://doi.org/10.1155/2017/6801023

Geisel, C. D. (2013). *Spacecraft orbit design in the circular restricted three-body problem using higher-dimensional poincaré maps*. Doctor of Philosophy (Aeronautics and Astronautics) thesis, Purdue University.

Gomez, G., & Masdemont, J. (2000). *Some zero cost transfers between libration point orbits*. In Point Orbits, AAS paper 00-177, AAS/AIAA Astrodynamics Specialist Conference, Florida.

Gordon, D. P. (2008). *Transfers To Earth-Moon L2 halo orbits using lunar proximity and invariant manifolds*. Master of Science (Aeronautics and Astronautics) thesis, Purdue University.

Haapala, A. F. (2014). *Trajectory design in the spatial circular restricted three-body problem exploiting higher-dimensional poincaré maps*. Doctor of Philosophy (Aeronautics and Astronautics) thesis, Purdue University.

Henninger, H. C., & Biggs, J. D. (2017). Near time-minimal earth to L1 transfers for low-thrust spacecraft. *Journal of Guidance, Control, and Dynamics, 40*, 2994–2999. https://doi.org/10.2514/1.G002373

Hénon, M. (1973). Vertical stability of periodic orbits in the restricted problem. *Astronomy and Astrophysics, 28*, 415–426.

Howell, K. C., Barden, B. T., Wilson, R. S., & Lo, M. W. (1998). Trajectory design using a dynamical systems approach with application to GENESIS. *Advances in the Astronautical Sciences, 97*, 1665–1684.

Howell, K. C., & Kakoi, M. (2006). Transfers between the Earth-Moon and Sun-Earth systems using manifolds and transit orbits. *Acta Astronautica, 59*, 367–380. https://doi.org/10.1016/j.actaastro.2006.02.010

Han, C., & Yin, J. (2011). Study of satellite relative motion in elliptical relative orbital elements. *Acta Aeronautica et Astronautica Sinica, 32*(12), 2244–2258.

Hyeraci, N., & Topputo, F. (2010). Method to design ballistic capture in the elliptic restricted three-body problem. *Journal of Guidance, Control, and Dynamics, 33*(6), 1814–1823.

Ilgen, M. R. (1993). Low thrust OTV guidance using Lyapunov optimal feedback control techniques. *Advances in the Astronautical Sciences, 85*(2), 1527–1545.

Jarabek, R. L. (2004) *Investigation of manifolds and optimized trajectories in the three-body problem*. Master of Science (Aerospace Engineering) thesis, University of Maryland.

Jia-Richards, O., & Lozano, P. C. (2021). Analytical guidance for circular orbit transfers with staging of space propulsion systems. *Acta Astronautica, 179*, 69–77.

Kim, M., &Hall, C. D. (2002). *Lyapunov and Halo Orbits about L_2*. Proceedings of the 2001 AAS/AIAA Astrodynamics Specialist Conference, Quebec City, Canada, also in, Advances in the Astronautical Sciences, 109, 349–366.

Kluever, C. A. (1993) *Optimal low thrust earth moon trajectories*. PhD Dissertation, Department of Aerospace Engineering and Engineering Mechanics, Iowa State University.

Koon, W. S., Lo, M. W., Marsden, J. E., & Ross, S. D. (2000). Heteroclinic connections between periodic orbits and resonance transitions in celestial mechanics. *Chaos, 10*(2), 427–469.

Koon, W. S., Lo, M. W., Marsden, J. E., & Ross, S. D. (2007). *Dynamical systems, the three-body problem and space mission design*. Springer.

Leeghim, H., Kim, D., & Turner, J. (2013). Solution for nonlinear three-dimensional intercept problem with minimum energy. *Mathematical Problems in Engineering, 2013*(435725), 8. https://doi.org/10.1155/2013/435725

Li, J. (2013) *Research on continuous thrust orbit maneuver optimization method*. Phd thesis, Beihang University.

Mengali, G., & Quarta, A. A. (2009). Escape from elliptic orbit using constant radial thrust. *Journal of Guidance, Control, and Dynamics, 32*(3), 1018–1022. https://doi.org/10.2514/1.43382

Mingotti, G., Topputo, F., & Bernelli-Zazzera, F. (2009). Low-energy, low-thrust transfers to the Moon. *Celestial Mechanics and Dynamical Astronomy, 105*, 61–74. https://doi.org/10.1007/s10569-009-9220-7

Musielak, Z. E., & Quarles, B. (2014). The three-body problem. *Reports on Progress in Physics*.

Ozimek, M. T., & Howell, K. C. (2010). Low-thrust transfers in the earth-moon system, including applications to libration point orbits. *Journal of Guidance, Control, and Dynamics, 33*, 533–549. https://doi.org/10.2514/1.43179

Patel, H. K., (2020) *Constrained motion analysis of spacecraft trajectory in restricted three body problem*. Master's Thesis in Aerospace Engineering, Embry-Riddle Aeronautical University. https://commons.erau.edu/edt/562

Petropoulos, A. E. (2005). Refinements to the Q-law for low-thrust orbit transfers. *Advances in the Astronautical Sciences, 120*(1), 963–982.

Poincaré, H. (1993). In D. Goroff (Ed.), *New methods of celestial mechanics*. American Institute of Physics.

Roa, J., Peláez, J., & Senent, J. (2016). Spiral Lambert's problem. *Journal of Guidance, Control, and Dynamics, 39*(10), 2250–2263.

Rogov, K. (2016). *Computation of libration point orbits and manifolds using collocation methods*. Master in Aerospace Science & Technology, Universitat Politècnica de Catalunya.

Shen, H., & Tsiotras, P. (2003). *Using Battin's method to obtain multiple-revolution Lambert's solutions*. AAS/AIAA Astrodynamics Specialist Conference, AAS-03-568.

Singh, S. K., Anderson, B. D., Taheri, E., & Junkins, J. L. (2021). Exploiting manifolds of $L1$ halo orbits for end-to-end Earth–Moon low-thrust trajectory design. *Acta Astronautica, 183*, 255–272.

Sundman, K. F. (1912–1913). Mémoire sur le problém des trois corps. *Acta Mathematica, 36*, 105–179.

Szebehely, V. G. (1967). *Theory of orbits – The restricted problem of three bodies*. Academic.

Roy, A. E. (2005). *Orbital motion*. Institute of Physics Publishing.

Topputo, F., Vasile, M., & Bernelli-Zazzera, F. (2005). Low Energy Interplanetary Transfers Exploiting Invariant manifolds of the restricted three body problem. *The Journal of Astronautical Sciences, 53*(4), 353–372.

Tsien, H. S. (1953). Take-off from satellite orbit. *Journal of the American Rocket Society, 23*(4), 233–236. https://doi.org/10.2514/8.4599

Yuan, R., Pingyuan, C., & Enjie, L. (2007). A low-thrust guidance law based on lyapunov feedback control and hybrid genetic algorithm. *Aircraft Engineering and Aerospace Technology, 79*(2), 144–149.

Zazzera, F. B., Topputo, F., & Massari. M. (2004). *Assessment of mission design including utilisation of libration points and weak stability boundaries*. Technical Report 03-4103b, European Space Agency, the Advanced Concepts Team, Available on line at www.esa.int/act. Accessed Nov, 2021.

Zeng, H., & Zhang, J. (2018). Modeling low-thrust transfers between periodic orbits about five libration points: manifolds and hierarchical design. *Acta Astronautica, 145*, 408–423. https://doi.org/10.1016/j.actaastro.2018.01.035

Zhang, C., Topputo, F., Bernelli-Zazzera, F., & Zhao Y.- S. (2015). Low-thrust minimum-fuel optimization in the circular restricted three-body problem. *Journal of Guidance, Control, and Dynamics, 38*(8), 1501–1510. https://doi.org/10.2514/1.G001080

Chapter 8
Missions to the Asteroids

8.1 Introduction

Asteroids have become extremely important members of our planetary community. Most residents on planet Earth now recognize the dangers of even a relatively small asteroid colliding with the Earth. The destruction will be enormous particularly if it strikes one of our population centres. Recent studies have indicated that such a possibility does exist, in principle, albeit quite remote.

Missions to asteroids have become particularly important due to the need for closely monitoring their orbits, for capturing these asteroids or deflecting these planetary bodies, if they are found to be on a collision course with the planet Earth. Yet missions to small planetary bodies within the solar system such as asteroids and comets face many dynamical challenges. These bodies are also of great scientific interest to physicists and astronomers. Although small near-Earth objects (NEOs) are within the range of accessible bodies in the solar system and are a natural target for low-cost space missions, the fact the spacecraft experiences relatively strong solar radiation pressure (SRP) forces en route, the irregular mass distribution of these objects that can affect the spacecraft's dynamics when in close proximity and the ever present gravitational effects of the Sun, throughout their orbital motion, are the major challenges that must be addressed in planning and executing missions to them. During the Japanese Hayabusa mission to the asteroid Itokawa (Kawaguchi, 2006), the spacecraft was required to maintain a quasi-stationary position in the asteroid-Sun-fixed frame, with station keeping requirement met by regular velocity increments. The frequency and magnitude of the velocity increments implies that the spacecraft's orbit is continually being corrected and therefore not following a single prescribed reference orbit. The fuel and power required to deliver the velocity increments limit the total life of mission.

When considering the orbital mechanics of a spacecraft in orbit about a small body such as a comet, stability of the spacecraft flying in close proximity of the celestial body in the presence of perturbations due to disturbances, as well as the

R. Vepa, *Space Vehicle Maneuvering, Propulsion, Dynamics and Control*,
https://doi.org/10.1007/978-3-031-65518-0_8

uncertainties due to body's geometry and mass distribution, is a challenging topic to consider. In addition, if it is further required to deliver a prescribed velocity impulse to the celestial object in a certain direction, the complexity of the problem is significantly greater. Prior to delivering the impulse, one may need to study in real time the optimal direction of the impulse, so as to deflect the object in a desired direction.

NEOs moving in resonant, Earth-orbit like orbits are potentially important, as there is always the possibility of a few of them colliding with the Earth at some point in the future. 3753 Cruithne, the Earth's first co-orbital asteroid was discovered in 1986. In 1997, 3753 Cruithne was found to move around the Earth in a horseshoe orbit. But it was found not to follow the same orbit as the Earth around the Sun and so cannot strictly be called co-orbital asteroid.

The object, 2002 AA29, found to be closely following Earth's orbit in a reversible horseshoe orbit. It is known to become a temporary second moon to the Earth from time to time, with the next rendezvous expected in 2600. Its orbit is special as it approaches the Earth on one side, as it did in 2003, and then in about 95 years sweeps to the other side of the Earth. Calculations show that in a time frame of 600 years, this asteroid will orbit the Earth, once a year for 50 years. In a sense its orbit can resemble that of a satellite. Behaviours such as the above involving large-amplitude oscillations of the eccentricities and inclinations could in principle be predicted by the Kozai-Lidov mechanism, which is a secular phenomenon in the case of co-orbiting bodies and involving a family of resonance type librations and bifurcations of the equilibrium states, leading to set of stable Kozai resonance states. The coupling of large oscillations of eccentricities and inclination with the continuous motion or libration of the argument of the perihelion is known as the Kozai resonance, Kozai (1962).

In the co-orbital region of the Earth, three types of orbits can occur: tadpole, horseshoe and quasi-satellite orbits. Moreover spatial tadpole or horseshoe orbits could merge with quasi-satellite orbits and transitions between different types of orbits could appear, in principle. Co-orbital asteroids (Brasser et al., 2004), with tadpole orbits, are also known as Trojans as they librate about the Trojan Lagrangian points of the Sun-Earth and third-body system. Horseshoe orbits enclose the L4, L3 and L5 points while tadpole orbits oscillate or librate only about the L4 or L5 Lagrangian points. Horseshoe orbits librate over much larger amplitude and consequently their shape resembles a horseshoe. Quasi-satellite orbits are co-orbital orbits with a high eccentricity so part of the orbit lies well beyond the planet's sphere of influence and could be unstable in the long term.

Another Earth co-orbital asteroid recently discovered is designated 2010 SO16 and is on a "horseshoe" class orbit. Since its discovery, eight other asteroids, 2010 TK7, 2013 BS45, 2013 LX28, 2014 OL339, 2015 SO2, 2015 XX169, 2015 YA and 2015 YQ1, have also been added to list of Earth co-orbiting satellites. Over 24 Earth co-orbital asteroids that have been discovered so far, only 2010 TK7 is a Trojan asteroid (Connors et al., 2011), and its orbit is a tadpole or Trojan orbit (Dvorak et al., 2012). While it is stable in the short term, its stability in the long term over periods ranging from 10,000 to a million years is questionable. 2016 HO3, an asteroid discovered on 27 April 2016, is possibly the most stable quasi-satellite of Earth. As it orbits the Sun, 2016 HO3 appears to circle around Earth as well. It is too distant to be a true satellite of Earth but is the best and most stable example of a quasi-satellite, a type of near-Earth object. They appear to orbit a point other than Earth itself, such as the orbital path of the NEO asteroid 3753 Cruithne. Other small natural objects in orbit around the Sun may enter orbit around Earth for a short amount of time, becoming temporary natural satellites. To date, the only confirmed example has been 2006 RH120 in Earth orbit during 2006 and 2007, though further instances are already predicted.

The increasingly high number of potentially hazardous asteroids discovered so far has led to the concept of asteroid deflection, defined by Izzo and Rathke (2006). The primary idea is to give the asteroid an impulse in a particular direction so as to deflect the asteroid from its current orbit and place it in a new orbit which is the preferred orbit of the asteroid in order to avoid possible collisions. Possibilities of collisions are found by evaluating the minimum orbit intersection distance. Phenomenon such as the Kozai resonance offer protection from possible collision with several of the asteroids as shown by Libert and Tsiganis (2009). Yet there are several others and the potential threat of a collision with an asteroid remains as discussed by Domingos and Winter (2005). For this reason, several researchers are considering the need and plausibility of being able to launch a mission to an asteroid, which would fly in close proximity to the asteroid and then deliver the impulse in a particular direction so as to deflect the asteroid from its current orbit and place it in another orbit. Based on the currently available methods of trajectory design, such a mission could take up to 3 years, before a satellite could arrive at an asteroid and fly alongside it in order to be able to deliver a precomputed impulse to it. There is also a need to minimize the fuel requirements in such a mission. Thus methods are being developed so as to reduce the overall fuel requirements to complete such mission within the constraints of currently available technologies.

8.2 Gravitational Potentials and Potential Modelling

The mutual gravitational potential, between two elements of mass, is integrated and the result is an integral over two massive bodies of arbitrary shape. The integrated mutual gravitational potential is thus expressed as

$$U = -\int_{M_1}\int_{M_2} \frac{G dm_1 dm_2}{|\rho_1 - \rho_2|}, \tag{8.1}$$

where G is the universal gravitational constant, M_1 and M_2 represent the two bodies, dm_1 and dm_2 are two mass elements in each body and ρ_1 and ρ_2 represent their position vector from an inertial origin of these two mass elements. The computation of the mutual potential is not trivial in this case. Current and past researchers have looked into methods and expressions to compute the mutual potential. One approach is to derive a polyhedron-based discretization and use polyhedron potential computations to model irregularly shaped bodies. Other methods involve the use of Lie group computations.

The problem is simplified by letting one of the bodies be a sphere while allowing the second body to have a general shape. In this case, the mutual potential is an integral over the two bodies, one of which is spherical and symmetrical, and written as

$$U = -GM_1 \int\limits_{M_2} \frac{dm_2}{|\rho_1 - \rho_2|}, \tag{8.2}$$

where M_1 here is the mass of the sphere and M_2 represents the general body.

This problem can be simplified by using an appropriate set of reference frames; in an inertial frame fixed at the centre of mass of the sphere, M_1 and a body fixed rotating frame fixed at the centre of mass of M_2. The mutual potential is time-invariant, which is independent of whether the system is in relative equilibrium or not.

Hence, for a frame fixed at the centre of mass of the non-spherical body (Fig. 8.1)

$$U = -GM_1 \int\limits_{M_2} \frac{dm_2}{|\rho_1 - \rho_2|} = -GM_1M_2 \int\limits_{M_2} \frac{d(m_2(\widetilde{\rho})/M_2)}{|\widetilde{\rho} + \mathbf{r}_b|} - GM_1M_2\widetilde{U}(\mathbf{r}_b). \tag{8.3}$$

where $\mathbf{r}_b$ is the relative position between the sphere and the other body in the body-fixed frame and $\widetilde{\rho}$ is the position vector of a mass element of the arbitrarily shaped body.

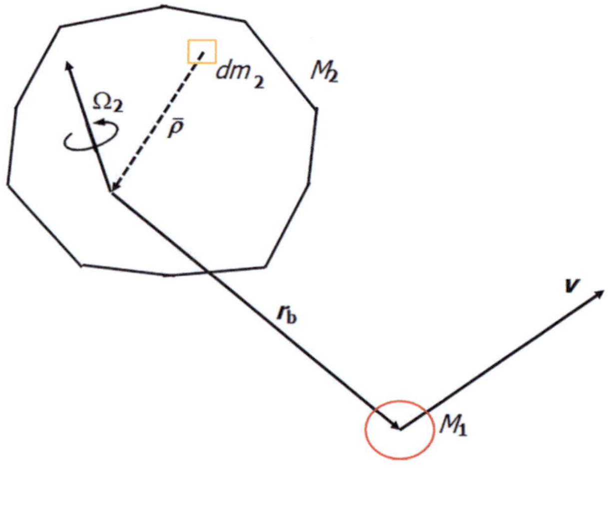

Fig. 8.1 Representation of the two-body problem, with an arbitrarily shaped body and sphere (one body is a sphere)

A general method of approximation is to express the general body potential using spherical harmonics, which simplifies if the body-fixed coordinate frame has its axes along the principal moments of inertia of the general body. The gravitational potential of the body is then calculated by enclosing it in a circumscribing spherical boundary called a "Brillouin sphere" of radius, R_0.

8.2.1 Spherical Harmonic Model

The potential $\widetilde{U}(\mathbf{r})$ is usually expressed in terms of spherical harmonic functions, in the form

$$\widetilde{U}(\mathbf{r})=\frac{1}{R_0}\sum_{n=0}^{\infty}\sum_{m=0}^{n}\left(\frac{R_0}{|\mathbf{r}|}\right)^{n+1}P_{nm}(\sin\delta)(C_{nm}\cos(m\lambda)+S_{nm}\sin(m\lambda)), \tag{8.4}$$

where the associated Legendre functions are obtained from the differentials of the Legendre functions and are given by

$$P_{nm}(x)=\left(1-x^2\right)^{\frac{m}{2}}\times\frac{d^m}{dx^m}P_n(x), P_n(x)=\frac{1}{2^n!}\times\frac{d^n}{dx^n}\left(x^2-1\right)^n, \tag{8.5}$$

are the associated Legendre functions, and C_{nm}, S_{nm} are the spherical harmonics coefficients, and δ and λ are the spherical coordinates of latitude and longitude in the general body. The latitude δ is measured from the x-y plane and longitude, and λ is measured from the x-axis and is taken to be positive towards east.

For convenience, let us introduce the abbreviations $t=\sin\delta$ and $u=\cos\delta$. The fully normalized associated Legendre functions $\overline{P}_{nm}(t)$, sometimes also called fully normalized harmonics, can be computed from the conventional associated Legendre functions $P_{nm}(t)$ by

$$\overline{P}_{nm}(t)=\sqrt{\frac{(2-\delta_{0m})(2n+1)(n-m)!}{(n+m)!}}P_{nm}(t), \tag{8.6}$$

The associated Legendre functions can be computed with the following recursive formulas:

$$\begin{gathered}P_{n+1,0}(t)=(2n+1)tP_{n,0}(t)-nP_{n-1,0}(t), P_{n,n}(t)=(2n-1)uP_{n-1,n-1}(t),\\ P_{n,m}(t)=P_{n-2,m}(t)+(2n-1)uP_{n-1,m-1}(t).\end{gathered} \tag{8.7}$$

With the starting values

$$P_{0,0}(t)=1, P_{1,0}(t)=t, P_{1,1}(t)=u, P_{2,0}(t)=(1/2)(3t^2-1),\\ P_{2,1}(t)=3ut, P_{2,2}(t)=3u^2. \tag{8.8}$$

The normalized coefficients are related to the un-normalized ones by the relation

$$\left[\overline{C}_{nm}, \overline{S}_{nm}\right]=\sqrt{\frac{(n+m)!}{(2-\delta_{0m})(2n+1)(n-m)!}}[C_{nm}, S_{nm}]. \tag{8.9}$$

It is also important to note that the associated Legendre functions are orthogonal functions which satisfy the orthogonality relations

$$\int_{-1}^{1} P_{nm}^2(x)dx=\frac{2}{2n+1}\frac{(n+m)!}{(n-m)!}, \quad \int_{-1}^{1} P_{nm}(x)P_{km}(x)dx \\ =\frac{2}{2n+1}\frac{(n+m)!}{(n-m)!}\delta_{nk}. \tag{8.10}$$

The potential $\widetilde{U}(\mathbf{r})$ may also be expressed in a perturbation form as

$$\widetilde{U}(\mathbf{r})=(1/|\mathbf{r}|)+R(\mathbf{r}),$$

$$R(\mathbf{r})=\frac{1}{R_0}\sum_{n=1}^{\infty}\sum_{m=0}^{n}\left(\frac{R_0}{|\mathbf{r}|}\right)^{n+1}P_{nm}(\sin\delta)(C_{nm}\cos(m\lambda)+S_{nm}\sin(m\lambda)). \tag{8.11}$$

Spherical harmonics are particularly suitable to describe small deviations from a sphere as they model the distribution of mass inside a sphere circumscribing the body.

An ellipsoid-sphere system has also been studied by Scheeres and Augenstein (2003) and by Scheeres (2004), investigating relative equilibria and stability. The potential of an ellipsoid can be written in terms of elliptic integrals. The exterior potential of an ellipsoid with the dimensions in the three principal x, y, z directions of the ellipsoid, α, β, γ, respectively, and uniform density ρ is expressed as

$$\widetilde{U}=G\rho\alpha\beta\gamma\int_{\lambda}^{\infty}\left(1-\frac{x^2}{\alpha^2+v}-\frac{y^2}{\beta^2+v}-\frac{z^2}{\gamma^2+v}\right)\frac{dv}{\Delta(v)}, \tag{8.12}$$

with

$$\Delta(v) = \sqrt{(\alpha^2 + v)\left(\beta^2 + v\right)(\gamma^2 + v)}, 0 < \gamma \leq \beta \leq \alpha, \tag{8.13}$$

where λ satisfies,

$$1 - \frac{x^2}{\alpha^2 + \lambda} - \frac{y^2}{\beta^2 + \lambda} - \frac{z^2}{\gamma^2 + \lambda} = 0. \tag{8.14}$$

The gradient is obtained from

$$\nabla \widetilde{U}(\mathbf{r}) = \frac{d\widetilde{U}(\mathbf{r})}{dx}\widehat{i} + \frac{d\widetilde{U}(\mathbf{r})}{dy}\widehat{j} + \frac{d\widetilde{U}(\mathbf{r})}{dz}\widehat{k}. \tag{8.15}$$

For purposes of differentiation, the potential $\widetilde{U}(\mathbf{r})$ is written in the form

$$\widetilde{U}(\mathbf{r}) = \frac{1}{R_0} \sum_{n=0}^{\infty} \sum_{m=0}^{n} (C_{nm} W_{nm} + S_{nm} V_{nm}). \tag{8.16}$$

With $m = 0$,

$$\frac{d\widetilde{U}(\mathbf{r})}{dx} = \frac{1}{R_0} \sum_{n=0}^{\infty} - C_{n0} V_{n+1,1}, \quad \frac{d\widetilde{U}(\mathbf{r})}{dy} = \frac{1}{R_0} \sum_{n=0}^{\infty} - C_{n0} W_{n+1,1},$$

$$\frac{d\widetilde{U}(\mathbf{r})}{dz} = \frac{1}{R_0} \sum_{n=0}^{\infty} - (n+1) C_{n0} V_{n+1,0} - S_{n0} W_{n+1,0}. \tag{8.17}$$

With $m > 0$,

$$\begin{aligned}
\frac{d\widetilde{U}(\mathbf{r})}{dx} &= \frac{1}{2R_0} \sum_{n=0}^{\infty} \sum_{m=0}^{n} - \left(C_{nm} V_{n+1,m+1} + S_{nm} W_{n+1m+1}\right) \\
&\quad + \frac{(n-m+2)!}{(n+m)!} \left(C_{nm} W_{n+1,m-1} + S_{nm} V_{n+1,m-1}\right) \\
\frac{d\widetilde{U}(\mathbf{r})}{dy} &= \frac{1}{2R_0} \sum_{n=0}^{\infty} \sum_{m=0}^{n} - \left(C_{nm} W_{n+1,m+1} - S_{nm} V_{n+1m+1}\right) \\
&\quad + \frac{(n-m+2)!}{(n+m)!} \left(-C_{nm} W_{n+1,m-1} + S_{nm} V_{n+1,m-1}\right) \\
\frac{d\widetilde{U}(\mathbf{r})}{dz} &= \frac{1}{R_0} \sum_{n=0}^{\infty} \sum_{m=0}^{n} - (n-m+1) C_{nm} V_{n+1,m} - S_{nm} W_{n+1m}.
\end{aligned} \tag{8.18}$$

The gradient of the gravitational potential may then be found.

8.2.2 MacCullagh's Moment of Inertia-Based Formula

Any aspherical body *S* can be considered as the grouping of all the contained masses within the volume element *dV*. Let *O* be the centre of mass of the body to which the coordinate system is attached. The gravitational potential *dU* is due to the mass element *dM* at a point *P* separated of a distance $\mathbf{q} \cdot \mathbf{s}$ is the distance between *O* and *dM* and *r* is the distance between *O* and *P* (see Fig. 8.2).

The gravitational potential *dU* is due to the mass element *dM* at a point *P*, $dU = -(G/q)dM$.

Referring to Fig. 8.2 and applying the law of cosines

$$q^2 = r^2 + s^2 - 2rs\cos\alpha, \tag{8.19}$$

and it follows that

$$\begin{aligned} dU &= -\frac{G}{q}dM = -\frac{G}{\sqrt{r^2 + s^2 - 2rs\cos\alpha}}dM \\ &= -\frac{G}{r\sqrt{1 + (s/r)^2 - 2(s/r)\cos\alpha}}dM. \end{aligned} \tag{8.20}$$

Expanding the denominator in a series

$$\begin{aligned} \left(1 + (s/r)^2 - 2(s/r)\cos\alpha\right)^{-1/2} &= 1 - \frac{1}{2}(s/r)^2 + (s/r)\cos\alpha \\ &+ \frac{3}{2}(s/r)^2\cos^2\alpha + O\left((s/r)^3\right)\cdots \end{aligned} \tag{8.21}$$

Thus

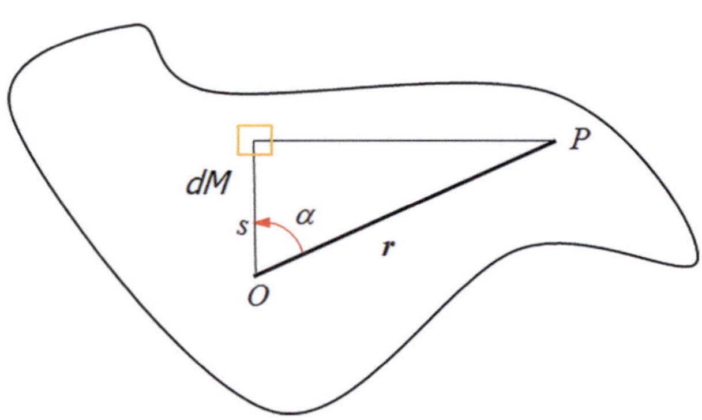

Fig. 8.2 Gravitational potential of an aspherical body

$$\left(1+(s/r)^2-2(s/r)\cos\alpha\right)^{-1/2}\approx 1+(s/r)\cos\alpha$$
$$+(1/2)(s/r)^2\left(3\cos^2\alpha-1\right). \tag{8.22}$$

It follows that

$$dU=-(G/r)\left(1+(s/r)\cos\alpha+(1/2)(s/r)^2\left(3\cos^2\alpha-1\right)\right)dM. \tag{8.23}$$

Hence, after integration one has

$$U=-(G/r)\int\limits_V\left(1+(s/r)\cos\alpha+(1/2)(s/r)^2\left(3\cos^2\alpha-1\right)\right)dM. \tag{8.24}$$

The gravitational potential U is expressed as

$$U=-\frac{G}{r}\int\limits_V dM-\frac{G}{r^2}\int\limits_V s\cos\alpha dM-\frac{G}{r^3}\int\limits_V s^2dM+\frac{3G}{2r^3}\int\limits_V s^2\left(1-\cos^2\alpha\right)dM, \tag{8.25}$$

or as

$$U=-\frac{G}{r}\int\limits_V dM-\frac{G}{r^2}\int\limits_V s\cos\alpha dM-\frac{G}{r^3}\int\limits_V s^2dM+\frac{3G}{2r^3}\int\limits_V s^2\sin^2\alpha dM. \tag{8.26}$$

In the above, the first integral is the gravitational potential of a point mass at O while the second integral is equal to zero, if it is assumed that the point O is at the centre of mass of the body, which must be known. The third integral is the half the sum of the moments of inertia of the body about the principal axes. The last integral is simply the moment of inertia about the axis OP. Hence the gravitational potential U is given by

$$U=-\frac{GM}{r}-\frac{G}{2r^3}\left(I_{xx}+I_{yy}+I_{zz}\right)+\frac{3G}{2r^3}I_{rr}. \tag{8.27}$$

The gravitational potential U is also expressed as

$$U=-\frac{GM}{r}\left(1+\frac{I_{xx}+I_{yy}+I_{zz}-3I_{rr}}{2Mr^2}\right). \tag{8.28}$$

MacCullagh's formula for the gravitational potential is symmetric by construction and related to the tri-axial ellipsoid representation. It can also be derived from the spherical harmonic representation of the potential by truncating the expansions to the second order.

8.2.3 Moment of Inertia Modelling

The moment of inertia of the body B about an axis OP is defined by the integral

$$I_{OP} = \int_V h^2 dm \tag{8.29}$$

where dm is an element of mass at an arbitrary point in the body with co-ordinates x, y, z and h is the perpendicular distance from the axis OP to the point. The integral is over the entire volume, V, of the body B. By applying Pythagoras theorem, it follows that

$$h^2 = (\mathbf{r}.\mathbf{r}) - (\mathbf{r}.\mathbf{e})^2, \tag{8.30}$$

where

$$\mathbf{r} = x\,i + y\,j + z\,k, \tag{8.31}$$

the position vector of the arbitrary point and

$$\mathbf{e} = e_x\,i + e_y\,j + e_z\,k, \tag{8.32}$$

is a unit vector in the direction of OP. Hence h^2 may also expressed as

$$h^2 = (\mathbf{r}.\mathbf{r})(\mathbf{e}.\mathbf{e}) - (\mathbf{r}.\mathbf{e})^2 = \mathbf{e}.[\mathbf{e}(\mathbf{r}.\mathbf{r}) - \mathbf{r}(\mathbf{r}.\mathbf{e})] = \mathbf{e}.(\mathbf{r} \times (\mathbf{e} \times \mathbf{r})). \tag{8.33}$$

The moment of inertia may be expressed as

$$I_{OP} = \int_V \left((\mathbf{r}.\mathbf{r})(\mathbf{e}.\mathbf{e}) - (\mathbf{r}.\mathbf{e})^2\right) dm. \tag{8.34}$$

Hence the moment of inertia of the body about the axis OP may be expressed as

$$I_{OP} = \int_V \left(\left(x^2 + y^2 + z^2\right)\left(e_x^2 + e_y^2 + e_z^2\right) - \left(xe_x + ye_y + ze_z\right)^2\right) dm. \tag{8.35}$$

To reduce the integral further, we define the six moments of inertia integrals as Table 8.1.

Table 8.1 Moment of inertia integrals

Integral	Moment of inertia (MI) reference axis	Integral	Product of inertia (PI) reference axes
$I_{xx} = \int_V (y^2 + z^2)dm$	MI about the Ox axis	$I_{xy} = \int_V xydm$	PI about the Ox, Oy axes
$I_{yy} = \int_V (x^2 + z^2)dm$	MI about the Oy axis	$I_{xz} = \int_V xzdm$	PI about the Ox, Oz axes
$I_{zz} = \int_V (x^2 + y^2)dm$	MI about the Oy axis	$I_{yz} = \int_V yzdm$	PI about the Oy, Oz axes

In terms of the integrals defined in Table 8.1, we may define the moment of inertia of the body about the axis OP as

$$I_{OP} = I_{xx}e_x^2 + I_{yy}e_y^2 + I_{zz}e_z^2 - 2I_{xy}e_xe_y - 2I_{xz}e_xe_z - 2I_{yz}e_ye_z, \tag{8.36}$$

which may also be expressed as the matrix product

$$I_{OP} = \begin{bmatrix} e_x & e_y & e_z \end{bmatrix} \begin{bmatrix} I_{xx} & -I_{xy} & -I_{xz} \\ -I_{xy} & I_{yy} & -I_{yz} \\ -I_{xz} & -I_{yz} & I_{zz} \end{bmatrix} \begin{bmatrix} e_x \\ e_y \\ e_z \end{bmatrix} = \begin{bmatrix} e_x & e_y & e_z \end{bmatrix} \mathbf{I} \begin{bmatrix} e_x \\ e_y \\ e_z \end{bmatrix}. \tag{8.37}$$

where

$$\mathbf{I} = \begin{bmatrix} I_{xx} & -I_{xy} & -I_{xz} \\ -I_{xy} & I_{yy} & -I_{yz} \\ -I_{xz} & -I_{yz} & I_{zz} \end{bmatrix}, \tag{8.38}$$

is the moment of inertia matrix.

For a spheroid having uniform density, the moment of inertia is that of an ellipsoid with an additional axis of symmetry. Given a description of a spheroid as having a major axis c, and minor axes and intermediate axes, $a \neq b$, the moments of inertia along these principal axes are in general

$$I_{cc} = (1/5)M\left(a^2 + b^2\right), I_{bb} = (1/5)M\left(a^2 + c^2\right), I_{aa} = (1/5)M\left(c^2 + b^2\right). \tag{8.39}$$

The mass is given by

$$M = (4/3)\rho\pi abc. \tag{8.40}$$

These formulas for a general ellipsoid can be derived from first principles and also extended to a double ellipsoidal model.

For a cylinder of radius R and length L

$$I_{cc} = (1/2)MR^2, I_{bb} = (1/12)M\left(L^2 + 3R^2\right) \text{ and } I_{aa} = (1/12)M\left(L^2 + 3R^2\right). \tag{8.41}$$

For regular tetrahedron (composed of four triangular faces, six straight edges and four vertex corners) of side s and mass M, the moment of inertia I about an axis through a vertex and the centre of mass and the mass are respectively given by

$$I = (1/20)Ms^2, M = \left(\rho/6\sqrt{2}\right)s^3. \tag{8.42}$$

For a sphere of radius r_2 with a spherical cavity with radius r_1, the moment of inertia about an axis passing through the centre of mass and the mass are respectively given by

$$I = \frac{2}{5}M\left(\frac{r_2^5 - r_1^5}{r_2^3 - r_1^3}\right) = \frac{8}{15}\rho\pi\left(r_2^5 - r_1^5\right), M = (4/3)\rho\pi\left(r_2^3 - r_1^3\right). \tag{8.43}$$

8.2.4 Gravitational Potential Models

Considering the general expression for the gravitational potential, because of rotational and latitudinal symmetries of the normal spheroid, only the zonal harmonics of even degrees are non-zero, for an ellipsoid. The general expression for the gravitational potential of an ellipsoid can be expressed in terms of elliptic integrals. Alternatively

$$\widetilde{U}(\mathbf{r}) = (1/R_0)\sum_{n=0}^{\infty}(R_0/|\mathbf{r}|)^{2n+1}C_{2n,0}P_{2n,0}(\sin\delta). \tag{8.44}$$

Another alternate expression is,

$$\widetilde{U}(\mathbf{r}) = \frac{GM}{|\mathbf{r}|}\left(1 - J_2\left(\frac{R_0}{|\mathbf{r}|}\right)^2 P_{2,0}(\sin\delta)\cdots - J_{2n}\left(\frac{R_0}{|\mathbf{r}|}\right)^{2n} P_{2n,0}(\sin\delta)\right). \tag{8.45}$$

Let $P_{n0}(x) \equiv P_n(x)$ denote the classical Legendre polynomial of degree n and

$$P_n^*(x) = \sqrt{\frac{2n+1}{2}} P_n(x) \tag{8.46}$$

its ortho-normalized version. Thus, with $\delta_{i,\,j}$ the Kronecker delta, the family $P_n^*(x)$ satisfies

$$\int_{-1}^{1} P_i^*(x) P_j^*(x) dx = \delta_{i,j}, i, j \geq 0. \tag{8.47}$$

The gravitational potential of a sphere of radius, R, is,

$$U = -\mu/r. \tag{8.48}$$

The gravitational potential of a disc of radius R at a point along its axis is

$$U = -\frac{2\mu}{R^2}\left(\sqrt{z^2+R^2} - z\right) = -2\pi G\rho\left(\sqrt{z^2+R^2} - z\right). \tag{8.49}$$

In the limit as $z \to \infty$, it reduces to, $U = -\mu/z$.

For a circular wire of radius, R, the gravitational potential at an arbitrary point is

$$U(\mathbf{r}) = \frac{2\mu}{p\pi} \int_0^{\pi/2} \frac{d\phi}{\sqrt{1-k^2\sin^2\phi}} = \frac{2\mu}{p\pi} K(k), \tag{8.50}$$

p and q, the shortest (q) and the longest (p) distances between the field point P and the wire and are

$$p^2 = (R+r)^2 + z^2, q^2 = (R-r)^2 + z^2, k = \sqrt{1 - q^2/p^2}. \tag{8.51}$$

The gravitational potential of a cylinder at a point along its axis is obtained by integrating along the length of the cylinder, the gravitational potential of a disc of radius R.

For a uniform circular disc, the gravitational potential in terms of the complete and incomplete elliptic integrals of first and second kinds is

$$\begin{aligned} U(x,y,z) &= 2G\sigma\left(-pE(k) - \frac{R^2-r^2}{p}K(k) + |z|\left(\frac{\pi}{2} + \frac{\pi}{2}\operatorname{sgn}(R-r)\right)\right) \\ &- 2G\sigma z \operatorname{sgn}(R-r)(E(k)F(\phi,k') + K(k)E(\phi,k') - K(k)F(\phi,k')), \end{aligned} \tag{8.52}$$

with

$$k' = \sqrt{1 - k^2}, \phi = \sin^{-1}(z/q), r^2 = x^2 + y^2, \tag{8.53}$$

where $\sigma = M/\pi R^2$ is the surface density of the disc; $E(k)$, $K(k)$, are the complete elliptic integrals of first and second kind, respectively; and $F(\phi, k)$, $E(\phi, k)$ are the incomplete elliptic integrals of first and second kind, respectively. The elliptic integrals are computed by using the methods provided by Carlson. The form of the potential function permits the rapid numerical evaluation of the potential function and its derivatives at a field point. The derivatives of the gravitation potential are

$$\frac{\partial}{\partial x}U(x,y,z) = -G\sigma\left(2x\frac{\sqrt{r^2+z^2+R^2+2Rr}}{r^2}\right)\left(\left(1-\frac{1}{2}k^2\right)K(k)-E(k)\right)$$

$$\frac{\partial}{\partial y}U(x,y,z) = -G\sigma\left(2y\frac{\sqrt{r^2+z^2+R^2+2Rr}}{r^2}\right)\left(\left(1-\frac{1}{2}k^2\right)K(k)-E(k)\right)$$

$$\begin{aligned}\frac{\partial}{\partial z}U(x,y,z) &= G\sigma\left(\frac{2x}{\sqrt{r^2+z^2+R^2+2Rr}}K(k)\right)\\ &-2G\sigma\left(\mathrm{sgn}(z)\left(\frac{\pi}{2}+\frac{\pi}{2}\,\mathrm{sgn}(R-r)\right)\right.\\ &\left.-\,\mathrm{sgn}(R-r)((E(k)-K(k))F(\phi,k')+K(k)E(\phi,k'))\right).\end{aligned} \tag{8.54}$$

8.2.5 Geoid Models

A new approach to modelling the gravitational potential between two bodies of arbitrary shape is presented by Conway (2016), based on the use of vector potentials. In this approach, the force, torque and mutual potential energy are formulated as surface integrals under the assumption of constant density. The force, torque and potential energy integrals are converted from volume integrals to surface integrals. The approach was recently applied to coplanar spheroids and thin disks by Wold and Conway (2021). Ho et al. (2021) have extended the work of Wold and Conway to the gravitational force, torque and potential energy to the dynamics between the two bodies, including the coupling between translational and rotational motion. The main feature of Conway's method is the representation of the force **F**, torque **M** and potential U of a solid body of uniform density ρ in a gravitational field $\mathbf{g}(\mathbf{r})$ in terms of surface integrals over the surface of the body.

The force of an extended body with constant density ρ, due to an external gravitational field $\mathbf{g}(\mathbf{r})$, can be expressed through a volume integral

$$\mathbf{F} = \rho \iiint_V \mathbf{g}(\mathbf{r}) dV, \tag{8.55}$$

where $\mathbf{r}$ is a position vector to a point in the body being integrated over. If the gravitational field is given by the gradient of a scalar potential $\Phi(\mathbf{r})$, so that $\mathbf{g}(\mathbf{r}) = \nabla \Phi(\mathbf{r})$, the divergence theorem can be applied to show that the force can be computed through a surface integral (Conway, 2016)

$$\mathbf{F} = \rho \iint_S \Phi(\mathbf{r}) \boldsymbol{n} \, dS. \tag{8.56}$$

It can also be shown (Conway, 2016) that the torque about the mass centre of an extended body is given by

$$\mathbf{M} = -\rho \iint_S \Phi(\mathbf{r}) \boldsymbol{n} \times \mathbf{r} \, dS. \tag{8.57}$$

The potential energy (Conway, 2016) may be expressed as

$$U = \frac{\rho}{3} \iint_S \left(\mathbf{r}\Phi(\mathbf{r}) - \frac{1}{2} |\mathbf{r}|^2 \mathbf{g}(\mathbf{r}) \right) \cdot \boldsymbol{n} \, dS. \tag{8.58}$$

In these formulas $\boldsymbol{n}$ is a unit normal to the body surface. Equations (8.56), (8.57) and (8.58), in terms of the surface integrals, are thus alternative expressions for the force, torque and potential between two bodies.

Analytical formulas for $\mathbf{g}(\mathbf{r})$ and $\Phi(\mathbf{r})$ in terms of special functions and integrals are given for spheroids and tri-axial ellipsoids by MacMillan (1930).

For a general ellipsoid with semi-axes $a > b > c$ and constant density ρ, the gravitational potential is given by

$$\Phi(\mathbf{r}) = \frac{3}{2} \frac{GM}{\sqrt{a^2 - c^2}} \widetilde{\Phi}(\mathbf{r}),$$

$$\begin{aligned} \widetilde{\Phi}(\mathbf{r}) = & \left(1 - \frac{x^2}{a^2 - b^2} + \frac{y^2}{a^2 - b^2}\right) F(\phi, k) \\ & + \left(\frac{x^2}{a^2 - b^2} - \frac{(a^2 - c^2) y^2}{(a^2 - b^2)(b^2 - c^2)} + \frac{z^2}{b^2 - c^2}\right) E(\phi, k) \\ & + \left(\frac{(c^2 + \lambda) y^2}{(b^2 - c^2)} - \frac{(b^2 + \lambda) y^2}{(b^2 - c^2)}\right) \frac{\sqrt{a^2 - c^2}}{\sqrt{(a^2 + \lambda)(b^2 + \lambda)(c^2 + \lambda)}}. \end{aligned} \tag{8.59}$$

where $F(\phi, k)$, $E(\phi, k)$ are the incomplete elliptic integrals of first and second kind respectively and λ is the largest root of Eq. (8.14) or

$$\frac{x^2}{\alpha^2+\lambda}+\frac{y^2}{\beta^2+\lambda}+\frac{z^2}{\gamma^2+\lambda}=1. \tag{8.60}$$

In Eq. (8.59), ϕ and k are defined as

$$\phi=\sin^{-1}\left(\sqrt{(a^2-c^2)/(a^2+\lambda)}\right), k=\sqrt{(a^2-b^2)/(a^2-c^2)}. \tag{8.61}$$

For an oblate spheroid with semi-axes $a = b > c$ the gravitational potential is given by

$$\begin{aligned}\Phi(\mathbf{r}) &= \frac{3}{2}GM\widetilde{\Phi}(\mathbf{r}),\\ \widetilde{\Phi}(\mathbf{r}) &= \frac{1}{\sqrt{a^2-c^2}}\left(1-\frac{x^2+y^2-2z^2}{2(a^2-c^2)}\right)\phi\\ &\quad+\frac{1}{2(a^2-c^2)}\left(\sqrt{c^2+\lambda}\frac{x^2+y^2}{a^2+\lambda}-\frac{2z^2}{\sqrt{c^2+\lambda}}\right),\end{aligned} \tag{8.62}$$

where λ still satisfies Eq. (8.60).

8.2.6 Geometrical Modelling

Given that the potential function in Eqs. (8.59) and (8.62) are only functions of the geometry, it is possible in principle to fit an ellipsoid to the geometry body. This is the approach adopted by Panou et al. (2020). There are other geometrical shapes that one could use to model an asteroid, so as to predict its gravitational field. In this section the methods discussed by Panou et al. (2020) and by Soler et al. (2020) and by Soler and Han (2020) will be briefly summarized.

The Cartesian equation of an ellipsoid, in an arbitrary reference system, may be expressed as

$$f(x,y,z)=\begin{bmatrix}x-x_c\\ y-y_c\\ z-z_c\end{bmatrix}^T\begin{bmatrix}w_{xx} & w_{xy} & w_{xz}\\ w_{xy} & w_{yy} & w_{yz}\\ w_{xz} & w_{yz} & w_{zz}\end{bmatrix}^{-1}\begin{bmatrix}x-x_c\\ y-y_c\\ z-z_c\end{bmatrix}=1. \tag{8.63}$$

where the elements of vector $[x_c\ y_c\ z_c]$ are the coordinates of the centre of the ellipsoid and the elements of symmetric weighting matrix

$$\begin{bmatrix} w_{xx} & w_{xy} & w_{xz} \\ w_{xy} & w_{yy} & w_{yz} \\ w_{xz} & w_{yz} & w_{zz} \end{bmatrix} = \begin{bmatrix} q_{xx} & q_{xy} & q_{xz} \\ q_{xy} & q_{yy} & q_{yz} \\ q_{xz} & q_{yz} & q_{zz} \end{bmatrix}^{-1} . \tag{8.64}$$

are functions of three rotation angles and the three ellipsoid semi-axes a, b and c. Thus, the mathematical model given by Eq. (8.63) includes nine parameters representing the spatial properties of an ellipsoid (the semi-axes and translation, rotation relative to the semi-axes). Equation (8.63) is equivalently written as

$$f(x,y,z) = \begin{bmatrix} x-x_c \\ y-y_c \\ z-z_c \end{bmatrix}^T \begin{bmatrix} q_{xx} & q_{xy} & q_{xz} \\ q_{xy} & q_{yy} & q_{yz} \\ q_{xz} & q_{yz} & q_{zz} \end{bmatrix} \begin{bmatrix} x-x_c \\ y-y_c \\ z-z_c \end{bmatrix} = \begin{bmatrix} x \\ y \\ z \end{bmatrix}^T \begin{bmatrix} q_{xx} & q_{xy} & q_{xz} \\ q_{xy} & q_{yy} & q_{yz} \\ q_{xz} & q_{yz} & q_{zz} \end{bmatrix} \begin{bmatrix} x \\ y \\ z \end{bmatrix}$$
$$-2\begin{bmatrix} x_c \\ y_c \\ z_c \end{bmatrix}^T \begin{bmatrix} q_{xx} & q_{xy} & q_{xz} \\ q_{xy} & q_{yy} & q_{yz} \\ q_{xz} & q_{yz} & q_{zz} \end{bmatrix} \begin{bmatrix} x \\ y \\ z \end{bmatrix} + \begin{bmatrix} x_c \\ y_c \\ z_c \end{bmatrix}^T \begin{bmatrix} q_{xx} & q_{xy} & q_{xz} \\ q_{xy} & q_{yy} & q_{yz} \\ q_{xz} & q_{yz} & q_{zz} \end{bmatrix} \begin{bmatrix} x_c \\ y_c \\ z_c \end{bmatrix} = 1. \tag{8.65}$$

It may also be written as

$$f(x,y,z) = \begin{bmatrix} x \\ y \\ z \end{bmatrix}^T \begin{bmatrix} q_{xx} & q_{xy} & q_{xz} \\ q_{xy} & q_{yy} & q_{yz} \\ q_{xz} & q_{yz} & q_{zz} \end{bmatrix} \begin{bmatrix} x \\ y \\ z \end{bmatrix} - \begin{bmatrix} q_x & q_y & q_z \end{bmatrix} \begin{bmatrix} x \\ y \\ z \end{bmatrix} = 1 - q_{xyz}, \tag{8.66}$$

with

$$\begin{bmatrix} q_x \\ q_y \\ q_z \end{bmatrix}^T = 2\begin{bmatrix} x_c \\ y_c \\ z_c \end{bmatrix}^T \begin{bmatrix} q_{xx} & q_{xy} & q_{xz} \\ q_{xy} & q_{yy} & q_{yz} \\ q_{xz} & q_{yz} & q_{zz} \end{bmatrix}, q_{xyz} = \begin{bmatrix} x_c \\ y_c \\ z_c \end{bmatrix}^T \begin{bmatrix} q_{xx} & q_{xy} & q_{xz} \\ q_{xy} & q_{yy} & q_{yz} \\ q_{xz} & q_{yz} & q_{zz} \end{bmatrix} \begin{bmatrix} x_c \\ y_c \\ z_c \end{bmatrix} . \tag{8.67}$$

Equation (8.66) may be normalized and the number of constants reduced to 9 instead of the 10 and expressed as

$$\begin{bmatrix} x \\ y \\ z \end{bmatrix}^T \begin{bmatrix} c_{xx} & c_{xy} & c_{xz} \\ c_{xy} & c_{yy} & c_{yz} \\ c_{xz} & c_{yz} & c_{zz} \end{bmatrix} \begin{bmatrix} x \\ y \\ z \end{bmatrix} - \begin{bmatrix} c_x & c_y & c_z \end{bmatrix} \begin{bmatrix} x \\ y \\ z \end{bmatrix} = 1, \tag{8.68}$$

The unknown coefficients are

$$\mathbf{C} = \begin{bmatrix} c_{xx} & c_{xy} & c_{xz} & c_{yy} & c_{yz} & c_{zz} & c_x & c_y & c_z \end{bmatrix}. \tag{8.69}$$

Equation (8.68) is

$$\begin{bmatrix} x^2 & xy & xz & y^2 & yz & z^2 & x & y & z \end{bmatrix} \mathbf{C} = 1.$$

The coefficient vector **C** can be obtained by the method of least squares, provided a minimum of ten points on the ellipsoid are known. Once as estimate of the coefficients defining the geometry of the ellipsoid are obtained, they could be improved either to model the geometry of the body itself or to improve the gravitational field of the body. Once the coefficient vector **C** is known, the coefficients themselves are transformed back into the geometric constants defining the size and shape of the tri-axial ellipsoid, as the coordinates of the origin of the ellipsoid with respect to the frame used as well as the semi-axes could be determined. The complete procedure has been given by Soler et al. (2020).

8.2.7 Polyhedron Model

Polyhedral shapes constructed two lists of data as in the case of finite element modelling of an arbitrarily shaped body. An ordered list of all the vertices expressed as vectors in the body-centred body-fixed frame and a second list of vertex number triplets that define each of the triangular facets or elements are used. Euler's formula provides a linear relation between the number of faces f, the number of edges e and the number of vertices v of a triangular faced polyhedron: $f = 2(v-2)$, $e = 3(v-2)$. Each polyhedron face with its edges subtends a unique solid angle at the centre of the body. Thus one may compute the gravitational potential of each face and of each edge and sum up for all edges and faces to construct the gravitational potential of the complete polyhedron. In order to compute the gravitational potential of this object, two elemental matrices are derived using the normal to each face and edge. These are each 3-by-3 matrices, one for each edge and face of the polyhedron.

To compute the gravitational potential from a position in space relative to the polyhedron barycentre, given the vectors from the field point to each vertex i, j, k of a given face, $\mathbf{r}_i$, $\mathbf{r}_j$, $\mathbf{r}_k$, the non-dimensional signed solid angle, ω_f covered by that triangular facet as seen from the field point is computed from

$$\omega_f = 2\tan^{-1}\left(\frac{\mathbf{r}_i \cdot \mathbf{r}_j \times \mathbf{r}_k}{r_i r_j r_k + r_i(\mathbf{r}_j \cdot \mathbf{r}_k) + r_j(\mathbf{r}_k \cdot \mathbf{r}_i) + r_k(\mathbf{r}_i \cdot \mathbf{r}_j)}\right). \tag{8.70}$$

The element along the edges arises as it has a logarithmic distribution of mass. In this case, given r_i and r_j the norms of the vectors from the field point to the two vertices of the edge taken into account, and the edge length e_{ij}, the potential is given by

$$L_e = \ln\left(\frac{r_i + r_j + e_{ij}}{r_i + r_j - e_{ij}}\right), \tag{8.71}$$

For a convex polygon, such as the triangle, the face normal vector is given by the cross product of vectors along the edge. Thus the un-normalized vector aligned with a convex polygons face is

$$\mathbf{n}_f = (\mathbf{r}_2 - \mathbf{r}_1) \times (\mathbf{r}_3 - \mathbf{r}_2) = \mathbf{r}_1 \times \mathbf{r}_2 + \mathbf{r}_2 \times \mathbf{r}_3 + \mathbf{r}_3 \times \mathbf{r}_1. \tag{8.72}$$

Upon normalization, the outward pointing surface normal vector of the face, f, $\widehat{\mathbf{n}}_f$ is obtained. Werner and Scheeres (1997) define a dyad $\mathbf{E}_{12} = \widehat{\mathbf{n}}_A \widehat{\mathbf{n}}_{12}^A + \widehat{\mathbf{n}}_A \widehat{\mathbf{n}}_{21}^B$, in terms of the two face and edge normal vectors associated with the edge P_1P_2 as shown in Fig. 8.3. Also shown in the figure is the polyhedral model of an arbitrarily shaped asteroid. Each edge of each face has an outward pointing edge normal vector $\widehat{\mathbf{n}}_e^f$ that is perpendicular to both the face normal vector $\widehat{\mathbf{n}}_f$ and the edge vector, $\widehat{e}_{12}$. The matrices $\mathbf{E}_e$ are defined similarly for each edge. As a matrix, $\mathbf{E}_{12}$ is defined as

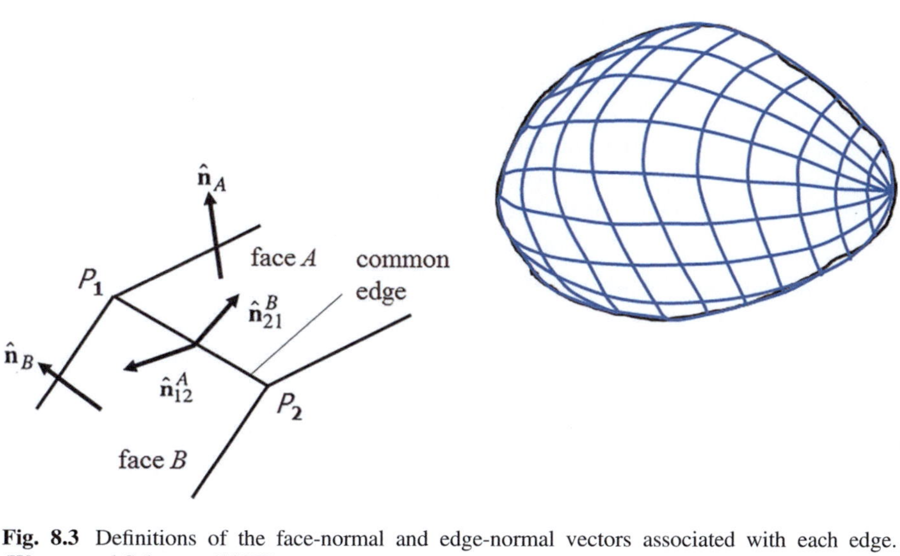

Fig. 8.3 Definitions of the face-normal and edge-normal vectors associated with each edge. (Werner and Scheeres (1997))

$\mathbf{E}_{12} = \widehat{\mathbf{n}}_A \left(\widehat{\mathbf{n}}_{12}^A \right)^T + \widehat{\mathbf{n}}_A \left(\widehat{\mathbf{n}}_{21}^B \right)^T$. Werner and Scheeres (1997) define the dyad $\mathbf{F}_f$ as the outer product of the face normal vector $\widehat{\mathbf{n}}_f$ with itself. Thus, the dyad, $\mathbf{F}_f = \widehat{\mathbf{n}}_f \widehat{\mathbf{n}}_f$, one for each face. In terms of matrices, $\mathbf{F}_f = \widehat{\mathbf{n}}_f \widehat{\mathbf{n}}_f^{T_f}$.

The gravitational potential and its gradient as derived by Werner and Scheeres (1997) are given by

$$\widetilde{U}(\mathbf{r}) = \frac{1}{2} G\rho \sum_{e \in edges} \mathbf{r}_e \cdot \mathbf{E}_e \cdot \mathbf{r}_e \cdot L_e - \frac{1}{2} G\rho \sum_{f \in faces} \mathbf{r}_f \cdot \mathbf{F}_f \cdot \mathbf{r}_f \cdot \omega_f, \tag{8.73}$$

$$\nabla \widetilde{U}(\mathbf{r}) = -G\rho \sum_{e \in edges} \mathbf{E}_e \cdot \mathbf{r}_e \cdot L_e + G\rho \sum_{f \in faces} \mathbf{F}_f \cdot \mathbf{r}_f \cdot \omega_f. \tag{8.74}$$

where G is the gravitational constant and ρ is the asteroid's density; $\mathbf{r}_e$ and $\mathbf{r}_f$ are body-fixed vectors from the field point to fixed points on the edge e and face f of the polyhedron, respectively; $\mathbf{E}_e$ and $\mathbf{F}_f$ are the face and edge related element matrices defined in terms of the edge normal and face normal vectors, respectively.

8.2.8 Discrete Element Models

Yu et al. (2019) proposed a finite element method (FEM) for computing the gravitational interactions between two arbitrarily shaped celestial bodies. Expressions for the gravitational potential, the force of attraction and torque acting on the orbiting body were derived by dividing the asteroid into several small tetrahedral elements. A finite element model of the binary near-Earth asteroid 65803 Didymos was developed and used to successfully predict the coupled spin-orbit motion of its two components.

In the method proposed by Yu et al. (2019), given one or more arbitrary rigid bodies, tetrahedral meshes are first generated to fill their volumes. Figure 8.4 illustrates an arbitrary pair of tetrahedral elements $\{c_k, c_m\}$ from two bodies, i.e. element c_k belongs to the first body and element c_m belongs to the second body. The basic geometry of the element pair is defined in the figure, and two or more coordinate systems are adopted:

(1) An inertially fixed world coordinate frame
(2) A body-fixed frame with its origin at the mass centre of each of the bodies, and the three axes of the frame coinciding with the minimum, medium and maximum principal inertia axes of the body

The absolute position vectors of the centroids of each of the bodies are reckoned or estimated from the origin of the world frame. Considering the fixed mesh as defined above, the four corner nodes of each tetrahedral element in each of the bodies have constant coordinates with respect to the corresponding body frame of the body to which the element belongs. The position of the ith corner node of a tetrahedral

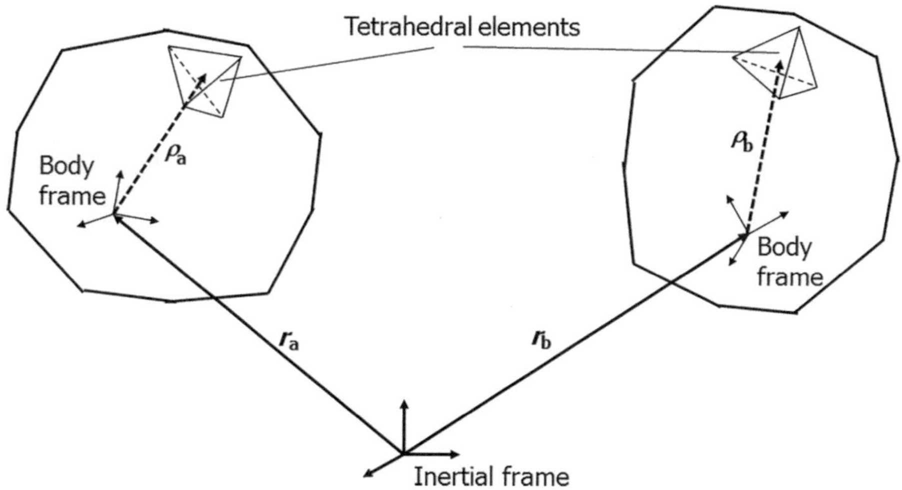

Fig. 8.4 An arbitrary pair of tetrahedral elements (from Yu et al., 2019)

element c_k is denoted as ρ_{ki}, while position of the jth corner node of a tetrahedral element c_m is denoted as ρ_{mj}. The generic position vector of a point inside an element within a body $\boldsymbol{a}$, with elements c_k, is denoted as $\rho_{\boldsymbol{a}}$, while the generic position vector of a point inside an element within a body $\boldsymbol{b}$, with elements c_m, is denoted as $\rho_{\boldsymbol{b}}$. Given a general form of any mutual field function $h(\rho_{\boldsymbol{a}}, \rho_{\boldsymbol{b}})$, the field integral over the volumes occupied by each of the elements c_k and c_m is determined by

$$H_{km} = \iiint_{c_m} \iiint_{c_k} h(\rho_{\boldsymbol{a}}, \rho_{\boldsymbol{b}}) dx_{\boldsymbol{a}} dy_{\boldsymbol{a}} dz_{\boldsymbol{a}} dx_{\boldsymbol{b}} dy_{\boldsymbol{b}} dz_{\boldsymbol{b}}. \tag{8.75}$$

The integral is evaluated by representing the general mutual field function in terms of the tetrahedral natural coordinates. A brief introduction to the tetrahedral natural coordinates is presented below.

The tetrahedral natural coordinates are very similar in principle to the triangular area coordinates. A typical tetrahedral element consists of four nodes, six corners and four faces. Fig. 8.5 shows a typical four-node tetrahedron. Its geometry is fully defined by specifying the location of the four corner nodes with respect to the local reference coordinate frame. When seen from the any node, the number convention followed for the remaining nodes is also shown in Fig. 8.5.

The Cartesian nodal coordinates of the nodes in the reference frame are respectively given by

$$[x_i \quad y_i \quad z_i], i = 1, 2, 3, 4. \tag{8.76}$$

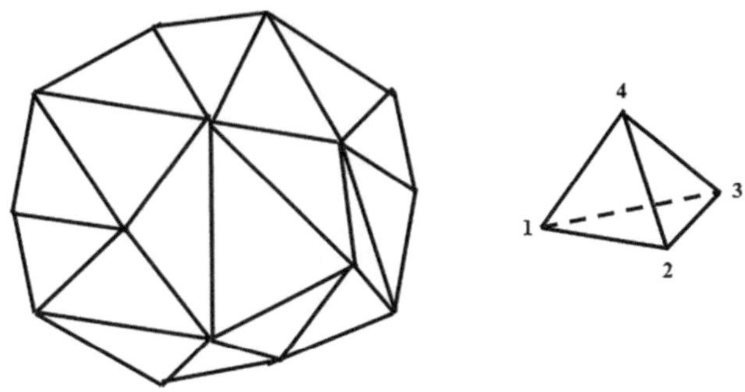

Fig. 8.5 A typical rigid body discretized into tetrahedral elements (left) and a typical tetrahedral element (right)

The numbering convention results in a positive volume. The location of the centroid is given by the mean of the nodal coordinates while the volume is given by the determinant

$$V=\frac{1}{6}\begin{vmatrix}1 & 1 & 1 & 1\\ x_1 & x_2 & x_3 & x_4\\ y_1 & y_2 & y_3 & y_4\\ z_1 & z_2 & z_3 & z_4\end{vmatrix}. \tag{8.77}$$

The natural coordinates of a point within a tetrahedral element are defined as each of the volumes induced by the internal point with each of the four faces of the element, normalized by the volume of the element. These are themselves tetrahedrons and hence the volumes can be obtained by using the volume formula cyclically. The set of local coordinates are now introduced and defined as

$$\boldsymbol{\zeta}^T=[\zeta_1 \quad \zeta_2 \quad \zeta_3 \quad \zeta_4]. \tag{8.78}$$

The value of ζ_i is one at corner i, is zero at the other three corners which make the opposite face and varies linearly in between. The sum of the four coordinates is equal to one:

$$\zeta_1+\zeta_2+\zeta_3+\zeta_4=1. \tag{8.79}$$

The coordinate ζ_1 is defined by

$$\zeta_1=\frac{1}{6V}\begin{vmatrix}1 & 1 & 1 & 1\\ x & x_2 & x_3 & x_4\\ y & y_2 & y_3 & y_4\\ z & z_2 & z_3 & z_4\end{vmatrix}. \tag{8.80}$$

The others are obtained by cyclic permutation. Thus

$$\zeta_i = (1/6V)(\alpha_i + \beta_i x + \gamma_i y + \delta_i z), \text{ or } \begin{bmatrix} \zeta_1 \\ \zeta_2 \\ \zeta_3 \\ \zeta_4 \end{bmatrix} = \frac{1}{6V} \begin{bmatrix} \alpha_1 & \beta_1 & \gamma_1 & \delta_1 \\ \alpha_2 & \beta_2 & \gamma_2 & \delta_2 \\ \alpha_3 & \beta_3 & \gamma_3 & \delta_3 \\ \alpha_4 & \beta_4 & \gamma_4 & \delta_4 \end{bmatrix} \begin{bmatrix} 1 \\ x \\ y \\ z \end{bmatrix}. \tag{8.81}$$

The values of α_1, β_1, γ_1, δ_1 are obtained as shown below as

$$\begin{aligned} \alpha_1 &= \begin{vmatrix} x_2 & x_3 & x_4 \\ y_2 & y_3 & y_4 \\ z_2 & z_3 & z_4 \end{vmatrix}, \beta_1 = -\begin{vmatrix} 1 & 1 & 1 \\ y_2 & y_3 & y_4 \\ z_2 & z_3 & z_4 \end{vmatrix}, \\ \gamma_1 &= \begin{vmatrix} 1 & 1 & 1 \\ x_2 & x_3 & x_4 \\ z_2 & z_3 & z_4 \end{vmatrix}, \delta_1 = -\begin{vmatrix} 1 & 1 & 1 \\ x_2 & x_3 & x_4 \\ y_2 & y_3 & y_4 \end{vmatrix}. \end{aligned} \tag{8.82}$$

The others are obtained by cyclic permutations of the index i. The transformation from the natural coordinates back to the local reference frame is given by

$$\begin{bmatrix} 1 \\ x \\ y \\ z \end{bmatrix} = \begin{bmatrix} 1 & 1 & 1 & 1 \\ x_1 & x_2 & x_3 & x_4 \\ y_1 & y_2 & y_3 & y_4 \\ z_1 & z_2 & z_3 & z_4 \end{bmatrix} \begin{bmatrix} \zeta_1 \\ \zeta_2 \\ \zeta_3 \\ \zeta_4 \end{bmatrix}. \tag{8.83}$$

Treating ζ_i, $i = 1, 2, 3$ as independent coordinates, the Jacobian determinant of the transformation from x, y and z to ζ_i, $i = 1, 2, 3$ is given by

$$J = \frac{1}{6V} \begin{vmatrix} \beta_1 & \gamma_1 & \delta_1 \\ \beta_2 & \gamma_2 & \delta_2 \\ \beta_3 & \gamma_3 & \delta_3 \end{vmatrix}. \tag{8.84}$$

Any function $f(x, y, z)$ that has values $f_i = f(x_i, y_i, z_i)$ $(i = 1, 2, 3, 4)$ at the four nodes of the tetrahedron may be linearly interpolated in terms of the natural coordinates as:

$$\tilde{f}(x, y, z) = \sum_{i=1}^{4} f_i \zeta_i = [f_1 \quad f_2 \quad f_3 \quad f_4][\zeta_1 \quad \zeta_2 \quad \zeta_3 \quad \zeta_4]^T. \tag{8.85}$$

The tetrahedral natural coordinates are now introduced for each set of elements to calculate the sextuple integral described in Eq. (8.75). Following Yu et al. (2019), and taking element c_k as an example, for a point located within the tetrahedral element, the natural coordinates are defined as

$$\zeta_{ki} = \frac{A_{ki}}{3V_k} \boldsymbol{n}_{ki} \cdot (\rho_{\boldsymbol{a}} - \rho_{ki}). \tag{8.86}$$

In Eq. (8.86), V_k is the volume of the tetrahedral element c_k, A_{ki} the area of the face opposite the ith corner and $\boldsymbol{n}_{ki}$ the unit normal vector of the opposite face, pointing toward the interior of the tetrahedral element. Thus, the natural coordinate ζ_{ki} represents the scaled volume of a "sub-tetrahedron" spanned by point at $\rho_{\boldsymbol{a}}$ and the face of the tetrahedral element opposite to the ith corner. The value of ζ_{ki} is one at the ith corner and zero at any point on the opposite face. The use of natural coordinates allows us to express the mutual integral in Eq. (8.75) in a compact form, independent of the Cartesian coordinates. Transforming the Cartesian coordinates in Eq. (8.75) to natural coordinates, the mutual field integral over the transformed domain is given by

$$H_{km} = \iiint_{c_m} \iiint_{c_k} h(\rho_{\boldsymbol{a}}, \rho_{\boldsymbol{b}}) J_k^{-1} J_m^{-1} d\zeta_{k1} d\zeta_{k2} d\zeta_{k3} d\zeta_{m1} d\zeta_{m2} d\zeta_{m3}. \tag{8.87}$$

To facilitate the evaluation of Eq. (8.87), the function $h(\rho_{\boldsymbol{a}}, \rho_{\boldsymbol{b}})$ is replaced by a bilinear interpolated function in terms of the values of the function at the nodes given by, $H_{ij}^{km} = h\left(\rho_{ki}, \rho_{mj}\right)$. Then the interpolated function is defined as

$$\widetilde{h}(\rho_{\boldsymbol{a}}, \rho_{\boldsymbol{b}}) = \sum_{i=1}^{4} \sum_{j=1}^{4} \zeta_{ki} \zeta_{mj} H_{ij}^{km}. \tag{8.88}$$

Introducing the approximation given by Eq. (8.88) to $h(\rho_{\boldsymbol{a}}, \rho_{\boldsymbol{b}})$ into Eq. (8.87), the latter equation may now be integrated exactly. Yu et al. (2019) give the first order approximation to Eq. (8.87) where $h(\rho_{\boldsymbol{a}}, \rho_{\boldsymbol{b}})$ is an arbitrary mutual field function as

$$H_{km} \approx \widetilde{H}_{km} = \frac{1}{576} J_k^{-1} J_m^{-1} \sum_{i=1}^{4} \sum_{j=1}^{4} H_{ij}^{km}. \tag{8.89}$$

Yu et al. (2019) apply Eq. (8.89) to derive approximations to mutual potential energy, force and torque. Choosing $h(\rho_{\boldsymbol{a}}, \rho_{\boldsymbol{b}})$ to be

$$h(\rho_a, \rho_b) = -G\frac{\sigma_a(\rho_a)\sigma_b(\rho_b)}{|r_b + \rho_b - r_a - \rho_a|}, \tag{8.90}$$

where $\sigma_a(\rho_a)$ describes the distribution of the density within the first body and $\sigma_b(\rho_b)$, the distribution of the density within the second. The corresponding interpolation of $h(\rho_a, \rho_b)$ is given by

$$\widetilde{h}(\rho_a, \rho_b) = \sum_{i=1}^{4}\sum_{j=1}^{4}\zeta_{ki}\zeta_{mj}U_{ij}^{km} = -G\sum_{i=1}^{4}\sum_{j=1}^{4}\zeta_{ki}\zeta_{mj}\frac{\sigma_a(\rho_{ki})\sigma_b(\rho_{mj})}{|r_b + \rho_{mj} - r_a - \rho_{ki}|}. \tag{8.91}$$

The bilinear approximation to the mutual potential energy function is given by

$$U_{km} \approx \widetilde{U}_{km} = \frac{1}{576}J_k^{-1}J_m^{-1}\sum_{i=1}^{4}\sum_{j=1}^{4}U_{ij}^{km}. \tag{8.92}$$

The mutual gravitational attraction force is derived by the same approach. Following Newton's third law of motion, the net gravitational force acting at the centroid of the first body is defined as f_{km} and the net force acting on the second body is $-f_{km}$. The function $h(\rho_a, \rho_b)$ is chosen as

$$h(\rho_a, \rho_b) = G\sigma_a(\rho_a)\sigma_b(\rho_b)\frac{\boldsymbol{r}_b + \boldsymbol{\rho}_b - \boldsymbol{r}_a - \boldsymbol{\rho}_a}{|r_b + \rho_b - r_a - \rho_a|^3}, \tag{8.93}$$

The corresponding interpolation of $h(\rho_a, \rho_b)$ is given by

$$\widetilde{h}(\rho_a, \rho_b) = \sum_{i=1}^{4}\sum_{j=1}^{4}\zeta_{ki}\zeta_{mj}f_{ij}^{km} = G\sum_{i=1}^{4}\sum_{j=1}^{4}\zeta_{ki}\zeta_{mj}\sigma_a(\rho_{ki})\sigma_b(\rho_{mj})\frac{\boldsymbol{r}_b + \boldsymbol{\rho}_{mj} - \boldsymbol{r}_a - \boldsymbol{\rho}_{ki}}{|r_b + \rho_{mj} - r_a - \rho_{ki}|^3}. \tag{8.94}$$

The bilinear approximation to the mutual gravitational force is given by

$$f_{km} \approx \widetilde{f}_{km} = \frac{1}{576}J_k^{-1}J_m^{-1}\sum_{i=1}^{4}\sum_{j=1}^{4}f_{ij}^{km}. \tag{8.95}$$

Similarly the gravitational torque on the first body due to the second is found by replacing $h(\rho_a, \rho_b)$ by,

$$h(\rho_a, \rho_b) = G\sigma_a(\rho_a)\sigma_b(\rho_b)\frac{\boldsymbol{\rho}_a \times (\boldsymbol{r}_b + \boldsymbol{\rho}_b - \boldsymbol{r}_a - \boldsymbol{\rho}_a)}{|r_b + \rho_b - r_a - \rho_a|^3}. \tag{8.96}$$

The corresponding approximation to the torque may then be found. Similarly the gravitational torque on the second body due to the first may also be found by the same approach.

8.2.9 The MASCON Model

The concept of modelling the gravitational potential of asteroids with a high number of concentrated point masses or MASCONS has been investigated and even used for several planned and other missions to asteroids. Collections of small spheres have also been used for the simulation of the so-called rubble pile asteroids letting the spheres free to move and relocate under their mutual gravity. Although it has been asserted that the use of a high number of point masses doesn't give a solution of the same quality as the same number of polyhedra, consequently a low number of masses have been used to model the gravitational potential due to the non-uniform mass distribution of the asteroid. The filling the complete volume of the asteroid with many tiny spheres is not really essential and need not be more efficient than with the pyramids of the polyhedra used for modelling. A large number of uniformly distributed point masses could be used to efficiently model the gravitational potential due to the non-uniform mass distribution of the asteroid by matching all the observed gravitational potential features of the asteroid, in question. The general form of the potential of a set of MASCONS is described by a summation of potentials due to each point mass in terms of the universal of gravitation constant, the values of the masses of the concentrated point masses and the location of these point masses relative to the origin of the rotating body-centred body-fixed frame. The expression is given by

$$U = -G\sum_{k=1}^{N} \frac{1}{|\mathbf{r}-\mathbf{s}_k|} m_k. \tag{8.97}$$

The gradient is obtained from

$$\nabla \widetilde{U}(\mathbf{r}) = \frac{d\widetilde{U}(\mathbf{r})}{dx}\widehat{i} + \frac{d\widetilde{U}(\mathbf{r})}{dy}\widehat{j} + \frac{d\widetilde{U}(\mathbf{r})}{dz}\widehat{k}, \tag{8.98}$$

with

$$\frac{d\widetilde{U}(\mathbf{r})}{dx} = G\sum_{k=1}^{N} \frac{x-x_k}{|\mathbf{r}-\mathbf{s}_k|^3} m_k,\ \frac{d\widetilde{U}(\mathbf{r})}{dy} = G\sum_{k=1}^{N} \frac{y-y_k}{|\mathbf{r}-\mathbf{s}_k|^3} m_k,\ \frac{d\widetilde{U}(\mathbf{r})}{dz} = G\sum_{k=1}^{N} \frac{z-z_k}{|\mathbf{r}-\mathbf{s}_k|^3} m_k. \tag{8.99}$$

One could add to this a sphere, a cylinder, an ellipsoid, a dumbbell or any other predefined body with a known mass and a gravitational potential. With a minimum of four masses, it is always possible to match the total mass and the three principal moments of inertia. With more masses and the locations chosen to match the properties of the asteroid, it is possible not only to match the total mass but also all the inertia components (6) of the body.

Rather than use concentrated masses an alternate strategy is to use, circular discs distributed along a predefined arbitrary line or axis. Such a discrete circular disc (DCD) model is expected to provide an extremely representative model for most asteroids, particular those which rotate about a central axis. The gravitational potential due to a disc of width Δz_1 and radius $R(z_1)$ located at a point, $z = z_1$ along the z-axis with its plane normal to it, induces a gravitational potential given by

$$\begin{aligned} U(x,y,z) &= 2G\rho\Delta z_1\left(-pE(k) - \frac{R^2-r^2}{p}K(k) + |z-z_1|\left(\frac{\pi}{2}+\frac{\pi}{2}\,\mathrm{sgn}(R-r)\right)\right) \\ &\quad - 2G\rho\Delta z_1(z-z_1)\ \mathrm{sgn}(R-r)(E(k)F(\phi,k') + K(k)E(\phi,k') - K(k)F(\phi,k')), \end{aligned} \tag{8.100}$$

with

$$\begin{aligned} p^2 &= (R+r)^2 + (z-z_1)^2, q^2 = (R-r)^2 + (z-z_1)^2, k \\ &= \sqrt{1-q^2/p^2}, k' = \sqrt{1-k^2}, \phi = \sin^{-1}((z-z_1)/q), r^2 = x^2+y^2. \end{aligned} \tag{8.101}$$

Thus with a N discs of width Δz_1 and radius $R(z_1)$ arrange sequentially, one has

$$\begin{aligned} &U(x,y,z) \\ &= 2G\sum_{i=1}^{N}\rho_i\Delta z_i\left(-p_iE(k_i) - \frac{R(z_i)^2-r^2}{p_i}K(k_i) + |z-z_i|\left(\frac{\pi}{2}+\frac{\pi}{2}\,\mathrm{sgn}(R(z_i)-r)\right)\right) \\ &\qquad - 2G\sum_{i=1}^{N}\rho_i\Delta z_i(z-z_i)\ \mathrm{sgn}(R(z_i)-r) \\ &\qquad \left(E(k_i)F\left(\phi_i,k_i'\right) + K(k_i)E\left(\phi_i,k_i'\right) - K(k_i)F\left(\phi_i,k_i'\right)\right), \end{aligned} \tag{8.102}$$

with

$$\begin{aligned} p_i^2 &= (R(z_i)+r)^2 + (z-z_i)^2, q_i^2 = (R(z_i)-r)^2 + (z-z_i)^2, k_i \\ &= \sqrt{1-q_i^2/p_i^2}, k_i' = \sqrt{1-k_i^2}\,\phi_i = \sin^{-1}((z-z_i)/q_i), r^2 = x^2+y^2. \end{aligned} \tag{8.103}$$

8.2.10 Other Models

One can conceive of several other models but those described in the earlier sections are generally adequate for modelling any asteroid or finite celestial body with a well-defined geometry, within the planetary system. There are many situations where the geometry may not be well defined or may have to be estimated as well, using pictures of views of the bodies in question, captured by a camera using methods as suggested by, for example, Passalis et al. (2007). It then reduces to the problem of the determination of the shape of the three dimensional objects from multiple camera images, which has been studied by several researchers.

Another important issue is the transformations from one model to the other, which is in general quite feasible.

8.3 Asteroid Equations of Motion

The motion of an asteroid around the Sun or the motion of a satellite around an asteroid should be treated locally in the context of the two-body theory. A typical two-body situation is illustrated in Fig. 8.6. On the other hand, considering that the asteroid itself is orbiting the Sun, as most asteroids are in heliocentric orbits, one must essentially also consider the three-body problem with a satellite in close proximity of an asteroid, which is itself orbiting the Sun. The typical three-body problem is illustrated in Fig. 8.7. M_1 is the mass of the central body, the Sun, and M_2 the mass of the asteroid.

8.3.1 Two-Body Problem: Equations of Motion

Let M_1 be the mass of the central body, the Sun, and M_2 be the mass of the asteroid. The asteroid is modelled as a tri-axial ellipsoid with three different major axes. Furthermore the distance between the Sun and the asteroid is chosen as the distance unit. If the mass ratio is defined as $\mu_1 = M_1/(M_1 + M_2)$, then relative to their centre of mass, each of the masses are located at

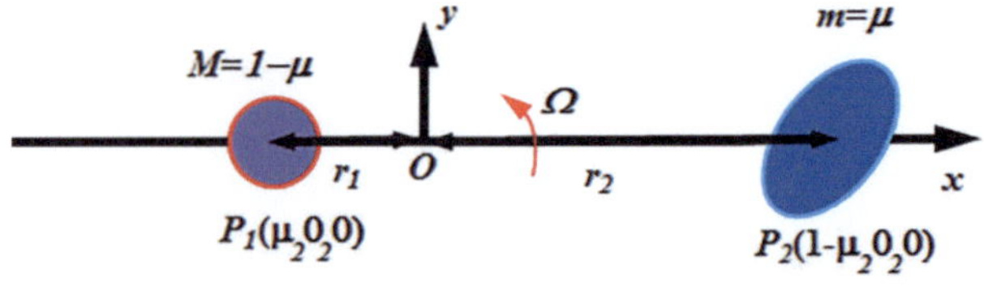

Fig. 8.6 Geometry of the full two-body problem for an ellipsoid-sphere system (one body is a sphere, depicting an asteroid in motion around the sun).

Fig. 8.7 The restricted full three-body problem

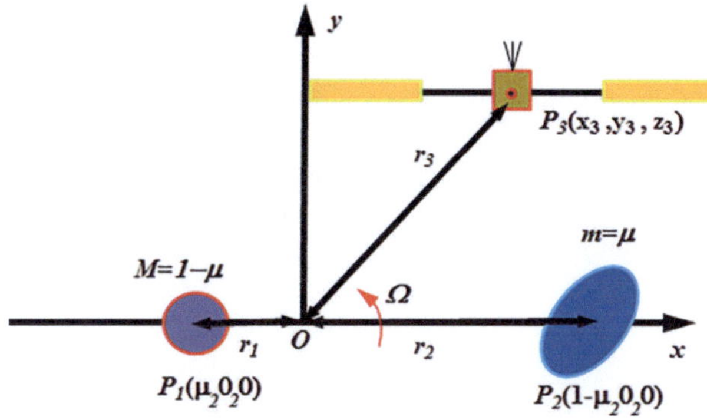

$$\mathbf{r}_1 = (1-\mu_1)\mathbf{r}_b, \mathbf{r}_2 = -\mu_1\mathbf{r}_b \text{ and } \mathbf{r}_b = \mathbf{r}_1 - \mathbf{r}_2. \tag{8.104}$$

The gravitational potential is further approximated as

$$U = \frac{\mu}{r}\left\{1 + \left(\frac{r_0}{r}\right)^2\left(C_{20}\left(1 + \frac{3}{2}\cos^2\delta\right) + 3C_{22}\cos^2\delta\cos(2\lambda)\right)\right\}, \tag{8.105}$$

with

$$C_{20} = -J_2 = \frac{1}{5r_0^2}\left(\gamma^2 - \frac{\alpha^2+\beta^2}{2}\right), C_{22} = \frac{1}{5r_0^2}\left(\frac{\alpha^2-\beta^2}{4}\right), \tag{8.106}$$

where μ is the gravitational parameter; λ and δ are the longitude and latitude angles, respectively; and r_0 is an arbitrary reference radius chosen to be equal to the largest semi-major axis of the ellipsoid.

In the body-fixed frame, the relative dynamics equation is defined by

$$\ddot{\mathbf{r}}_b + 2\boldsymbol{\Omega}\times\dot{\mathbf{r}}_b + \dot{\boldsymbol{\Omega}}\times\mathbf{r}_b + \boldsymbol{\Omega}\times(\boldsymbol{\Omega}\times\mathbf{r}_b) = G(M_1+M_2)\partial\widetilde{U}/\partial\mathbf{r}_b, \tag{8.107}$$

and the rotational dynamics of the ellipsoid is described by

$$\widetilde{\mathbf{I}}\cdot\dot{\boldsymbol{\Omega}} + \boldsymbol{\Omega}\times\widetilde{\mathbf{I}}\cdot\boldsymbol{\Omega} = -GM_1\mathbf{r}_b\times\partial\widetilde{U}/\partial\mathbf{r}_b, \tag{8.108}$$

where $\boldsymbol{\Omega}$ is the angular velocity of the ellipsoid and its moment of inertia matrix is given by $\widetilde{\mathbf{I}}$. To simplify the dynamics and its computation, one may normalize all distances by the maximum dimension of the ellipsoid and the mean orbital motion of the two-body system at the radius equal to the normalising distance. So, let $\mathbf{r} = \mathbf{r}_b/\alpha$ and $n^2 = G(M_1 + M_2)/\alpha^3$. Hence, the normalized angular velocity of the ellipsoid is $\boldsymbol{\omega} = \boldsymbol{\Omega}/n$. In the body-fixed frame, with normalized coordinates, the relative dynamics is defined by

$$\ddot{\mathbf{r}} + 2\boldsymbol{\omega} \times \dot{\mathbf{r}} + \dot{\boldsymbol{\omega}} \times \mathbf{r} + \boldsymbol{\omega} \times (\boldsymbol{\omega} \times \mathbf{r}) = \partial U / \partial \mathbf{r}, \tag{8.109}$$

and the rotational dynamics of the ellipsoid is described by

$$\mathbf{I} \cdot \dot{\boldsymbol{\omega}} + \boldsymbol{\omega} \times \mathbf{I} \cdot \boldsymbol{\omega} = -\mu_1 \mathbf{r} \times \partial U / \partial \mathbf{r}_b, \tag{8.110}$$

which may be written explicitly as

$$\begin{bmatrix} I_{xx} & -I_{xy} & -I_{xz} \\ -I_{xy} & I_{yy} & -I_{yz} \\ -I_{xz} & -I_{yz} & I_{zz} \end{bmatrix} \begin{bmatrix} \dot{\omega}_x \\ \dot{\omega}_y \\ \dot{\omega}_z \end{bmatrix} + \begin{bmatrix} 0 - \omega_z & & \omega_y \\ \omega_z & 0 & -\omega_x \\ -\omega_y & \omega_x & 0 \end{bmatrix}$$
$$\times \begin{bmatrix} I_{xx} & -I_{xy} & -I_{xz} \\ -I_{xy} & I_{yy} & -I_{yz} \\ -I_{xz} & -I_{yz} & I_{zz} \end{bmatrix} \begin{bmatrix} \omega_x \\ \omega_y \\ \omega_z \end{bmatrix} + \mu_1 \mathbf{r} \times \frac{\partial U}{\partial \mathbf{r}_b} = \mathbf{0}. \tag{8.111}$$

The corresponding gravitational potential may be expressed as

$$U = \frac{3}{4} \int_{\lambda}^{\infty} \left(1 - \frac{\overline{x}^2}{1+v} - \frac{\overline{y}^2}{\beta^2+v} - \frac{\overline{z}^2}{\gamma^2+v}\right) \frac{dv}{\Delta(v)}, \Delta(v)$$
$$= \sqrt{(1+v)(\beta^2+v)(\gamma^2+v)}, 0 < \gamma \leq \beta \leq \alpha, \tag{8.112}$$

where $0 < \gamma \leq \beta \leq 1$, while λ satisfies

$$1 - \frac{\overline{x}^2}{1+\lambda} - \frac{\overline{y}^2}{\beta^2+\lambda} - \frac{\overline{z}^2}{\gamma^2+\lambda} = 0. \tag{8.113}$$

It is assumed that the asteroid rotates around its axis of the largest moment of inertia and that the z-axis of the body-fixed frame points to the axis along which the moment of inertia of the asteroid is the largest. The normalized principal moments of inertia are

$$I_{xx} = \frac{1}{5}\left(\beta^2 + \gamma^2\right), I_{yy} = \frac{1}{5}\left(1 + \gamma^2\right), I_{zz} = \frac{1}{5}\left(1 + \beta^2\right). \tag{8.114}$$

The scalar equations on motion are

$$\ddot{\overline{x}} - 2\omega_z \dot{\overline{y}} - \dot{\omega}_z \overline{y} - \omega_z^2 \overline{x} = \frac{\partial U}{\partial \overline{x}}, \ddot{\overline{y}} + 2\omega_z \dot{\overline{x}} + \dot{\omega}_z \overline{x} - \omega_z^2 \overline{y} = \frac{\partial U}{\partial \overline{y}}, \ddot{\overline{z}}$$
$$= \frac{\partial U}{\partial \overline{z}}, \dot{\omega}_z = -\frac{\mu_1}{I_{zz}} \left(\overline{x} \frac{\partial U}{\partial \overline{y}} - \overline{y} \frac{\partial U}{\partial \overline{x}}\right). \tag{8.115}$$

8.3.2 Planar Equations of Motion

The position of the sphere relative to the ellipsoid in the plane is denoted as $\mathbf{q} = \mathbf{r}$, and the inertial velocity is

$$\mathbf{p} = \dot{\mathbf{r}} + \boldsymbol{\omega} \times \mathbf{r}. \tag{8.116}$$

The kinetic energy and the total energy are respectively expressed as

$$T = \frac{1}{2}\mathbf{p} \cdot \mathbf{p} + \frac{1}{2\mu_1} I_{zz}\omega_z^2, E = \frac{1}{2}\mathbf{p} \cdot \mathbf{p} + \frac{1}{2\mu_1} I_{zz}\omega_z^2 - U(\mathbf{q}). \tag{8.117}$$

The momentum integral is

$$K = (1/\mu_1)I_{zz}\omega_z + \overline{z} \cdot (\mathbf{q} \times \mathbf{p}). \tag{8.118}$$

Hence one can obtain the rotation rate as

$$\omega_z = \mu_1(K - \overline{z} \cdot (\mathbf{q} \times \mathbf{p}))/I_{zz}. \tag{8.119}$$

Thus the energy integral or the Hamiltonian is

$$E = H(\mathbf{q}, \mathbf{p}) = (1/2)\mathbf{p} \cdot \mathbf{p} + (\mu_1/2I_{zz})(K - \overline{z} \cdot (\mathbf{q} \times \mathbf{p}))^2 - U(\mathbf{q}). \tag{8.120}$$

The Hamiltonian equations of motion are

$$\dot{\mathbf{q}} = \partial H(\mathbf{q}, \mathbf{p})/\partial \mathbf{p}, \dot{\mathbf{p}} = -\partial H(\mathbf{q}, \mathbf{p})/\partial \mathbf{q}. \tag{8.121}$$

In component form, they are

$$\begin{aligned}
\dot{q}_x &= \dot{\overline{x}} = \frac{\partial H(\mathbf{q}, \mathbf{p})}{\partial p_x} = p_x + \frac{\mu_1 \overline{y}}{I_{zz}}(K - \overline{x}p_y + \overline{y}p_x), \\
\dot{q}_y &= \dot{\overline{y}} = \frac{\partial H(\mathbf{q}, \mathbf{p})}{\partial p_y} = p_y - \frac{\mu_1 \overline{x}}{I_{zz}}(K - \overline{x}p_y + \overline{y}p_x), \\
\dot{p}_x &= -\frac{\partial H(\mathbf{q}, \mathbf{p})}{\partial q_x} = \frac{\mu_1 p_y}{I_{zz}}(K - \overline{x}p_y + \overline{y}p_x) + \frac{\partial U}{\partial \overline{x}}, \\
\dot{p}_y &= -\frac{\partial H(\mathbf{q}, \mathbf{p})}{\partial q_y} = -\frac{\mu_1 p_x}{I_{zz}}(K - \overline{x}p_y + \overline{y}p_x) + \frac{\partial U}{\partial \overline{y}}.
\end{aligned} \tag{8.122}$$

Given that the planar potential can be expressed as

$$U = \frac{3}{4} \int_{\lambda}^{\infty} \left(1 - \frac{\overline{x}^2}{1+v} - \frac{\overline{y}^2}{\beta^2+v}\right) \frac{dv}{\Delta(v)},$$

$$\frac{\partial U}{\partial \overline{x}} = \frac{3}{4} \int_{\lambda}^{\infty} \frac{\partial}{\partial \overline{x}} \left(1 - \frac{\overline{x}^2}{1+v} - \frac{\overline{y}^2}{\beta^2+v}\right) \frac{dv}{\Delta(v)} = -\frac{3\overline{x}}{2} \int_{\lambda}^{\infty} \left(\frac{1}{1+v}\right) \frac{dv}{\Delta(v)}$$

$$\frac{\partial U}{\partial \overline{y}} = \frac{3}{4} \int_{\lambda}^{\infty} \frac{\partial}{\partial \overline{y}} \left(1 - \frac{\overline{x}^2}{1+v} - \frac{\overline{y}^2}{\beta^2+v}\right) \frac{dv}{\Delta(v)} = -\frac{3\overline{y}}{2} \int_{\lambda}^{\infty} \left(\frac{1}{\beta^2+v}\right) \frac{dv}{\Delta(v)}. \quad (8.123)$$

Furthermore

$$\frac{\partial U}{\partial \overline{z}} = -\frac{3\overline{z}}{2} \int_{\lambda}^{\infty} \left(\frac{1}{\gamma^2+v}\right) \frac{dv}{\Delta(v)}. \quad (8.124)$$

A simple change in the variables leads to the integrals

$$\frac{\partial U}{\partial \overline{x}} = -\frac{3\overline{x}}{2} \int_{0}^{\infty} \left(\frac{1}{1+v+\lambda}\right) \frac{dv}{\Delta(v+\lambda)} \equiv -\overline{x} R_{j\alpha},$$

$$\frac{\partial U}{\partial \overline{y}} = -\frac{3\overline{y}}{2} \int_{0}^{\infty} \left(\frac{1}{\beta^2+v+\lambda}\right) \frac{dv}{\Delta(v+\lambda)} \equiv -\overline{y} R_{j\beta},$$

$$\frac{\partial U}{\partial \overline{z}} = -\frac{3\overline{z}}{2} \int_{0}^{\infty} \left(\frac{1}{\gamma^2+v+\lambda}\right) \frac{dv}{\Delta(v+\lambda)} \equiv -\overline{z} R_{j\gamma}. \quad (8.125)$$

The potential can be expressed as

$$U = \frac{3}{4} \int_{0}^{\infty} \left(1 - \frac{\overline{x}^2}{1+v+\lambda} - \frac{\overline{y}^2}{\beta^2+v+\lambda} - \frac{\overline{z}^2}{\gamma^2+v+\lambda}\right) \frac{dv}{\Delta(v+\lambda)}$$

$$\equiv \frac{1}{2} \left(3R_{j0} - \overline{x}^2 R_{j\alpha} - \overline{y}^2 R_{j\beta} - \overline{z}^2 R_{j\gamma}\right), \quad (8.126)$$

with

$$R_{j0} = \frac{1}{2} \int_{0}^{\infty} \frac{dv}{\Delta(v+\lambda)}. \quad (8.127)$$

The energy integral is

$$E = \frac{1}{2}\mathbf{p} \cdot \mathbf{p} + \frac{1}{2\mu_1} I_{zz}\omega_z^2 - \frac{1}{2}\left(3R_{j0} - \overline{x}^2 R_{j\alpha} - \overline{y}^2 R_{j\beta} - \overline{z}^2 R_{j\gamma}\right). \tag{8.128}$$

Under equilibrium conditions, one could obtain the rotation speed in terms of the centrifugal force component.

8.3.3 Cartesian Coordinates

Relative to a rotating body-fixed coordinate frame, the equations of motion of a particle can be expressed by

$$\begin{aligned}
&\ddot{\overline{x}} + 2\omega_y\dot{\overline{z}} - 2\omega_z\dot{\overline{y}} - \dot{\omega}_z\overline{y} + \dot{\omega}_y\overline{z} + \omega_y\omega_x\overline{y} + \omega_z\omega_x\overline{z} - \left(\omega_y^2 + \omega_z^2\right)\overline{x} = \partial U/\partial \overline{x},\\
&\ddot{\overline{y}} + 2\omega_z\dot{\overline{x}} - 2\omega_x\dot{\overline{z}} + \dot{\omega}_z\overline{x} - \dot{\omega}_x\overline{z} + \omega_y\omega_x\overline{x} + \omega_y\omega_z\overline{z} - \left(\omega_x^2 + \omega_z^2\right)\overline{y} = \partial U/\partial \overline{y},\\
&\ddot{\overline{z}} - 2\omega_y\dot{\overline{x}} + 2\omega_x\dot{\overline{y}} + \dot{\omega}_x\overline{y} - \dot{\omega}_y\overline{x} + \omega_y\omega_z\overline{y} + \omega_z\omega_x\overline{x} - \left(\omega_x^2 + \omega_y^2\right)\overline{z}\\
&\quad = \partial U/\partial \overline{z}.
\end{aligned} \tag{8.129}$$

When the rotation speeds are constants

$$\begin{aligned}
\ddot{\overline{x}} + 2\omega_y\dot{\overline{z}} - 2\omega_z\dot{\overline{y}} + \omega_y\omega_x\overline{y} + \omega_z\omega_x\overline{z} - \left(\omega_y^2 + \omega_z^2\right)\overline{x} &= \partial U/\partial \overline{x},\\
\ddot{\overline{y}} + 2\omega_z\dot{\overline{x}} - 2\omega_x\dot{\overline{z}} + \omega_y\omega_x\overline{x} + \omega_y\omega_z\overline{z} - \left(\omega_x^2 + \omega_z^2\right)\overline{y} &= \partial U/\partial \overline{y},\\
\ddot{\overline{z}} - 2\omega_y\dot{\overline{x}} + 2\omega_x\dot{\overline{y}} + \omega_y\omega_z\overline{y} + \omega_z\omega_x\overline{x} - \left(\omega_x^2 + \omega_y^2\right)\overline{z} &= \partial U/\partial \overline{z}.
\end{aligned} \tag{8.130}$$

Furthermore when $\omega_x = 0$

$$\begin{aligned}
&\ddot{\overline{x}} + 2\omega_y\dot{\overline{z}} - 2\omega_z\dot{\overline{y}} - \left(\omega_y^2 + \omega_z^2\right)\overline{x} = \partial U/\partial \overline{x},\ \ \ddot{\overline{y}} + 2\omega_z\dot{\overline{x}} + \omega_y\omega_z\overline{z} - \omega_z^2\overline{y}\\
&\quad = \partial U/\partial \overline{y},\ \ \ddot{\overline{z}} - 2\omega_y\dot{\overline{x}} + \omega_y\omega_z\overline{y} - \omega_y^2\overline{z} = \partial U/\partial \overline{z}.
\end{aligned} \tag{8.131}$$

Finally when both $\omega_x = \omega_y = 0$

$$\ddot{\overline{x}} - 2\omega_z\dot{\overline{y}} - \omega_z^2\overline{x} = \partial U/\partial \overline{x}, \ddot{\overline{y}} + 2\omega_z\dot{\overline{x}} - \omega_z^2\overline{y} = \partial U/\partial \overline{y}, \ddot{\overline{z}} = \partial U/\partial \overline{z}. \tag{8.132}$$

The Hamiltonian is

$$\mathbf{H} = (1/2)\dot{\mathbf{r}} \cdot \dot{\mathbf{r}} - (1/2)(\boldsymbol{\omega} \times \mathbf{r}) \cdot (\boldsymbol{\omega} \times \mathbf{r}) - U(\mathbf{r}). \tag{8.133}$$

The total mechanical energy is

$$\mathbf{E} = (1/2)(\dot{\mathbf{r}} + \boldsymbol{\omega} \times \mathbf{r}) \cdot (\dot{\mathbf{r}} + \boldsymbol{\omega} \times \mathbf{r}) - U(\mathbf{r}). \tag{8.134}$$

The zero velocity curves follow from

$$(1/2)(\boldsymbol{\omega} \times \mathbf{r}) \cdot (\boldsymbol{\omega} \times \mathbf{r}) + U(\mathbf{r}) = \text{constant}. \tag{8.135}$$

The normalized angular velocity of the orbiting body is $\boldsymbol{\omega} = \boldsymbol{\Omega}/n$. In the body-fixed frame, with normalized coordinates, the relative dynamics is defined by

$$\ddot{\mathbf{r}} + 2\boldsymbol{\omega} \times \dot{\mathbf{r}} + \dot{\boldsymbol{\omega}} \times \mathbf{r} + \boldsymbol{\omega} \times (\boldsymbol{\omega} \times \mathbf{r}) = \partial U/\partial \mathbf{r}. \tag{8.136}$$

The rotational dynamics of the body is described by

$$\mathbf{I} \cdot \dot{\boldsymbol{\omega}} + \boldsymbol{\omega} \times \mathbf{I} \cdot \boldsymbol{\omega} = -\mu_1 \mathbf{r} \times \partial U/\partial \mathbf{r}_b. \tag{8.137}$$

8.3.4 Conditions of Equilibrium

The conditions for the equilibrium state are

$$\boldsymbol{\omega} \times (\boldsymbol{\omega} \times \mathbf{r}) = \partial U/\partial \mathbf{r}, \boldsymbol{\omega} \times \mathbf{I} \cdot \boldsymbol{\omega} = -\mu_1 \mathbf{r} \times \partial U/\partial \mathbf{r}_b. \tag{8.138}$$

If one lets

$$\mathbf{g} = \boldsymbol{\omega} \times (\boldsymbol{\omega} \times \mathbf{r}) - \partial U/\partial \mathbf{r}, \mathbf{h} = \boldsymbol{\omega} \times \mathbf{I} \cdot \boldsymbol{\omega} + \mu_1 \mathbf{r} \times \partial U/\partial \mathbf{r}_b, \tag{8.139}$$

the conditions of equilibrium are

$$\mathbf{g} = \boldsymbol{\omega} \times (\boldsymbol{\omega} \times \mathbf{r}) - \partial U/\partial \mathbf{r} = 0, \mathbf{h} = \boldsymbol{\omega} \times \mathbf{I} \cdot \boldsymbol{\omega} + \mu_1 \mathbf{r} \times \partial U/\partial \mathbf{r}_b = 0. \tag{8.140}$$

Thus

$$\begin{gathered}
-\left(\omega_y^2 + \omega_z^2\right)\bar{x} - (\dot{\omega}_z - \omega_y\omega_x)\bar{y} + (\dot{\omega}_y + \omega_z\omega_x)\bar{z} - \partial U/\partial \bar{x} = 0, \\
(\dot{\omega}_z + \omega_y\omega_x)\bar{x} - (\omega_x^2 + \omega_z^2)\bar{y} - (\dot{\omega}_x - \omega_y\omega_z)\bar{z} - \partial U/\partial \bar{y} = 0, \\
(\omega_z\omega_x - \dot{\omega}_y)\bar{x} + (\dot{\omega}_x + \omega_y\omega_z)\bar{y} - \left(\omega_x^2 + \omega_y^2\right)\bar{z} - \partial U/\partial \bar{z} = 0.
\end{gathered} \tag{8.141}$$

In matrix form

$$\begin{bmatrix} -\left(\omega_y^2+\omega_z^2\right) & -\left(\dot{\omega}_z-\omega_y\omega_x\right) & \left(\dot{\omega}_y+\omega_z\omega_x\right) \\ \left(\dot{\omega}_z+\omega_y\omega_x\right) & -\left(\omega_x^2+\omega_z^2\right) & -\left(\dot{\omega}_x-\omega_y\omega_z\right) \\ -\left(\dot{\omega}_y-\omega_z\omega_x\right) & \left(\dot{\omega}_x+\omega_y\omega_z\right) & -\left(\omega_x^2+\omega_y^2\right) \end{bmatrix} \begin{bmatrix} \overline{x} \\ \overline{y} \\ \overline{z} \end{bmatrix} = \begin{bmatrix} \partial U/\partial \overline{x} \\ \partial U/\partial \overline{y} \\ \partial U/\partial \overline{z} \end{bmatrix}. \tag{8.142}$$

Alternately in the case of an ellipsoid

$$\begin{bmatrix} -\left(\omega_y^2+\omega_z^2\right) & -\left(\dot{\omega}_z-\omega_y\omega_x\right) & \left(\dot{\omega}_y+\omega_z\omega_x\right) \\ \left(\dot{\omega}_z+\omega_y\omega_x\right) & -\left(\omega_x^2+\omega_z^2\right) & -\left(\dot{\omega}_x-\omega_y\omega_z\right) \\ -\left(\dot{\omega}_y-\omega_z\omega_x\right) & \left(\dot{\omega}_x+\omega_y\omega_z\right) & -\left(\omega_x^2+\omega_y^2\right) \end{bmatrix} \begin{bmatrix} \overline{x} \\ \overline{y} \\ \overline{z} \end{bmatrix} = -\begin{bmatrix} \overline{x}R_{j\alpha} \\ \overline{y}R_{j\beta} \\ \overline{z}R_{j\gamma} \end{bmatrix}. \tag{8.143}$$

The above is solved for $\overline{x}$, $\overline{y}$, $\overline{z}$.
From

$$\boldsymbol{\omega}\times\mathbf{I}\cdot\boldsymbol{\omega}+\mu_1\mathbf{r}\times\partial U/\partial\mathbf{r}_b=\mathbf{0}, \tag{8.144}$$

if one lets

$$\begin{bmatrix} L & M & N \end{bmatrix}^T=\mu_1\mathbf{r}\times\partial U/\partial\mathbf{r}_b. \tag{8.145}$$

then the condition for rotational equilibrium is

$$\boldsymbol{\omega}\times\mathbf{I}\cdot\boldsymbol{\omega}+\mu_1\mathbf{r}\times\partial U/\partial\mathbf{r}_b=\boldsymbol{\omega}\times\mathbf{I}\cdot\boldsymbol{\omega}+\begin{bmatrix} L & M & N \end{bmatrix}^T=\mathbf{0}. \tag{8.146}$$

However, since

$$\mathbf{I}=\begin{bmatrix} I_{xx} & -I_{xy} & -I_{xz} \\ -I_{xy} & I_{yy} & -I_{yz} \\ -I_{xz} & -I_{yz} & I_{zz} \end{bmatrix}, \tag{8.147}$$

the rotational dynamics is given by

$$\begin{bmatrix} I_{xx} & -I_{xy} & -I_{xz} \\ -I_{xy} & I_{yy} & -I_{yz} \\ -I_{xz} & -I_{yz} & I_{zz} \end{bmatrix} \begin{bmatrix} \dot{\omega}_x \\ \dot{\omega}_y \\ \dot{\omega}_z \end{bmatrix} + \begin{bmatrix} 0 - \omega_z & & \omega_y \\ \omega_z & 0 & -\omega_x \\ -\omega_y & \omega_x & 0 \end{bmatrix}$$
$$\times \begin{bmatrix} I_{xx} & -I_{xy} & -I_{xz} \\ -I_{xy} & I_{yy} & -I_{yz} \\ -I_{xz} & -I_{yz} & I_{zz} \end{bmatrix} \begin{bmatrix} \omega_x \\ \omega_y \\ \omega_z \end{bmatrix} + \mu_1 \mathbf{r} \times \frac{\partial U}{\partial \mathbf{r}_b} = \mathbf{0}. \tag{8.148}$$

When $\dot{\omega}_x = 0$, $\dot{\omega}_y = 0$ and $\dot{\omega}_z = 0$, we obtain for rotational equilibrium

$$\begin{bmatrix} \left(I_{zz} - I_{yy}\right)\omega_y\omega_z + I_{xy}\omega_x\omega_z - I_{xz}\omega_x\omega_y + I_{yz}\left(\omega_z^2 - \omega_y^2\right) \\ (I_{xx} - I_{zz})\omega_x\omega_z + I_{yz}\omega_x\omega_y - I_{xy}\omega_y\omega_z + I_{xz}\left(\omega_x^2 - \omega_z^2\right) \\ \left(I_{yy} - I_{xx}\right)\omega_x\omega_y + I_{xy}\left(\omega_y^2 - \omega_x^2\right) - I_{yz}\omega_x\omega_z + I_{xz}\omega_y\omega_z \end{bmatrix} + \begin{bmatrix} L \\ M \\ N \end{bmatrix} = \mathbf{0}. \tag{8.149}$$

The equation for rotational equilibrium may be expressed as

$$\begin{bmatrix} \left(I_{zz} - I_{yy}\right) & I_{xy} & -I_{xz} \\ -I_{xy} & (I_{xx} - I_{zz}) & I_{yz} \\ I_{xz} & -I_{yz} & \left(I_{yy} - I_{xx}\right) \end{bmatrix} \begin{bmatrix} \omega_y\omega_z \\ \omega_x\omega_z \\ \omega_x\omega_y \end{bmatrix}$$
$$+ \begin{bmatrix} 0 & -I_{yz} & I_{yz} \\ I_{xz} & 0 & -I_{xz} \\ -I_{xy} & I_{xy} & 0 \end{bmatrix} \begin{bmatrix} \omega_x^2 \\ \omega_y^2 \\ \omega_z^2 \end{bmatrix} + \begin{bmatrix} L \\ M \\ N \end{bmatrix} = \mathbf{0}. \tag{8.150}$$

The above can be solved for ω_x, ω_y and ω_z.

8.3.5 Stability of Equilibrium

The equations of motion are

$$\ddot{\mathbf{r}} + 2\boldsymbol{\omega} \times \dot{\mathbf{r}} + \dot{\boldsymbol{\omega}} \times \mathbf{r} + \mathbf{g} = \mathbf{0}, \mathbf{I} \cdot \dot{\boldsymbol{\omega}} + \mathbf{h} = 0. \tag{8.151}$$

The equations motion may now be expanded about the equilibrium state. To this end, the following are defined, where the equilibrium state is defined with the subscript "*e*".

Thus let

$$\widetilde{x} = x - x_e, \widetilde{y} = y - y_e, \widetilde{z} = z - z_e, \alpha = \omega_x - \omega_{x_e}, \beta = \omega_y - \omega_{y_e}, \gamma = \omega_z - \omega_{z_e}. \tag{8.152}$$

Furthermore let

$$\mathbf{g}=\begin{bmatrix} g_1 \\ g_2 \\ g_3 \end{bmatrix}=\begin{bmatrix} g_{1x} & g_{1y} & g_{1z} \\ g_{2x} & g_{2y} & g_{2z} \\ g_{3x} & g_{3y} & g_{3z} \end{bmatrix}\begin{bmatrix} \widetilde{x} \\ \widetilde{y} \\ \widetilde{z} \end{bmatrix}, \mathbf{h}=\begin{bmatrix} h_1 \\ h_2 \\ h_3 \end{bmatrix}=\begin{bmatrix} h_{1\alpha} & h_{1\beta} & h_{1\gamma} \\ h_{2\alpha} & h_{2\beta} & h_{2\gamma} \\ h_{3\alpha} & h_{3\beta} & h_{3\gamma} \end{bmatrix}\begin{bmatrix} \alpha \\ \beta \\ \gamma \end{bmatrix},$$
$$\mathbf{I}^{-1}=[\mathbf{i}_1 \quad \mathbf{i}_2 \quad \mathbf{i}_3]=\begin{bmatrix} \widetilde{\mathbf{i}}_1 \\ \widetilde{\mathbf{i}}_2 \\ \widetilde{\mathbf{i}}_3 \end{bmatrix}. \tag{8.153}$$

Hence, it follows that the translational perturbation equations are

$$\begin{aligned} &\ddot{\widetilde{x}}-2\omega_z\dot{\widetilde{y}}-\dot{\omega}_z\widetilde{y}+g_{1x}\widetilde{x}+g_{1y}\widetilde{y}+g_{1z}\widetilde{z}=0, \\ &\ddot{\widetilde{y}}+2\omega_z\dot{\widetilde{x}}+\dot{\omega}_z\widetilde{x}+g_{2x}\widetilde{x}+g_{2y}\widetilde{y}+g_{2z}\widetilde{z}=0, \\ &\ddot{\widetilde{z}}+g_{3x}\widetilde{x}+g_{3y}\widetilde{y}+g_{3z}\widetilde{z}=0. \end{aligned} \tag{8.154}$$

The rotational perturbation equations are

$$\begin{aligned} \begin{bmatrix} \dot{\alpha} \\ \dot{\beta} \\ \dot{\gamma} \end{bmatrix} &= -[\mathbf{i}_1 \quad \mathbf{i}_2 \quad \mathbf{i}_3]\begin{bmatrix} h_{1\alpha} & h_{1\beta} & h_{1\gamma} \\ h_{2\alpha} & h_{2\beta} & h_{2\gamma} \\ h_{3\alpha} & h_{3\beta} & h_{3\gamma} \end{bmatrix}\begin{bmatrix} \alpha \\ \beta \\ \gamma \end{bmatrix} \\ &= -\begin{bmatrix} i_{11} & i_{12} & i_{13} \\ i_{21} & i_{22} & i_{23} \\ i_{31} & i_{32} & i_{33} \end{bmatrix}\begin{bmatrix} h_{1\alpha} & h_{1\beta} & h_{1\gamma} \\ h_{2\alpha} & h_{2\beta} & h_{2\gamma} \\ h_{3\alpha} & h_{3\beta} & h_{3\gamma} \end{bmatrix}\begin{bmatrix} \alpha \\ \beta \\ \gamma \end{bmatrix}. \end{aligned} \tag{8.155}$$

The rotational perturbation equations may be decomposed as

$$\begin{aligned} \begin{bmatrix} \dot{\alpha} \\ \dot{\beta} \\ \dot{\gamma} \end{bmatrix} &= -[\mathbf{i}_1 \quad \mathbf{i}_2 \quad \mathbf{i}_3]\left\{\begin{bmatrix} h_{1\alpha} \\ h_{2\alpha} \\ h_{3\alpha} \end{bmatrix}\alpha+\begin{bmatrix} h_{1\beta} \\ h_{2\beta} \\ h_{3\beta} \end{bmatrix}\beta+\begin{bmatrix} h_{1\gamma} \\ h_{2\gamma} \\ h_{3\gamma} \end{bmatrix}\gamma\right\} \\ &= -[\mathbf{i}_1 \quad \mathbf{i}_2 \quad \mathbf{i}_3]\begin{bmatrix} h_{1\alpha}\alpha+h_{1\beta}\beta+h_{1\gamma}\gamma \\ h_{2\alpha}\alpha+h_{2\beta}\beta+h_{2\gamma}\gamma \\ h_{3\alpha}\alpha+h_{3\beta}\beta+h_{3\gamma}\gamma \end{bmatrix}, \end{aligned} \tag{8.156}$$

or as

$$\begin{bmatrix} \dot{\alpha} \\ \dot{\beta} \\ \dot{\gamma} \end{bmatrix} = -[[\mathbf{i}_1 \quad \mathbf{i}_2 \quad \mathbf{i}_3] \cdot \mathbf{h}_\alpha \quad [\mathbf{i}_1 \quad \mathbf{i}_2 \quad \mathbf{i}_3] \cdot \mathbf{h}_\beta \quad [\mathbf{i}_1 \quad \mathbf{i}_2 \quad \mathbf{i}_3] \cdot \mathbf{h}_\gamma] \begin{bmatrix} \alpha \\ \beta \\ \gamma \end{bmatrix}. \tag{8.157}$$

Hence, the rotational perturbation equations may be expressed as

$$\begin{aligned} &\dot{\alpha} + \left[\tilde{\mathbf{i}}_1 \cdot \mathbf{h}_\alpha \quad \tilde{\mathbf{i}}_1 \cdot \mathbf{h}_\beta \quad \tilde{\mathbf{i}}_1 \cdot \mathbf{h}_\gamma\right][\alpha \quad \beta \quad \gamma]^T = 0, \\ &\dot{\beta} + \left[\tilde{\mathbf{i}}_2 \cdot \mathbf{h}_\alpha \quad \tilde{\mathbf{i}}_2 \cdot \mathbf{h}_\beta \quad \tilde{\mathbf{i}}_2 \cdot \mathbf{h}_\gamma\right][\alpha \quad \beta \quad \gamma]^T = 0, \\ &\dot{\gamma} + \left[\tilde{\mathbf{i}}_3 \cdot \mathbf{h}_\alpha \quad \tilde{\mathbf{i}}_3 \cdot \mathbf{h}_\beta \quad \tilde{\mathbf{i}}_3 \cdot \mathbf{h}_\gamma\right][\alpha \quad \beta \quad \gamma]^T = 0. \end{aligned} \tag{8.158}$$

In matrix form, the linearized equations about the equilibrium points are

$$\begin{bmatrix} \ddot{\tilde{x}} \\ \ddot{\tilde{y}} \\ \ddot{\tilde{z}} \end{bmatrix} + 2\omega_z \begin{bmatrix} 0 & -1 & 0 \\ 1 & 0 & 0 \\ 0 & 0 & 0 \end{bmatrix} \begin{bmatrix} \dot{\tilde{x}} \\ \dot{\tilde{y}} \\ \dot{\tilde{z}} \end{bmatrix} + \dot{\omega}_z \begin{bmatrix} 0 & -1 & 0 \\ 1 & 0 & 0 \\ 0 & 0 & 0 \end{bmatrix} \begin{bmatrix} \tilde{x} \\ \tilde{y} \\ \tilde{z} \end{bmatrix} + \begin{bmatrix} g_{1x} & g_{1y} & g_{1z} \\ g_{2x} & g_{2y} & g_{2z} \\ g_{3x} & g_{3y} & g_{3z} \end{bmatrix}$$

$$\times \begin{bmatrix} \tilde{x} \\ \tilde{y} \\ \tilde{z} \end{bmatrix} = \begin{bmatrix} 0 \\ 0 \\ 0 \end{bmatrix},$$

$$\begin{bmatrix} \dot{\alpha} \\ \dot{\beta} \\ \dot{\gamma} \end{bmatrix} + \begin{bmatrix} \tilde{\mathbf{i}}_1 \cdot \mathbf{h}_\alpha & \tilde{\mathbf{i}}_1 \cdot \mathbf{h}_\beta & \tilde{\mathbf{i}}_1 \cdot \mathbf{h}_\gamma \\ \tilde{\mathbf{i}}_2 \cdot \mathbf{h}_\alpha & \tilde{\mathbf{i}}_2 \cdot \mathbf{h}_\beta & \tilde{\mathbf{i}}_2 \cdot \mathbf{h}_\gamma \\ \tilde{\mathbf{i}}_3 \cdot \mathbf{h}_\alpha & \tilde{\mathbf{i}}_3 \cdot \mathbf{h}_\beta & \tilde{\mathbf{i}}_3 \cdot \mathbf{h}_\gamma \end{bmatrix} \begin{bmatrix} \alpha \\ \beta \\ \gamma \end{bmatrix} = \begin{bmatrix} 0 \\ 0 \\ 0 \end{bmatrix}. \tag{8.159}$$

Assuming the coefficients are constants, the corresponding characteristic polynomials are

$$\det\left| s^2\mathbf{I} + 2\omega_z s \begin{bmatrix} 0 & -1 & 0 \\ 1 & 0 & 0 \\ 0 & 0 & 0 \end{bmatrix} + \dot{\omega}_z \begin{bmatrix} 0 & -1 & 0 \\ 1 & 0 & 0 \\ 0 & 0 & 0 \end{bmatrix} + \begin{bmatrix} g_{1x} & g_{1y} & g_{1z} \\ g_{2x} & g_{2y} & g_{2z} \\ g_{3x} & g_{3y} & g_{3z} \end{bmatrix} \right| = 0, \tag{8.160}$$

and

$$\det\left|s\mathbf{I}+\begin{bmatrix}\widetilde{\mathbf{i}}_1\cdot\mathbf{h}_\alpha & \widetilde{\mathbf{i}}_1\cdot\mathbf{h}_\beta & \widetilde{\mathbf{i}}_1\cdot\mathbf{h}_\gamma\\ \widetilde{\mathbf{i}}_2\cdot\mathbf{h}_\alpha & \widetilde{\mathbf{i}}_2\cdot\mathbf{h}_\beta & \widetilde{\mathbf{i}}_2\cdot\mathbf{h}_\gamma\\ \widetilde{\mathbf{i}}_3\cdot\mathbf{h}_\alpha & \widetilde{\mathbf{i}}_3\cdot\mathbf{h}_\beta & \widetilde{\mathbf{i}}_3\cdot\mathbf{h}_\gamma\end{bmatrix}\right|=0. \tag{8.161}$$

The dynamic stability of the equilibrium solution can be ascertained by considering the roots of the characteristic polynomial which must not lie in the right half of the complex plane. Tumbling asteroids are not expected to be stable in rotation. A good example is 4179 Toutatis (Scheeres et al., 1998). Nazari et al. (2014) have designed a novel observer-based feedback controller so the satellite hovers over the tumbling asteroid 4179 Toutatis in the body-fixed frame. An extended Kalman filter was used to process range measurements from a set of ground stations, to estimate the spacecraft's state vector and the gravitational parameters of the asteroid assuming a second degree and order gravity field model. The estimated states are used in the optimal feedback control laws which were designed using two alternate methods: a time-dependent linear quadratic regulator and a linear quadratic regulator using a time-invariant system obtained by using the Liapunov-Floquet transformation of the system dynamics.

8.4 Asteroid Trajectories: Application Examples

Several researchers are considering the plausibility of being able to rapidly launch a mission to an asteroid, which would fly in close proximity of the asteroid either to grab a sample from the asteroid's surface or to deliver an impulse in a particular direction so as to deflect the asteroid from its current orbit. A selection of these are briefly discussed here. Goossens et al. (2021) considered the robust determination of the mass and shape of the asteroid (101955) Bennu while Scheeres et al. (2016) investigated the geophysical environment around the same asteroid. Fukushima (2017) considered the computation of the gravitational field of a general finite body and its application to the asteroid 433 EROS. Colagrossi et al. (2015) successfully applied the MASCON approach to study dynamical evolution of the trajectory about asteroids 433 EROS, 216 Kleopatra and 4179 Toutatis and also compared these results with those obtained by other methods. Yu and Baoyin (2012) studied the equilibrium points, the stability and topological cases of equilibrium points, the topological structure of zero velocity curves and other features of the asteroid 216 Kleopatra while Shepard et al. (2018) revised the shape model of the same asteroid. Tracking and following an asteroid, in close proximity, is the subject of the paper by Vepa and Shaheed (2021) where the problem of synthesizing an optimal trajectory to a NEO such as the asteroid 2010 TK7 is considered. Chai et al. (2019) have reviewed the optimization techniques that are most suitable for spacecraft flight trajectory design. Sánchez et al. (2018) have discussed the design of asteroid retrieval trajectories, the strategies for asteroid retrieval, the different dynamical models and the trajectory design methodologies. Morimoto et al. (2004) have discussed the preliminary design of two sample return missions to the asteroids and back, with mission times ranging from 7 to 14 years with an impulsive velocity increment ranging from 2.6 to 22.0.4 km/s. They estimated the total fuel required, the electric and chemical propulsion system masses, the structural mass, the power system and bus masses, the masses of the Earth return capsule and sampling mechanism and other payload masses with a margin of 5%. Luo and Zhou (2014) proposed a trajectory design method for asteroid rendezvous mission using multi-objective optimization techniques. Watanabe et al. (2017) have discussed the Hayabusa2 mission scheduled to land on the asteroid 162173 Ryugu, in some detail. Youtao et al. (2019) have presented trajectory design and optimization to the asteroid 2010TK7 as the main target, an Earth-2016QL44-Venus-Earth-2014UM192-2010TK7, including two flybys past two other asteroids 2016QL44 and 2014UM192, two gravity assist flybys past Venus and Earth and two trajectory manoeuvres, involving several legs.

8.5 Three-Body Problem: Vector Equations of Motion in Rotating Coordinates with a Point Mass Model

The three-body problem was illustrated in Fig. 8.7. The masses are assumed to be point masses. The dynamics of satellite's position relative to the centre of mass of an asteroid and the Sun are expressed in rotating coordinate system as

$$\ddot{\widetilde{\boldsymbol{\rho}}}+2\boldsymbol{\Omega}\times\dot{\widetilde{\boldsymbol{\rho}}}+\dot{\boldsymbol{\Omega}}\times\widetilde{\boldsymbol{\rho}}+\boldsymbol{\Omega}\times(\boldsymbol{\Omega}\times\widetilde{\boldsymbol{\rho}})=\partial\widetilde{U}_{12}/\partial\widetilde{\boldsymbol{\rho}}, \tag{8.162}$$

where

$$\widetilde{U}_{12}=\frac{GM_1}{|\widetilde{\boldsymbol{\rho}}-(1-\mu_1)\mathbf{r}_b|}+GM_2\widetilde{U}(\widetilde{\boldsymbol{\rho}}+\mu_1\mathbf{r}_b),\mu_1=M_1/(M_1+M_2). \tag{8.163}$$

To simplify the dynamics and its computation, one may normalize all distances by the maximum dimension of the ellipsoid and the mean orbital motion of the two-body system at the radius equal to the normalising distance. So, let $\boldsymbol{\rho}=\widetilde{\boldsymbol{\rho}}/\alpha$ and $n^2 = G(M_1 + M_2)/\alpha^3$. Hence, the normalized angular velocity of the ellipsoid is $\boldsymbol{\omega}=\boldsymbol{\Omega}/n$. In the body-fixed frame, with normalized coordinates, the relative dynamics is defined by

$$\ddot{\boldsymbol{\rho}}+2\boldsymbol{\omega}\times\dot{\boldsymbol{\rho}}+\dot{\boldsymbol{\omega}}\times\boldsymbol{\rho}+\boldsymbol{\omega}\times(\boldsymbol{\omega}\times\boldsymbol{\rho})=\partial U_{12}/\partial\boldsymbol{\rho}, \tag{8.164}$$

with

$$U_{12} = \frac{\mu_1}{|\boldsymbol{\rho} - (1-\mu_1)\mathbf{r}|} + (1-\mu_1)U(\boldsymbol{\rho} + \mu_1 \mathbf{r}). \tag{8.165}$$

If $\boldsymbol{\rho} = x\hat{i} + y\hat{j} + z\hat{k}$, the scalar equations on motion are

$$\begin{aligned}
\ddot{\bar{x}} - 2\omega_z\dot{\bar{y}} - \dot{\omega}_z\bar{y} - \omega_z^2\bar{x} &= -\mu_1(\bar{x} - (1-\mu_1)|\mathbf{r}|)/r_1^3 - (1-\mu_1)(\bar{x} + \mu_1|\mathbf{r}|)R_{j\alpha},\\
\ddot{\bar{y}} + 2\omega_z\dot{\bar{x}} + \dot{\omega}_z\bar{x} - \omega_z^2\bar{y} &= -\mu_1\left(\bar{y}/r_1^3\right) - (1-\mu_1)\bar{y}R_{j\beta},\\
\ddot{\bar{z}} &= -\mu_1\left(\bar{z}/r_1^3\right) - (1-\mu_1)\bar{z}R_{j\gamma},
\end{aligned} \tag{8.166}$$

with $r_1^2 = (\bar{x} - (1-\mu_1)|\mathbf{r}|)^2 + \bar{y}^2 + \bar{z}^2$.

The Jacobi integral is given by

$$J = (1/2)\left(\dot{\bar{x}}^2 + \dot{\bar{y}}^2 + \dot{\bar{z}}^2\right) + \left(\omega_z^2/2\right)\left(\bar{x}^2 + \bar{y}^2\right) - U = C. \tag{8.167}$$

The equilibrium points satisfy the relations

$$\begin{aligned}
&\dot{\omega}_z\bar{y} + \omega_z^2\bar{x} = \mu_1(\bar{x} - (1-\mu_1)|\mathbf{r}|)/r_1^3 + (1-\mu_1)(\bar{x} + \mu_1|\mathbf{r}|)R_{j\alpha}, \dot{\omega}_z\bar{x}\\
&\quad - \omega_z^2\bar{y} = -\mu_1\left(\bar{y}/r_1^3\right) - (1-\mu_1)\bar{y}R_{j\beta}, \mu_1\left(\bar{z}/r_1^3\right) + (1-\mu_1)\bar{z}R_{j\gamma} = 0.
\end{aligned} \tag{8.168}$$

When ω_z is known, the $\bar{x}$, $\bar{y}$ and $\bar{z}$ Cartesian coordinates based model, defined by Eq. (8.166), can be used as a basis for identifying the dynamics of the asteroid with the further assumption that $R_{j\alpha}$, $R_{j\beta}$ and $R_{j\gamma}$ are slowly varying and with the dynamics driven by noise. Thus

$$\dot{R}_{j\alpha} = w_1, \dot{R}_{j\beta} = w_2, \dot{R}_{j\gamma} = w_3. \tag{8.169}$$

In the above, w_1, w_2 and w_3 are assumed to be process noise inputs, each with a predefined covariance.

8.5.1 Three-Body Problem: Derivation of the Equations of Motion with a Finite Second Body

The three-body problem was illustrated in Fig. 8.7. Let the position vectors of the three bodies in an inertial frame be, $\mathbf{r}_1$, $\mathbf{r}_2$ and $\mathbf{r}_3$. Only the bodies 1 and 3 are assumed to be either spherical or point masses while body 2 is assumed to be arbitrarily shaped. Define

$$\mathbf{r}_{cm} = \frac{1}{m_1 + m_2 + m_3} \sum_{i=1}^{3} m_i \mathbf{r}_i, \mathbf{r}_{cm_{12}} = \frac{m_1 \mathbf{r}_1 + m_2 \mathbf{r}_2}{m_1 + m_2}, \mathbf{r} = \mathbf{r}_2 - \mathbf{r}_1, \widetilde{\boldsymbol{\rho}} = \mathbf{r}_3 - \mathbf{r}_{cm_{12}}. \tag{8.170}$$

Thus

$$\mathbf{r}_3 - \mathbf{r}_1 = \mathbf{r}_3 - \mathbf{r}_{cm_{12}} + \mathbf{r}_{cm_{12}} - \mathbf{r}_2 + \mathbf{r}_2 - \mathbf{r}_1 = \widetilde{\boldsymbol{\rho}} + \mathbf{r}_{cm_{12}} - \mathbf{r}_2 + \mathbf{r}, \tag{8.171}$$

which reduces to

$$\mathbf{r}_3 - \mathbf{r}_1 = \widetilde{\boldsymbol{\rho}} + \frac{m_1 \mathbf{r}_1 + m_2 \mathbf{r}_2}{m_1 + m_2} - \mathbf{r}_2 + \mathbf{r} = \widetilde{\boldsymbol{\rho}} + \frac{m_2 \mathbf{r}}{m_1 + m_2}, \tag{8.172}$$

and

$$\mathbf{r}_3 - \mathbf{r}_2 = \widetilde{\boldsymbol{\rho}} + \frac{m_2 \mathbf{r}}{m_1 + m_2} + \mathbf{r}_1 - \mathbf{r}_2 = \widetilde{\boldsymbol{\rho}} + \frac{m_2 \mathbf{r}}{m_1 + m_2} - \mathbf{r} = \widetilde{\boldsymbol{\rho}} - \frac{m_1 \mathbf{r}}{m_1 + m_2}. \tag{8.173}$$

For two bodies, in terms of an external force potential, V

$$\mathbf{F} = -\frac{G m_1 m_2}{|\mathbf{r}|^2} \left(\frac{\mathbf{r}}{|\mathbf{r}|} \right), m_2 \ddot{\mathbf{r}} = \mathbf{F} = m_2 \frac{\partial V}{\partial \mathbf{r}}, V = \frac{G m_1}{\mathbf{r}}. \tag{8.174}$$

Let $V_i(\mathbf{r}) \equiv Gm_i/|\mathbf{r}|$. Hence, it follows that

$$V_1(\mathbf{r}_{21}) = \frac{G m_1}{|\mathbf{r}_{21}|}, V_2(\mathbf{r}_{12}) = \frac{G m_2}{|\mathbf{r}_{12}|} + R(\mathbf{r}_{12}), V_3(\mathbf{r}_{23}) = \frac{G m_3}{|\mathbf{r}_{23}|}. \tag{8.175}$$

The derivatives of the potential functions defined in Eq. (8.175) are

$$\begin{aligned} \frac{\partial V_1(\mathbf{r}_{21})}{\partial \mathbf{r}_{21}} &= \frac{\partial V_1(\mathbf{r}_{21})}{\partial |\mathbf{r}_{21}|} \frac{\partial |\mathbf{r}_{21}|}{\partial \mathbf{r}_{21}} = \frac{\partial V_1(\mathbf{r}_{21})}{\partial |\mathbf{r}_{21}|} \frac{\mathbf{r}_{21}}{|\mathbf{r}_{21}|} \\ &= -\frac{G m_1 \mathbf{r}_{21}}{|\mathbf{r}_{21}|^3}, \frac{\partial V_1(\mathbf{r}_{31})}{\partial \mathbf{r}_{31}} = -\frac{G m_1 \mathbf{r}_{31}}{|\mathbf{r}_{31}|^3}. \end{aligned} \tag{8.176}$$

Similarly

$$\begin{aligned} \frac{\partial V_2(\mathbf{r}_{12})}{\partial \mathbf{r}_{12}} &= -\frac{G m_2 \mathbf{r}_{12}}{|\mathbf{r}_{12}|^3} + \frac{\partial R(\mathbf{r}_{12})}{\partial \mathbf{r}_{12}}, \frac{\partial V_3(\mathbf{r}_{23})}{\partial \mathbf{r}_{23}} \\ &= -\frac{G m_3 \mathbf{r}_{23}}{|\mathbf{r}_{23}|^3}, \frac{\partial V_2(\mathbf{r}_{32})}{\partial \mathbf{r}_{32}} = -\frac{G m_2 \mathbf{r}_{32}}{|\mathbf{r}_{32}|^3} + \frac{\partial R(\mathbf{r}_{32})}{\partial \mathbf{r}_{32}}. \end{aligned} \tag{8.177}$$

The position vectors satisfy the equations

$$\ddot{\mathbf{r}}_1 = \frac{\partial V_2(\mathbf{r}_{12})}{\partial \mathbf{r}_{12}} + \frac{\partial V_3(\mathbf{r}_{13})}{\partial \mathbf{r}_{13}}, \ddot{\mathbf{r}}_2 = \frac{\partial V_1(\mathbf{r}_{21})}{\partial \mathbf{r}_{21}} + \frac{\partial V_3(\mathbf{r}_{23})}{\partial \mathbf{r}_{23}},$$
$$\ddot{\mathbf{r}}_3 = \frac{\partial V_2(\mathbf{r}_{32})}{\partial \mathbf{r}_{32}} + \frac{\partial V_1(\mathbf{r}_{31})}{\partial \mathbf{r}_{31}}. \tag{8.178}$$

The relative position vector $\mathbf{r}$ satisfies,

$$\ddot{\mathbf{r}} = \ddot{\mathbf{r}}_2 - \ddot{\mathbf{r}}_1 = \frac{\partial V_1(\mathbf{r}_{21})}{\partial \mathbf{r}_{21}} + \frac{\partial V_3(\mathbf{r}_{23})}{\partial \mathbf{r}_{23}} - \frac{\partial V_2(\mathbf{r}_{12})}{\partial \mathbf{r}_{12}} - \frac{\partial V_3(\mathbf{r}_{13})}{\partial \mathbf{r}_{13}}. \tag{8.179}$$

It follows that

$$\ddot{\mathbf{r}} = \ddot{\mathbf{r}}_2 - \ddot{\mathbf{r}}_1 = -\frac{G(m_1+m_2)\mathbf{r}_{21}}{|\mathbf{r}_{21}|^3} - \frac{Gm_3}{|\mathbf{r}_{23}|^3}\mathbf{r}_{23} + \frac{Gm_3}{|\mathbf{r}_{13}|^3}\mathbf{r}_{13} - \frac{\partial R(\mathbf{r}_{12})}{\partial \mathbf{r}_{12}}. \tag{8.180}$$

Equation (8.180) may be expressed as

$$\ddot{\mathbf{r}} = -\frac{G(m_1+m_2)\mathbf{r}}{|\mathbf{r}|^3} + \frac{Gm_3}{|\mathbf{r}_{32}|^3}\mathbf{r}_{32} - \frac{Gm_3}{|\mathbf{r}_{13}|^3}\mathbf{r}_{31} - \frac{\partial R(\mathbf{r}_{12})}{\partial \mathbf{r}_{12}}. \tag{8.181}$$

or as

$$\ddot{\mathbf{r}} = -\frac{G(m_1+m_2)\mathbf{r}}{|\mathbf{r}|^3} + \frac{Gm_3}{|\mathbf{r}_{32}|^3}\mathbf{r}_{32} - \frac{Gm_3}{|\mathbf{r}_{13}|^3}\mathbf{r}_{31} - \frac{\partial R(\mathbf{r}_{12})}{\partial |\mathbf{r}_{12}|}\frac{\mathbf{r}_{12}}{|\mathbf{r}_{12}|}. \tag{8.182}$$

Furthermore

$$\ddot{\mathbf{r}}_{cm_{12}} = \frac{m_1\ddot{\mathbf{r}}_1 + m_2\ddot{\mathbf{r}}_2}{m_1+m_2}$$
$$= -\frac{1}{m_1+m_2}\left(\frac{Gm_1m_2\mathbf{r}_{12}}{|\mathbf{r}_{12}|^3} + \frac{Gm_1m_3\mathbf{r}_{13}}{|\mathbf{r}_{13}|^3} + \frac{Gm_1m_2\mathbf{r}_{21}}{|\mathbf{r}_{21}|^3} + \frac{Gm_2m_3\mathbf{r}_{23}}{|\mathbf{r}_{23}|^3} - m_1\frac{\partial R(\mathbf{r}_{12})}{\partial \mathbf{r}_{12}}\right), \tag{8.183}$$

so

$$\ddot{\mathbf{r}}_{cm_{12}} = \frac{1}{m_1+m_2}\left(\frac{Gm_1m_3\mathbf{r}_{32}}{|\mathbf{r}_{32}|^3} + \frac{Gm_2m_3\mathbf{r}_{31}}{|\mathbf{r}_{31}|^3} + m_1\frac{\partial R(\mathbf{r}_{12})}{\partial \mathbf{r}_{12}}\right). \tag{8.184}$$

Equation (8.184) may be expressed as

$$\ddot{\mathbf{r}}_{cm_{12}} = \frac{1}{m_1 + m_2}\left(-m_3\ddot{\mathbf{r}}_3 + m_3\frac{\partial R(\mathbf{r}_{32})}{\partial \mathbf{r}_{32}} + m_1\frac{\partial R(\mathbf{r}_{12})}{\partial \mathbf{r}_{12}}\right). \tag{8.185}$$

Since

$$\ddot{\mathbf{r}}_3 = \frac{\partial V_2(\mathbf{r}_{32})}{\partial \mathbf{r}_{32}} + \frac{\partial V_1(\mathbf{r}_{31})}{\partial \mathbf{r}_{31}} = -\frac{Gm_2\mathbf{r}_{32}}{|\mathbf{r}_{32}|^3} - \frac{Gm_1\mathbf{r}_{31}}{|\mathbf{r}_{31}|^3} + \frac{\partial R(\mathbf{r}_{32})}{\partial \mathbf{r}_{32}}, \tag{8.186}$$

$$\widetilde{\boldsymbol{\rho}} = \mathbf{r}_3 - \mathbf{r}_{cm_{12}} \text{ or } \widetilde{\boldsymbol{\rho}} = \frac{m_1}{m_2 + m_1}\mathbf{r}_{31} + \frac{m_2}{m_2 + m_1}\mathbf{r}_{32}. \tag{8.187}$$

It also follows that

$$\ddot{\widetilde{\boldsymbol{\rho}}} = \ddot{\mathbf{r}}_3 - \ddot{\mathbf{r}}_{cm_{12}} = \ddot{\mathbf{r}}_3 - \frac{1}{m_1 + m_2}\left(-m_3\ddot{\mathbf{r}}_3 + m_3\frac{\partial R(\mathbf{r}_{32})}{\partial \mathbf{r}_{32}} + m_1\frac{\partial R(\mathbf{r}_{12})}{\partial \mathbf{r}_{12}}\right). \tag{8.188}$$

Thus

$$\ddot{\widetilde{\boldsymbol{\rho}}} = \frac{1}{m_1 + m_2}\left(\left(\sum m_i\right)\ddot{\mathbf{r}}_3 - m_3\frac{\partial R(\mathbf{r}_{32})}{\partial \mathbf{r}_{32}} - m_1\frac{\partial R(\mathbf{r}_{12})}{\partial \mathbf{r}_{12}}\right). \tag{8.189}$$

Hence the equation of relative motion is

$$\ddot{\widetilde{\boldsymbol{\rho}}} = -G\frac{\left(\sum m_i\right)}{m_1 + m_2}\left(\frac{m_2\mathbf{r}_{32}}{|\mathbf{r}_{32}|^3} + \frac{Gm_1\mathbf{r}_{31}}{|\mathbf{r}_{31}|^3}\right) + \frac{\partial R(\mathbf{r}_{32})}{\partial \mathbf{r}_{32}} - \frac{m_1}{m_1 + m_2}\frac{\partial R(\mathbf{r}_{12})}{\partial \mathbf{r}_{12}}, \tag{8.190}$$

where

$$\mathbf{r}_{31} = \widetilde{\boldsymbol{\rho}} + \frac{m_2}{m_1 + m_2}\mathbf{r}, \mathbf{r}_{32} = \widetilde{\boldsymbol{\rho}} - \frac{m_1}{m_1 + m_2}\mathbf{r}. \tag{8.191}$$

It is now possible to generalize the equation for relation motion for $\widetilde{\boldsymbol{\rho}}$ to the case of a rotating frame of reference.

8.5.2 Equations of Circular Restricted Motion in Rotating Coordinates

The dynamics of satellite's position relative to the centre of mass of an asteroid and the Sun are expressed in rotating coordinate system as

$$\ddot{\widetilde{\boldsymbol{\rho}}}+2\boldsymbol{\Omega}\times\dot{\widetilde{\boldsymbol{\rho}}}+\dot{\boldsymbol{\Omega}}\times\widetilde{\boldsymbol{\rho}}+\boldsymbol{\Omega}\times(\boldsymbol{\Omega}\times\widetilde{\boldsymbol{\rho}})=F_{3b},$$

$$F_{3b}=-G\frac{\left(\sum m_i\right)}{m_1+m_2}\left(\frac{m_2\mathbf{r}_{32}}{|\mathbf{r}_{32}|^3}+\frac{Gm_1\mathbf{r}_{31}}{|\mathbf{r}_{31}|^3}\right)+\frac{\partial R(\mathbf{r}_{32})}{\partial\mathbf{r}_{32}}-\frac{m_1}{m_1+m_2}\frac{\partial R(\mathbf{r}_{12})}{\partial\mathbf{r}_{12}}. \tag{8.192}$$

If $\boldsymbol{\rho}=x\widehat{i}+y\widehat{j}+z\widehat{k}$, $\mathbf{r}=r\widehat{i}+0\widehat{j}+0\widehat{k}$, $\boldsymbol{\Omega}=0\widehat{i}+0\widehat{j}+\omega_z\widehat{k}$, $\mu_1=m_1/(m_1+m_2)$, and let,

$$\begin{aligned}\mathbf{r}_1\equiv\mathbf{r}_{31}=\widetilde{\boldsymbol{\rho}}+(1-\mu_1)\mathbf{r}=(x+(1-\mu_1)r)\widehat{i}+y\widehat{j}+z\widehat{k},\mathbf{r}_2\\ \equiv\mathbf{r}_{32}=\widetilde{\boldsymbol{\rho}}-\mu_1\mathbf{r}=(x-\mu_1r)\widehat{i}+y\widehat{j}+z\widehat{k}.\end{aligned} \tag{8.193}$$

Assuming r is constant (circular motion of asteroid relative to the primary), the equations of motion in rotating or synodic coordinates are

$$\begin{aligned}\ddot{x}-2\omega_z\dot{y}-\dot{\omega}_zy-\omega_z^2x=\partial U_{3b}/\partial x=f_{3b,x},\\ \ddot{y}+2\omega_z\dot{x}+\dot{\omega}_zx-\omega_z^2y=\partial U_{3b}/\partial y=f_{3b,y},\ddot{z}=\partial U_{3b}/\partial z=f_{3b,z},\end{aligned} \tag{8.194}$$

where

$$\begin{bmatrix}\partial U_{3b}/\partial x\\ \partial U_{3b}/\partial y\\ \partial U_{3b}/\partial z\end{bmatrix}=\begin{bmatrix}f_{3b,x}\\ f_{3b,y}\\ f_{3b,z}\end{bmatrix}=-\frac{G\left(\sum m_i\right)(1-\mu_1)}{\left(\sqrt{(x-\mu_1r)^2+y^2+z^2}\right)^3}\begin{bmatrix}x-\mu_1r\\ y\\ z\end{bmatrix}$$
$$-\frac{G\left(\sum m_i\right)\mu_1}{\left(\sqrt{(x+(1-\mu_1)r)^2+y^2+z^2}\right)^3}\begin{bmatrix}x+(1-\mu_1)r\\ y\\ z\end{bmatrix}+\frac{\partial R(\mathbf{r}_{32})}{\partial\mathbf{r}_{32}}-\mu_1\frac{\partial R(\mathbf{r}_{12})}{\partial\mathbf{r}_{12}}. \tag{8.195}$$

In general the potential $V_2(\mathbf{r})$ may be expressed as

$$V_2(\mathbf{r})=\frac{Gm_i}{|\mathbf{r}|}+R(\mathbf{r}),\text{ and }\frac{\partial R(\mathbf{r})}{\partial\mathbf{r}}=\frac{R(\mathbf{r})}{\partial|\mathbf{r}|}\frac{\mathbf{r}}{|\mathbf{r}|}, \tag{8.196}$$

with

$$\frac{\partial R(\mathbf{r}_{32})}{\partial \mathbf{r}_{32}} - \mu_1 \frac{\partial R(\mathbf{r}_{12})}{\partial \mathbf{r}_{12}} = \frac{\partial R(\mathbf{r}_{32})}{\partial |\mathbf{r}_{32}|} \frac{\mathbf{r}_{32}}{|\mathbf{r}_{32}|} - \mu_1 \frac{\partial R(\mathbf{r}_{12})}{\partial |\mathbf{r}_{12}|} \frac{\mathbf{r}_{12}}{|\mathbf{r}_{12}|}$$
$$= \frac{\partial R(\mathbf{r}_{32})}{\partial |\mathbf{r}_{32}|} \frac{\mathbf{r}_{32}}{|\mathbf{r}_{32}|} + \mu_1 \frac{\partial R(\mathbf{r}_{12})}{\partial |\mathbf{r}_{12}|} \frac{\mathbf{r}_{21}}{|\mathbf{r}_{21}|}. \tag{8.197}$$

Hence, since $\mathbf{r}_{32} = \widetilde{\boldsymbol{\rho}} - \mu_1 \mathbf{r} = (x - \mu_1 r)\widehat{i} + y\widehat{j} + z\widehat{k}$

$$\begin{bmatrix} \partial U_{3b}/\partial x \\ \partial U_{3b}/\partial y \\ \partial U_{3b}/\partial z \end{bmatrix} = \begin{bmatrix} f_{3b,x} \\ f_{3b,y} \\ f_{3b,z} \end{bmatrix} = \frac{1}{|\widetilde{\boldsymbol{\rho}} - \mu_1 \mathbf{r}|} \left(\frac{\partial R(\widetilde{\boldsymbol{\rho}} - \mu_1 \mathbf{r})}{\partial |\widetilde{\boldsymbol{\rho}} - \mu_1 \mathbf{r}|} - \frac{G(\sum m_i)(1-\mu_1)}{\left(\sqrt{(x-\mu_1 r)^2 + y^2 + z^2}\right)^2} \right) \begin{bmatrix} x - \mu_1 r \\ y \\ z \end{bmatrix}$$
$$- \frac{G(\sum m_i)\mu_1}{\left(\sqrt{(x + (1-\mu_1) r)^2 + y^2 + z^2}\right)^3} \begin{bmatrix} x + (1-\mu_1) r \\ y \\ z \end{bmatrix} + \mu_1 \frac{\partial R(-\mathbf{r})}{\partial |\mathbf{r}|} \begin{bmatrix} 1 \\ 0 \\ 0 \end{bmatrix}. \tag{8.198}$$

Furthermore, U_{3b} is,

$$U_{3b} = R(\widetilde{\boldsymbol{\rho}} - \mu_1 \mathbf{r}) + \mu_1 \frac{\partial R(-\mathbf{r})}{\partial |\mathbf{r}|} (\widetilde{\boldsymbol{\rho}} \cdot i) - \frac{G(\sum m_i)(1-\mu_1)}{|\widetilde{\boldsymbol{\rho}} - \mu_1 \mathbf{r}|} - \frac{G(\sum m_i)\mu_1}{|\widetilde{\boldsymbol{\rho}} + (\mathbf{1} - \mu_1)\mathbf{r}|}. \tag{8.199}$$

The Jacobi integral is given by

$$J = \frac{1}{2}\left(\dot{\bar{x}}^2 + \dot{\bar{y}}^2 + \dot{\bar{z}}^2\right) + \frac{\omega_z^2}{2}\left(\bar{x}^2 + \bar{y}^2\right) - U_{3b} = \dot{C}. \tag{8.200}$$

The zero velocity curves are given by

$$\frac{\omega_z^2}{2}\left(\bar{x}^2 + \bar{y}^2\right) - U_{3b} = C. \tag{8.201}$$

8.5.3 Conditions of Equilibrium of the Circular Restricted Motion in Rotating Coordinates

Equilibrium is characterized by two equations. The first or vector equation in rotating coordinates is

$$\ddot{\mathbf{r}} + 2\boldsymbol{\Omega}\times\dot{\mathbf{r}} = -\dot{\boldsymbol{\Omega}}\times\mathbf{r} - \boldsymbol{\Omega}\times(\boldsymbol{\Omega}\times\mathbf{r}) - \frac{G(m_1+m_2)\mathbf{r}}{|\mathbf{r}|^3} + \frac{Gm_3}{|\mathbf{r}_{32}|^3}\mathbf{r}_{32} - \frac{Gm_3}{|\mathbf{r}_{13}|^3}\mathbf{r}_{31} + \frac{\partial R(\mathbf{r}_{12})}{\partial|\mathbf{r}_{12}|}\frac{\mathbf{r}_{21}}{|\mathbf{r}_{21}|} = 0. \tag{8.202}$$

The above equation may be expressed as

$$\left(\frac{G(m_1+m_2)\mathbf{r}}{|\mathbf{r}|^2} - \frac{\partial R(\mathbf{r}_{12})}{\partial|\mathbf{r}_{12}|}\right)\frac{\mathbf{r}}{|\mathbf{r}|} + \dot{\boldsymbol{\Omega}}\times\mathbf{r} + \boldsymbol{\Omega}\times(\boldsymbol{\Omega}\times\mathbf{r}) = \left(\frac{Gm_3}{|\mathbf{r}_{32}|^3}\mathbf{r}_{32} - \frac{Gm_3}{|\mathbf{r}_{13}|^3}\mathbf{r}_{31}\right). \tag{8.203}$$

Physically it represents the equilibrium of inertial forces (centrifugal and Coriolis) and gravitational forces.

The second or the set of equations in component form are

$$\ddot{x} - 2\omega_z\dot{y} = \dot{\omega}_z y + \omega_z^2 x + f_{3b,x} = 0, \ddot{y} + 2\omega_z\dot{x} = -\dot{\omega}_z x + \omega_z^2 y + f_{3b,y} = 0, \ddot{z} = f_{3b,z} = 0. \tag{8.204}$$

Assuming $\dot{\omega}_z = 0$, the equilibrium equations are

$$\omega_z^2 x + f_{3b,x} = 0, \omega_z^2 y + f_{3b,y} = 0, f_{3b,z} = 0. \tag{8.205}$$

When the second body is a completely spherical body

$$V_2(\mathbf{r}) = Gm_i/|\mathbf{r}|, R(\mathbf{r}) \equiv 0, \tag{8.206}$$

the equilibrium equations reduce to the classical, three-body equilibrium equations.

8.5.4 *Stability of the Equilibrium Solutions*

The study of the stability of the equilibrium solutions or equilibrium points proceeds in exactly the same way as in the case of the two-body problem considered in Sect. 8.3. However, in this case we have tacitly assumed that the equations of motion of the orientation states of the spacecraft or third body are decoupled from the equations of motion of the spacecraft along its orbit. The equations motion may now be expanded about the equilibrium state, by perturbing only the translational motion states about their equilibrium values and treating the others are parameters. Thus, one may define the perturbations, $\widetilde{x}=x-x_e$, $\widetilde{y}=y-y_e$, $\widetilde{z}=z-z_e$, and linearize the equations of motion in terms the perturbation states. The stability of the equilibrium points may then be ascertained from a study of the linearized equations of motion in terms the perturbation states.

8.6 Application to a Spacecraft Orbiting the Sun in the Vicinity of an Asteroid

When considering the motion of a spacecraft orbiting the SUN in the vicinity of an Asteroid, is essential to consider various regimes of the motion. First there is the case of the spacecraft orbiting the SUN, far away from the asteroid. However, the case when the asteroid is absent is considered first.

8.6.1 *Solar Radiation Pressure (Based on a Flat Plate Model)*

For a spacecraft in a heliocentric orbit, the solar radiation pressure potential is combined with the gravitational potential of the Sun, and the two together are expressed as

$$V_s=(\mu_s-\beta)/r_s, \tag{8.207}$$

where μ_s is the gravitational parameter for the Sun, r_s is the distance of the Sun from the spacecraft and β is the force parameter of the solar radiation pressure that is proportional to inverse of the spacecraft's mass to area ratio and computed from

$$\beta=G_s/B_s=\left(10^8/B_s\right)km^3/s^2, \tag{8.208}$$

with B_s equal to the spacecraft's mass to area ratio in *kg*/m^2. Typically for a spacecraft, with a total surface area from the front of the spacecraft equal to 25 m^2 and a mass of 1000 *kg*, its value is 40 *kg*/m^2.

8.6.2 Gravitational Potential with an Asteroid Present

Considering the presence of the asteroid, irrespective of whether it is close to or far from the spacecraft in a heliocentric orbit, described in a Sun-centred inertial, space-fixed frame, the spacecraft's equations of motion can be expressed as

$$\begin{aligned}\ddot{\mathbf{r}}_s &= \partial V_{sc}/\partial \mathbf{r}_s, V_{sc} = V_s + V_a + \frac{\mu_a}{r_a^2}\widehat{\mathbf{r}}_a \cdot \mathbf{r}_s, V_s = \frac{\mu_s - \beta}{r_s}, V_a \\ &\equiv \widetilde{U}(\mathbf{r}_s - \mathbf{r}_a) = \frac{\mu_a}{|\mathbf{r}_s - \mathbf{r}_a|} + R(\mathbf{r}_s - \mathbf{r}_a).\end{aligned} \tag{8.209}$$

In the above, $\mathbf{r}_s$ is the position vector of the spacecraft relative to the Sun, $\mathbf{r}_a$ is the position vector of the asteroid relative to the Sun, $\widehat{\mathbf{r}}_a$ is a unit vector in direction of $\mathbf{r}_a$, μ_a is the gravitational parameter of the asteroid and $R(\mathbf{r}_s - \mathbf{r}_a)$ is the perturbation gravitational potential of the asteroid due to its non-spherical geometry. The Sun exerts a force of the spacecraft in the direction of the Sun, while the asteroid exerts a force on the spacecraft. In the expression for the force potential, V_{sc}, the first two terms give the potential function due the Sun and asteroid, respectively; the last term is the potential due to the component of the force exerted by the asteroid on the spacecraft.

The equations of motion of a spacecraft in rotating coordinates may now be suitably modified.

8.7 Asteroid Trajectory Examples

There are several examples of the application of the equations of motion of a spacecraft in the vicinity of an asteroid in a heliocentric orbit. In this section three such application examples are briefly considered.

8.7.1 De-tumbling Control of an Asteroid: Yo-Yo De-spin

In a Yudintsev and Aslanov (2017), a method for de-tumbling a space debris object such as a small asteroid, using a modified yo-yo mechanism that does not contribute to the inertia properties of the space debris or asteroid. A mathematical model of the spatial motion of the space object with attached yo-yo mechanism is developed. Several examples in the literature demonstrate the possibility of using mechanisms to de-spin a space object, such as Kang et al. (2019). Several methods of active debris removal (ADR) are proposed by several researchers. One such method employs a tethered space tug for rendezvous, docking and towing of dangerous objects to a graveyard orbit or into the atmosphere for burn up (Kang et al., 2021).

The ability to "capture" and safely de-tumble an asteroid depends on its attitude motion and dynamics. The attitude motion of an ADR system must be synchronized to the target's so that the capturing operation is conducted with relatively low relative motion. The ADR system is equipped with rocket thrusters, and once rendezvous docking and synchronization are achieved, the de-tumbling can begin. One approach is to use a yo-yo mechanism as proposed by Yudintsev and Aslanov (2017) or a tethered space tug (Kang & Zhu, 2023). Primarily these systems try and increase the moment of inertia of the tumbling body about an axis of maximum rotation to slow the objects rotational speed. Further reduction in the orbital speed or a change in the orbit is obtained by applying appropriate velocity impulses to the object after it has been acquired by the ADR system.

8.7.2 Asteroid Deflection

One approach to deflect an asteroid from its known orbit is to irradiate the asteroid by a train of short duration LASER (light amplification by stimulated emission of radiation) pulses. The recoil impulse transferred to the target asteroid by each laser pulse in the train is then estimated using the area matrix approach. The sum of all the impulses transferred gives the total velocity increment delivered to the target. This approach is based on the assumption that impulse due to a single laser pulse is exactly opposite to the ablation recoil due to the pulse. Thus the impulse delivered is equal to the magnitude ablation recoil impulse. It is assumed the ablation is in the direction of the local surface normal.

Thus one could estimate the impulse delivered by irradiating the asteroid with a single short duration laser pulse. The right duration and intensity are determined to provide the right magnitude of momentum so as to obtain a change in the velocity eventually to deflect the asteroid into a hyperbolic escape orbit, so it does not collide with the planet Earth.

Asteroids of different shapes would have different characteristic attitude motions. It is natural to expect that the momentum transfer is likely to vary depending on the shape and size of the asteroid. To establish the methodology and the design equations, an arbitrary shaped asteroid of known geometry is assumed.

The LASER incident energy is the LASER beam energy which comes in contact with the object is expressed

$$E_{inc} = F_{inc} A, \tag{8.210}$$

where E_{inc} is the incident energy, A is illuminated area and F_{inc} is the laser fluence or the energy of the number of photons per unit cross-sectional area with which a material is irradiated. The intensity of the laser beam is determined by the laser fluence which varies depending on the specification of the laser. Assuming that the ablation vector is parallel to the local normal, the impulse imparted onto a surface area element is expressed as

$$d(m\Delta\overline{v}) = C_m F_{inc} \left| e_{inc} \cdot \vec{n} \left| dA \right| - \vec{n} \right| = C_m F_{inc} e_{inc} \cdot \left(\vec{n}\,\vec{n} \right) dA. \tag{8.211}$$

In the above equation, $\vec{n}$ denotes the unit normal vector of the object area element dA and e_{inc} is the unit vector in the direction of the laser beam. Furthermore, C_m is the momentum coupling coefficient which gives an indication of the transferred impulse magnitude per unit of laser energy and is given by

$$C_m \equiv m\Delta v / E_{inc}, \tag{8.212}$$

in units of Ns/J, Δv is the velocity increment due to the incident laser energy and m is the mass of the target object. $\left(\vec{n}\,\vec{n} \right)$ is obtained by transforming the direction cosine matrix in body-fixed coordinates, to the appropriate reference frame. The total impulse is the sum of all the element impulses calculated as

$$m\Delta\overline{v} = \iint_S C_m F_{inc} e_{inc} \cdot \left(\vec{n}\,\vec{n} \right) dA, \tag{8.213}$$

where S represents the surface area of the region irradiated by the laser beam.

8.7.3 Co-orbital Dynamics of Two Spacecraft

The governing equations of motion considered in the study of co-orbital motion are in the framework of the circular restricted three-body problem (CR3BP). In the CR3BP, two celestial bodies of masses M_i are considered, known as primaries, which revolve around their common centre of mass under their mutual gravitational attraction describing Keplerian circular orbits. Third and fourth bodies, i.e. spacecraft, labelled $j = 1, 2$, respectively, are considered with comparatively negligible masses $m_j < < M_i$, such that their motions are driven by gravitational interaction with the primaries, and the primaries are not affected by the presence of the spacecraft. The motion of the spacecraft is conveniently studied in a co-rotating, synodic frame S, defined as follows: the x_s axis points from M_1 to M_2, following the latter in its relative circular motion around M_1; the z_s axis is set perpendicular to the orbital plane of the primaries (along the angular momentum of the system); and the y_s axis is mutually perpendicular to the other two. The origin of S is located at the barycentre of the system. When realized in the synodic frame, the primaries' position vectors occupy fixed positions along the x_s axis. The unit vectors $\{\mathbf{i}, \mathbf{j}, \mathbf{k}\}$ are defined, respectively, along each of the coordinate axes of the synodic frame. This is illustrated in Fig. 8.8, which depicts the definition of the synodic frame with respect to the inertial frame. The latter is arbitrarily defined to be aligned with S at the reference epoch $t_0 = 0$.

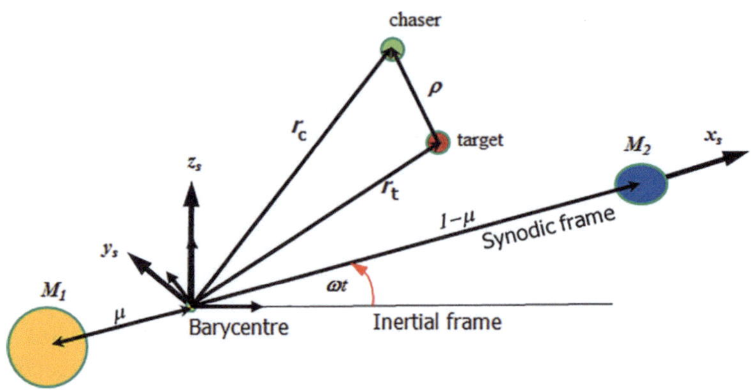

Fig. 8.8 Relative co-orbital motion in the CR3BP

In the following, dimensionless coordinates are used, where the distance between the primaries is used as the characteristic length, the orbital period of the primaries around their common barycentre is taken as the characteristic time, and the masses of the primaries are referred to as the total mass of the system. In these dimensionless coordinates, the primaries revolve at one radian per unit of dimensionless time, and the reduced mass of M_2 can be defined as

$$\mu = M_2/(M_1 + M_2). \tag{8.214}$$

The primaries are set at a unit length apart from each other, so their position vectors, constant when realized in the synodic frame, can be written as

$$\mathbf{R}_1 = -\mu\mathbf{i}, \mathbf{R}_2 = (1-\mu)\mathbf{i}. \tag{8.215}$$

Under the aforementioned setup, the motion of the spacecraft is governed by the following second-order ordinary differential equation

$$\ddot{\mathbf{r}} + 2\boldsymbol{\omega}\times\dot{\mathbf{r}} + \boldsymbol{\omega}\times(\boldsymbol{\omega}\times\mathbf{r}) = -(1-\mu)\frac{\mathbf{r}-\mathbf{R}_1}{\|\mathbf{r}-\mathbf{R}_1\|^3} - \mu\frac{\mathbf{r}-\mathbf{R}_2}{\|\mathbf{r}-\mathbf{R}_2\|^3}, \tag{8.216}$$

where $\mathbf{r}$ is the dimensionless spacecraft position vector, as measured by an observer in the synodic frame, and the overhead dot $(\dot{\cdot})$ indicates derivatives with respect to the non-dimensional time. The vector $\boldsymbol{\omega}$ denotes the dimensionless angular velocity vector of the synodic frame S with respect to the inertial frame. Again, note that, in non-dimensional units, $\boldsymbol{\omega} = \mathbf{k}$.

In this section the formulation given by Cuevas del Valle et al. (2023) is tacitly adopted. The relative co-orbital motion between two spacecraft, termed target and chaser, following the usual nomenclature, are now considered. Both vehicles are modelled as point masses, and the mutual gravity interaction is ignored. The target spacecraft is assumed to be cooperative and passive (i.e. with no orbital manoeuvring capabilities), and its trajectory is also assumed to be accurately known.

Let vectors $\mathbf{r}_t$ and $\mathbf{r}_c$ denote the positions of the target and chaser spacecraft, respectively, so their relative position vector is defined as $\boldsymbol{\rho} = \mathbf{r}_c - \mathbf{r}_t$. Differentiating twice, with respect to time, and including the appropriate inertial terms, the following governing equations are obtained:

$$\begin{aligned}
\ddot{x} - 2\dot{y} - x &= (1-\mu)\left(\frac{\xi+\mu}{\|\mathbf{r}_t - \mathbf{R}_1\|^3} - \frac{x+\xi+\mu}{\|\boldsymbol{\rho} + \mathbf{r}_t - \mathbf{R}_1\|^3}\right) \\
&\quad + \mu\left(\frac{\xi-1+\mu}{\|\mathbf{r}_t - \mathbf{R}_2\|^3} - \frac{x+\xi-1+\mu}{\|\boldsymbol{\rho} + \mathbf{r}_t - \mathbf{R}_2\|^3}\right) + u_x, \\
\ddot{y} + 2\dot{x} - y &= (1-\mu)\left(\frac{\eta}{\|\mathbf{r}_t - \mathbf{R}_1\|^3} - \frac{y+\eta}{\|\boldsymbol{\rho} + \mathbf{r}_t - \mathbf{R}_1\|^3}\right) \\
&\quad + \mu\left(\frac{\eta}{\|\mathbf{r}_t - \mathbf{R}_2\|^3} - \frac{y+\eta}{\|\boldsymbol{\rho} + \mathbf{r}_t - \mathbf{R}_2\|^3}\right) + u_y, \\
\ddot{z} &= (1-\mu)\left(\frac{\zeta}{\|\mathbf{r}_t - \mathbf{R}_1\|^3} - \frac{z+\zeta}{\|\boldsymbol{\rho} + \mathbf{r}_t - \mathbf{R}_1\|^3}\right) \\
&\quad + \mu\left(\frac{\zeta}{\|\mathbf{r}_t - \mathbf{R}_2\|^3} - \frac{z+\zeta}{\|\boldsymbol{\rho} + \mathbf{r}_t - \mathbf{R}_2\|^3}\right) + u_z,
\end{aligned} \tag{8.217}$$

where $\{x, y, z\}$ are the synodic frame coordinates of the relative position vector $\boldsymbol{\rho} = \mathbf{r}_c - \mathbf{r}_t$, $\{\xi, \eta, \zeta\}$ are the synodic frame coordinates of the target spacecraft position vector $\mathbf{r}_t$ and the control acceleration $\mathbf{u} = \mathbf{u}_x\mathbf{i} + \mathbf{u}_y\mathbf{j} + \mathbf{u}_z\mathbf{k}$ is introduced. In deriving the above equations, it is assumed that the force of attraction between the chaser and target may be ignored.

When the force of attraction of the target body is not negligible, the equations of motion, now including the mass ratio of the target body are, but still ignoring its mass relative to that of the primary and secondary

$$\ddot{x} - 2\dot{y} - x = F_x, \ddot{y} + 2\dot{x} - y = F_y, \ddot{z} = F_z, \tag{8.218}$$

$$\begin{aligned}
F_x &= (1-\mu)\left(\frac{\xi+\mu}{\|\mathbf{r}_t - \mathbf{R}_1\|^3} - \frac{x+\xi+\mu}{\|\boldsymbol{\rho} + \mathbf{r}_t - \mathbf{R}_1\|^3}\right) \\
&\quad + \mu\left(\frac{\xi-1+\mu}{\|\mathbf{r}_t - \mathbf{R}_2\|^3} - \frac{x+\xi-1+\mu}{\|\boldsymbol{\rho} + \mathbf{r}_t - \mathbf{R}_2\|^3}\right) - \mu_t \frac{x}{|\boldsymbol{\rho}|^3} + u_x,
\end{aligned}$$

$$F_y = (1-\mu)\left(\frac{\eta}{\|\mathbf{r}_t - \mathbf{R}_1\|^3} - \frac{y+\eta}{\|\boldsymbol{\rho} + \mathbf{r}_t - \mathbf{R}_1\|^3}\right) + \mu\left(\frac{\eta}{\|\mathbf{r}_t - \mathbf{R}_2\|^3} - \frac{y+\eta}{\|\boldsymbol{\rho} + \mathbf{r}_t - \mathbf{R}_2\|^3}\right) - \mu_t \frac{y}{|\boldsymbol{\rho}|^3} + u_y,$$

$$F_z = (1-\mu)\left(\frac{\zeta}{\|\mathbf{r}_t - \mathbf{R}_1\|^3} - \frac{z+\zeta}{\|\boldsymbol{\rho} + \mathbf{r}_t - \mathbf{R}_1\|^3}\right) + \mu\left(\frac{\zeta}{\|\mathbf{r}_t - \mathbf{R}_2\|^3} - \frac{z+\zeta}{\|\boldsymbol{\rho} + \mathbf{r}_t - \mathbf{R}_2\|^3}\right) - \mu_t \frac{z}{|\boldsymbol{\rho}|^3} + u_z. \tag{8.219}$$

where $\mu_t = M_t/(M_1 + M_2)$.

The relative motion equations are closed, assuming the target spacecraft ephemerides are known. Interestingly, the scalar equations can be compactly written into a first-order, control-affine nonlinear system

$$\dot{\mathbf{s}} = \mathbf{f}(\mu, \mathbf{s}, \mathbf{r}_t) + \mathbf{u}, \tag{8.220}$$

where $\mathbf{s} = \{\boldsymbol{\rho}, \dot{\boldsymbol{\rho}}\}^T$ is the state vector accounting for the Cartesian relative position and velocity of the chaser with respect to the target.

Encke's formulation of the dynamics of an orbiting body could also be adopted to model the co-orbital dynamics, which is most appropriate for numerical orbit propagation

$$\ddot{\boldsymbol{\rho}} + 2\boldsymbol{\omega} \times \dot{\boldsymbol{\rho}} + \boldsymbol{\omega} \times (\boldsymbol{\omega} \times \boldsymbol{\rho}) = \sum_{i=1}^{2} \frac{\mu_i}{\|\mathbf{r}_t - \mathbf{R}_i\|^3} \times \left(f(q_i)(\mathbf{r}_t - \mathbf{R}_i) + (1 + f(q_i))\boldsymbol{\rho}\right) + \mu_3 \frac{\boldsymbol{\rho}}{|\boldsymbol{\rho}|^3}, \tag{8.221}$$

with

$$q_i = -\frac{2(\mathbf{r}_t - \mathbf{R}_i)\cdot\boldsymbol{\rho} + \boldsymbol{\rho}\cdot\boldsymbol{\rho}}{\|\mathbf{r}_t - \mathbf{R}_i + \boldsymbol{\rho}\|^2}, f(q_i) = q_i \frac{3(1+q_i) + q_i^2}{1 + (1+q_i)^{3/2}}. \tag{8.222}$$

In the above, $\mu_3 = GM_t$ and $\mu_i = GM_i$, $i = 1, 2$, and $\boldsymbol{\omega} = (\mu_1 + \mu_2)\boldsymbol{k}$ is the dimensionless angular velocity vector of the synodic frame S with respect to the inertial frame. Furthermore, the distance between the primary and secondary masses is normalized to unity.

There are several applications of the equations of co-orbital dynamics.

Exercises

1. A tiny asteroid 2020 QG passed above the Southern Indian Ocean at an altitude of 2950 km in August 2020 with a velocity of 12.3 km per second, at the point of closest approach to the planet, Earth. The turning angle as it passed by the Earth was 45 degrees.

   (a) Assume the orbit is Keplerian and determine the eccentricity of the orbit.
   (b) The semi-major axis of the orbit (assume the radius of the Earth is 6400 km).
   (c) The excess velocity and the total energy of the spacecraft as it approached the planet.
   (d) The velocity increment the satellite acquires as it swings by the planet Earth.
   (e) The excess velocity of the satellite as it departed the planet Earth.

2. A large asteroid—labelled (153201) 2000 WO107—swept past Earth in late November 2020. The semi-major axis of its elliptic heliocentric orbit was 1.3636×10^{08} km. The orbit eccentricity was 0.7806 and the true anomaly was 185.26°.

   (a) Assume the orbit is Keplerian and calculate its position or distance from the Sun.
   (b) Estimate the velocity of the asteroid as it swept past the Earth; ($\mu_{Sun} = 1327 \times 10^8 \text{km}^3/\text{s}^2$).
   (c) Given that the inclination was 7.77°, the Longitude of Ascending Node was 69.25° and the Argument of Periapsis was 213.72°, obtain the transformation to convert the orbit position and velocity to a heliocentric Cartesian frame. Clearly explain the derivation of the transformation. Hence transform the position, the radial component of the velocity and the tangential component of velocity to obtain the position and velocity vector of the asteroid in the heliocentric Cartesian frame.

3. The semi-major axis of the asteroid 944 Hidalgo in a heliocentric orbit was 5.7388 AU (1 AU = 1.496×10^{08} km, $\mu = 1.32712 \times 10^{11}$ km^3/s^2) on a particular day in the year 2023. The orbit eccentricity was 0.6611 and the true anomaly was 225.0815°.

   (a) Assume the orbit is Keplerian and calculate its position or distance from the Sun in km.
   (b) Estimate the velocity of the asteroid at that particular position in km/s. Obtain the radius of the sphere of influence of the asteroid, if the mass of the asteroid is 8.4×10^{15}kg (μ for the asteroid is 560641.2 km^3/s^2)?
   (c) Given that the orbital inclination was 42.5231°, the longitude of ascending node was 21.4255° and the argument of periapsis was 56.7263°, obtain the transformation to convert the orbit position and velocity to a heliocentric Cartesian frame. Clearly explain the derivation of the transformation. Hence transform the position, the radial component of the velocity and the tangential component of velocity to obtain the position and velocity vector of the asteroid in the heliocentric Cartesian frame.

4. For the heliocentric asteroid, 10370 Hylonome, the following orbital elements are given: eccentricity = 0.2482; semi-major axis = 25.1017 in AU (Astronomical Unit), where 1 AU = 149597870.7 km, orbital inclination = 4.1412°, longitude of the ascending node = 178.0727°, argument of the periapsis = 6.5554° and mean anomaly = 60.5576°. The gravitation constant for a heliocentric orbit is 132,712,440,018 km^3/s^2.

   (a) If the unit of distance is taken to be 1 AU, determine the unit of time in days such that the gravitation constant for a heliocentric orbit is unity.
   Also compute the orbital period of the asteroid in years.
   (b) In the units, for which the gravitation constant for a heliocentric orbit is unity, determine the distance of closest approach to the central body, the distance when the asteroid is farthest from the central body and the corresponding velocities.
   (c) In the units, for which the gravitation constant for a heliocentric orbit is unity, compute the velocity of the asteroid when it is exactly 25 AU from the central body, in its orbit.
   In the units, for which the gravitation constant for a heliocentric orbit is unity, compute the lowest velocity that must be imparted to the asteroid, so it escapes from the Sun's gravitational field.
   (d) An approximate value for the eccentric anomaly is given as $E = 74.238°$. Obtain a better approximation for the eccentric anomaly.

5. Consider the requirement to transfer a satellite, initially in a circular equatorial parking orbit about the Earth at a radius of 8200 km that is to be placed in the same orbit as an asteroid. The semi-major axis and the eccentricity of the asteroid's orbit are, respectively, given by 0.9115 AU and 0.7807. (1 AU = 1.496×10^{08} km). The inclination and the argument of the periapsis of the asteroid's orbit are, respectively, given by 7.77° and 213.72°.

   (a) Draw a diagram of the asteroid's orbit indicating clearly where this orbit intersects with the ecliptic plane and the two distances (i.e. as the asteroid approaches the Sun and then where it is moving away from the sun) from the Sun at these points. Calculate the Sun-asteroid distances at these points of interception.
   (b) Calculate the minimum values of Δv (in km/s) that would be required to take the satellite from its parking orbit about the Earth so that it could intersect the orbit of the asteroid at one of the two locations where the asteroid would cross the ecliptic plane. In this calculation you should explain clearly each step of your calculations with appropriate diagrams as well as state the appropriate equations.
   (c) Calculate the magnitude of the additional Δv that would be required to change the velocity of the spacecraft as it intersects the asteroid's orbit, so that it could fly in the same orbital plane and along the same orbit as the asteroid. You should explain in words as well as writing down the relevant equations as to what manoeuvre would be performed and state the optimal manoeuvre.

(d) How does the total, minimum Δv compare if one adopts a direct nonplanar transfer from the parking orbit, to a point flying alongside the asteroid, assuming that the asteroid is located at a nodal point?

6. Reconsider exercise (4) if the asteroid is the asteroid 162173 Ryugu. Consider the requirement to transfer a satellite, initially in a circular equatorial parking orbit about the Earth at a radius of 8200 km that is to be to be placed in the same orbit as the asteroid 162173 Ryugu. The semi-major axis and the eccentricity of the asteroid's orbit are, respectively, given by 1.1896 AU and 0.1902 (1 AU = 1.496×10^{08} km). The inclination and the argument of the periapsis of the asteroid's orbit are, respectively, given by 5.8837° and 211.43°. The asteroid's longitude of the ascending node is 251.62°. For this asteroid repeat parts 5(a)–5(d).
7. A small asteroid—labelled 2023 BU—swept past Earth on 27 January 2023. As it flew past the Earth, in its heliocentric orbit, at a distance of 9967 km from the Earth, its orbital characteristics were altered. The pre-flyby and post flyby orbital characteristics are listed below, with the distances in AU (Astronomical Unit), where 1 AU = 149597870.7 km

   Pre-flyby: 2022-10-25 period = 358.9 days
   Aphelion = 1.05, Perihelion = 0.926, a = 0.988, e = 0.063, incl. = 2.35°
   Post-flyby: 2023-02-25 period = 425.4 days
   Aphelion = 1.23, Perihelion = 0.984, a = 1.11, e = 0.111, incl. = 3.75°.

   (a) Determine the equivalent Δv acquired by the asteroid at its perihelion, during the flyby past the Earth, both in the orbital plane and normal to it. Hence compute the magnitude of the total Δv acquired by the asteroid 2023 BU when it was at its perihelion. Explain all your steps.
   (b) Determine the magnitude of the velocity of the asteroid as it approached the Earth (pre-flyby) in its heliocentric orbit ($\mu = 1.327 \times 10^{11}$ km^3/s^2). Explain all your steps.
   (c) Given that the post-flyby inclination was 3.75°, the longitude of ascending node was 125.5° and the argument of perihelion was 355.8°, obtain the transformation to convert the orbit position and velocity to a heliocentric Cartesian frame. Clearly explain the derivation of the transformation. Hence transform the position, the radial component of the velocity and the tangential component of velocity to obtain the position and velocity vector of the asteroid when it is at the perihelion, in the heliocentric Cartesian frame.

8. For the heliocentric asteroid, 944 Hidalgo, the following orbital elements are given: eccentricity = 0.6611; semi-major axis = 5.7388 in AU (Astronomical Unit), where 1 AU = 149597870.7 km, orbital inclination = 42.5231°, longitude of the ascending node = 21.4255°, argument of the periapsis = 56.7263° and mean anomaly = 301.3931°. The gravitation constant for a heliocentric orbit is 132,712,440,018 km^3/s^2.

(a) If the unit of distance is taken to be 1 AU, determine the unit time in days such that the gravitation constant for a heliocentric orbit is unity. Explain all your steps.

Also compute the orbital period of the asteroid in years. Explain all your steps.

(b) In the units, for which the gravitation constant for a heliocentric orbit is unity, determine the distance of closest approach to the central body, the distance when the asteroid is farthest from the central body and the corresponding velocities. Explain all your steps.

(c) In the units, for which the gravitation constant for a heliocentric orbit is unity, compute the velocity of the asteroid when it is exactly 5 AU from the central body, in its orbit. Explain all your steps.

In the units, for which the gravitation constant for a heliocentric orbit is unity, compute the lowest velocity that must be imparted to the asteroid, so it escapes from the Sun's gravitational field. Explain all your steps.

(d) An approximate value for the eccentric anomaly is given as $E = 263.7569°$. Obtain a better approximation for the eccentric anomaly. Explain all your steps.

References

Brasser, R., Innanen, K. A., Connors, M., Veillet, C., Wiegert, P., Mikkola, S., & Chodas, P. W. (2004). Transient co-orbital asteroids. *Icarus, 171*, 102–109.

Chai, R., Savvaris, A., Tsourdos, A., Chai, S., & Xia, Y. (2019). A review of optimization techniques in spacecraft flight trajectory design. *Progress in Aerospace Sciences, 109*, 100543.

Colagrossi, A., Ferrari, F., Lavagna, M., & Howell, K. (2015). Dynamical Evolution about asteroids with high fidelity gravity field and perturbations modelling. *AAS*, 15–637.

Connors, M., Wiegert, P., & Veillet, C. (2011). Earth's Trojan asteroid, research letter. *Nature, 475*, 481–483.

Conway, J. T. (2016). Vector potentials for the gravitational interaction of extended bodies and laminas with analytical solutions for two disks. *Celestial Mechanics and Dynamic Astronomy, 125*, 161–194. https://doi.org/10.1007/s10569-016-9679-y

Cuevas del Valle, S., Urrutxua, H., & Solano-López, P. (2023). Optimal floquet station keeping under the relative dynamics of the three-body problem. *Aerospace, 10*, 393. https://doi.org/10.3390/aerospace10050393

Domingos, R. C., & Winter, O. C. (2005). Possibility of collision between co-orbital asteroids and the Earth. *Journal of Computational and Applied Mathematics, 24*(1) (Petrópolis Jan./Apr. 2005), 99–107.

Dvorak, R., Lhotka, C., & Zhou, L. (2012). The Orbit of 2010 TK7: Possible regions of stability for other Earth Trojan asteroids. *Astronomy and Astrophysics, A & A, 541*(A127), 1–10. https://doi.org/10.1051/0004-6361:201118374

Fukushima, T. (2017). Precise and fast computation of the gravitational field of a general finite body and its application to the gravitational study of asteroid EROS. *The Astronomical Journal, 154*, 145. https://doi.org/10.3847/1538-3881/aa88b8

Goossens, S., et al. (2021). Mass and shape determination of (101955) bennu using differenced data from multiple OSIRIS-REx mission phases. *The Planetary Science Journal, 2*, 219. https://doi.org/10.3847/PSJ/ac26c4

Ho, A., Wold, M., Conway, J. T., & Poursina, M. (2021). Extended two-body problem for rotating rigid bodies. *Celestial Mechanics and Dynamical Astronomy, 133*(8). https://doi.org/10.1007/s10569-021-10034-8

Izzo, D., & Rathke, A. (2006). *The asteroid deflection formula for Earth quasi-co-orbiting asteroids*. ACT Technical Report, ACT-RPT-4100-DI-ADFECA—August 2006, 1–3.

Kang, J., & Zhu, Z. (2023). Passivity-based model predictive control for tethered despin of massive space objects by small space tug. *IEEE Transactions on Aerospace and Electronic Systems, 59*(2), 1239–1248.

Kang, J., Zhu, Z. H., Wang, W., Wang, C., & Li, A. (2019). Dynamics and de-spin control of massive target by single tethered space tug. *Chinese Journal of Aeronautics, 32*(3), 653–659.

Kang, J., Zhu, Z. H., & Santaguida, L. F. (2021). Analytical and experimental investigation of stabilizing rotating uncooperative target by tethered space tug. *IEEE Transactions on Aerospace and Electronic Systems, 57*(4), 2426–2437.

Kawaguchi, J. (2006). *Hayabusa, summary of guidance, navigation and control achievement in its proximity phase*. 2006 AIAA/AAS Astrodynamics Specialist Meeting, Keystone, Colorado, August 2006. Paper AIAA-2006-6533.

Kozai, Y. (1962). Secular perturbations of asteroids with high inclination and eccentricity, 1962. *The Astronomical Journal, 67*, 591–598.

Libert, A.-S., & Tsiganis, K. (2009). Kozai resonance in extrasolar systems. *Astronomy and Astrophysics, A & A, 493*, 677–686. https://doi.org/10.1051/0004-6361:200810843

Luo, Y.-Z., & Zhou, L.-N. (2014). Asteroid rendezvous mission design using multi-objective particle swarm optimization. *Mathematical Problems in Engineering, 2014.*, ID 823659, 13 pages. https://doi.org/10.1155/2014/823659

MacMillan, W. D. (1930). *The theory of the potential: Theoretical mechanics*. McGraw-Hill Book Company, Incorporated, (Dover Publications, 1958). 469 pages.

Morimoto, M., Yamakawa, H., Yoshikawa, M., Abe, M., & Yano, H. (2004). Trajectory design of multiple asteroid sample return missions. *Advances in Space Research, 34*, 2281–2285.

Nazari, M., Wauson, R., Critz, T., Butcher, E. A., & Scheeres, D. J. (2014). Observer-based body-frame hovering control over a tumbling asteroid. *Acta Astronautica, 102*, 124–139.

Panou, G., Korakitis, R., & Pantazis, G. (2020). Fitting a triaxial ellipsoid to a geoid model. *Journal of Geodetic Science, 10*, 69–82. https://doi.org/10.1515/jogs-2020-0105

Passalis, G., Sgouros, N., Athineos, S., & Theoharis, T. (2007). Enhanced Reconstruction of 3D Shape and Texture from Integral Photography Images. *Applied Optics, 46*(22), 5311–5320. https://doi.org/10.1364/ao.46.005311

Sánchez, J.-P., Neves, R., & Urrutxua, H. (2018). Trajectory design for asteroid retrieval missions: A short review. *Frontiers in Applied Mathematics and Statistics, 4*, 44. https://doi.org/10.3389/fams.2018.00044

Scheeres, D. J., Ostro, S. J., Hudson, R. S., DeJong, E. M., & Suzuki, S. (1998). Dynamics of orbits close to asteroid 4179Toutatis. *Icarus, 132*, 53–79.

Scheeres, D. J. & Augenstein, S. (2003). *Spacecraft motion about binary asteroids, Astrodynamics, Part II, Advances in the Astronautical Sciences Series, 116* (pp. 991–1010). Univelt, San Diego, AAS Paper 03-564.

Scheeres, D. J. (2004). Stability of relative equilibria in the full two-body problem. *Annals of the New York Academy of Sciences, 1017,* 81–94.

Scheeres, D. J., Hesar, S. G., Tardivel, S., Hirabayashi, M., Farnocchia, D., McMahon, J. W., Chesley, S. R., Barnouin, O., Binzel, R. P., Bottke, W. F., Daly, M. G., Emery, J. P., Hergenrother, C. W., Lauretta, D. S., Marshall, J. R., Michel, P., Nolan, M. C., & Walsh, K. J. (2016). The geophysical environment of Bennu. *Icarus, 276*, 116–140.

Shepard, M. K., et al. (2018). A revised shape model of asteroid (216) Kleopatra. *Icarus, 311*, 197–209.

Soler, T., & Han, J. Y. (2020). Determination of the parameters of the triaxial earth ellipsoid as derived from present-day geospatial techniques. *GPS Solutions, 24*(4), 117. https://doi.org/10.1007/s10291-020-01033-7

Soler, T., Han, J.-Y., & Huang, C. J. (2020). Estimating the variance-covariance matrix of the parameters of a fitted triaxial ellipsoid. *Journal of Surveying Engineering, 146*(2), 04020003.

Vepa, R., & Shaheed, M. H. (2021). Optimal trajectory synthesis for spacecraft asteroid rendezvous. *Symmetry, 13*, 1403. https://doi.org/10.3390/sym13081403

Watanabe, S., Tsuda, Y., Yoshikawa, M., Tanaka, S., Saiki, T., & Nakazawa, S. (2017). Hayabusa2 mission overview. *Space Science Reviews, 208*(2017), 1–14. https://doi.org/10.1007/s11214-017-0377-1

Werner, R. A., & Scheeres, D. J. (1997). Exterior gravitation of a polyhedron derived and compared with harmonic and mascon gravitation representations of asteroid 4769 Castalia. *Celestial Mechanics and Dynamical Astronomy, 65*(3), 313–344. https://doi.org/10.1007/BF00053511. ISSN 0923-2958

Wold, M., & Conway, J. T. (2021). The planar two-body problem for spheroids and disks. *Celestial Mechanics and Dynamic Astronomy, 133*(6), 27. https://doi.org/10.1007/s10569-021-10023-x

Youtao, G., Xi, L., Yuming, P., Bo, X., & Tanran, Z. (2019). Trajectory optimization of multiple asteroids exploration with asteroid 2010TK7 as main target. *Advances in Space Research, 63*(2019), 432–442.

Yu, Y., & Baoyin, H. (2012). Orbital dynamics in the vicinity of asteroid 216 Kleopatra. *Astronomy Journal, 143*(2012), 62–70.

Yu, Y., Cheng, B., Hayabayashi, M., Michel, P., & Baoyin, H. (2019). A finite element method for computational full two-body problem: I. The mutual potential and derivatives over bilinear tetrahedron elements. *Celestial Mechanics and Dynamical Astronomy, 131*, 51. https://doi.org/10.1007/s10569-019-9930-4

Yudintsev, V., & Aslanov, V. (2017). Detumbling space debris using modified yo-yo mechanism. *Journal of Guidance, Control, & Dynamics, 40*(3), 713–720.

Chapter 9
Physics of Plasmas

9.1 Plasma and Physical Laws

Charged particles are produced by ionization of a gas, by various means, which is equivalent to adding or removing an electron from the atoms of the gas. Thus ions (positively or negatively charged), electrons and electrically neutral particles coexist and form what is known as a plasma. A plasma is then a collection of the various charged particles that will move freely in response to applied or self-generated field to the entire collection. Thus a plasma is an ionized gas, consisting of a collection of charged (and neutral particles), which as a group are quasi-neutral and exhibit a collective behaviour governed by Maxwell's laws of electromagnetism. The physics and dynamics of plasmas is a vast subject which has been discussed by many authors (Holt & Haskell, 1965; Wiesemann, 2023; Lieberman & Lichtenberg, 1994; Shang & Surzhikov, 2018; Bittencourt, 2004; Goldston & Rutherford, 1995; Colonna & D'Angola, 2016). It is only being summarized in this chapter while focussing on those concepts that are important for space propulsion.

Briefly, Maxwell's laws of electromagnetism are Lorentz law, which gives rise to force due to the interaction between a current density $\vec{j}$ and the magnetic field $\vec{B}$, $\vec{F} = \vec{j} \times \vec{B}$, Gauss's and Faraday's laws, which define the electric field intensity $\vec{E}$ in terms of the charge density ρ, and its relationship to the magnetic field, $\varepsilon \nabla \cdot \vec{E} = \rho$, $\nabla \times \vec{E} = -\partial \vec{B} / \partial t$, and equations defining the magnetic field, including $\nabla \cdot \vec{B} = 0$, when no magnetic dipoles exist and Ampere's law, $\nabla \times \vec{B} = \mu_0 \left(\vec{j} + \varepsilon_0 \partial \vec{E} / \partial t \right)$. These will be discussed in some depth in a later section.

The most natural way of generating plasma is to subject the gas to extreme source of heat and heat it to a very high temperature. Energized electrons with a very high speed would collide with the atoms and break away the electrons in the outermost shells of the atom out of their orbits. The degree of ionization, in such a situation,

R. Vepa, *Space Vehicle Maneuvering, Propulsion, Dynamics and Control*,
https://doi.org/10.1007/978-3-031-65518-0_9

depends on temperature of the gas. A general relation between the degree of ionization and temperature may be established based on the statistical description of the plasma dynamics, which is assumed to be in state of thermodynamic equilibrium.

9.2 Electrostatics of Plasma

The electrostatic case refers to the special situation when the magnetic field $\vec{B}$ is either non-existent or when it is a constant. In this case one has $\nabla \times \vec{E} = -\partial \vec{B}/\partial t = 0$. It follows that the electric field can be obtained from an electric potential function, $\vec{E} = -\nabla \phi$. The relationship between the electric field intensity $\vec{E}$ and the charge density ρ, reduces to

$$\varepsilon \nabla^2 \phi = -\rho. \tag{9.1}$$

In a thermodynamically generated plasma, equilibrium is established between the rate of ionization that depends on the temperature T and rate of recombination of the ions that depends on the charge density ρ.

9.2.1 *Quasi-neutrality of the Plasma*

One of the main characteristic of plasma is its capacity to remain electrically neutral. So it follows that $|q_e n_e| \approx |q_i n_i|$. An imbalance on the local charge density will create an electrostatic force that will restore the charge equilibrium. So, $n_e \approx n_i$ and $|q_e| \approx |q_i|$. Thus the quasi-neutrality of the plasma is a state that it actively tries to achieve following a disturbance, by readjusting the local charge distribution. There is an important consequence.

9.2.2 *Plasma Oscillations*

Consider a cold and stationary plasma with $n_e \approx n_i = n_0$. In an electric field directed along the x-axis on a single electron with a mass and a charge, $F_{el,x} = m_e a_x = m_e d^2x/dt^2 = q_e E_x$. By Gauss's theorem, $\varepsilon_0 E_x = -n_e q_e x$. Hence, it follows

$$m_e \frac{d^2x}{dt^2} = -\frac{n_e q_e^2}{\varepsilon_0} x \Rightarrow \frac{d^2x}{dt^2} = -\frac{n_e q_e^2}{m_e \varepsilon_0} x = -\omega_{np}^2 x. \tag{9.2}$$

Thus, when the plasma is dominated by electrons, the natural frequency of plasma oscillations ω_{np}, also known as the Langmuir frequency (Bohm & Gross, 1949), satisfies

$$\omega_{np}^2 = n_e q_e^2 / m_e \varepsilon_0. \tag{9.3}$$

Quasi-neutrality provides a mathematical way of stating that even though the particles making up a plasma consist of free electrons and ions, their overall charge densities cancel each other in equilibrium. So if n_e and n_i are, respectively, the number densities of electrons and ions with charge state Z, then when these are locally balanced, $n_e \approx Zn_i$. In a similar manner to electron dominated plasmas, if the motion of the plasma is dominated by ions, the ion plasma frequency can be found to be

$$\Omega_{np}^2 = Zn_i q_e^2 / m_i \varepsilon_0. \tag{9.4}$$

Charged particles in an electric thruster generally have a Maxwellian distribution of the velocity, which is the most probable distribution of velocities for a group of particles in thermal equilibrium. In one dimension, the Maxwellian velocity distribution function is

$$f(v) = \sqrt{\frac{m}{2\pi k_B T_e}} \exp\left(\frac{m}{2k_B T_e} v^2\right). \tag{9.5}$$

where m is the mass of the particle, k_B is Boltzmann's constant and the width of the distribution is characterized by the equilibrium temperature T_e. The average kinetic energy of a particle in the Maxwellian distribution in one dimension is

$$\overline{E} = \frac{\int\limits_{-\infty}^{\infty} (1/2) m v^2 f(v) dv}{\int\limits_{-\infty}^{\infty} f(v) dv}. \tag{9.6}$$

By using the probability distribution function and integrating by parts, the average energy per particle in each dimension is

$$\overline{E} = (1/2) k_B T_e. \tag{9.7}$$

To know how fast the electrons and ions actually move, for equal ion and electron temperatures ($T_e = T_i$), given that the Boltzmann constant is k_B, one has in three dimensions, from the thermal kinetic energies

$$\overline{E} = (1/2) m_e \overline{v}_e^2 = (1/2) m_i \overline{v}_i^2 = (3/2) k_B T_e. \tag{9.8}$$

Thus it follows even for the lightest ions; the ion velocity is only 0.0233 of the electron velocity and hence almost stationary relative to an electron. An approximation of the number density of the ions is then by

$$n_i \approx n_0 = N_A \rho_m / A_n, \tag{9.9}$$

where N_A is the Avogadro number, ρ_m is the mass density of the plasma and A_n is the atomic number. In thermal equilibrium, the electron number density follows a Boltzmann distribution and is related to the ion number density, in terms of the potential function of the disturbance by

$$n_e = n_i \exp(q_e \phi / k_B T_e), \tag{9.10}$$

The potential function satisfies Gauss's law and is given by

$$\nabla^2 \phi = -\rho/\varepsilon_0 = -(q_e/\varepsilon_0)(n_i - n_e), \tag{9.11}$$

Combining Eqs. (9.10) and (9.11), and using spherical coordinates with spherical symmetry, Eq. (9.11) is

$$\nabla^2 \phi \equiv \frac{1}{r^2} \frac{d}{dr}\left(r^2 \frac{d\phi}{dr}\right) = -\frac{q_e}{\varepsilon_0} n_i \left(1 - \exp\left(\frac{q_e \phi}{k_B T_e}\right)\right), \tag{9.12}$$

Assuming that $\phi \to 0$ as $r \to \infty$, an approximate solution of the equation is given by

$$\phi = \frac{1}{4\pi r \varepsilon_0} \exp\left(-\frac{r}{\lambda_D}\right). \tag{9.13}$$

The characteristic length-scale λ_D scaling the distance in the exponential factor is known as the Debye length and is given by

$$\lambda_D = \sqrt{\varepsilon_0 k_B T_e / \left(n_e q_e^2\right)}. \tag{9.14}$$

The Debye length is a fundamental property of nearly all plasmas of interest and depends equally on the plasma's temperature and density. The Debye sphere is a sphere of radius λ_D. The total number of particles contained in a Debye sphere is generally expected to be very large, so one can expect the particles to behave collectively in a predictable way. It is often important to compare the Debye length λ_D to the de Broglie wavelength λ_B, which is defined as the ratio of Planck's constant and the thermal momentum of an electron

$$\lambda_B = \bar{h}/m_e v_{th}, \quad \frac{1}{2} m_e v_{th}^2 = k_B T_e \Rightarrow \lambda_B = \bar{h}/\sqrt{2 m_e k_B T_e}. \tag{9.15}$$

Hence, using Eq. (9.3) for ω_{np}

$$\frac{\lambda_D}{\lambda_B} = \sqrt{2}\frac{k_B T_e}{\hbar\omega_{np}}. \tag{9.16}$$

9.2.3 Saha Ionization Equation

Consider a single nucleus, if it has a single bound electron, then it could be considered to be an atom. Without the electron, it is an ion. A plasma is a quasi-neutral gas of charged particles such as ions and electrons showing a collective behaviour. Based on the Boltzmann distribution, the probability electron is in the energy state ε_k is

$$w_k = A\exp(\varepsilon_k/T). \tag{9.17}$$

where A is a constant. Using the principles enunciated by Bohr, and the definition of the partition function representing the ratio of the quantum volume to the volume of an electron, it can be shown that the probability w_a of a proton having a bound electron is

$$w_a = \frac{\exp\left(I_p/T\right)}{\exp\left(I_p/T\right) + 1/\left(\lambda_B^3 n_e\right)}. \tag{9.18}$$

where λ_B is a De Broglie wavelength of electrons for thermal motion, I_p is the ionization potential of the lowest (for a hydrogen atom) energy level (13.6 eV) and n_e is the number of electrons per unit volume of the plasma. The corresponding probability for ions is

$$w_i = 1 - w_a = \frac{1/\left(\lambda_B^3 n_e\right)}{\exp\left(I_p/T\right) + 1/\left(\lambda_B^3 n_e\right)}. \tag{9.19}$$

Then, the relative ion density is

$$\frac{n_i}{n_a} = \frac{w_i}{w_a} = \frac{1}{\lambda_B^3 n_e}\exp\left(-I_p/T\right) \Rightarrow \frac{n_e n_i}{n_a} = \frac{1}{\lambda_B^3}\exp\left(-I_p/T\right). \tag{9.20}$$

Electrons, ions and atoms could have several internal degrees of freedom, which affect their partition functions. For instance, electrons have two spin states. Consequently they are assigned thus the statistical weight of 2, because each of the energy levels can be occupied by two electrons. Therefore, in the last equation we should replace n_e with n_e/g_e, where g_e is the statistical weight of the electron.

Taking into account the statistical weights for ions and atoms, the relative ion density is expressed as

$$\frac{n_e n_i}{n_a} = \frac{g_e g_i}{g_a \lambda_B^3} \exp\left(-I_p/T\right). \tag{9.21}$$

If the degree of ionization is defined as

$$\alpha = \frac{n_e}{n_{a0}} = \frac{n_i}{n_{a0}}, \tag{9.22}$$

where n_{a0} is the total number of atoms and ions, then $n_a = (1-\alpha)\ n_{a0}$ and the equation for the degree of ionization is given by

$$\frac{\alpha^2}{(1-\alpha)} = \frac{g_e g_i}{n_{a0} g_a \lambda_B^3} \exp\left(-I_p/T\right). \tag{9.23}$$

This is the Saha equation. The Saha ionization equation, also known as Saha-Langmuir equation, describes the ionization state of a plasma or ionized gas relating it to the temperature and the pressure. The equilibrium is achieved between the ionization rate (that depends on the temperature) and recombination rate (that depends on the charge density). In general, the Saha ionization equation may be expressed as

$$\frac{n_i}{n_n} = \left(\approx 2.4 \times 10^{21}\right) = \frac{T^{3/2}}{n_i} e^{\left(-\frac{E_i}{kT}\right)}. \tag{9.24}$$

with $I_p = E_i/k$. Because of the nature of the temperature distribution on the Earth's surface, the plasma particles present near the Earth's surface are relatively very low.

9.3 The Hall Effect or the Hall Voltage

The mechanisms for momentum transfer in a plasma are governed by the Lorentz force. In the absence of a physical wall, the charges deflected by the applied magnetic field accumulate at the two faces of a conductor. A net voltage is established which is known as the Hall voltage.

$$\mathbf{F} = q(\mathbf{E} + \mathbf{v} \times \mathbf{B}) = 0 \Rightarrow \mathbf{E} = -\mathbf{v} \times \mathbf{B} \Rightarrow V = \mathbf{E}/d. \tag{9.25}$$

The voltage induces a current in the direction mutually perpendicular to both **E** and **B**. Furthermore since

$$\mathbf{F} = q(\mathbf{E} + \mathbf{v} \times \mathbf{B}) = 0 \Rightarrow \mathbf{v} \equiv v_e = -\mathbf{E} \times \mathbf{B}/B^2. \tag{9.26}$$

The ions are accelerated by the electrostatic field **E** and the ions are in a quasi-neutral plasma. An equal and opposite force is applied to the free electrons in that zone due to applied radial magnetic field **B**, drift azimuthally (so normal to the **E** × **B** direction) with a drift velocity, enough to make the force equal and opposite. Thus in polar coordinates, with an applied radial magnetic field and an electric field in the forward direction, the azimuthal electron drift velocity is

$$v_{e,\theta} = -E_x/B_r. \tag{9.27}$$

The corresponding Hall Current is given by

$$\boldsymbol{j}_{\text{Hall}} = -qn_e v_{e,\theta} = qn_e \boldsymbol{E} \times \boldsymbol{B}/B^2. \tag{9.28}$$

Given a Hall current in the plasma, the magnetic force density, the Lorentz force acting on the plasma is

$$\boldsymbol{f} = \boldsymbol{j}_{\text{Hall}} \times \boldsymbol{B} = -qn_e v_e = qn_e \left(\boldsymbol{E} \times \boldsymbol{B}/B^2\right) \times \boldsymbol{B} = qn_e \boldsymbol{E}. \tag{9.29}$$

The motion of the ions happen in a quasi-neutral plasma so there's no Child-Langmuir space charge (or accumulation of space charge) or any saturated current limitation.

9.4 Force Density on an Ion Stream in an Axial Electric Field: Magnitude of the Force Density

If the force acting on each charged particle is averaged in time over periods longer than characteristic atomic or molecular orbital times and summed over the particles contained in a volume, ΔV, where ΔV is large compared with atomic or molecular dimensions, then one can divide this total averaged force by ΔV to obtain the force density given by

$$\mathbf{F} = \rho\mathbf{E} + \mathbf{j} \times \mathbf{B}. \tag{9.30}$$

The magnitude of the force density on a stream of ions in an electric field is given by

$$|\mathbf{F}| = \left|\vec{F}\right| = \frac{1}{2}\varepsilon_0 E_a^2, \tag{9.31}$$

where

$$E_a = \frac{4V}{3d}, \text{ and } \varepsilon_0 = 8.85 \times 10^{-12} \frac{\text{Farads}}{m}. \tag{9.32}$$

The average force is then give as

$$F_{\text{ave}} = \frac{1}{2}qE_a = \frac{1}{2}\frac{C}{x}E_a^2 = \frac{1}{2}\varepsilon_0 E_a^2. \tag{9.33}$$

9.5 Probabilistic Nature of Plasma Velocities: Plasma Velocities in an Electric Potential

It was already observed in Sect. (9.2.2) that a particle of a gas that is in thermal equilibrium has a distribution of velocities in three dimensions. The most appropriate distribution is the Maxwellian distribution. In three dimensions, the Maxwellian velocity distribution function is

$$f(u) = f_0 \exp\left(\frac{m}{2k_B T_e}u^2\right), \tag{9.34}$$

where the constant f_0 is given by

$$f_0 = n(m/2\pi k_B T_e)^{3/2}. \tag{9.35}$$

In Eq. (9.35), the quantity n is the particle density and is related to the distribution function by the relation

$$n^{-1} = \int_{-\infty}^{\infty} f(u)du. \tag{9.36}$$

In the presence of an external electric potential ϕ

$$f(u) = f_0 \exp\left(\frac{(mu^2/2) - q\phi}{k_B T_e}\right), \tag{9.37}$$

Integration of the velocity distribution gives the Boltzmann relation

$$n_e = n_0 \exp\left(\frac{q\phi}{k_B T_e}\right). \tag{9.38}$$

The Boltzmann relation relates the electron density with a potential to the same in the zero-potential region.

9.5.1 Average Kinetic Energy and Velocities

It has already been stated that in one and in three dimensions, respectively

$$\overline{E}=\frac{1}{2}k_BT_e,\ \overline{E}=\frac{3}{2}k_BT_e. \tag{9.39}$$

It can also be shown that the mean, RMS and most probable velocities are all different but proportional to $\sqrt{k_BT_e/m}$.

The mean velocity is

$$\overline{v}=\sqrt{8k_BT_e/\pi m}. \tag{9.40}$$

The RMS velocity is

$$v_{\mathrm{RMS}}=\sqrt{3k_BT_e/m}. \tag{9.41}$$

The most probable velocity is

$$\widetilde{v}=\sqrt{2k_BT_e/m}. \tag{9.42}$$

9.5.2 Collisions Between Charged Particles and Debye Shielding

Collisions mediate the transfer of energy and momentum between various species in a plasma and allow a treatment of highly ionized plasma as a single conducting fluid with resistivity determined by electron-ion collisions. Collisions that conserve total kinetic energy are called elastic. Some examples are atom-atom, electron-atom, ion-atom (charge exchange), etc. In inelastic collisions, there is some exchange between potential and kinetic energies of the system. Examples are electron-impact ionization/excitation and collisions with surfaces. Both types of collision processes are considered in this section. The dominant process in a plasma is elastic Coulomb collisions between charged particles.

Slowly inserting a single additional particle or "test" particle with a given charge into an initially unperturbed, spatially uniform neutral plasma, with equal ion and electron densities, will perturb the equilibrium, as the inserted particle would interact with each of the particles in the plasma. Consequently, particles having the same polarity as the inserted particle will be slightly repelled whereas particles of the opposite polarity will be slightly attracted. The net resulting small displacements resulting from these repulsions and attractions will contribute to a small but finite change in the potential in the plasma. The change in the potential is due to the cumulative sum of the inserted particle's potential and the potentials of the plasma particles that have displaced slightly in response to the inserted particle. The net resulting small displacement of plasma particles is called shielding or screening of the inserted particle because the displacement tends to reduce the effectiveness of the inserted particle's field which can be shown by using Poisson's equation.

The electro-static force between two charged particles is

$$F = q_1 q_2 / \left(4\pi\varepsilon_0 r^2\right). \tag{9.43}$$

So, in theory as $r \to 0$, $F \to \infty$. A fundamental characteristic of the plasma is its ability to shield out electric potentials that are applied to it. This is known as Debye shielding. In a plasma, however, the Debye shielding limits the range of the force so that an effective cross section can be found.

Ions are generally very much heavier than electrons. In the case of argon, the ion is almost 80,000 times heavier than an electron. Consequently electrons are much more mobile than ions and can only be contained by the application of a potential. This potential is usually referred to as the plasma potential and is established so as to generate an equal flux of electrons and ions near a wall in steady state. Within the interior of the plasma, the potential is approximately constant, but nearer the walls, within a few Debye lengths, it begins to change rapidly. This region is referred to as a plasma sheath.

The potential at a distance r due to a single charge is $\phi = q/(4\pi\varepsilon_0 r)$. When immersed in a plasma, a positive charge attracts negative charges and repels positive ions. As a result, the number density of ions and electrons will vary slightly around a charge band yet will be in charge balance for large r (Gibbon, 2016). The effect of this screening will be to exponentially damp over a distance comparable to the Debye length: $\overline{\phi} = \phi \exp(-r/\lambda_D)$ with λ_D given by Eq. (9.14) (from Poisson's equation for the potential ϕ.) The Debye length λ_D, given by Eq. (9.14), may be interpreted as a probabilistic measure of the shielding distance or thickness of the sheath. As the density increases, the Debye length decreases, since each layer of plasma contains more electrons. The Debye length increases with increasing $k_B T_e$. In fact, without thermal agitation, the charge cloud would collapse to an infinitely thin layer. To define the layer, the temperature of the electrons is used because they're more mobile than the ions, so they do the shielding by moving so as to create a surplus or a deficit of negative charge.

One could relate the quasi-neutrality to the Debye shielding. If the dimensions of a system are much larger than the Debye length, then whenever local concentrations of charge arise or external potentials are introduced into the system, these are shielded out in a distance less than the dimension, leaving the bulk of the plasma free of large electric potentials or fields. Outside of the sheath on the wall or on an obstacle within which Debye shielding is effective, the gradient of the field is very small, quasi-neutrality follows, and a small charge imbalance would give rise to potentials of the order of $k_B T_e$.

9.5.3 *Collective Behaviour of Plasma Particles*

An ionized gas can be considered to be a plasma if it's dense enough, i.e. if its dimensions are much greater than the Debye length (and the time scales are much greater than $1/\omega_{np}$). The definition of Debye shielding is only valid, from a statistical point of view, only if there are enough particles in the charge cloud. This number can be estimated with a parameter called "Plasma Parameter"—Λ, the number of particles inside a Debye sphere. Important characteristic of the plasma is its collective behaviour. Depending on the density, a plasma behaves sometimes like a fluid where the collective effects are "preponderant" and, sometimes, like a collection of individual particles. The collective behaviour of the plasma is also due to the collision cross section between electrons and neutral particles being much lower than the collision cross section between electrons and ions. Thus, the frequency of the collisions with the neutral particles should be low enough. Thus, the frequency of the collisions with the neutral particles should be low enough so that the dynamics of the motion of the plasma is governed primarily by the electromagnetic forces and not by the hydrodynamic forces. Over the corresponding domain, it is the magneto-hydro dynamic equations that are important.

9.6 Motion of a Single Charged Particle in an EM Field

The dynamics of plasmas under the action of magnetic and inertial force is of some importance (Gerwin, 1966; Goldston & Rutherford, 1995). Consider the motion of a single charged particle in an electromagnetic field. There are two cases to consider: (i) in the presence of only the magnetic field and (ii) in the presence of both the electric and magnetic fields. In the first case, with only the magnetic field present

$$m\frac{d\vec{v}}{dt} = \vec{F} = q\left(\vec{E} + \vec{v} \times \vec{B}\right) = q\left(\vec{v} \times \vec{B}\right). \tag{9.44}$$

If magnetic field is uniaxial, $\vec{B} = B\boldsymbol{k}$, where $\boldsymbol{k} = \vec{B}/B = \boldsymbol{b}$, the equations of motion satisfied by the velocity components are

$$m\frac{d}{dt}\begin{bmatrix} v_x \\ v_y \\ v_z \end{bmatrix} = qB\begin{bmatrix} v_y \\ -v_x \\ 0 \end{bmatrix}. \tag{9.45}$$

The in-plane velocity components satisfy

$$\frac{d^2}{dt^2}\begin{bmatrix} v_x \\ v_y \end{bmatrix} = \frac{qB}{m}\frac{d}{dt}\begin{bmatrix} v_y \\ -v_x \end{bmatrix} = -\left(\frac{qB}{m}\right)^2\begin{bmatrix} v_x \\ v_y \end{bmatrix}. \tag{9.46}$$

These are the equations of two uncoupled simple harmonic oscillators with the same natural frequency acting in two mutually perpendicular directions, resulting in a circular motion. The natural frequency, known as the cyclotron frequency, is given by $\omega_{nc} = qB/m$. Both velocity components are sinusoidal but out of phase by 90°. Thus the resulting motion is a rotation in a circle and in plane normal to the applied **B**-field. The circle's radius is known as the Larmor radius and is

$$r_L = v_{xy}/\omega_{nc} = v_{xy}m/B|q|. \tag{9.47}$$

The magnetic field generated by the circular rotation of the charged particle opposes the applied field.

In the out-of-plane or the Z direction

$$m\frac{dv_z}{dt} = 0. \tag{9.48}$$

The solution for the motion is

$$\begin{bmatrix} x \\ y \\ z \end{bmatrix} = \begin{bmatrix} x_0 \\ y_0 \\ z_0 \end{bmatrix} + \left[r_L \sin(\omega_{nc}t) \pm r_L \cos(\omega_{nc}t) v_{z0}t \right]. \tag{9.49}$$

Thus there is a net drift in the direction of the applied magnetic field.

9.6.1 With an Applied Electric Field and General Motion

In this case

$$m\frac{d\vec{v}}{dt} = \vec{F} = q\left(\vec{E} + \vec{v} \times \vec{B}\right). \tag{9.50}$$

With uniform electric field and $E_y = 0$

$$m\frac{d}{dt}\begin{bmatrix} v_x \\ v_y \\ v_z \end{bmatrix} = q\begin{bmatrix} E_x \\ 0 \\ E_z \end{bmatrix} + qB\begin{bmatrix} v_y \\ -v_x \\ 0 \end{bmatrix}. \tag{9.51}$$

For the in-plane and out-of-plane velocity components, respectively

$$\frac{d^2}{dt^2}\begin{bmatrix} v_x \\ v_y \end{bmatrix} = \frac{q}{m}\frac{d}{dt}\begin{bmatrix} E_x \\ 0 \end{bmatrix} + \frac{qB}{m}\frac{d}{dt}\begin{bmatrix} v_y \\ -v_x \end{bmatrix} = -\left(\frac{q}{m}\right)^2 B\begin{bmatrix} 0 \\ E_x \end{bmatrix} - \left(\frac{qB}{m}\right)^2\begin{bmatrix} v_x \\ v_y \end{bmatrix}, \quad m\frac{dv_z}{dt} = qE_z. \tag{9.52}$$

The result is a helical or growing screw motion. The centre of the Larmor circle drifts, with the velocity

$$v_{\text{drift}} = \boldsymbol{E} \times \boldsymbol{B}/B^2. \tag{9.53}$$

Considering the general motion, given by Eq. (9.50), the component of this equation parallel to the magnetic field is

$$mdv_{\|}/dt = qE. \tag{9.54}$$

In the normal direction

$$v_{\perp} = \frac{\vec{E} \times \vec{B}}{B^2} + v_0 \frac{qB\omega_{nc}}{m} \exp(j(\omega_{nc}t + \gamma)). \tag{9.55}$$

The first term represents the drift and arises from

$$\vec{E} + \vec{v} \times \vec{B} = 0 \Rightarrow \vec{E} \times \vec{B} = -\vec{v} \times \vec{B} \times \vec{B} = vB^2 - \vec{B}\left(\vec{v} \cdot \vec{B}\right) = v_{\perp}B^2$$
$$\Rightarrow v_{\perp} = \vec{E} \times \vec{B}/B^2. \tag{9.56}$$

The equations derived can be generalized for motion under an applied force. In particular the drift velocity of the Larmor orbit centre (LOC) can be expressed as

$$v_{\text{drift}} = v_{\text{LOC}} = \boldsymbol{E} \times \boldsymbol{B}/B^2 = (\boldsymbol{E}q) \times \boldsymbol{B}/qB^2 = \boldsymbol{F} \times \boldsymbol{B}/qB^2. \tag{9.57}$$

In the special case of a gravitational force

$$v_{\text{LOC}} = \boldsymbol{F} \times \boldsymbol{B}/qB^2 = m\boldsymbol{g} \times \boldsymbol{B}/qB^2. \tag{9.58}$$

Unlike, earlier the drift velocity is now a function of both the mass and charge on the particle!

9.7 Plasma Instability, Drifts due to Non-uniform Magnetic Fields and Curvature

The dynamic nature of magnetically confined plasmas is to be expected since the pressure gradient, which is maintained by the magnetic Lorentz force through the balance of forces can act as a source of free energy that can be released through the plasma drift motion, which could be considered to be a form of instability. The pressure gradient-driven drift modes in magnetically confined plasmas are briefly considered in this section. For example, under a gravitational force, ions and electrons will drift in opposite directions, causing a net current

$$\boldsymbol{j} = n(m+M)\boldsymbol{g} \times \boldsymbol{B}/B^2. \tag{9.59}$$

The magnitude of the LOC drift velocity is negligible. Although the lines of force are curved, there's an effective rotational motion due to the centrifugal force that is not negligible, is independent of the mass and will create a plasma instability called "gravitational instability".

In the case of a non-uniform magnetic field, the gradient of the field causes the Larmor radius to be different at the bottom of the orbit and at the top. It results in a drift, in opposite directions for ions and electrons, perpendicular to both the field and its gradient. The drift is proportional to both the position and velocity. Given that

$$\vec{F} = q\left(\vec{E} + \vec{v} \times \vec{B}\right), \quad F_y = -qv_x \vec{B}_y = -qv_{xy}\cos(\omega_{nc}t)\left(B_0 \pm r_L \cos(\omega_{nc}t)\frac{\partial B}{\partial y}\right) \tag{9.60}$$

Averaging the expansion

$$|v_{\text{LOC}}| = \left|\boldsymbol{F} \times \boldsymbol{B}/qB^2\right| = F_y/qB = \left(\mp v_{xy} r_L/2B\right)\partial B/\partial y. \tag{9.61}$$

Thus the gradient drift velocity is therefore given by

$$\vec{v}_{\nabla B} = \mp\left(qv_{xy} r_L/2B\right)\frac{\boldsymbol{B} \times \nabla \boldsymbol{B}}{qB^2}. \tag{9.62}$$

in opposite directions for ions and electrons, transverse to the applied non-uniform magnetic field.

When the force acting on the particle is centrifugal force and the magnetic field is axi-symmetric, i.e. the lines of force are circular

$$\vec{v}_{\text{LOC}} = \frac{mv_c^2}{R_c}\frac{\boldsymbol{R}_c}{R_c} \times \frac{\boldsymbol{B}}{qB^2} = \frac{mv_c^2}{q}\frac{\boldsymbol{R}_c \times \boldsymbol{B}}{R_c^2 B^2}. \tag{9.63}$$

Further when $\boldsymbol{B}$ varies as $\boldsymbol{R}_c/R_c^2$ and it follows that

$$\frac{\nabla \boldsymbol{B}}{\boldsymbol{B}} = -\frac{\boldsymbol{R}_c}{R_c^2}, \text{ then } \vec{v}_{\nabla B} = \frac{mv_{xy}^2}{2qB^2}\frac{\boldsymbol{R}_c \times \boldsymbol{B}}{R_c^2}, \quad \vec{v}_{\text{LOC}} + \vec{v}_{\nabla B} = \frac{m}{qB^2}\frac{\boldsymbol{R}_c \times \boldsymbol{B}}{R_c^2}\left(v_c^2 + \frac{v_{xy}^2}{2}\right). \tag{9.64}$$

9.7.1 Axially Varying Magnetic Fields, Radial Component of the Magnetic Field and Components of the Lorentz Force

Consider a magnetic field directed in the axial z direction with a magnitude that varies along z. It is assumed that the field is axially symmetric, with the angular component in the transverse plane equal to zero and its variation with the angular axis also equal to zero. There a non-zero radial component B_r that creates a force which can trap a particle in a magnetic field. Since there can exist no free magnetic dipoles, $\nabla \cdot \vec{B} = 0$, and the gradient operator in cylindrical coordinates is given by

$$\nabla = i\left(\frac{\partial}{\partial r} + \frac{1}{r}\right) + j\frac{1}{r}\frac{\partial}{\partial \theta} + k\frac{\partial}{\partial z}, \tag{9.65}$$

it follows that

$$\frac{\partial B_r}{\partial r} + \frac{1}{r}B_r = \frac{1}{r}\frac{\partial (rB_r)}{\partial r} = -\frac{\partial B_z}{\partial z}. \tag{9.66}$$

Integrating the preceding equation after making a simplifying assumption

$$B_r = -\frac{1}{r}\int_0^r r\frac{\partial B_z}{\partial z}dr = -\frac{r}{2}\frac{\partial B_z}{\partial z}\bigg|_{r=0}. \tag{9.67}$$

The Lorentz force and its components are respectively given by

$$\vec{F} = q\left(\vec{v} \times \vec{B}\right), \quad \begin{bmatrix} F_r \\ F_\theta \\ F_z \end{bmatrix} = q\begin{bmatrix} 0 & -v_z & v_\theta \\ v_z & 0 & -v_r \\ -v_\theta & v_r & 0 \end{bmatrix}\begin{bmatrix} B_r \\ B_\theta \\ B_z \end{bmatrix}$$
$$= q\begin{bmatrix} v_\theta B_z \\ v_z B_r - v_r B_z \\ -v_\theta B_r \end{bmatrix}. \tag{9.68}$$

In the z directions, $F_z = F_{\|}$

$$F_z = -qv_\theta B_r = \frac{qrv_\theta}{2}\frac{\partial B_z}{\partial z}\bigg|_{r=0} = \frac{qv_\theta^2}{2\omega_{nc}}\frac{\partial B_z}{\partial z}\bigg|_{r=0} = -\frac{mv_\perp^2}{2B}\frac{\partial B_z}{\partial z}\bigg|_{r=0} = -\mu_m\frac{\partial B_z}{\partial z}\bigg|_{r=0}, \tag{9.69}$$

with the magnetic moment $\mu_m = mv_\perp^2/2B$.

For the component $v_{\|}$ it can be shown that

$$mv_{\|}\frac{dv_{\|}}{dt}=v_{\|}F_{\|}=\frac{d}{dt}\left(\frac{1}{2}mv_{\|}^{2}\right)=-\mu_{m}v_{\|}\frac{dB}{dz}=-\mu_{m}\frac{dB}{dt}. \tag{9.70}$$

From the principle of energy conservation

$$\frac{d}{dt}\left(\frac{1}{2}mv_{\|}^{2}+\frac{1}{2}mv_{\perp}^{2}\right)=0=-\mu_{m}\frac{dB}{dt}+\frac{d}{dt}(\mu_{m}B)\Rightarrow\frac{d\mu_{m}}{dt}=0. \tag{9.71}$$

It follows that μ_m is the invariant, lowest order. It is the basis for one of the main schemes for plasma confinement: the magnetic mirror.

9.7.2 Magnetic Mirrors and Plasma Confinement

Returning to Eq. (9.50), the scalar product of the equation of motion with the velocity gives

$$m\frac{d\vec{v}\cdot\vec{v}}{dt}=q\vec{v}\cdot\left(\vec{E}+\vec{v}\times\vec{B}\right)=q\vec{v}\cdot\vec{E}. \tag{9.72}$$

This equation governs the evolution of the particle energy during its motion. Averaging the preceding equation one can show, the rate of change of the total kinetic energy satisfies

$$\frac{d(KE)}{dt}=q\vec{v}\cdot\vec{E}+\mu_{m}\frac{\partial B}{\partial t}. \tag{9.73}$$

If the electric field is obtained from a potential, so $\vec{E}=-\nabla\cdot\phi$ it can be shown that the total particle energy is a constant.

A charged particle neither gains nor loses energy as it moves around in non-time-varying electromagnetic fields. Because both the magnetic moment and total energy are constants of the motion, the equation for the total energy may be rearranged for the particle velocity components along or orthogonal to the magnetic field lines. While charged particles can still drift along or opposite the directions of the magnetic field lines in regions where the total energy exceeds the sum of the magnetic and electric energy and the kinetic energy along the electric field lines, the particles are excluded from these regions otherwise (because imaginary velocities are not feasible). Charged particles can still reverse direction at those points on magnetic field lines where the total energy remains conserved. Such points are termed bounce points or mirror points. Such a device capable of confining a collision-less or a very high temperature plasma could be physically constructed only by confining a hot plasma using a magnetic field. This technique is called magnetic confinement. Magnetic confinement involves the confinement of stabilized plasma over a long period of time.

9.7.3 Non-uniform Electric Fields, Drift Instabilities and Time-Varying Fields

In the case electric field is not uniform

$$m\frac{d\vec{v}}{dt} = \vec{F} = q\left(\vec{E}(x) + \vec{v} \times \vec{B}\right). \tag{9.74}$$

With $E_y = 0$, from Eq. (9.52) and $E_x = E_0 \cos(kx)$

$$\frac{d^2}{dt^2}\begin{bmatrix} v_x \\ v_y \end{bmatrix} = \pm\omega_{nc}\frac{1}{B}\frac{d}{dt}\begin{bmatrix} E_x \\ 0 \end{bmatrix} - \omega_{nc}^2\frac{1}{B}\begin{bmatrix} 0 \\ E_x \end{bmatrix} - \omega_{nc}^2\begin{bmatrix} v_x \\ v_y \end{bmatrix}. \tag{9.75}$$

One is interested primarily in the drift velocity and so the oscillatory component is got rid of by averaging over a period, which results in

$$v_\perp = \frac{\vec{E} \times \vec{B}}{B^2}\left(1 - \frac{r_L^2}{4}k^2\right) = \frac{\vec{E} \times \vec{B}}{B^2}\left(1 - \frac{r_L^2}{4}\frac{\nabla^2 E_x}{E_x}\right). \tag{9.76}$$

The additional term in $v_\perp$ is known as the "finite Larmor radius effect". It is representative of a "drift instability" that makes the electric field drive an instability in the plasma. Also the grad-B drift is a similar effect, but since in the grad-B drift it is proportional to $\nabla\vec{B}$ while in this case it is proportional to $\nabla^2\vec{E}$, the non-uniformity of the electric field is more important, for large gradients. Drift instabilities belong to a general class of instabilities also known as "micro-instabilities". The effects of curvature of the magnetic field on the drift-dissipative instability of a non-uniform plasma are such that the curvature of the magnetic field stabilizes the instability produced by particle collisions.

A time variation of the electric field leads to another type of drift. Since the drift is in opposite direction for ions and electrons, it will result in a polarization current, hence known as a polarization drift. Time variation of the magnitude of magnetic fields lead to variations in the period of gyration and of the direction of the drift, and this variation is known as inertial drift.

9.7.4 Guiding Centre Drifts: Summary

One may summarize all of the drift expressions (Goldston & Rutherford, 1995):

$$\text{Electric field:}\quad v_\perp = \vec{E} \times \vec{B}/B^2, \tag{9.77a}$$

$$\text{General force:}\quad v_{\mathrm{LOC}} = \boldsymbol{F} \times \boldsymbol{B}/qB^2, \tag{9.77b}$$

$$\text{Gravitational force:}\quad v_{\mathrm{LOC}} = m\boldsymbol{g} \times \boldsymbol{B}/qB^2, \tag{9.77c}$$

$$\text{Non-uniform electric field:}\quad v_{\perp} = \frac{\vec{E} \times \vec{B}}{B^2}\left(1 - \frac{r_L^2}{4}\frac{\nabla^2 E_x}{E_x}\right), \tag{9.77d}$$

$$\text{Grad B field:}\quad \vec{v}_{\nabla B} = \frac{\pm v_{\perp} r_L}{2B}\frac{\boldsymbol{B} \times \nabla \boldsymbol{B}}{B^2}, \tag{9.77e}$$

$$\text{Curvature drift:}\quad v_{\mathrm{LOC}} = \left(mv_{\parallel}^2/q\right)\boldsymbol{R}_c \times \boldsymbol{B}/\left(R_c^2 B^2\right), \tag{9.77f}$$

$$\text{Curved vacuum field:}\quad \vec{v}_{\mathrm{LOC}} + \vec{v}_{\nabla B} = \frac{m}{qB^2}\frac{\boldsymbol{R}_c \times \boldsymbol{B}}{R_c^2}\left(v_{\parallel}^2 + \frac{v_{\perp}^2}{2}\right), \tag{9.77g}$$

$$\text{Polarization drift:}\quad v_P = \frac{1}{B\omega_{nc}}\frac{d\vec{E}}{dt}. \tag{9.77h}$$

9.8 Plasma as a Fluid: Plasma Modelling and Electromagnetic Fields

Computer modelling of plasmas involves, with increasing detail, ideal magneto-hydrodynamic (MHD) models at the lowest level, where (MHD) the plasma is modelled as one fluid coupled with the associated electromagnetic (EM) fields and at the highest level of abstraction one has the particle-in-cell (PIC) simulations, where the electron and ion motions are permitted to dynamically evolve in accordance with the applied Lorentz forces in the EM fields.

The full set of Maxwell's equations are applied to define the electromagnetic fields and their effects. In between PIC and MHD, there exist a several range of descriptions for the plasma, including (i) gyro-kinetic models, where the guiding centre position of each particle is followed in space and time, instead of solving the whole gyro-motion of particles, hybrid models of plasma dynamics, where usually electrons are treated as a neutralizing massless fluid while ions are treated as kinetic particles, with some ion species, occasionally, being treated as a fluid and multi-fluid models for the plasma, where different plasma species (or different populations of the same species) are treated as interacting fluids coupled with EM fields. The associated electromagnetic fields may be

(i) Purely electrostatic or
(ii) Fully electromagnetic

Maxwell's equations could be different for vacuum or in a specific medium. They are briefly reviewed in the next section.

9.8.1 Maxwell's Equations and the Complete Equations of Fluid Flow

In the following, ρ is the electric charge density, $\boldsymbol{J}$ is the electric current density, the electric field is $\boldsymbol{E}$, and the magnetic field (flux density) is $\boldsymbol{B}$.

$$\text{Gauss's law:} \quad \nabla \cdot \mathbf{D} = \rho, \tag{9.78a}$$

$$\text{Gauss's law:} \quad \nabla \cdot \mathbf{B} = 0, \tag{9.78b}$$

$$\text{Faraday's law:} \quad \nabla \times \mathbf{E} = -\frac{\partial \mathbf{B}}{\partial t}, \tag{9.78c}$$

$$\text{Amphere's law:} \quad \nabla \times \mathbf{H} = \mathbf{J} + \frac{\partial \mathbf{D}}{\partial t}. \tag{9.78d}$$

The electric displacement is

$$\mathbf{D} = \varepsilon_r \varepsilon_0 \mathbf{E}. \tag{9.79}$$

The magnetic intensity is

$$\mathbf{H} = \mathbf{B}/\mu_0 \mu_r. \tag{9.80}$$

The basic equations of motions characterizing any fluid flow are the continuity and momentum equations. In the absence of body forces, the equations of continuity and momentum may be expressed, respectively, as

$$\frac{\partial \rho}{\partial t} + \frac{\partial (\rho u_j)}{\partial x_j} = 0. \tag{9.81}$$

$$\frac{\partial (\rho u_i)}{\partial t} + \frac{\partial (\rho u_i u_j)}{\partial x_j} = -\frac{\partial p}{\partial x_i} + \frac{\partial \tau_{ji}}{\partial x_j} + \vec{F}. \tag{9.82}$$

The variables are the Cartesian components of the local flow velocity vector (u_j), the local pressure (p), the corresponding density (ρ), the viscous stress tensor (τ_{ji}) and body forces, $\vec{F}$. The independent variable are the Cartesian coordinates (x_j) corresponding to (x, y, z) and the time t. It is also important to consider the equation for the energy.

9.8.2 The Magneto-Gas Dynamic Equations, Equations of Charge-Neutral Fluid Flow in Vector Form

The magneto-gas dynamic equations comprise Maxwell's and the gas dynamic equations. Maxwell's equations are

$$\begin{gathered}\varepsilon\nabla\cdot\vec{E}=\rho=\varepsilon\sigma,\ \ \nabla\cdot\vec{B}=0,\ \ \nabla\times\vec{E}=-\partial\vec{B}/\partial t,\ \ \nabla\times\vec{B}=\mu_0\vec{j}\\ +(1/c^2)\partial\vec{E}/\partial t,\ \ c=1/\sqrt{\mu_0\varepsilon_0}.\end{gathered}\tag{9.83}$$

The gas dynamic equations

$$\begin{gathered}\frac{\partial\rho}{\partial t}+\nabla\cdot(\rho\mathbf{u})=0,\ \ \frac{\partial(\rho\mathbf{u})}{\partial t}+\nabla\cdot(\rho\mathbf{uu})=-\nabla p+\sigma\mathbf{E}+\mathbf{j}\times\mathbf{B}+\nabla\cdot\bar{\bar{\tau}},\\ \frac{\partial e}{\partial t}+\nabla\cdot\{\mathbf{u}\nabla\cdot(e+p)\}=\mathbf{E}\cdot\mathbf{J}+\nabla\cdot(\mathbf{u}\cdot\bar{\bar{\tau}}),\ \ e=\frac{\rho\mathbf{u}\cdot\mathbf{u}}{2}+\frac{p}{\gamma-1}.\end{gathered}\tag{9.84}$$

Partially eliminating the current density and using Poynting's theorem and Ohm's law, the momentum and energy may be expressed as

$$\begin{gathered}\frac{\partial(\rho\mathbf{u})}{\partial t}+\nabla\cdot\left(\rho\mathbf{uu}-\frac{\mathbf{BB}}{\mu_0}\right)=-\nabla\left(p+\frac{\mathbf{BB}}{\mu_0}\right)+\nabla\cdot\bar{\bar{\tau}},\\ \frac{\partial}{\partial t}\left(e+\frac{\mathbf{B}\cdot\mathbf{B}}{2\mu_0}\right)+\nabla\cdot\left\{\mathbf{u}\nabla\cdot\left(e+p+\frac{\mathbf{B}\cdot\mathbf{B}}{\mu_0}\right)-\left(\frac{\mathbf{u}\cdot\mathbf{B}}{\mu_0}\right)\mathbf{B}\right\}\\ =\nabla\cdot(k_{th}\nabla T)+\nabla\cdot(\mathbf{u}\cdot\bar{\bar{\tau}})-\nabla\cdot\left(\mathbf{J}\times\frac{\mathbf{B}}{\sigma\mu_0}\right).\end{gathered}\tag{9.85}$$

The preceding equation now also includes the specific internal energy e and the heat flux due to conduction $k_{th}\nabla T$ and could be generalized to include all the internal heat sources.

9.8.3 Single Fluid Model: The MHD Equations

Magneto-hydrodynamics (MHD) (Callen, 2006) describes the "slow" evolution of plasma consisting of electrons and protons where the evolution on time scales longer than those on which individual particles are important or on which the electrons and ions might evolve independent of one another. The magneto-hydrodynamic (MHD) model is a "single-fluid" description of plasma with a single-temperature approximation and is used to describe plasma (charge-neutral) flow dynamics. The resistive MHD equations comprise a hydrodynamic part described by the compressible

Navier-Stokes equations with magnetic Lorentz force and Joule heating source terms, coupled with Maxwell's electromagnetic equations. The entire system is closed by an equation of state and a generalized Ohm's law.

9.9 Kinetic Theory of Plasma

In developing the MHD equations, the kinetic theory is invoked which is the simplest description of plasma. The velocity distribution of all particles is assumed to be Maxwellian everywhere and only dependent on the temperature. Each species is not treated independently. The kinetic theory explains how kinetic energy of the particles as a group is dissipated resulting in the increase in temperature. When time scales are long enough to let electrons and ions move, the quasi-neutrality condition is valid so the particle densities of ions and electrons are equal. This is called the "plasma approximation". It is valid just for low-frequency applications. For high-frequency electronic waves, this is not suitable and $\vec{E}$ must be determined with Maxwell's Equations.

9.9.1 The Complete Set of MHD Equations

The first equation in the continuum model of plasma dynamics is mass continuity in the plasma flow. The second set of equations relate to the momentum continuity in the plasma flow, characterized by the material derivative of the momentum vector and driven by the gradient of the total pressure and includes the divergence of the dissipative stress tensor. The third equation is the energy equation stating that the energy generation rate equals the divergence of the heat flux, Ohmic heating, momentum loss of the particles (collisions) and heat exchanged between the particles. They are

$$\frac{\partial \rho}{\partial t} + \nabla \cdot (\rho \mathbf{u}) = 0, \tag{9.86}$$

where $\mathbf{u}$ is the velocity vector

$$\begin{aligned} &\frac{\partial (\rho \mathbf{u})}{\partial t} + \nabla \cdot \left(\rho \mathbf{u}\mathbf{u} - \frac{\mathbf{B}\mathbf{B}}{\mu_0} \right) = -\nabla (p_{\text{tot}}) + \nabla \cdot \overline{\overline{\tau}}, \\ &\frac{\partial e_{\text{tot}}}{\partial t} + \nabla \cdot \left\{ \mathbf{u} \nabla \cdot (e + p_{\text{tot}}) - \left(\mathbf{u} \cdot \frac{\mathbf{B}}{\mu_0} \right) \mathbf{B} \right\} = \nabla \cdot (k_{th} \nabla T) \\ &+\nabla \cdot (\mathbf{u} \cdot \overline{\overline{\tau}}) - \nabla \cdot \left(\mathbf{J} \times \frac{\mathbf{B}}{\sigma \mu_0} \right). \end{aligned} \tag{9.87}$$

with $p_{\text{tot}} = p + \mathbf{B} \cdot \mathbf{B}/\mu_0$, $e = \rho\mathbf{u} \cdot \mathbf{u}/2 + p/(\gamma - 1)$ and $e_{\text{tot}} = e + \mathbf{B} \cdot \mathbf{B}/2\mu_0$. A slight change in notation is necessary to avoid confusion with other variables. The variables are the magnetic field $\mathbf{B}$, the current density $\mathbf{J}$, the permeability of free space μ_0, the conductivity k_{th} and dissipative stress tensor, $\overline{\overline{\tau}}$.

The final subset of equations arise from combining the magnetic induction equation with Faraday's and Ampere's laws. The magnetic field evolution may be expressed as

$$\frac{\partial \mathbf{B} \nabla \times \mathbf{B} = \mu_0 \mathbf{J},}{\partial t} = -\nabla \times \mathbf{E} = \frac{1}{\sigma \mu_0} \nabla^2 \mathbf{B} + \nabla(\mathbf{Bu} - \mathbf{uB}) = \nabla \cdot \overleftrightarrow{\mathcal{E}}_{\text{res}} + \nabla(\mathbf{Bu} - \mathbf{uB}), \tag{9.88}$$

$\overleftrightarrow{\mathcal{E}}_{\text{res}}$ is the resistive diffusion tensor, satisfying

$$\nabla \cdot \overleftrightarrow{\mathcal{E}}_{\text{res}} = -\nabla \times \left(\overline{\overline{\eta}} \cdot \frac{\nabla \times \mathbf{B}}{\mu_0} \right), \tag{9.89}$$

$\mathbf{J}$ is related to the electric and magnetic fields, $\mathbf{E}$ and $\mathbf{B}$, by the generalized Ohm's law

$$\left(\eta \mathbf{I} - \frac{1}{n_e q} \mathbf{B} \times \right) \mathbf{J} = \mathbf{E} + \mathbf{u} \times \mathbf{B} + \frac{1}{n_e q} (n_e k T_e), \tag{9.90}$$

The preceding equations are in terms of the electron temperature T_e, the electron number density n_e, the Boltzmann constant k, the electron charge q, the resistivity η and the full resistivity tensor $\overline{\overline{\eta}}$, which includes the dispersive Hall effect. The momentum equation contains the electromagnetic body force/unit volume $\mathbf{J} \times \mathbf{B}$, written as the divergence of the Maxwell stress tensor (body forces and heat sources are absent).

The final set of dynamic equations could be written in conservative form and are

$$\frac{\partial \rho}{\partial t} + \nabla \cdot (\rho \mathbf{u}) = 0,$$

$$\frac{\partial (\rho \mathbf{u})}{\partial t} + \nabla \cdot \left(\rho \mathbf{uu} - \frac{\mathbf{BB}}{\mu_0} \right) + \nabla \left(p + \frac{\mathbf{BB}}{\mu_0} \right) = \nabla \cdot \overline{\overline{\tau}},$$

$$\frac{\partial \mathbf{B}}{\partial t} + \nabla(\mathbf{uB} - \mathbf{Bu}) = \frac{1}{\sigma \mu_0} \nabla^2 \mathbf{B},$$

$$\frac{\partial}{\partial t}\left(e+\frac{\mathbf{B}\cdot\mathbf{B}}{2\mu_0}\right)+\nabla\cdot\left\{\mathbf{u}\nabla\cdot\left(e+p+\frac{\mathbf{B}\cdot\mathbf{B}}{\mu_0}\right)-\left(\frac{\mathbf{u}\cdot\mathbf{B}}{\mu_0}\right)\mathbf{B}\right\}$$
$$=\left(\frac{1}{\sigma\mu_0}\frac{|\nabla\times\mathbf{B}|^2}{\mu_0}\right)+\nabla\cdot(k_{th}\nabla T)+\nabla\cdot(\mathbf{u}\cdot\bar{\bar{\tau}}). \quad (9.91)$$

The first and last are scalar equations, while others are vector equations.

9.9.2 Two Species Model: Maxwell's Equations

Consider, for simplicity, a plasma with only two species—ions and electrons—the charge and current densities will be respectively given by

$$\sigma=q_i n_i+q_e n_e,\quad \vec{j}=\boldsymbol{j}=v_i q_i n_i+v_e q_e n_e. \quad (9.92)$$

Ignoring both the collisions and the viscosity (and replacing the total stress tensor by the scalar pressure), the complete set of fluid equations are given below. The first four equations are given by Maxwell's equations which are

$$\varepsilon_0\nabla\cdot\mathbf{E}=n_i q_i+n_e q_e,\quad \nabla\cdot\mathbf{B}=0,\quad \nabla\times\mathbf{E}=-\partial\mathbf{B}/\partial t,\quad \nabla\times\mathbf{B}=\mathbf{v}_i n_i q_i$$
$$+\mathbf{v}_e n_e q_e+\varepsilon_0\partial\mathbf{E}/\partial t. \quad (9.93)$$

The simplified fluid equations for each of the two species, where *j* is not an index are

$$\frac{\partial n_j}{\partial t}+\nabla\cdot(n_j\mathbf{v}_j)=0,\quad m_j n_j\frac{\partial\mathbf{v}_j}{\partial t}+(\mathbf{v}_j\cdot\nabla)\mathbf{v}_j=n_j q_j(\mathbf{E}+\mathbf{v}_j\times\mathbf{B})$$
$$-\nabla p_j,\quad p_j=C_j n_j^{\gamma_j-1},\quad j=i(\text{ion})\text{ or }e(\text{electron}). \quad (9.94)$$

9.10 The Boltzmann and Vlasov Equation: Electron Distributions

The Boltzmann and Vlasov equations describe the evolution of the system of charged particles in a force field, over short and long ranges, respectively. The Vlasov equation can be considered to be the Boltzmann equation without collisions. Given a multi-particle system, one could introduce the Distribution Function for a species α defined as

$$f_\alpha(\vec{r},\vec{v})d\vec{r}\,d\vec{v}=dN(\vec{r},\vec{v}). \quad (9.95)$$

It is the number of particles in the volume element, in the six dimensional phase space. Thus, one has the Boltzmann equation

$$\frac{\partial f}{\partial t} + \vec{v} \cdot \nabla_{\vec{r}}(f) + \frac{q\left(\mathbf{E} + \vec{v} \times \mathbf{B}\right)}{m} \cdot \nabla_{\vec{v}}(f) = \left(\frac{\partial f}{\partial t}\right)_{\text{collisions}}. \tag{9.96}$$

From the Boltzmann equation for the evolution of the probability distribution, one can set the right-hand side of Eq. (9.96) to zero, in the absence of collisions to get the Vlasov equation:

$$\frac{\partial f}{\partial t} + \vec{v} \cdot \nabla_{\vec{r}}(f) + \frac{q\left(\mathbf{E} + \vec{v} \times \mathbf{B}\right)}{m} \cdot \nabla_{\vec{v}}(f) = 0. \tag{9.97}$$

In the case of electrons which could be assume to have reached steady state and with $\mathbf{B} = 0$

$$m\vec{v} \cdot \nabla_{\vec{r}}(f) + q\mathbf{E} \cdot \nabla_{\vec{v}}(f) = 0. \tag{9.98}$$

A solution is given by

$$f_e = f_{e0} \exp(q\phi/kT_e) \quad \text{and} \quad n_e = n_{e0} \exp(q\phi/kT_e), \tag{9.99}$$

where the electron temperature is T_e, the electron number density is n_e, the Boltzmann constant is k and the electron charge is q.

9.11 Mobility and Diffusion: Ambi-polar Diffusion and Plasma Decay

In a weakly ionized plasma, both electron and ion are participating species, and the majority of collisions are between charged particles and neutral particles. Electrons and ions have different viscosity and diffusion coefficients. The corresponding flux will be initially different but eventually become the same because of the electric field between them. Thus, the electron which has higher mobility and diffusion rate due to lower mass will slow down while ion which has lower mobility and diffusion rate will speed up.

In a real plasma, the diffusion processes lead to collisions which attenuate the pressure gradients. If a stationary process is considered and the convective terms are neglected, the momentum equation with collisions and a collision frequency ν is

$$mn\frac{\partial \mathbf{v}}{\partial t} + (\mathbf{v}\cdot\nabla)\mathbf{v} = mn\frac{\partial \mathbf{v}}{\partial t} = nq\mathbf{E} - \nabla p - mn\nu\mathbf{v},$$
$$\Rightarrow \mathbf{v} = \pm\frac{q}{m\nu}\mathbf{E} - \frac{kT_e}{m\nu}\frac{\nabla n}{n} = \pm\mu_D\mathbf{E} - D\frac{\nabla n}{n}, \tag{9.100}$$

In the above, μ_D is the mobility and D is the diffusion coefficient.

Consider now a plasma created in a finite volume that decays by diffusion. Electrons and ions will recombine somewhere, so the density after recombination will be zero. Continuity is given by

$$\frac{\partial n}{\partial t} + \nabla\cdot(n\mathbf{v}) = \frac{\partial n}{\partial t} + \nabla\cdot\Gamma = 0. \tag{9.101}$$

where Γ is the flux. For quasi-neutrality, $\Gamma_i = \Gamma_e = \Gamma$, with

$$\Gamma = n\mu_{D_i}\mathbf{E} - D_i\nabla n = -n\mu_{D_e}\mathbf{E} - D_e\nabla n. \tag{9.102}$$

Equation (9.102) leads to

$$\mathbf{E} = \frac{(D_i - D_e)}{\left(\mu_{D_i} + \mu_{D_e}\right)}\frac{\nabla n}{n}, \quad \Gamma = -\frac{\mu_{D_i}D_e + D_i\mu_{D_e}}{\left(\mu_{D_i} + \mu_{D_e}\right)}\nabla n = -D_{adc}\nabla n. \tag{9.103}$$

In Eq. (9.103) D_{adc} is the Ambi-polar diffusion coefficient. When an ion and an electron collide, they have a finite probability of recombining into a neutral atom. To conserve momentum, a third body must be present. If the third body could be an emitted photon, or a particle, the loss of plasma or plasma decay is by diffusion and recombination.

9.11.1 Bohm Diffusion, Momentum Gain and Plasma Resistivity

The classical diffusion process does not apply to confined plasmas because of the imperfection of the confinement. A diffusion process is formularized, which represents a much higher rate of diffusion than in classical diffusion and found to work much better in practice, due to Debye shielding. It is known as Bohm diffusion. Plasma that diffuses with this process decays exponentially with time rather than linearly with time.

The diffusion of plasma across a magnetic field was believed to follow the Bohm diffusion scaling. This predicted that the rate of diffusion was linear with temperature and inversely linear with the strength of the confining magnetic field. The rate predicted by Bohm diffusion is much higher than the rate predicted by classical diffusion, which develops from a random walk within the plasma. The classical model scaled inversely with the square of the magnetic field.

Including the effects of the momentum gain of the ion fluid caused by collisions with electrons, and vice versa, with $j = i$(ion) or e(electron)

$$m_j n \partial \mathbf{v}_j / \partial t = nq\left(\mathbf{E} + \mathbf{v}_j \times \mathbf{B}\right) - \nabla p_j - \nabla \cdot \vec{\pi}_j + P_{j\bar{j}}. \tag{9.104}$$

where $P_{i,\ e}$ and $P_{e,\ i}$ are momentum gain of the ion fluid caused by collisions with electrons, and vice versa.

Furthermore $P_{e,\ i} = \ - P_{i,\ e}$, $\nabla \cdot \vec{\pi}_j \approx 0$ and, given the resistivity η

$$P_{e,i} = mn(\mathbf{v}_i - \mathbf{v}_e)\mathbf{v}_{ei} = \eta q^2 n^2 (\mathbf{v}_i - \mathbf{v}_e). \tag{9.105}$$

From Eq. (9.105), one has

$$\mathbf{v}_{ei} = \eta n q^2 / m \Rightarrow \eta = m_e \mathbf{v}_{ei} / n q^2. \tag{9.106}$$

Furthermore, with $\mathbf{B} = 0$, $nq\mathbf{E} = P_{e,\ i}$ and with, $\mathbf{j} = qn(\mathbf{v}_i - \mathbf{v}_e)$

$$P_{e,i} = nq\mathbf{E} = \eta nq\mathbf{j} \Rightarrow \mathbf{E} = \eta \mathbf{j}. \tag{9.107}$$

It is a form of Ohm's law and η represents a specific resistivity of a material. Moreover, $\mathbf{E} = \mathbf{V}/L = R\mathbf{i}/L$, $R = \eta L/A$. Thus

$$\mathbf{E} = \eta \mathbf{i}/A = \eta \mathbf{j}. \tag{9.108}$$

9.12 Quasi-one-dimensional Model and Transformation of Variables

Considering a one-dimensional duct, the generic MHD equations reduce to

$$\partial(\rho A)/\partial t + \partial(\rho A u)/\partial x = 0, \tag{9.109}$$

$$\partial(\rho A u)/\partial t + \partial\left(\rho A u^2\right)/\partial x + A \partial p / \partial x + A B_y \partial\left(B_y/\mu_0\right)/\partial x = 0, \tag{9.110}$$

$$\frac{\partial B_y}{\partial t} + \frac{\partial}{\partial x}\left(B_y u\right) = \frac{1}{\sigma \mu_0} \frac{\partial^2 B_y}{\partial x^2}. \tag{9.111}$$

$$A \frac{\partial}{\partial t}\left(e + \frac{B_y^2}{2\mu_0}\right) + \frac{\partial}{\partial x}\left(h_t u A + \frac{B_y^2}{\mu_0} u A\right) = \left(\frac{1}{\sigma} J_y^2 A\right), \tag{9.112}$$

with $e = \rho u^2/2 + p/(\gamma - 1)$, $h_t = p + e = \frac{\rho u^2}{2} + \frac{\gamma p}{\gamma - 1}$ and $p + e + B_y^2/\mu_0 = \rho u^2/2 + \gamma p/(\gamma - 1) + B_y^2/\mu_0$.

Changing the variables, the equations may be expressed as a set of coupled equations as

$$\frac{\partial}{\partial t}\begin{bmatrix} \rho \\ \rho u \\ B_y \\ e+\dfrac{B_y^2}{2\mu_0} \end{bmatrix} = \begin{bmatrix} 1 & 0 & 0 & 0 \\ u & \rho & 0 & 0 \\ 0 & 0 & 1 & 0 \\ \dfrac{1}{2}u^2 & \rho u & \dfrac{B_y}{\mu_0} & \dfrac{1}{\gamma-1} \end{bmatrix} \frac{\partial}{\partial t}\begin{bmatrix} \rho \\ u \\ B_y \\ p \end{bmatrix} \equiv L\frac{\partial}{\partial t}\begin{bmatrix} \rho \\ u \\ B_y \\ p \end{bmatrix}. \tag{9.113}$$

So, it follows that

$$\frac{\partial}{\partial t}\begin{bmatrix} \rho \\ u \\ B_y \\ p \end{bmatrix} + \begin{bmatrix} u & \rho & 0 & 0 \\ 0 & u & \dfrac{B_y}{\rho\mu_0} & \dfrac{1}{\rho} \\ 0 & B_y & u & 0 \\ 0 & \gamma p & 0 & u \end{bmatrix} \frac{\partial}{\partial x}\begin{bmatrix} \rho \\ u \\ B_y \\ p \end{bmatrix} = L^{-1}\begin{bmatrix} -\dfrac{\rho u}{A}\dfrac{dA}{dx} \\ -\dfrac{\rho u^2}{A}\dfrac{dA}{dx} \\ \dfrac{1}{\sigma\mu_0}\dfrac{\partial^2 B_y}{\partial x^2} \\ \dfrac{1}{\sigma}J_y^2 - \dfrac{h_t u}{A}\dfrac{dA}{dx} \end{bmatrix}. \tag{9.114}$$

To define the characteristic polynomial

$$\begin{vmatrix} u-\lambda & \rho & 0 & 0 \\ 0 & u-\lambda & \dfrac{B_y}{\rho\mu_0} & \dfrac{1}{\rho} \\ 0 & B_y & u-\lambda & 0 \\ 0 & \gamma p & 0 & u-\lambda \end{vmatrix} = 0. \tag{9.115}$$

Two of the roots are u while the other two are

$$\lambda_{3,4} = u \mp \sqrt{\frac{\gamma p}{\rho} + \frac{B_y^2}{\rho\mu_0}} = u \mp \sqrt{c_{ac}^2 + c_{alf}^2}. \tag{9.116}$$

The acoustic wave speed, the Alfven wave speed (or the velocity of propagation of magnetic disturbances) and the magneto acoustic wave speed are respectively given by

$$c_{ac} = \sqrt{\frac{\gamma p}{\rho}},\quad c_{alf} = \sqrt{\frac{B_y^2}{\rho\mu_0}},\quad c_{mac} = \sqrt{c_{ac}^2 + c_{alf}^2}. \tag{9.117}$$

The steady solution satisfies

$$\partial(\rho A u)/\partial x = 0, \tag{9.118}$$

$$\partial\left(\rho A u^2\right)/\partial x + A\partial p/\partial x + AB_y\partial\left(B_y/\mu_0\right)/\partial x = 0, \tag{9.119}$$

$$\frac{\partial}{\partial x}\left(B_y u\right) = \frac{1}{\sigma\mu_0}\frac{\partial^2 B_y}{\partial x^2}. \tag{9.120}$$

$$\frac{\partial}{\partial x}\left(h_t uA + \frac{B_y^2}{\mu_0}uA\right) = \left(\frac{1}{\sigma}J_y^2 A\right), \tag{9.121}$$

with

$$e = \rho u^2/2 + p/(\gamma-1),\ \ h_t = p + e = \frac{\rho u^2}{2} + \frac{\gamma p}{\gamma - 1} \quad \text{and} \quad p + e + B_y^2/\mu_0 = \rho u^2/2 + \gamma p/(\gamma-1) + B_y^2/\mu_0.$$

They are solved subject to the boundary conditions.

9.13 Applications

Generally the complete dynamic models stated earlier in this chapter are used to simulate the two-dimensional axisymmetric or the three-dimensional fluid plasma equations governing the flow through the thruster duct and the exhaust nozzle. For each combination of plasma fluid, magnetic field and charge state, the ion continuity and momentum equations are solved numerically. Simultaneously, the electric potential and electron currents are solved using a combination of the generalized Ohm's law and current/charge conservation equations while the electron fluid energy equation is solved to determine the corresponding temperature (in electron-volts). These equations are particularly useful in simulating the flow field through magnetically guided plasma flow fields to generate the thrust developed. The magnetic nozzle is a convergent/divergent magnetic field used to guide the plasma flow and transform the non-directed kinetic energy to the directed one. The magnetic nozzle acts functionally like a physical de Laval nozzle but incorporates plasma-field interactions to constrain and accelerate the plasma flow, rather than a walled physical nozzle, thus avoiding all direct contact of the plasma with walls of the nozzle. The shape of the magnetic nozzle can be easily controlled to control both the magnitude and direction thrust vector without using the mechanical actuation. Thus a completely controllable convergent/divergent magnetic field is used in a magnetic nozzle to expand and accelerate a stream of ions and electrons in a plasma into vacuum in space for generating thrust to propel a spacecraft. The stream of ions and electrons in the plasma is allowed to expand along the diverging magnetic field lines, which are applied by a combination of an array of permanent magnets and several a solenoidal current loops. The presence of the solenoidal current loops allows for the complete control of the magnetic field lines, within limits. Although many aspects of magnetic nozzle performance can be accurately captured with a quasi-one-dimensional model, accurate modelling of the two-dimensional plasma flow in the nozzle is required to deduce the thrust, efficiency and specific impulse of the thruster and to capture the features induced by the nozzle geometry. Moreover the two-dimensional plasma flow model is also inadequate for long thrusters which must be modelled as three-dimensional plasma flows.

Exercises

1. Compare the typical thermal speed of a Xenon atom leaving a 300-K surface with the speed of an ion beam from a 300-V Hall thruster. Also determine the mean velocity, the RMS velocity and the most probable velocity.
2. (a) For a plasma (quasi-neutral) with a particle density of 10^{18} m^{-3}, dominated by electrons (and charge of an electron: 1.60217662 × 10^{-19} C, electron charge to mass ratio = 1.75822 × 10^{11}C/kg, permittivity of free space = 8.8542 × 10^{-12}F/m), estimate the natural frequency of plasma oscillations. Show that for an electron-dominated plasma, the natural frequency of plasma oscillations can be expressed as $\omega_{np}^2 = n_e q_e^2 / m_e \varepsilon_0 = 324\pi^2 n_e$.

   (b) Estimate the de Broglie wavelength, the ratio of the Debye length and the de Broglie wavelength and the Debye length at an electron temperature of 5 eV (5* 11604.5250061657 °K)
3. The current density of ions entering the sheath at the edge of the plasma can be found from the number density and the ion velocity at the sheath edge. The electron number density at the sheath edge may be assumed to be 60% of the number density at the centre of the plasma. The ions are assumed to fall through a potential in the plasma equivalent to 50% of the plasma electron temperature T_e, before entering the sheath. Thus estimate current density of ions entering the sheath at the edge of the plasma, if the Xenon plasma electron temperature is 5 eV and the number density is 10^{18} m^{-3} and the ion collection area is 100 cm^2.

   If instead of Xenon plasma, the plasma is a Krypton plasma. How does the current change?
4. A proton is moving in a magnetic field of strength $B = 5 \times 10^{-3}$ T at a speed of $v = 1\times 10^6$ m/s. The angle between the proton's velocity vector and the magnetic field is $\theta = 30°$.

   Estimate the Larmor radius and the cyclotron frequency.

5. Estimate the Lorentz force, on a charged particle with a charge $q = 2 \times 10^{-6}$ C (Coulombs), is moving with a velocity $v_x = 4$ m/s in the x-direction (in a Cartesian frame) in the presence of an electric field $E_y = 4$ N/C in the y-direction and a magnetic field $B_z = 0.5$ T in the z direction. The charged particle velocity, the electric field and magnetic field are in mutually perpendicular directions, x, y and z.
6. A 2-mm-diameter cylindrical probe 5 mm long in a Xenon plasma with $T_e = 3$ eV collects 1 mA of ion saturation current. What is the average plasma density?
7. When the applied magnetic field lines are curved and the particle has a velocity along the field, another drift occurs. In a frame fixed at the guiding centre, a centrifugal force appears because the plasma is rotating about the centre of curvature. Hence derive the equation for the curvature drift velocity.
8. Generalize the notion of the $\boldsymbol{E} \times \boldsymbol{B}$ drift to the case when a particle experiences a force and hence derive an equation for the gradient drift. Show that this drift velocity, like in the case of gravitational drift, depends on the sign of the charged particle and so it results in a net current.

References

Bittencourt, J. A. (2004). *Fundamentals of plasma physics, Third Edition*. Springer Science & Business Media.

Bohm, D., & Gross, E. P. (1949). Theory of plasma oscillations. A origin of medium-like behaviour. *Physics Review, 75*(12), 1851–1864.

Callen, J. D. (2006). *Fundamentals of plasma physics, 2006*. https://www.scribd.com/document/85976601/fundamentals-of-plasma-physics-2006

Colonna, G., & D'Angola, A. (2016). *Plasma modeling, methods and applications*. Institute of Physics Publishing.

Gerwin, R. A. (1966). Inertial effects in the diffusion of a plasma across a magnetic field, Phd Thesis, Electrical Engineering, Technische Hogeschool Eindhoven. https://doi.org/10.6100/IR72949

Gibbon, P. (2016). Introduction to plasma physics. In *Proceedings of the CAS-CERN Accelerator School: Plasma wake acceleration, Geneva, Switzerland*, 23–29 November 2014, edited by B. Holzer, CERN-2016-001, CERN, Geneva, 2016.

Goldston, R. J., & Rutherford, P. H. (1995). *Introduction to plasma physics*. Institute of Physics Publishing.

Holt, E. H., & Haskell, R. E. (1965). *Foundations of plasma dynamics* (2nd ed.). Macmillan.

Lieberman, M. A., & Lichtenberg, A. J. (1994). *Principles of plasma discharges and material processing*. Wiley-Interscience.

Shang, J. J. S., & Surzhikov, S. T. (2018). *Plasma dynamics for aerospace engineering*. Cambridge University Press, ISBN:9781108292566, https://doi.org/10.1017/9781108292566

Wiesemann, K. (2023). *A short introduction to plasma physics*. https://cds.cern.ch/record/1693043/files/p85.pdf, Accessed November 15, 2023.

Chapter 10
Space Vehicle Electric Propulsion

10.1 Introduction to Electric Propulsion

In a spacecraft, on-board propulsion systems are required for orbit transfer manoeuvres, for example, from LEO to GEO, LEO to HEO or HEO to a heliocentric solar orbit, inter-planetary orbit transfers, drag force and other disturbing torque compensation, orbit- and station-keeping attitude control manoeuvres including attitude hold and acquire as well as several other manoeuvres and slewing operations. Typical requirements for the total velocity increment over a mission could exceed 10–12 km/s.

The rocket equation was derived in Chap. 5. It relates the ratio of the final or empty mass to the initial mass to the ratio of the velocity increment delivered to a spacecraft to the propellant ejection or exhaust speed. It is expressed as

$$m_E/m_0 = \left(m_0 - m_p\right)/m_0 = 1/R = \exp(-\Delta v_m/c_e). \tag{10.1}$$

The rocket equation shows there are two options to achieve a large velocity increment: either to burn a substantial amount of propellant or to use a thruster able to generate a high velocity of exhaust. The second option permits a bigger a bigger and heavier payload. For this reason, it is generally preferable. The necessary velocity increments to perform a range of manoeuvres that must be completed to accomplish a mission can be computed from orbital mechanics. It is also imperative that the maximum possible velocity increment must be achieved with the minimum use of the propellant. The propellant consumption should be reduced in order to deliver a higher payload mass to the intended destination. This in turn implies the velocity of the exhaust must be much larger than the velocity increment. Figure 10.1 shows the evolution of the empty-to-initial-mass ratio with the velocity of the exhaust for increasing velocity increments. A high exhaust speed allows to drastically bring the mass fraction to near unity and thereby reduce the propellant fraction, which would permit a larger payload or increased duration of flight, making the mission much more cost-effective. Both liquid and solid propellant–based chemical

R. Vepa, *Space Vehicle Maneuvering, Propulsion, Dynamics and Control*,
https://doi.org/10.1007/978-3-031-65518-0_10

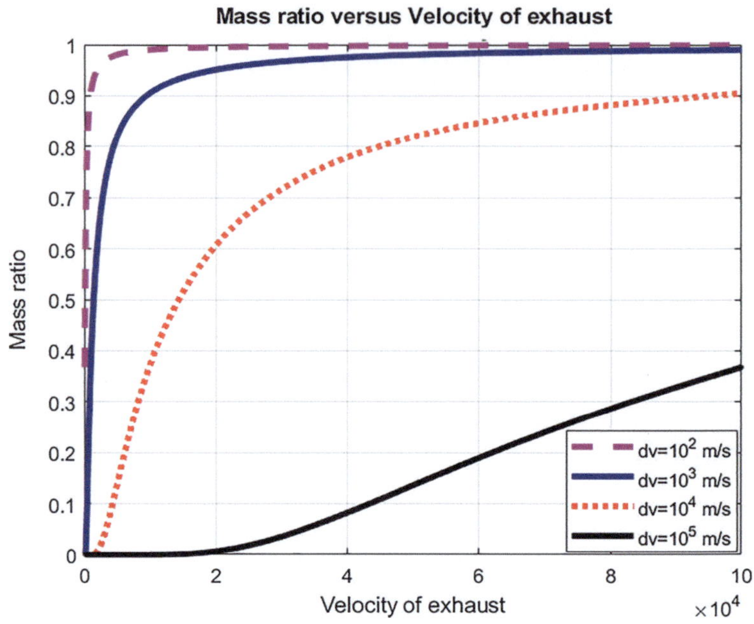

Fig. 10.1 Mass ratio from the rocket equation as a function of exhaust velocity for a number of different velocity increments for completing a mission

thrusters have limited capacity to store energy related to the energy per unit of mass stored in the propellant. From a practical viewpoint, this limits the velocity of the exhaust to a maximum attainable value of about 5000 m/s.

As seen in Fig. 10.1, as the velocity of exhaust increases, the mass ratio increases as well, for all increments in velocity. If this ratio is as close to unity as possible, the spacecraft would require as little propellant as possible. There is also the lowest possible mass ratio that a chemical rocket can be designed for that must be considered which is of the order of 0.1. The theoretical limit on the exhaust speed for chemical propulsion systems implies that for velocity increments greater than 10 km/s, only a very low mass ratio is achievable and almost all of the rocket mass must be the propellant. Thus, with chemical propulsion systems, velocity increments greater than 10 km/s are practically not achievable and there is indeed a fundamental limitation for a number of missions. Electric propulsion (EP) allows for higher exhaust speeds, and thus, much larger mass ratios are possible when a whole class of electric thrusters (ET) are used. This then facilitates much larger payloads to be carried on missions, so allowing for a number of other missions.

With their specific impulse 1.3–30 times greater than in chemical propulsion, electric propulsion possesses several notable advantages. For satellites in orbit, the mass of fuel or structures containing that fuel are upwards of 50% of the total mass in orbit. The greater efficiency of electric propulsion can drastically reduce the volume needed and mass of that propellant. Satellites are lighter, fitted with smaller thrusters and consequent of lower costs. The reduction in fuel mass allows for additional payload, making it a more useful satellite with a longer range, while avoiding the need for combustion (no oxidizer needed as well). By using easily stored inert propellants such as xenon, EP has increased safety margins. Quite unlike in chemical propulsion, in electrical propulsion, the energy is independently supplied from an external source, and as a consequence, the propellant and energy necessary to create momentum are not related. The power can be generated from several types of independent power sources such as photovoltaic (PV) solar panels, batteries, a radioisotope thermoelectric generator (RTG) or even nuclear power generator. The total power available on-board the spacecraft, which must be carried on-board, directly determines the thrust level, as shown by the equation

$$T/P_{\text{input}} = 2\eta/c_e = 2\eta/\left(gI_{\text{sp}}\right). \tag{10.2}$$

The mass of the energy source of an electrical rocket does not decrease in time. It always corresponds to a dry mass. As a consequence, the power-to-mass ratio or the specific power delivered by the power source is a critical parameter. From Eq. (5.99),

$$\Delta v_m = c_e \ln\left\{\frac{(1+K_t)\frac{v_c^2}{c_e^2}+1}{\left(\frac{m_p}{m_0}+K_t\right)\frac{v_c^2}{c_e^2}+1}\right\} = c_e \ln\left\{\frac{(1+K_t)\frac{2\eta t_b h_s}{c_e^2}+1}{\left(\frac{m_p}{m_0}+K_t\right)\frac{2\eta t_b h_s}{c_e^2}+1}\right\}. \tag{10.3}$$

In Eq. (10.3), h_s is the power-to-mass ratio or the specific power delivered by the power source. The velocity increment is plotted as a function of the velocity of exhaust for several values of the power-to-mass ratio or the specific power delivered by the power source h_s in Fig. 10.2. Typical values of the specific power h_s for PV solar panels, batteries, RTGs and nuclear power generators are shown in Table 10.1. A review of solar energy for space applications by Verduci et al. (2022) discusses the applications of PV cells to solar arrays operating in space. More information on batteries may be found in the DoITPoMS Teaching and Learning Package (2023a) and and in Aifantis, Hackney and Kumar (Eds.) (2010). More information on fuel cells may be found in the DoITPoMS Teaching and Learning Package (2023b). From the table, one could conclude that batteries and fuel cells combined with solar panels provide the best power source for electric propulsion applications. RTGs and nuclear power generators have significant disadvantages due to radiation hazards.

Figure 10.2 clearly shows the spacecraft velocity increment does not increase monotonically with the velocity of exhaust. There is an optimum and the optimum increases with the specific power. Beyond the optimum, the velocity increment delivered by the thruster degreases rapidly with increasing velocity of exhaust. The power-to-mass ratio of the power unit is clearly a key element to obtain the maximum velocity increment and to optimize the mission.

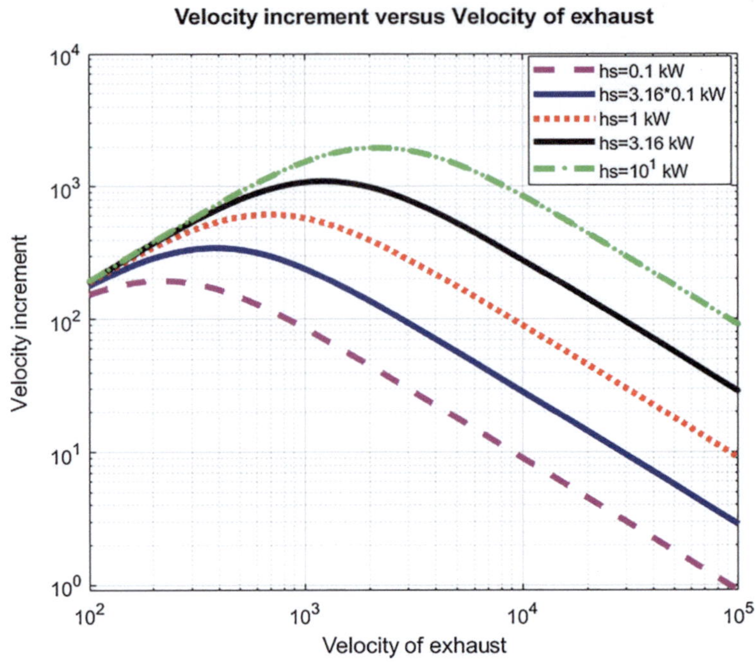

Fig. 10.2 Velocity increment versus the exhaust velocity for several values of the power-to-mass ratio, h_s

Table 10.1 Specific power values for various power sources

Power sources⇨	Solar panels	Batteries (and fuel cells)	RTG	Nuclear power generators
Typical power (in kW)	1–20	0.1–250	0.1–1	100–1000
Nominal max efficiency (%)	35	96	7	30
Specific power, h_s (kW/kg)	0.15–0.3	0.1–1.2	0.005	0.005–0.05
Degradation	High	Low	Medium	Low
Application	Earth orbits	Inter-planetary	Inter-planetary	Inter-planetary

10.2 Electric Propulsion (EP) Basics and Schematics

An electric propulsion thruster is used for spacecraft manoeuvring and orbit control. Charged particles are produced by electric discharge, then attracted and accelerated; i.e. momentum is transferred to them via electric or electromagnetic fields. The thrust force is equivalent to the momentum exhausted by the device, and the spacecraft can thus be propelled in space. In an electric propulsion device, atoms or molecules of the input propellant, which can be in a solid, liquid or gaseous state, are converted first into ions before being accelerated to high velocity and ejected outwards. Only a small fraction of the stored propellant is transformed into ions to produce thrust.

Thus, an essential parameter in determining the thrust is the mass or propellant utilization efficiency is given by Eq. (5.80) which is

$$\eta_m = \frac{\dot{m}_{\text{ions}}}{\dot{m}_f} \equiv \frac{\dot{m}_i}{\dot{m}_f} = \frac{i_b M_{\text{ions}}}{q\dot{m}_f}, \tag{10.4}$$

where $\dot{m}_i$ is the ion mass flow rate, i_b is the ion current in the exhaust beam, q is the elementary charge of the ion, M_{ions} is the propellant atomic mass and $\dot{m}_f$ is the mass flow rate of the propellant. It is assumed the exhaust beam only contains singly charged ions, although the propellant utilization efficiency can easily be corrected to account for multiply charged ion species. Hence, for ionic exhaust thrusters in which the propellant is partially ionized, the thrust is $T = c_e\dot{m}_i = \eta_m c_e \dot{m}_f$ and the specific impulse given by I_{sp} should be effectively multiplied by η_m. In fact, when $\eta_m = 1$, it could be interpreted in two ways in electric propulsion. Thus, either the propellant flow is fully ionized, which demonstrates a high efficiency of ion generation, or the thrust production mechanism does not depend on the acceleration of ions. The propellant could be electrically heated and expanded into the vacuum.

As η_m is a quantity that is difficult to accurately determine, the thrust efficiency of any electrical thruster is defined in terms of the actual thrust produced and is given by Eq. (5.16) which is

$$\eta_{th} = \frac{P_e}{P_{\text{in}}} = \frac{(1/2)\dot{m}_f c_e^2}{P_{\text{in}}} = \frac{(1/2)\dot{m}_e c_e^2}{P_{\text{in}}} = \frac{(1/2)T^2}{\dot{m}_e P_{\text{in}}}, \tag{10.5}$$

where P_{in} is the input electrical power. Equation (10.5) may also be expressed as

$$\frac{T}{P_{\text{in}}} = \frac{2\eta_{th}}{gI_{\text{sp}}}. \tag{10.6}$$

The thrust delivered is limited by the available on-board power and inversely proportional to the specific impulse. Electrical thrusters generally are highly specific impulse devices and deliver a low thrust although they are capable of operating for long periods of time.

However, EP has several limitations; low mass throughput and total impulse because high power demands per kg of propellant reduces the practical mass flow rate. It is not possible to compensate for this in the operational lifetime as it is too short. Electric thrusters (ET) produce very low thrust, from μN to tens of Ns of force instead of the mega-Ns produced by chemical thrusters. EP systems are only partially scalable. They require extensive testing under vacuum conditions.

EPs need complex electronics and power supplies. The on-board power generating equipment can be quite large, and as specific impulse increases, the power conditioning mass will grow quite substantially. For current EP technology, suitable for missions that require velocity increments of ~10 km/s, the ideal main ET, has an approximate specific impulse equal to 3000 s.

EP-propelled spacecraft spend a long time in the Van Allen belts which could result in radiation damage. Electromagnetic compatibility (EMC) could be a problem. Electrical components of ion thrusters and the thrusters themselves are sources and sinks of electromagnetic radiation that can interact with other electronics of the satellite. Noble gases are expensive resource which are depleting. Most EPs need neutralizers. Neutralizers are among the most failure-prone components.

10.2.1 Electric Propulsion Performance

In the case of electric propulsion, one may assume that the power plant power supplied is given by $P_s = m_s h_s$, where h_s is the specific power of the power plant. Here, we have made the assumption that the entire structural mass is equal to the propulsion system mass. The power P_s is related to the actual power converted into useful power by the jet as $P_T = \eta P_s$ where η is the overall mechanical power conversion efficiency. Earlier in Chap. 5, an expression for the Δv ratio $\Delta v/v_c$ was found in terms of the payload mass ratio m_p/m_0 and the exhaust velocity ratio c_e/v_c.

$$\frac{\Delta v_m}{v_c} = \frac{c_e}{v_c} \ln\left(\frac{1 + \frac{c_e^2}{v_c^2}}{\frac{m_p}{m_0} + \frac{c_e^2}{v_c^2}}\right). \tag{10.7}$$

Here, c_e is the relative exhaust velocity and v_c is the characteristic velocity of the electric propulsion system

$$v_c^2 = 2\eta t_b h_s. \tag{10.8}$$

For $c_e < v_c$, the rocket behaves like a chemical rocket. Optimum values exist for c_e and v_c, corresponding to the exhaust velocity ratio when the velocity increment ratio is a maximum. In Fig. 10.3, Eq. (10.7) is plotted versus c_e/v_c for several values of the payload ratio m_p/m_0. The maxima on each plot almost lie along a straight line.

A plot of the maximum of $\Delta v/v_c$ with the exhaust velocity ratio c_e/v_c is shown in Fig. 10.4. The plot of the maximum of $\Delta v/v_c$ with the payload ratio m_p/m_0 is shown in Fig. 10.5. Both these can be approximated by cubic splines.

Over typical values of interest, $0.1 < c_e/v_c < 5$ and payload ratios of $0 < m_p/m_0 < 0.6$, the maximum of $\Delta v/v_c$ obtained from Eq. (10.7) may be approximated as

$$\left(\frac{\Delta v}{v_c}\right)_{\max} = 0.7988 - 1.6187\left(\frac{m_p}{m_0}\right) + 1.8118\left(\frac{m_p}{m_0}\right)^2 - 1.1936\left(\frac{m_p}{m_0}\right)^3, \tag{10.9}$$

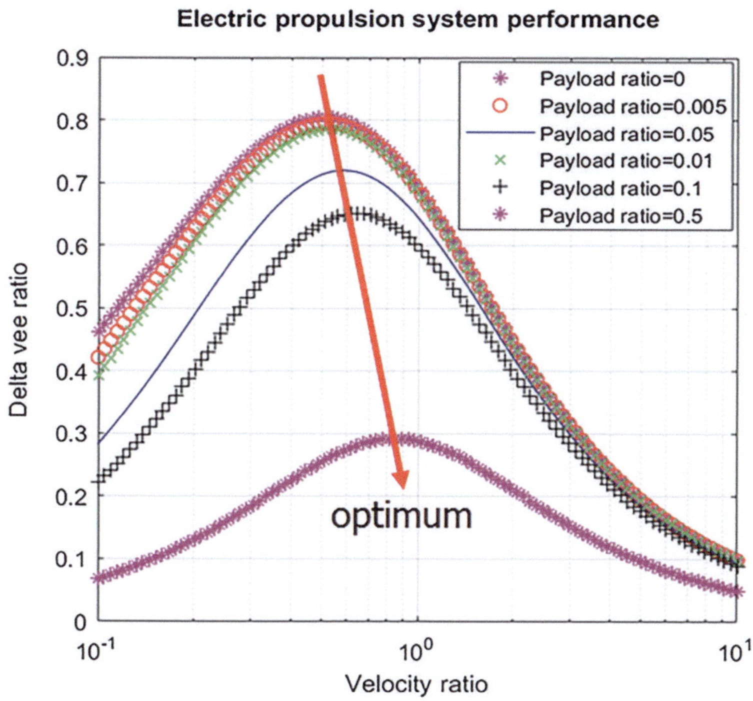

Fig. 10.3 Plots of $\Delta v/v_c$ versus c_e/v_c for several values of the payload ratio m_p/m_0

$$\left(\frac{\Delta v}{v_c}\right)_{\max} = 0.6044 + 2.1042\left(\frac{c_e}{v_c}\right) - 4.0684\left(\frac{c_e}{v_c}\right)^2 + 1.3625\left(\frac{c_e}{v_c}\right)^3. \tag{10.10}$$

The inverse relationships are

$$\frac{m_p}{m_0} = 0.9966 - 1.9724\left(\frac{\Delta v}{v_c}\right)_{\max} + 0.9268\left(\frac{\Delta v}{v_c}\right)^2_{\max} - 0.0188\left(\frac{\Delta v}{v_c}\right)^3_{\max}, \tag{10.11}$$

$$\frac{c_e}{v_c} = 1.0252 - 0.7091\left(\frac{\Delta v}{v_c}\right)_{\max} + 0.5155\left(\frac{\Delta v}{v_c}\right)^2_{\max} - 0.5398\left(\frac{\Delta v}{v_c}\right)^3_{\max}. \tag{10.12}$$

Thus, for a given optimum $\Delta v/v_c$ ratio, the corresponding the exhaust velocity ratio c_e/v_c is obtained from Eq. (10.12) and the payload ratio from Eq. (10.11). When these ratios are altered, the corresponding corrected values of the maximum of $\Delta v/v_c$ are obtained from Eqs. (10.9) and (10.10).

A typical numerical example illustrates the application of Eqs. (10.7)–(10.12). With typical values for h_s = 10–50 Watts/kg, the value for η is assumed to be 0.5. The burn time t_b is taken to be 100 days. Thus, for h_s = 50 Watts/Kg,

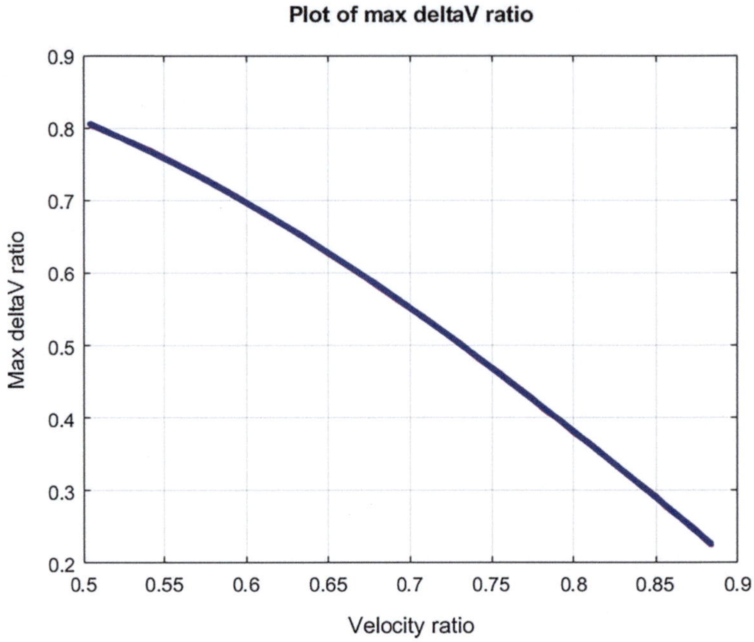

Fig. 10.4 Plot of the maximum of $\Delta v/v_c$ with the exhaust velocity ratio c_e/v_c

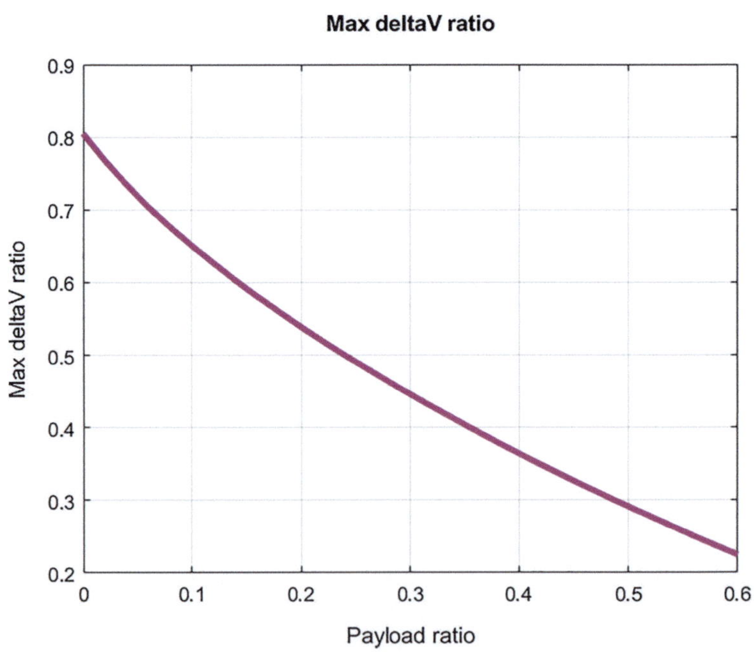

Fig. 10.5 Plot of the maximum of $\Delta v/v_c$ with the payload ratio m_p/m_0

$v_c = \sqrt{2\eta t_b h_s} = 20.785\text{km}$. For a payload ratio of $m_p/m_0 = 0.1$, $(\Delta v/v_c)_{\max} = 0.6538544$ and the corresponding c_e/v_c is 0.6310. The maximum available Δv for the mission is then 13.59 km/s. Even with, $h_s = 10$ Watts/Kg, $v_c = \sqrt{2\eta t_b h_s} = 9.2953\text{km}$. The maximum available Δv for the mission is then 6.0778 km/s.

In contrast, for a chemical rocket, with $I_{sp} = 400$ s, for a payload ratio $m_p/m_0 = 0.05$, Δv is limited to only 4.6 km/s.

As the thrust generated per unit power plant power goes down rapidly as the I_{sp} is increased, the highest possible I_{sp} is not the optimum for an electric thruster. Since the thrust satisfies

$$T = \dot{m} v_e = \frac{2P_T}{v_e} = \frac{2\eta P_w}{I_{sp} g_0}, \tag{10.13}$$

it follows that the thrust generated per unit power plant power is given by the relation

$$\frac{T}{P_w} = \frac{2\eta}{I_{sp} g_0}. \tag{10.14}$$

This relation is illustrated in Fig. 10.6.

To establish the optimum I_{sp} for minimum total mass, both propulsive efficiency and the power plant mass, m_{ppw}, need to be considered. Considering the thrust,

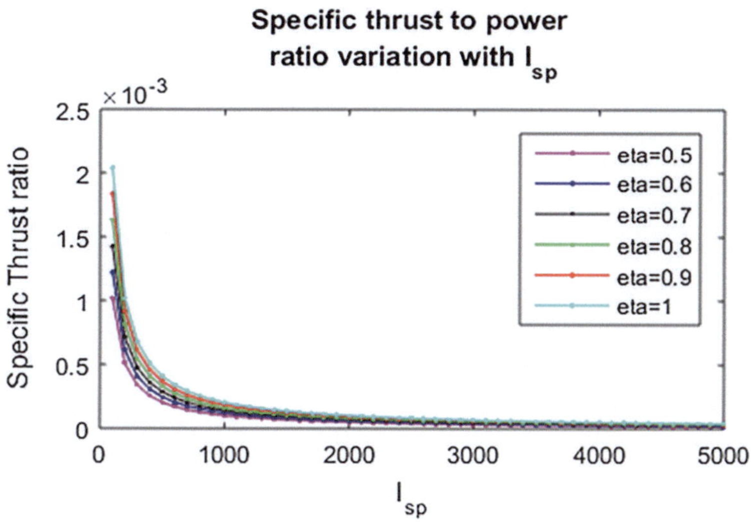

Fig. 10.6 Variation of the thrust generated per unit power plant power with I_{sp}

$$T = \dot{m} v_e = \frac{2P_w}{v_e} = \frac{2m_{\text{ppw}} h_s}{I_{\text{sp}} g} \Rightarrow m_{\text{ppw}} = \frac{T I_{\text{sp}} g}{2h_s}. \tag{10.15}$$

Since it was assumed that the structure mass is equal to the power plant mass in its entirety, $m_s = m_{\text{ppw}}$, assuming a constant fuel burn rate and using Eq. (5.83),

$$m_{\text{fuel}} = \frac{T t_b}{v_e} = \frac{T t_b}{I_{\text{sp}} g} \Rightarrow m_s = m_{\text{fuel}} \frac{c_e^2}{2\eta_T t_b h_s} = m_{\text{fuel}} \frac{I_{\text{sp}}^2 g^2}{v_c^2} = \frac{T t_b I_{\text{sp}} g}{v_c^2}. \tag{10.16}$$

Thus, the total mass can be approximated by

$$m_{\text{Total}} \approx m_{\text{fuel}} \left(\frac{I_{\text{sp}}^2 g^2}{v_c^2} + 1 \right). \tag{10.17}$$

The optimum I_{sp} for minimum total mass is

$$\left(I_{\text{sp}}\right)_{\text{opt}} = \frac{v_c}{g} = \frac{\sqrt{2h_s \eta_T t_b}}{g}. \tag{10.18}$$

In Fig. 10.7, the plot shows the mass variations with I_{sp} where the masses are for a given thrust level = 1 Mn and a burn time $t_b = 100$ days. The optimum I_{sp} is equal to 1895. The typical I_{sp} delivered by various chemical and electric thrusters is compared in Table 10.2.

In the preceding design equations, it was assumed that the structure mass is equal to the power plant mass in its entirety, $m_s = m_{\text{ppw}}$. If this is not so and the ratio of the

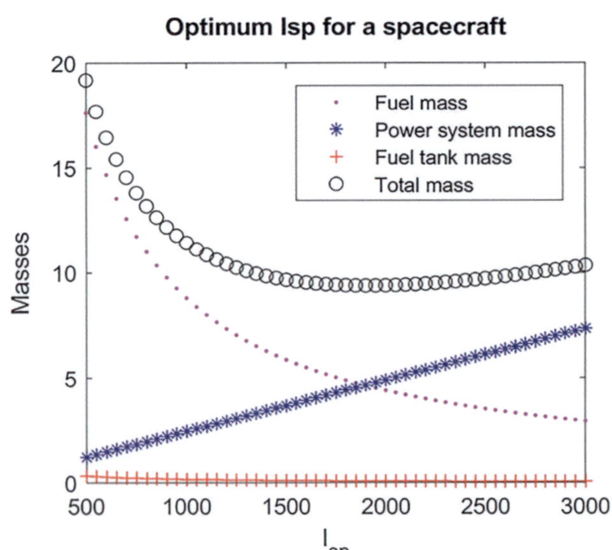

Fig. 10.7 Plot shows the mass variations with I_{sp} where the masses are for a given thrust level and burn time t_b

Table 10.2 Typical specific impulse I_{sp} delivered by various chemical and electric thrusters

Class	Thruster	I_{sp}	Efficiency %	Propellant	Input power kW	Thrust
Cold gas	Cold gas	50–75		Various		
Chemical	Mono	150–225		N_2H_4, H_2O_2		
	Bi-prop	300–450		Various		
Electro-thermal	Resistojet	300	65–90	N_2H_4 mono-prop	0.5–1	1–5 μN
	Arcjet	500–600	25–45	N_2H_4 mono-prop	0.9–2.2	2.5–10 μN
Electrostatic	Ion	2500–3600	40–80	Xenon gas	0.4–4.3	30 mN
	Hall	1500–2000	35–60	Xenon gas, Bi	1.5–4.5	10–1500 mN
	RFIT	3400–3700		Various		3.5 mN
	FEEP	5000–8000		Liquid Cs, In or Hg		2.5–10 μN
	Colloidal	1000–2000	60–70	Plexi glass, Teflon, polymide, silicon organics		
	Helicon double layer	10,000		Ar, Kr, Xe, He, H gas		1 mN
Electromagnetic	PPT	850–1200	7–13	Teflon	<0.2	
	MPD	1000–10,000		H, Ar, Na, Al gas		20–200 N
	IPT	1000–10,000				1 mN–100 N

power delivered by the thruster to the power plant mass is defined as α while the fraction of the power plant mass of the structure mass is defined as f_s, then the power generated by the thruster can be expressed as

$$P_w = m_{\text{ppw}}\alpha = (f_s m_s)\alpha = \alpha f_s m_s = h_s m_s \Rightarrow h_s = \alpha f_s. \tag{10.19}$$

Thus, all of the design equations can be easily modified when the ratio of the power delivered by the thruster to the power plant mass and the fraction of the power plant mass of the structure mass are known.

10.2.2 Electric Propulsion Types and Overviews

Electric thruster may be classified on the basis of how the ionized or neutral gas is produced by the electrical power source and how propellant is accelerated to generate the exhaust.

Thus, the electric thrusters are classified into three broad categories:

Electro-thermal propulsion: The propellant is either heated or disassociated (ionized) in a chamber. It is the expansion of the hot neutral or ionized gas through a nozzle (see for example Sutton and Biblarz, 2016) to convert the thermal energy within the chamber into kinetic energy at the exhaust, to deliver the thrust to the spacecraft. Typical examples of electro-thermal propulsion are resistance-heated (resistojet) thrusters (typical values of specific impulse I_{sp}, in the range of 300–500 s), electric arc and jet (or nozzle) thruster (arcjet), (I_{sp} = 500–700 s) and inductively or radiofrequency-heated jet thruster (I_{sp} = 400–600 s).

Electrostatic propulsion: The power source is used to generate the ions by ionizing the propellant. Ions are subsequently accelerated by means of electrodes or by any other method of application of an electric field. Typical examples of electrostatic ion propulsion include several types of ion and plasma thrusters (I_{sp} = 2500–10,000 s).

Electromagnetic propulsion: The power source is used to generate the ions by ionizing the propellant as in electrostatic thrusters. However, the ionized propellant is then accelerated under the combined action of both an electric and a magnetic field. Typical examples of electromagnetic thrusters are thrusters which are based on an exclusive electromagnetic field–accelerated propulsion (I_{sp} = 800–1200 s), Hall effect ion thrusters (I_{sp} = 1000–2800 s), magnetoplasmadynamic ion thrusters (I_{sp} = 6000–8000 s) and helicon thrusters (I_{sp} = 6000–10,000 s).

A more general classification of electric propulsion–based thrusters is displayed in Fig. 10.8 (Fortescue et al., 2011), while in Fig. 10.9, electro-thermal propulsion, electrostatic propulsion and electromagnetic propulsion–based thrusters are further classified into sub-systems (Heister et al., 2019).

In the webpage maintained by SatSearch (Anon, 2024a, CubeSat thrusters and small satellite propulsion systems), a premier search engine and marketplace for the global space industry, several commercial space thrusters are identified for various space applications. These include both chemical and electric propulsion systems. The chemical propulsion–based thrusters can deliver from 0.1 up to 500 N of thrust, while the electric propulsion–based thrusters provide from very low thrust to about 200 N of thrust. These include several patented technologies based on a synergy of principles discussed earlier in this chapter and in the next. However, the datasheets for these thrusters are available via this web page, and the reader is strongly advised to compare the performance of these industry standard thrusters with the flight heritage. The Orbital Propulsion Centre of the Ariane Group (Anon, 2024b, Space Propulsion Systems for Satellites and Spacecraft, Orbital Propulsion Centre) also

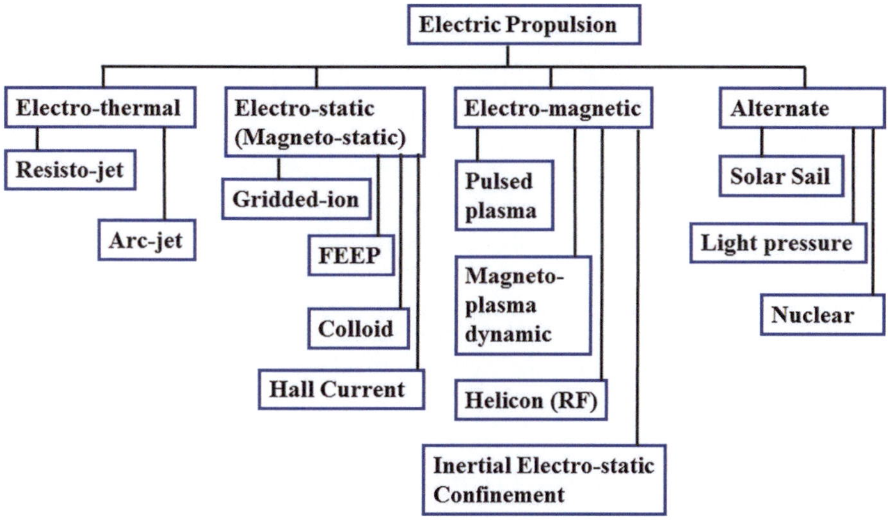

Fig. 10.8 General classification of electric propulsion–based thrusters

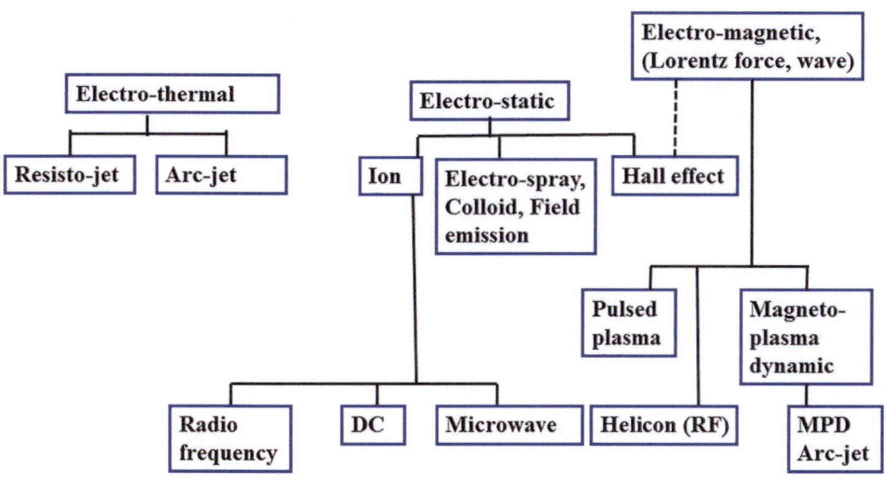

Fig. 10.9 Sub-classification of electric propulsion–based thrusters

maintains a webpage, of a complete range of monopropellant, bi-propellant and electric ion propulsion systems for various space applications. Typical applications of the propulsion systems include applications for orbital satellites and spacecraft; inter-planetary spacecraft and probes; thrusters for the control of launch and re-entry vehicles, including ascent roll control and stabilization of light to heavy launch vehicles; launch vehicle upper stage attitude, orbital and roll control; as well as for the automated resupply missions to the International Space Station. The webpage provides informative brochures on several patented technologies. Another extremely useful compendium has been assembled by Hsu et al. (2023). The webpage of important manufacturers of satellite systems such as Dawn Aerospace and Kongsberg NanoAvionics also provides considerable useful information on company proprietary modular propulsion systems for space applications, as do several other sites of manufactures. A comprehensive list of space systems manufacturers is provided in Appendix 10.2. Several webpages maintained by NASA and ESA generally provide vast amounts of useful proprietary information for spacecraft and space systems designers including principles of operation, systems selection principles and criteria, performance characteristics, integration data and material properties. Material properties of various components used in space systems can be found in webpages like Anon (2024e), ASM Specifications, ASM Aerospace Specification Metals Inc. as well as other systems providers.

10.2.3 Physics of Electric Propulsion

It is possible to show the benefits of electric propulsion, by re-considering the energy delivered to the exhaust by thermal means and comparing it to the energy delivered to it by non-thermal means as in the case of electric propulsion. From the rocket equation, the payload fraction increases exponentially with exhaust or exit velocity. The nozzle flow process converts the internal energy-driven random molecular motion to directed exhaust flow. From Chap. 5,

$$v_e^2 = \frac{2\gamma R_0 T_c\left(1 - (p_e/p_c)^{\gamma - 1/\gamma}\right)}{(\gamma - 1)m_w}. \tag{10.20}$$

From the preceding relations, it could be observed that there is a relationship between the exit velocity and temperature of the flow. Considering the underlying physical processes, for a perfect gas in thermodynamic equilibrium,

$$p = \rho RT, R = R_0/m_w, R_0 = kN_A. \tag{10.21}$$

R_0 is the universal gas constant, R is gas constant for a specific gas, m_w is the molecular weight, N_A is Avogadro's number and k is Boltzmann's constant.

As the scalar pressure is directly proportional to the force, it relates to the omni-directional momentum flux of the molecules across a surface, in the mean. It can be shown that, in terms of the temperature,

$$p = \rho\bar{c}^2/3 = \rho RT = \rho(kN_A/m_w)T. \tag{10.22}$$

where $\bar{c}^2$ is the mean thermal speed of the molecules ($N_A = 6.02 \times 10^{26}$ Kg mol, $k = 1.38 \times 10^{-23}$J/ $^\circ$ K). The mass of a molecule is m_w/N_A. Hence, the translational kinetic energy of the gas is

$$(1/2)m\overline{c}^2 = 3(1/2)kT_{\text{Translation}}. \tag{10.23}$$

In addition to pure translation, the internal energy of spherical real gas molecules is made up of contributions from rotations, due the molecules' moment of inertia and the internal mode of rotational excitations, vibrations due to the elastic nature of the bonds between atoms and excitations as well as electronic excitation which can lead to ionization where energy can only be exchanged by molecular collisions (radiation due to photonic collisions is also possible).

Considering molecular collisions, if the molecules have diameters d_p and d_q, a collision cross-section is defined as

$$\sigma_{pq} = \left(d_p + d_q\right)^2 \pi/4. \tag{10.24}$$

Define the mean diameter as $d_{pq} = (d_p + d_q)/2$, and it follows that

$$\sigma_{pq} = d_{pq}^2 \pi/4. \tag{10.25}$$

If one lets the relative rotation speed of the molecules be c_{rpq} and consider a test particle "p" move through a volume of "q" stationary particles of number density n_p, the mean value of the relative speed is $\overline{\sigma_{pq} c_{rpq}}$. Thus, the collision frequency ν_{pq} for the test particle "p" is

$$\nu_{pq} = n_p \overline{\sigma_{pq} c_{rpq}}. \tag{10.26}$$

Summing over all particle types to get the total number of collisions per unit time,

$$N_c = (1/2) \sum n_p n_q \overline{\sigma_{pq} c_{rpq}}. \tag{10.27}$$

The factor half is to avoid counting species twice while summing up. Energy exchange rates will depend on the square of the number density. Hence, as the density drops, the reaction rates also drop by a square law.

Collision processes can themselves be considered as a "reaction", while an ionization process is essential for forming a plasma. There are two important collision/ionization processes to be considered, which are collision ionization and photo-ionization. In the case of collision ionization, if $\widetilde{X}$ is a particle in an excited state, then

$$\widetilde{X} + A = X + A^+ + e^-. \tag{10.28}$$

The above reaction results in the generation of an electron which is characterized by a high speed and an associated kinetic energy. The reverse process requires a three-particle or three-body recombination, while reverse reaction rate is proportional to n^3.

In the case of photo-ionization, a photon has an energy of magnitude, $h\upsilon$, where h is known as Planck's constant and υ is the frequency ($h = 6.6 \times 10^{-34}$Joule sec), *t*he reverse reaction is a radiative recombination and involves radiation of quanta of energy. Moreover, the resultant atom can be in an excited state as energy is conserved. Both forward and reverse photo-ionization processes are often associated with electric propulsion.

In the case of electro-thermal propulsion, ionization is by thermal means, *e*very non-thermal forward reaction process takes away energy from the translational internal energy although other forms of energy could increase. There is a fall in the temperature of translation $T_{\text{Translation}}$. Non-thermal reverse process can however increase the translational internal energy. Increased internal energy results in overall higher temperature and exit velocity. Thus, propellant energization by non-thermal means (electrical means) can result in higher exit velocity.

10.3 Electro-thermal Propulsion: Concepts

Electric propulsion also relies on the principle of the conservation of momentum, and as a consequence, a force or thrust is generated and applied on a space vehicle following the ejection of the flow of ionized particles with high kinetic energy. The specialist feature of electric thrusters that distinguishes them from the others is that they achieve high propellant exhaust speeds as a large amount of energy, from externally stored fuels, can be transferred to the propellant or ionized particulate exhaust. The large exhaust velocity is coupled with the relatively low consumption of the fuel, which could be considered as its primary advantage. The other advantages of electric propulsion thrusters are their long operation times and relative ease of controllability. Most electric thrusters are quite easily controllable. The prime disadvantage of electric thrusters is their very low thrust level as a consequence of the power source limitation. In the case of electro-thermal thrusters, where the propellant is heated to a very high temperature and subsequently ionized albeit partially, the hot neutral or ionized gas is passed through a nozzle that facilitates the conversion of the thermal energy into kinetic energy, thereby imparting momentum to the exhaust and generating the thrust on the spacecraft.

10.3.1 Resistojets

The propellant is heated either via heating elements powered by electricity followed by the expansion through a nozzle as in a "resistojet" characterized by resistance heating of single propellant. The typical thrust levels are from 0.1 to a few Newtons. Alternately, the heating is done via an electric discharge in the gaseous propellant as in an "arcjet" or by heating the propellant via radio-frequency heating using microwaves, as in a "microwave jet" (as in a microwave oven). The typical thrust levels

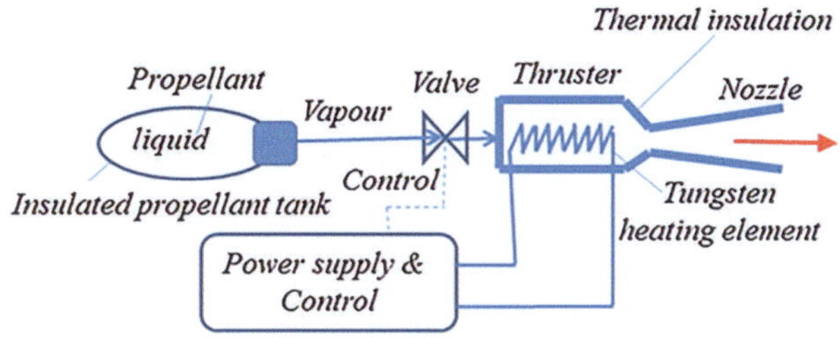

Fig. 10.10 Principle of operation of a resistojet thruster

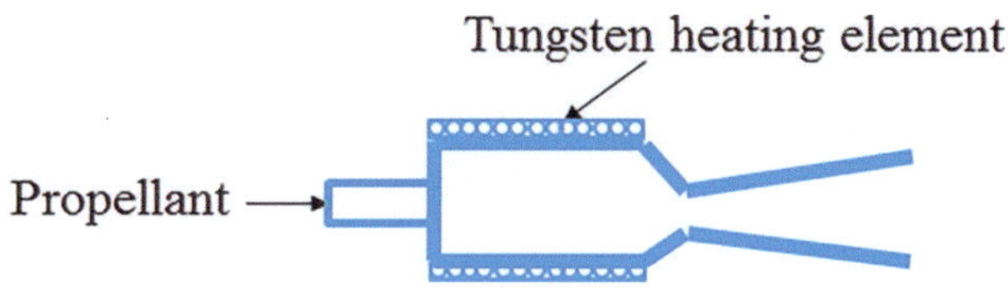

Fig. 10.11 Alternative arrangement of the heating element

are from 0.1 to 0.3 Newtons. Thus, the simplest forms of electrically augmented propulsion for spacecraft are resistance-heated and arc-heated thrusters, generally called resistojets and arcjets, respectively. As with all forms of electric propulsion, electric energy added from the spacecraft increases the specific impulse over that achievable with a chemical reaction and expansion through a nozzle. With electro-thermal devices, the added energy is imparted to the propellant as thermal energy. In the specific case of the resistojet, energy is added to the propellant with a resistance heater. A typical resistojet thruster is illustrated in Fig. 10.10.

The entire jet is encapsulated in thermal insulation, and the chamber is lined with a catalytic bed. An alternate arrangement of the heating element is also used. It is illustrated in Fig. 10.11.

In a conventional resistojet, the amount of energy added to the flow is limited by the maximum working temperature of the materials used in its construction

$$v_e^2 = 2C_p(T_c - T_e). \tag{10.29}$$

In Eq. (10.29), the exhaust velocity v_e is limited by T_c where T_c is the maximum temperature, achieved due to dissociation; T_e is the equilibrium temperature, which is generally quite low; and C_p is the specific heat capacity of the gas at constant pressure and is the amount of heat that must be added per unit mass to increase its temperature by $1°\ K$. The dissociation of propellant molecules on the surface of a metallic hot filament releases energy, which is converted to kinetic energy. Ionization is generally not present, and the absence of recombination, a relatively slow process that is proportional to n^3, means T_c is not also high enough, thus limiting the exhaust velocity v_e. Resistojets are limited by the fact that the expansion of the exhaust is a non-equilibrium process and the losses in the kinetic energy of exhaust severely limit the exhaust velocity. The performance losses are generally referred to as "frozen flow" losses and are associated with constant γ and non-equilibrium nozzle flow. Resistojets are also limited by thermal limitation of the filament materials (rhenium, platinum, tungsten, tantalum, molybdenum). There are also heat losses in the chamber, where the pressure is also limited. The typical I_{sp} is 800 s for a H_2 thruster and 340 s for a hydrazine thruster. The expansion in the nozzle is intermediate between two extremes of "frozen" flow and "equilibrium" flow. In the case of frozen flow, the exit conditions are assumed to be constant, while in the case of equilibrium flow, the expansion is based on local conditions of temperature and pressure but using equilibrium relations. The fast chemical reactions within the chamber influence the local conditions in the latter case. The latter case results in higher performance due to recovery of the chemical energy from the decomposition of complex molecular species in the chamber.

Performance can be assessed relative to the frozen flow. Thus, the frozen flow efficiency is given by

$$\eta_f = \left(h_c - h_e'\right)/h_c. \tag{10.30}$$

In Eq. (10.30), h_e' is the frozen state enthalpy and from equilibrium considerations, $h_c = C_p T_c$. If the efficiency for a number of thrusters is known from tests, $h_e' = \left(1 - \eta_f\right)h_c$ may be estimated. A typical set of performance curves are shown in Fig. 10.12.

Frozen flow losses may be avoided using longer nozzles to aid recovery. The increased time for conditions to reach equilibrium increases mass flow, increases the chamber pressure, reduces the level of dissociation/ionization and increases the mass flow. It is important that the most appropriate propellants are selected so as to maintain ease of storage, as well as storage over long times, and to meet the specific power requirements.

Not only the specific heat capacity of propellant C_p should be high but also η_f needs to be high, which in turn is dependent on molecular structure and ease with which dissociation takes place. Thermal conductivity of the propellant must be high to get energy uniformly into the gas. But this could lead to high heat losses to walls

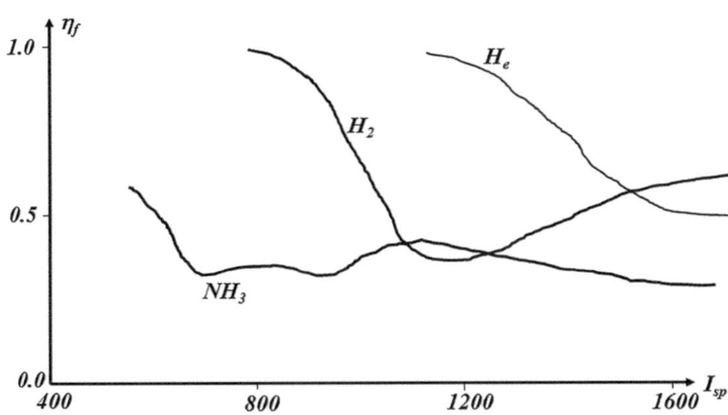

Fig. 10.12 Typical performance curves for resistojet thrusters

Table 10.3 Typical values of C_p (kJ/kg °K) of propellants at two temperatures

Fuel	C_p at 1000°K	C_p at 3000°K
H_2	15	18.4
H_e	5.2	5.2
NH_3	3.2	4.5
N_2H_4	2.8	

and hence a compromise or trade-off is essential. Some examples of propellants are hydrogen (H_2) which has a good C_p and good η_f but requires cryogenic storage and cannot be stored for long times, ammonia (NH_3) which has a low C_p and η_f but is erosive and hydrazine (N_2H_4) which decomposes exothermically but requires simple storage. Xenon (Xe) also has been used although it is prohibitively expensive and has a very low C_p (C_p = 0.16 kJ/kg °K). Typical values of C_p of propellants at two temperatures are given in Table 10.3.

For a *resistojet*, the thrust (T) produced by the propellant at a stagnation pressure depends on stagnation number density of propellant (n_o) in m^{-3}, stagnation temperature (T_s) and the probability (χ) of a molecule exiting the expansion slot area (A_o). The specific impulse ($I_{\rm sp}$) is a function of the stagnation temperature and the mass of the expended propellant (m_e). Thus,

$$T = A_o \chi n_o k_b T_s / 2, I_{sp} = \sqrt{\pi k_b T_s / 2m_e} / g, \tag{10.31}$$

where k_b is the Boltzmann constant and g is the acceleration due to gravity at sea level.

10.3.2 Arcjet Principles

In the case of an arcjet, an arc is generated directly through the propellant between a cathode and an anode. An anode (+ve) [cathode (-ve)] is the electrode from which the conventional current enters (leaves) a polarized device. In the case of all electro-thermal thrusters, the augmented specific impulse is limited by the maximum temperature tolerable by key components over the necessary life of the thruster. The maximum thrust is a function of input electrical power and is limited by the efficiency with which the electrical energy is coupled into and contained within the flow. A typical arcjet thruster is illustrated in Fig. 10.13.

An electrical arc (stream of energetic electrons, from ionization) across an electrode gap heats the propellant that is then expanded in a nozzle. The typical currents is I ~ 100 Amps, the voltage is V = 20 Volts, and the typical specific impulse is $I_{\rm sp}$~600–900 s. Arc discharge provides the energy increase that raises the temperature of the gas beyond what is achievable by a resistojet. Arc initiation across a gap is dependent on the voltage, the gap gas pressure and type of gas. The breakdown condition occurs when a high voltage is applied across an electrode gap and is a condition that is aided by secondary electrons which induce further

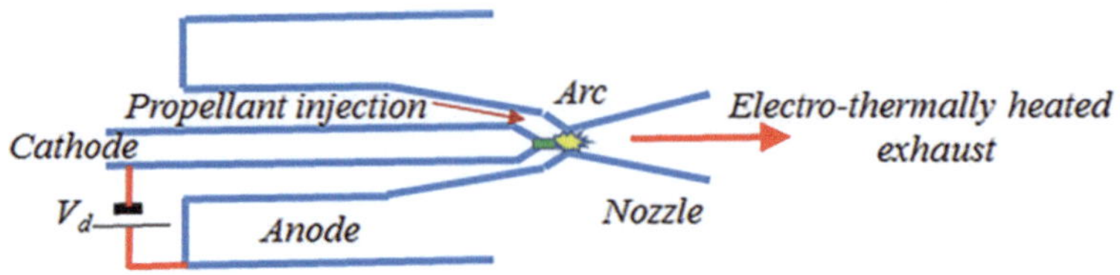

Fig. 10.13 Principle of operation of an arcjet thruster

ionization of the gas. Traditionally, Paschen's curves have been used to obtain the breakdown voltage for gaseous ionization between two electrodes. Paschen's curves describe the breakdown voltages as a function of the electrode spacing or electrode gap, the operating pressure and composition of the gas. The mathematical formulation of Paschen's curve is derived from Townsend's description of the basic charge generation processes including electron impact ionization (the α processes) and secondary electron emission from the cathode due primarily to ion bombardment (the γ processes), with the assumption that other bombardment processes play an insignificant role. The pre-breakdown current j_{pbd} satisfies a relation involving positive feedback due to avalanche propagation and secondary emission, in terms of ionization α and secondary electron emission γ_i events that take the form of an exponential relation

$$\begin{aligned} j_{\text{pbd}} + j_{\text{pbd}}\gamma_i &= (j_0 + j_{\text{pbd}}\gamma_i)\exp(\alpha \times \text{gap}) \\ \Rightarrow j_{\text{pbd}} &= \frac{j_0 \exp(\alpha \times \text{gap})}{1 - \gamma_i(\exp(\alpha \times \text{gap}) - 1)}. \end{aligned} \tag{10.32}$$

In Eq. (10.32), j_0 is a background photo-generated cathode emission current. The breakdown criterion is then based on the mathematical condition where the current increases to infinity due to a self-sustaining discharge or

$$1 - \gamma_i(\exp(\alpha \times \text{gap}) - 1) = 0. \tag{10.33}$$

The definition of the ionization constant α, based on a similarity law and experimental data, takes the form

$$\alpha = Ap\exp\left(-\frac{B \times p \times \text{gap}}{V}\right). \tag{10.34}$$

Using this breakdown criterion and the definition of α, the breakdown voltage V_b is a function of the product of the gap and pressure and takes the following form:

$$V_b = \left(\frac{B \times p \times \text{gap}}{\ln(p \times \text{gap}) + \ln(A/\ln(1 + 1/\gamma_i))}\right). \tag{10.35}$$

An explicit expression for the effective secondary electron emission coefficient γ_i that includes ion-enhanced field emission may be used to simply this formula. The obtained numerical values for the breakdown voltage, for a fixed pressure of $p = 760$ Torr or for a fixed electrode gap of gap = 1 μm, are then fitted by analytical expressions (Radmilović-Radjenović et al., 2013) to give

$$V_b = a + b \times \mathrm{gap}^c,\ V_b = m \times p^n. \tag{10.36}$$

Equations similar to Eqs. (10.36) are the basis of the traditional Paschen breakdown curves, in terms of the constants for a gas. The typical plots known as Paschen curves show the breakdown voltages as a function of the product of the pressure in Torrs (1 Torr = 1 mm of Hg = 133.3 Pa) and the electrode gap in cm. Thus, the breakdown voltages for He, Ne, Ar, H_2 and N_2 while ionization (in the arc) at temperatures of 10,000–20,000 °K with typically high power inputs could then be obtained.

The discharge, following an applied DC voltage V_D across a pair of electrodes with a small gap that develops between the electrodes following breakdown, has a number of behavioural regions, depending on the discharge current I_D level, as modelled in Fig. 10.14. Starting from a low discharge current and exponentially increasing voltage, the discharge regions are the Townsend region, normal glow, abnormal glow and the arc region. The Townsend region is characterized by typical non-linear, exponential behaviour. The glow discharge begins after the Townsend region, which is primarily characterized by a dark current. Both the initial development of the Townsend region and the change from the peak of the Townsend region to the normal glow plateau region can exhibit hysteresis-type behaviours. In the plateau of the normal glow region, as the ionizing discharge current is increased, the

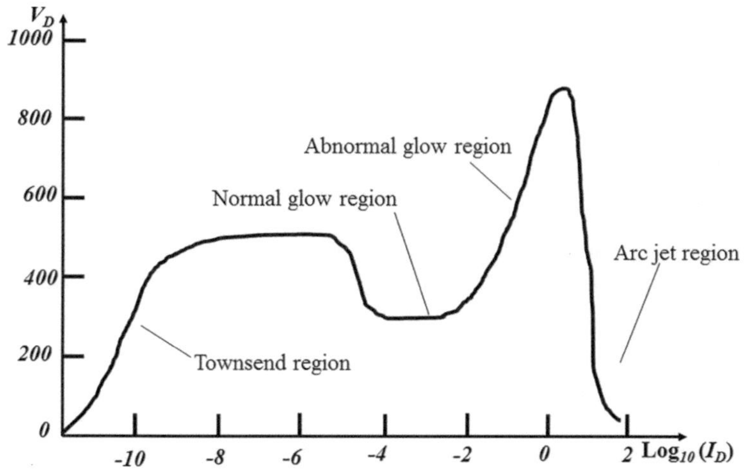

Fig. 10.14 Sketch of the typical behaviour of the discharge voltage versus the discharge current in an arcjet

coverage of the glow also increases till the current begins to saturate. The abnormal glow region follows and is the region used in generating plasmas. At higher discharge currents, the resulting emission is known as the thermionic emission region. Further increases in the current, which is characterized by a negative slope or resistance, result in the arc.

Clearly, there are several regions characterizing the voltage–current behaviour in a typical gas discharge. Initially, there is a low current discharge which could be due to random events. However, the processes change significantly through the regions as the current is increased. The arc discharge region is characterized by an unstable negative local resistance, where with the overall resistance decreasing, the conductivity is on the increase. Thus, there is a need for a stabilizing ballast resistance to restrict current and the resulting temperature at the electrodes. Ionization follows in the discharge within the arcjet region, and charged species are generated at the electrodes.

10.3.3 Emissions and Reactions in a Plasma Region

At the normal room temperatures, the electrons in the conduction band of a metal are not sufficiently energized to leave the surface of the metal. When the metals are heated above a certain critical temperature, the electrons are sufficiently energized due the large amplitude vibrations within the lattice which is sufficient for them to cross the energy barrier and leave the metal. The emission of electrons that are emitted from the surface of a heated metal is known as thermionic emission. Thermally induced flow of charge carriers from a surface or over a potential energy barrier results in the emission of ions at anode and electrons at cathode. The current density J depends on a material constant, the barrier work function W_f and the temperature, and the resulting thermionic emission is defined by the Richardson–Dushman equation given by

$$J = J_0 (T/T_0)^2 e^{-W_f/kT}. \tag{10.37}$$

The work function of a metal is the minimum external energy required for it to emit free electrons from the surface of a metal. The energy required for the electrons to cross the energy barrier is reduced when an external field is applied as it aids the electron in crossing the barrier. The effective work function of the cathode where the electrons are emitted decreases and is given by

$$W_{f_E} = W_f - \sqrt{\varepsilon^3 E}. \tag{10.38}$$

Field emission results due to the ejection of electrons from the surface of the metal when it is primarily subjected to a high electric field and could be the principal factor in breakdown initiation across small gaps. Field emission occurs at temperatures that are lower than that required for thermionic emission. The region close to the electrodes is the fall region or the region of the plasma sheath, due to the sudden drop in the potential, and in this region, the electric field $E = dV/dx$ is large. Thus, a secondary emission is induced and this is called field emission. Field electron emission is the emission of electrons induced by an electrostatic field and requires the application of very high field, often of the order of a million volts per millimetre. Thus, an applied field changes the nature of the potential barrier so it can be overcome, and this gives rise to field emission.

Field emission can be modelled by the Fowler–Nordheim equation. The Fowler–Nordheim equation predicts the current density due to field emission, generally from metals that, due to closely spaced electrodes, thus experience only the high surface electric fields over small emission areas and no thermal inputs. The Fowler–Nordheim-type equations are therefore approximate theoretical equations derived to describe the local current density J emitted from the internal electron states in the conduction band of a bulk metal. The emission current density J for some small uniform region of an emitting surface is usually expressed as a function J of the local work function ϕ and only the local barrier field E that characterizes the small region. The Fowler–Nordheim equation is given by

$$J = (A\beta E/t(y))^2 e^{-B\phi^{3/2}\nu(y)/\beta E}/\phi. \tag{10.39}$$

In Eq. (10.39), A and B are constants, and β is the field enhancement factor, while $t(y)$ and $\nu(y)$ represent corrections that include the influence of a tunnelling barrier shape on the exponential term.

Once a stream of electrons and ions are generated, the laws of chemical equilibrium may be applied to the charge balance between ions, electrons and neutral particles. Given a reaction of two constituents A and B generating products C and D, the reaction is represented by the general equation:

$$aA + bB \rightarrow cC + dD. \tag{10.40}$$

The rate of reaction is expressed by the equation

$$\text{rate} = k[A]^m[B]^n, \tag{10.41}$$

where m and n are constants. The proportionality constant k is called the rate constant; its value is characteristic of the reaction and the reaction conditions. In the rate equation, [G] is the concentration of the constituent “G”. A given reaction has a particular rate constant value under a given set of conditions, such as temperature, pressure and solvent; varying the temperature or the solvent usually changes the value of the rate constant. The value of k, however, does not change as the reaction progresses under a given set of conditions. When a chemical reaction reaches a state of equilibrium, the concentration of the reactants and the products remains the same over time.

The law of chemical equilibrium is a relation stating that in a reaction mixture at equilibrium, there is a condition (given by the equilibrium constant, k_c) relating the concentrations of the reactants and products. For the bi-directional reaction, when the reverse reaction is also feasible,

$$aA + bB \leftrightarrow cC + dD. \tag{10.42}$$

From the rate law, the equilibrium constant, k_c, is

$$k_c = [C]^c[D]^d/[A]^a[B]^b. \tag{10.43}$$

The law of chemical equilibrium may be applied to the charge balance between ions, electrons and neutral particles:

$$n_0 \leftrightarrow i^+ + e^-. \tag{10.44}$$

The rate constant is

$$S(T) = n_i \times n_e/n_0. \tag{10.45}$$

The rate constant for the charge balance equation is the Saha function and from Eq. (9.21) is given by

$$S(T) = \frac{n_i \times n_e}{n_0} = \frac{g_e g_i}{g_0 \lambda_B^3} \exp(-I_p/T). \tag{10.46}$$

In Eq. (10.46), I_p is the ionization potential of the lowest energy level. The Saha equation may also be written as Eq. (9.24) and is

$$S(T) = \frac{n_i}{n_n} = (\approx 2.4 \times 10^{21}) = \frac{T^{3/2}}{n_i} e^{\left(-\frac{E_i}{kT}\right)} = 2\frac{q_i}{q_n}\left(\frac{2\pi m_e kT}{h^2}\right)^{3/2} e^{\left(-\frac{E_i}{kT}\right)}. \tag{10.47}$$

The Saha ionization equation, also known as the Saha–Langmuir equation, relates the ionization state of a gas, in thermal equilibrium with the recombination state, to the temperature and pressure. In the Saha function (10.46), E_i is the ionization potential in electron Volts or "eV" (13.6 eV for H_2, 12.1 eV for Xe, 15.75 eV for Ar); m_e is the mass of the electron, $9.10938356 \times 10^{-31}$ kg; and $q_{n,\ i}$ is the ground state degeneracy (degenerate energy levels) of the neutral atom or ion ($q_{n,\ i} = 2$ for H_2, $q_{n,\ i} = 1$ for H^+).

10.3.4 Estimating the Current Density in a Plasma Region

For each species and for the total respectively, the pressures are

$$p_j = n_j kT, p = \sum p_j. \tag{10.48}$$

For most plasmas away from a boundary or wall of a reaction chamber, there is no net charge; so, from the neutrality of the net charge, it follows that $n_e = n_i$. Furthermore, from Eq. (10.48),

$$p = \sum p_j = \sum n_j kT = nkT, n = \sum n_j. \tag{10.49}$$

Combining with the Saha equation,

$$S(T) = \frac{n_i \times n_e}{n_0} = \frac{n_e^2}{n - 2 \times n_e}, \tag{10.50}$$

results in the quadratic equation for the number of electrons and the number of ions ($n_e = n_i$) in the reaction chamber which is

$$n_e^2 + 2 \times n_e S(T) - nS(T) = (n_e + S(T))^2 - S(T)(S(T) + n) = 0. \tag{10.51}$$

The preceding equation is a quadratic with one positive root for a particular pressure, p. Thus, one can estimate the number of electrons or ions in a confined plasma.

For an idealized thruster model, the thruster discharge chamber has a pre-defined volume that fully encloses the plasma, produced by ionization of neutral gas by the plasma electrons. The ions from the plasma flow only to an electrode or grid within the confines of the discharge chamber, with a current given by the Bohm current. For plasmas with an imposed electric field, E, the force of the electron is due to the imposed electric field, $F_e = -eE$. The force due to the applied field is balanced by a drag force which originates due to the collision processes between particles:

$$F_{de} = m_e \nu_e \cdot v_e. \tag{10.52}$$

To estimate the force, one must know both the collision frequency and the collision velocity of the electrons. The collision frequency is the total electron collision frequency. As the speed of the electron quite significantly exceeds the speed of the neutral particles and ions, $v_e >> v_p, v_n$, the ions and neutrals may be considered to be at rest. The collision frequency is defined as that for which the electron momentum is completely lost. Consequently, there are appropriate collision cross-sections for this, both for the ions and for the neutral particles which are respectively given by σ_{ei} and σ_{e0}. Hence, the collision frequency can be estimated and expressed as

$$\nu_e = n_i \overline{\sigma_{ei} c_{re}} + n_o \overline{\sigma_{eo} c_{re}}. \tag{10.53}$$

The mean speed of a particle (electron) or electron thermal velocity is

$$c_e = \sqrt{8 k_B T_e / \pi m_e}. \tag{10.54}$$

The total electron collision frequency considered in the present model consists of electron–neutral, ion collisions, electron–wall collisions and anomalous plasma collisions which were identified as Bohm diffusion in the preceding chapter, where the diffusion of plasma across a magnetic field was conjectured to follow the Bohm diffusion scaling. The Bohm diffusion predicted that the rate of diffusion was linear with temperature and inversely linear with the strength of the confining magnetic field. The rate predicted by Bohm diffusion is much higher than the rate predicted by classical diffusion, which develops from a random walk within the plasma. Bohm surmised that these fluctuating fields were caused by anomalous plasma collisions. The classical model scaled inversely with the square of the magnetic field. From the force on the electron due to the electric field and the balancing drag force,

$$F_e = -eE = m_e \nu_e \cdot v_e \Rightarrow v_e = -(e/m_e \nu_e) E = -\mu_e E. \tag{10.55}$$

In Eq. (10.55), the electron mobility is defined as $\mu_e = e/m_e \nu_e$ and the electron velocity is proportional to $-\mu_e E$. The current density is simply the product of the charge, electron velocity and the charge density:

$$J = -e n_e v_e = e n_e \mu_e E = \kappa E. \tag{10.56}$$

Goebel and Katz (2008) and Jahn (1968) derive similar relationships. The current density–electric field relation, $J = \kappa E$, is quite similar to Ohm's law. Thus, κ may be identified as the conductivity of the plasma,

$$\kappa = e n_e \mu_e = e^2 n_e / m_e \nu_e. \tag{10.57}$$

The units of the plasma conductivity are Siemens/*m* (*Si*/*m*). Approximate relationships may be established for the conductivity of the plasma (Moisan & Pelletier, 2012).

If it is assumed that the plasma is in equilibrium, the Maxwell distribution for the species velocity is most appropriate to establish an equation for the mean velocity. The Maxwell distribution function is

$$f_e = n_e \left(\frac{m_e}{2\pi k_B T_e} \right)^{3/2} \exp\left(-\frac{m_e}{2k_B T_e} c_e^2 \right). \tag{10.58}$$

c_e is the microscopic (individual) electron velocity of thermal origin. The mean velocity of the electron could be estimated as

$$\overline{c}_e = \sqrt{8k_B T_e/\pi m_e} = 6210\sqrt{T_e}. \tag{10.59}$$

The number of collisions per unit time between a particle and its "target" particles is referred to as the collision frequency. The ratio of the particle's mean speed and the mean free path is used to estimate the collision frequency. The collision frequency may also be estimated from the mean velocity of the electron and the collision cross-sections. Models are available for the collision cross-sections. The ion–electron values for the collision cross-sections are far greater than the neutral particle–electron values and of the order of 10^6, due the Coulombic attraction. However, the ion–electron cross-sections will decrease with the temperature as there is less time for interaction—but they are still dominant.

As the plasma becomes fully ionized, an approximate relation for the conductivity is

$$\kappa = 0.153\sqrt{T^3}/\log\Lambda \text{ in}(Si/m) \text{ with } T \text{ in}^\circ K, \tag{10.60}$$

where $\Lambda = 11.35 + 2 \ln T - \ln P/2$. In Eq. (10.60), P is in *atm* and T is the electron temperature in °K. For hydrogen, the temperature is about 8000 °K. Thus,logΛ is almost a constant. There are important consequences of the relation for the conductivity. There is a very strong dependence of the conductivity of the plasma on temperature. The Ohmic heating (with an equivalent circuit with resistance R and voltage V, the heat dissipation is V^2/R) is simply given by κE^2, where κ is the conductivity and E replaces V . The dissipation results in heating with an increase in conductivity (κ) followed by an increase in temperature.

In an arcjet, the arc becomes highly localized into a filament which carries all the current and all heating—potentially an unstable process, requiring stabilization which is frequently achieved by swirling the flow of the propellant. Hence, the process of heating the propellant is one in which the propellant flows through the arc with temperatures exceeding 10,000 °K to about 100,000 °K. Moreover, the net exhaust flow is neutral, and consequently, there is no net exchange of charge between the thruster exhaust and the environment.

10.3.5 Microwave Jet Thrusters

Microwave jet thrusters offer a highly specific impulse and significant lifetime increases when compared to resistojets and arcjets. The electrodeless design is

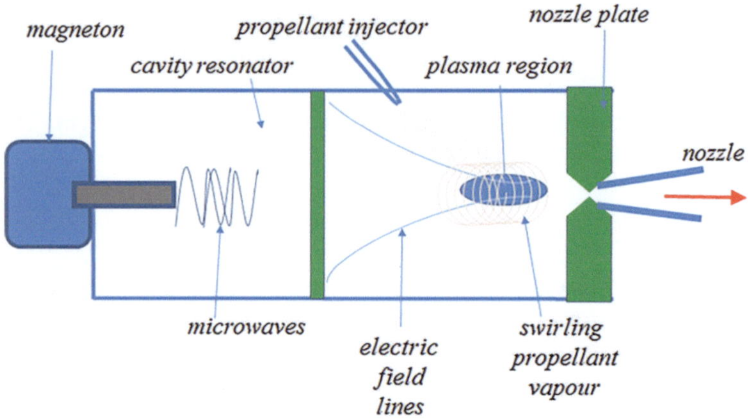

Fig. 10.15 Principle of operation of a microwave jet thruster

completely scalable and is based on the use of a magnetron microwave valve and a cavity resonator. Microwave jet thrusters create free-floating plasma discharges within a cavity resonator, heating the propellant and providing reasonable specific impulse with relatively low power consumption. Choice of a range of propellant enables a wide range of specific impulse (above 300 s) and thrust-to-power ratio combinations offering high I_{sp} performance or high thrust. The power consumption depends on the frequency of operation that can be varied, leading to thrusters able to operate at low power (30 Watts) or at much higher power (1 kW). A typical microwave jet thruster for spacecraft applications is illustrated in Fig. 10.15.

The principle of operation is based on a resonant cavity chamber which is a cylindrical waveguide shortened by two conducting endplates. One endplate houses a nozzle and the other a magnetron valve for generating microwaves, along the cavity axis. The cavity could be partitioned by dielectric plates so the rear of the cavity housing the nozzle could be maintained at a pressure different from the cavity where the microwaves are generated. Most of the power supplied is used to generate the waves which are generating the plasma. Ohmic heating of the walls is minimized relative to the power used for plasma generation. The propellant is injected into the nozzle section, and a free-floating region of stabilized plasma is formed by the high electric energy density at the nozzle entrance. The propellant is injected from a side of the cavity to generate a swirl vortex flow inside the chamber, so the radial pressure gradient maintains the plasma along the cavity axis. The plasma then heats the surrounding propellant which is then expelled and exhausted through a choked nozzle.

Exercises

1. (a) The following propulsion technologies may be used on-board a satellite for orbit maintenance. By identifying a typical propellant for each system, discuss the advantages and disadvantages when using that propellant, the typical performance obtained and any issues regarding storage of the propellant.

   (i) Monopropellant systems, (ii) bi-propellant system, (iii) an electrostatic propulsion system

   (b) An arcjet thruster is chosen as the propulsion system to raise the altitude of a satellite in low Earth orbit. The thruster consumes a gaseous propellant at a rate of 700 mg/s and has an exhaust velocity Ve = 9000 m/s. Calculate the specific impulse and thrust.
   (c) The arcjet in part (b) is powered by solar arrays and associated power conditioning (the power subsystem), which have a specific power efficiency of 15 kg/kW. The thruster converts 70% of the electrical power to kinetic power. Calculate the mass of the power subsystem required. Comment on your answer.
   (d) This arcjet is used to raise a satellite orbit from a low earth orbit to an orbit at geostationary altitude. Explain why this manoeuvre cannot be analysed as a Hohmann transfer and identify the principal changes that will arise between these mission scenarios. Identify thruster control requirements for the arcjet solution.

2. (a) Describe the key features and operating principles of two different electro-thermal propulsion systems, including the nature of propellants that may be used and the thrust and power level achievable in such devices.

   Identify what is meant by frozen flow efficiency and why it is particularly important in propellant selection for an electro-thermal thruster.
   (b) A spacecraft in circular orbit about the Earth has an electro-thermal propulsion unit, which provides a thrust tangentially to the orbit in the plane of the orbit. Under these conditions, the rate of change in semi-major axis may be written:

$$da/dt = 2f\sqrt{a^3/\mu},$$

   where f is the thrust per unit mass acting on the satellite.

   If it is assumed that the thrust is constant and the satellite is initially in a circular orbit of semi-major axis r_1, this shows that the equivalent Δv required to raise the orbit from r_1 to r_2 may be written in the form.

$$\Delta v/v_1 = 1 - \sqrt{r_1/r_2}.$$

   where v_1 is the initial orbital velocity of the satellite. Make clear any assumptions that you have made in your derivation.

(c) An electro-thermal thruster has a specific impulse of 1000 s when operating at a thrust of 10 N. This unit is used to raise the orbit of a 500 kg mass spacecraft in low Earth orbit from an altitude of 300 km to an altitude of 400 km. Calculate the mass of propellant used by the thruster for this orbit raising manoeuvre, assuming that there is negligible change in the spacecraft mass during the manoeuvre.

3. (a) Describe, with the aid of diagrams, the components and layout of two forms of on-board propulsion: chemical bi-propellant and electro-thermal propulsion systems. In your answer, you should provide an indication of the performance achieved.
(b) Chemical bi-propellant and electro-thermal propulsion systems both suffer from frozen flow losses. Identify the detailed causes of such losses and define frozen flow efficiency. What methods may be used to reduce the magnitude of such losses in an electro-thermal thruster?
(c) An electro-thermal propulsion system is required to undertake an orbit raising task in 100 h. If the power system specific power is 40 W/kg and the propellant has a specific heat capacity c_p of 2.8 kJ/kg, calculate the temperature the propellant must reach in order to achieve optimum performance if frozen flow losses are ignored. (d) What type of electro-thermal propulsion system is capable of achieving the performance required in part (c), and do you believe in such a system it is reasonable to ignore frozen flow losses?

4. A hydrazine electro-thermal thruster operating at a temperature of 1800 K has a frozen flow efficiency of 0.4. Calculate the loss of specific impulse (I_{sp}) in this thruster relative to the ideal performance of the device. The specific heat capacity for hydrazine may be assumed constant, having a value of 2.8×10^3 J/kg.

Appendix 10.1

Basic Units Used in Plasma Physics

Charge of an electron: $1.60217662 \times 10^{-19}$ Coulombs.

Electron volt: the electron volt (eV) is a unit of energy equal to $1.60217662 \times 10^{-19}$ Joule (J). It is the amount of energy gained by the charge of a single electron moved across an electric potential difference of one volt.

Speed of light: $c = 2.99792458 \times 10^8$ m/s.

Mass of an electron: 1.782662×10^{-36} kg (eV/c^2).

Thus, one could write the momentum in units of eV/c.

Appendix 10.2

List of Manufacturers and Some Useful Websites

In this appendix, a list of manufacturers of space propulsions systems and thrusters (Anon 2024a, b, c, d, e) are provided which also provide considerable useful information on company proprietary modular propulsion systems for space applications. Using the names of these manufacturers in a search engine, the reader can access the manufacturer's website and find all the relevant datasheets and information associated with their products, as released to the public by the Company.

Chemical Propulsion Systems Manufacturers

AAC Clyde Space
Benchmark Space Systems
Busek Co. Inc.
CU Aerospace LLC
Dawn Aerospace
ECAPS (Ecological Advanced Propulsion Systems)
IHI Aerospace
Laboratory of Advanced Jet Propulsion (LAJP)
Rafael Ltd.
Rubicon Space Systems

Electric Propulsion Systems Manufacturers

Ad Astra Rocket
Advanced Technology Institute (ATI)
Airbus Defence and Space
ALIENA Pte Ltd
Applied Ion Systems
Astra (https://astra.com/)
Aurora Propulsion Technologies
Bellatrix Aerospace Pvt Ltd
Benchmark Space Systems
Bradford Space
Busek Co. Inc.
CU Aerospace LLC
D-Orbit
ENPULSION
ExoTerra Resource LLC
Exotrail

IENAI SPACE
Orbital Astronautics Ltd
Pale Blue
Phase Four
Rafael Ltd.
Seran Systems
SITAEL S.p.A.
Space Electric Thruster Systems (SETS)
SteamJet Space Systems
T4i (Technology for Propulsion and Innovation) (Italy, University of Padua)
Tethers Unlimited Inc.
ThrustMe

Space Systems Manufacturers

AAC Clyde
ACCION (USA)
Aerojet (USA)
Aerojet Rocketdyne
Aerospace Innovation (Germany)
Aerotech (Apogee Components)
AFRL/USC (USA)
Alameda AASC Space (Alameda Applied Sciences Corporation) USA
Aliena (SGP)
Alta Space, now (SITAEL) (Italy)
Applied Ion Systems
Applied Science Corporation (USA)
Ariane Group (Airbus/Safran)
Asher Space Research Institute (ASRI) (Israel)
ASTRA (USA)
Aurora (Finland)
Austrian Research Center (ARC)/Austria Institute of Technology (AIT) Aerospace Technology (Mechatronic System technik) (Austria)
Ball Aerospace (USA)
Bellatrix Aerospace (India)
Boeing (USA)
Bradford Engineering
BUSEK
Celestron
ClydeSpace
COMAT Propulsion (France)
Composite Technology Development, Inc. (USA)
CSSAR (Applied Research of Chinese Academy of Sciences)
Dawn Aerospace (Netherlands/New Zealand)

Delfi Space
Deutsches Zentrum für Luft- und Raumfahrt (DLR)
D-Orbit (Italy)
DSSSP (USA)
Earth Observant (USA)
ECAPS (Ecological Advanced Propulsion Systems)
EMC (USA)
ESA (European Space Agency)
Espace
Exoterra (USA)
EXOTRAIL (France)
Experimental Propulsion Laboratories (EPL)/CRP Windform Manufacturer (USA)
Fakel (Russia)
FOTEC/ENPULSION (Austria), ENPULSION Inc. (USA)
French National Center of Scientific Research, LAAS CNRS (France)
Froeberg Aerospace (USA)
Frontier Aerospace (USA)
GreenDelta/DLR
Honeywell/Atlantic Research Corporation (USA)
Howe Industries (USA)
Hypernova (South Africa, ZAF)
Ienai Space (ESP)
Industrial Solid Propulsion (ISP) (USA)
JAXA (Japan Aerospace Exploration Agency)
Los Alamos LANL (USA)
LPMI (Laboratoire de Physique des Millieux Ionises) (France)
Magdrive (UK)
Marotta
Mars Space Limited (UK)
MICRO SPACE (Italy/Singapore)
Miles Space (USA)
MMA Design (USA)
MOE Key Lab for micro and nano electromechanical systems, Northwestern Polytechnical University (China)
Momentus (USA)
MOOG Inc.
Morpheus Space
Nanjing University of Science and Technology (China)
NanoAvionics (Lithuania)
NanoSpace AB (Sweden) now GOMSPACE (Denmark)
National Aeronautics and Space Agency (NASA)
NASA Goddard (USA)
NASA JPL
NASA Marshall
National Natural Science Foundation of China (NNSFC)

National Space Organization, Taiwan (Taiwan)
Neumann Space Australia (Australia)
Next Generation Space Tech. Research Assoc. (NESTRA) (Japan)
Northrop Grumman
Northwestern Polytechnical University
ONERA (FRA)
Orbion Space Technology (USA)
Orbital ATK
Orbital Technologies Corporation (ORBITEC)
Pacific Scientific (PACSCI EMC)
Pacific Scientific (PacSci)
Pale Blue (JPN)
Parabilis (USA)
Patch Electronics (Japan)
Phase Four
Planetary Society (USA)
Plasma Processes (Rubicon) (USA)
Plasma Processes (USA)
Plasmos
Prime Lightworks (USA)
Rafael Advanced Defence Systems
RCC Kurchatov Institute/Fakel
ROCCOR (USA)
ROSCOSMOS
Space Dynamics Laboratory (SDL)
Space Systems, Ltd. (UK/Russia)
SteamJet
Stellar Exploration (USA)
Surrey Satellite Technologies (SSTL)
Surrey Space Center (UK)
SystematIC BV
T4i (Technology for Propulsion and Innovation) (Italy, University of Padua)
Tethers Unlimited (USA)
The Aerospace Corporation (USA)
ThrustME (France)
TransAstra
TRW (Thompson Ramo Wooldridge Inc., Northrop Grumman)
VACCO (USA)
Ventions, LLC (USA), now Astra Space (USA)
Vestigo Aerospace
Yuki Precision/Takasago (Japan)
Zenno Astronautics (New Zealand)

Academic Institutions (Universities) Involved in Space Research

Tsinghua University
TU Delft
TU Dresden
UCLA/JPL (USA)
Universidad Carlos III de Madrid (ESP)
University of Arizona
University of Illinois Urbana Champaign/Alameda
University of Malta (Malta)
University of Michigan (USA)
University of Missouri (USA)
University of Padua (Italy)
University of Seville (Spain)
University of Southampton, AVS UK Ltd. (UK)
University of Stuttgart (DEU)
University of Texas at Austin (USA)
University of Tokyo
University of Toronto Space Flight Laboratory
University of Toronto, Institute for Aerospace Studies (UTIAS) Space Flight Laboratory (SFL) (Canada)
University of Twente
University of Ulsan (South Korea)
University: Australian National University (Australia)
University: BUAA (Beijing University of Aeronautics and Astronautics)
University: California Polytechnic State University, Cal Poly CubeSat Lab (USA)
University: Chinese Academy of Sciences
University: Cornell University (USA)
University: Cranfield University
University: Delft university of technology (Netherlands)
University: George Washington University (USA)
University: Georgia Tech (USA)
University: Justus Liebig University (DEU)
University: Korea Advanced Institute of Science and Technology (KAIST) (South Korea)
University: Lanzhou Institute of Physics (LIP)
University: MIT
University: National Aerospace University (KhAI) (Ukraine)
University: National University of Singapore
University: Naval Research Laboratory (NRL) (USA)
University: Osaka Sangyo University (OIT) (Japan)
University: Penn State
University: Purdue University
University: RAPPEL lab at the Royal Military College of Canada

University: Research Institute of Applied Mechanics and Electrodynamics of Moscow Aviation Institute (RIAME)
University: Royal Military College of Canada (Canada)
University: Shanghai Engineering Centre for Microsatellites
University: Shizuoka University
University: SPMI (Shanghai Spaceflight Power Machinery Institute)
University: Technical University of Madrid (UPM) (ESP)
University: Western Michigan University (USA)
US Naval Academy
Utah State University (USA)
Utah State University USU (USA)

References

Aifantis, K. E., Hackney, S. A., & Kumar, R. V. (Eds.). (2010). *High density energy lithium batteries*. Wiley-VCH.

Anon. (2024a). *CubeSat thrusters and small satellite propulsion systems*. https://blog.satsearch.co/2019-07-10-cubesat-thrusters-and-in-space-propulsion-products-for-small-satellites-on-the-global-marketplace, Accessed 01-06-2024.

Anon. (2024b). *Space propulsion Systems for Satellites and Spacecraft, orbital propulsion Centre*. Ariane Group. https://www.space-propulsion.com/spacecraft-propulsion/propulsion-systems/index.html, Accessed 01-06-2024.

Anon. (2024c). *Green propulsion for any satellite*. Dawn Aerospace. https://www.dawnaerospace.com/green-propulsion, Accessed 01-06-2024.

Anon. (2024d). *SmallSat propulsion systems EPSS*. Kongsberg NanoAvionics. https://nanoavionics.com/cubesat-components/cubesat-propulsion-system-epss/, Accessed 01-06-2024.

Anon. (2024e). *ASM specifications*. ASM Aerospace Specification Metals. https://asm.matweb.com/search/SpecificMaterial.asp?bassnum=NINC34, Accessed 01-06-2024.

DoITPoMS Teaching and Learning Packages. (2023a). *Batteries, © DoITPoMS*. University of Cambridge. https://www.doitpoms.ac.uk/tlplib/batteries/index.php, Accessed November, 2023.

DoITPoMS Teaching and Learning Packages. (2023b). *Fuel-Cells, © DoITPoMS*. University of Cambridge. https://www.doitpoms.ac.uk/tlplib/fuel-cells/index.php, Accessed November, 2023.

Fortescue, P. W., Swinerd, G. S., & Stark, J. P. W. (2011). *Spacecraft systems engineering* (4th ed.). Wiley.

Goebel, D. M., & Katz, I. (2008). *Fundamentals of electric propulsion: Ion and hall thrusters* (JPL space science and technology series) (1st ed.). Wiley.

Heister, S. D., Anderson, W. E., Pourpoint, I. L., & Cassidy, R. J. (2019). *Rocket propulsion*. Cambridge University Press.

Hsu, A., Dragnea, H., Schilling, J. & Doumitt, A. (2023). 2023 Small Satellite Propulsion Technologies Compendium, The Aerospace Corporation, DISTRO A: OTR-2024-00338, Version 1.41.

Jahn, R. G. (1968). *Physics of electric propulsion*. McGraw Hill Book.

Moisan, M., & Pelletier, J. (2012). *Physics of collisional plasmas* (pp. 387–411). Springer Science +Business Media. https://doi.org/10.1007/978-94-007-4558-2

Radmilović-Radjenović, M., Radjenović, B., Klas, M., Bojarov, A., & Matejčik, Š. (2013). The breakdown mechanisms in electrical discharges: The role of the field emission effect in direct current discharges in micro-gaps. *Acta Physica Slovaca, 63*(3), 105–205.

Sutton, G. P., & Biblarz, O. (2016). *Rocket Propulsion Elements, Ninth Edition*. Wiley.

Verduci, R., Romano, V., Brunetti, G., Nia, N. Y., Di Carlo, A., D'Angelo, G., & Ciminelli, C. (2022). Solar energy in space applications: Review and technology perspectives. *Advanced Energy Materials, 12*, 2200125(1-24).

Chapter 11
Space Vehicle Electro-static and Electro-magnetic Propulsion

11.1 Introduction

In the preceding chapter, electric propulsion systems were considered where the plasma was used to heat the propellant before it was expended via a nozzle. In this chapter, electric propulsion systems for spacecraft would be considered which actually expel the plasma and ionic flows themselves via nozzles. The gridded ion engine (GIE), or ion thruster and a Hall current thruster (HCT), is an example of plasma or ion accelerator in which the thrust generation is achieved through electro-static or steady electro-magnetic interactions in generating the plasma. In these types of thrusters, the creation of the propellant ions by ionization and subsequent acceleration processes of the ions and electrons or the plasma are done in physically separated regions and could be independently controlled.

Plasma is known to be ionized gases where positively charged ions of the gas, negatively charged electrons and neutral particles carrying no charge can coexist in an equilibrium state but not bonded together, for example, the noble gases helium, neon, xenon, argon, radon and krypton, particularly amenable to ionization at relatively low voltages. Electric propulsion thrusters generate thrust by creating and expelling plasma at high velocities, resulting in plasma plumes in the exhaust. Plasma plumes expand into vacuum and can have varying properties. Typical electro-static, magnetically guided plasma plumes are characterized by ion velocities of 10–100 km/s with electron energies, from 0 to 20 eV, and electron densities varying from 0 to 10^{15} to 10^{18} /m^3. The Mach numbers of the ions are estimated from $M_i = u_i/\sqrt{E_e/m_i}$. A typical simulated plasma plume takes the form of a divergent electro-static, electro-magnetically guided nozzle and is illustrated in Fig. 11.1. The plume regions, within the divergent half angles, are defined by different levels of the ejected plasma current and contain almost 90% of the generated plasma.

The typical electron units used in describing the physics of plasmas in ionic propulsion thrusters are summarized in Table 11.1.

R. Vepa, *Space Vehicle Maneuvering, Propulsion, Dynamics and Control*,
https://doi.org/10.1007/978-3-031-65518-0_11

Fig. 11.1 Typical simulated plasma plume

Table 11.1 Typical set electron units used in plasma physics

Measurement	Unit	SI value of unit
Energy	eV	$1.6021766208(98) \times 10^{-19}$ J
Mass	eV/c^2	1.782662×10^{-36} kg
Momentum	eV/c	5.344286×10^{-28} kg·m/s
Temperature	eV/k$_B$	$1.1604505(20) \times 10^4$ °K
Time	ħ/eV	6.582119×10^{-16} s
Distance	(ħ × c)/eV	1.97327×10^{-7} m
Proton mass	kg	$1.67262190 \times 10^{-27}$ kg

11.2 Plasma Behaviour in Electric and Magnetic Fields

In electric ion propulsion, the charged particles are usually produced in a plasma, and the main source of thrust results from the positively charged ions. From the thrust equation, the ionic thrust maybe written as

$$T_i = v_{\text{iex}} \dot{m}_i, \tag{11.1}$$

where m_i is the ionized propellant mass and v_{iex} is the ion exhaust velocity. If an ion is accelerated by a net voltage, V_b, then, by conservation of energy, the resulting ionic velocity is, v_i,

$$v_i = \sqrt{2qV_b/M_i}, \tag{11.2}$$

where q is the ion charge and M_i is the mass of the ion. Hence, the thrust T_{ib} produced by a zero-divergence mono-energetic ion beam accelerated by a potential V_b before ejection will be

$$T_{ib} = v_i M_i I_b / q = I_b \sqrt{2V_b M_i / q}, \tag{11.3}$$

where I_b is the ion beam current. The thrust is proportional to the beam current, the square root of the ion mass and acceleration potential and inversely proportional to the square root of the ion charge.

Sources of electron emissions include free electrons that are emitted from the surface of a metal when external heat energy is applied which is the thermionic emission. It is a function of both the work function and temperature and is defined by the Richardson–Dushman equation, given by Eq. (10.37). It applies to metals when they are heated to a very high temperature. Field emission refers to free electrons which are emitted under a strong applied electric field which is related to the Schottky effect. Field electron emission occurs both in metals and fluids when they are subjected to a very strong electric field, even at low temperatures, and is defined by the Fowler–Nordheim Eq. (10.39). Emissions due to radio frequency excitation or microwave heating to high temperatures are also possible sources of electron emissions.

11.2.1 Electro-static/Electro-magnetic Acceleration

The electro-static Coulomb force between two or more charged bodies is the force between them due to Coulomb's law. If the particles are both positively and negatively charged, the force is repulsive, and if they are of opposite charge, it is attractive. The law is given by

$$\mathbf{F} = \frac{q_1 q_2}{|\mathbf{r}|^3} \frac{\mathbf{r}}{4\pi\varepsilon_0}, \tag{11.4}$$

where ε_0 is the permittivity of free space, $8.85418782 \times 10^{-12}$ Farads(F)/m.

A particle with a charge q and a speed $\mathbf{v}$ in an electric field $\mathbf{E}$ and a magnetic field $\mathbf{B}$ experiences a force $\mathbf{F}$. This is known as a Lorentz force. It is given by

$$\mathbf{F} = q(\mathbf{E} + \mathbf{v} \times \mathbf{B}). \tag{11.5}$$

The force is a vector and normal to the velocity $\mathbf{v}$ and magnetic field $\mathbf{B}$ vectors. The electric field accelerates the particle. The magnetic field controls the direction of its trajectory. A magnetic field that is perpendicular to the velocity $\mathbf{v}$ will constrain the particle to circular trajectory in a plane that is perpendicular to the field.

Consider the motion of a charged particle of mass m, charge q, in a circular trajectory under the influence of a magnetic field. The circular velocity is assumed to be $\mathbf{v}$. In the radial direction, the force due to a transverse magnetic field $\mathbf{B}$ is $\mathbf{B}\mathbf{v}q$. This is balanced by the centrifugal force $m\mathbf{v} \cdot \mathbf{v}/r$. So,

$$\mathbf{B}\mathbf{v}q = m\mathbf{v} \cdot \mathbf{v}/r \Rightarrow r = m\mathbf{v}/\mathbf{B}q. \tag{11.6}$$

The radius, r, is known as the *Larmor* radius. Also $\omega_c = \mathbf{v}/r = \mathbf{B}q/m$ is the *cyclotron frequency*.

11.2.2 Generation of a Plasma at a Source

Both gridded ion engines and Hall current thrusters require a source of electrons to initiate the generation of the plasma and/or ion beam. The process of generation is always a hollow cathode which is quite similar to the discharge in an arc jet thruster. However, electrons are principally emitted as a result of thermionic emission from a hot cathode material as given by the Richardson–Dushman equation. With the anode at a positive potential, the current density for thermionic emission at an electron emitter cathode could be expressed as

$$J = A(1 - r)T^2 \exp(-\phi/kT). \tag{11.6}$$

where $A = 4\pi mk^2e/h^3 \sim 120$Amperes/(cm ° K)^2 is a fundamental constant, r is the reflection coefficient of the emitted electrons, T is the temperature and ϕ is the work function of the surface. The Richardson–Dushman equation is modified by the presence of an electric field, F, and is defined by the Schottky effect equation:

$$J = A(1 - r)T^2 \exp\left(-\left(\phi - \sqrt{\Delta\phi F}\right)/kT\right). \tag{11.7}$$

The emitter of the heated cathode requires a low work function and is frequently based upon barium oxide possibly in matrix form with added metallic components, e.g. tungsten. The outer electrode (the keeper) is held at a positive voltage (~200 V) with respect to the cathode. The design of a heated cathode is described by Pedrini et al. (2017). Figure 11.2 illustrates the design of a thermionic emission–based heated cathode, with a lanthanum hexaboride electron emitter, tungsten spring and graphite sleeve encased in a tube and heated by a tungsten/rhenium filament in a boron nitride insulator, protected by a tantalum heat shield, within an electrode keeper made of refractory materials (graphite, titanium alloys).

11.2.3 *Child–Langmuir Law*

When the anode voltage is small, electrons travel slowly in vacuum. Therefore, their density, $n = J/(ev)$, will be large (v is the electron velocity and e its charge.) This

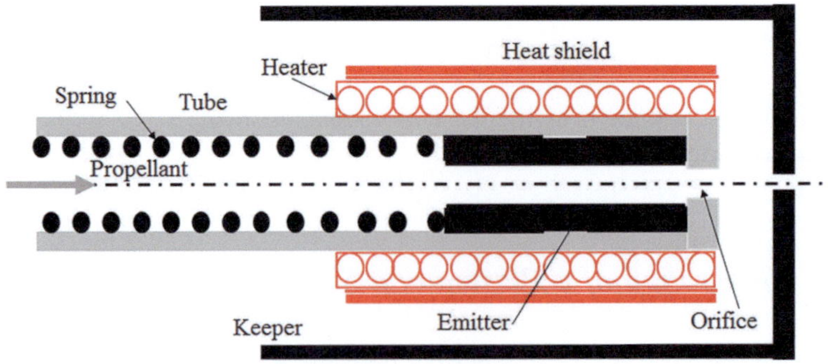

Fig. 11.2 Typical design of a heated cathode

density, called the "space charge", induces an electro-static potential by Poisson's law that is negative in front of the cathode and, therefore, presents a barrier, in other words, an electron that exits the solid with a cloud of electrons in front of the surface which may return it to the cathode. The space charge is responsible for J increasing initially as $V_a^{3/2}$ (Child–Langmuir law) and saturating at large V_a to the value predicted by the Richardson–Dushman equation.

The current density j_{scl} given by the space charge limited flow is

$$j_{\text{scl}} = \rho(x)v(x). \tag{11.8}$$

where $\rho(x)$ and $v(x)$ are the density and velocity of the electrons, respectively. From the principle of energy conservation, $v(x)$ is given by

$$(1/2)mv^2(x) = q \times V(x). \tag{11.9}$$

Due to the space charge inside the gap, the potential $V(x)$ must obey Poisson's equation:

$$\nabla^2 V(x) = -\rho(x)/\varepsilon_0. \tag{11.10}$$

Eliminating $\rho(x)$ and $v(x)$ using Eqs. (11.8) and (11.9), Eq. (11.10) reduces to a second-order non-linear differential equation for the potential in one dimension given by

$$\nabla^2 V(x) = -\rho(x)/\varepsilon_0 = -j_{\text{scl}}/\varepsilon_0 v(x) = (-j_{\text{scl}}/\varepsilon_0)\left(\sqrt{m_i/2qV(x)}\right). \tag{11.11}$$

Multiplying both sides by $dV(x)/dx$ and integrating, it can be shown that

$$\left(\frac{dV(x)}{dx}\right)^2 - \left(\frac{dV(x)}{dx}\bigg|_{x=0}\right)^2 = \frac{4j_{\text{scl}}}{\varepsilon_0}\sqrt{\frac{m_i}{2q}}\sqrt{V(x)}. \tag{11.12}$$

If one sets the electric field at the cathode to zero, taking the square root of Eq. (11.11) and integrating the electric field at the space charge limited condition, the potential for the space charge limited condition is

$$V(x) = \left(\frac{3x}{2}\right)^{4/3}\left(\frac{j_{\text{scl}}}{\varepsilon_0}\sqrt{\frac{m_i}{2q}}\right)^{2/3} + C. \tag{11.13}$$

By applying the boundary condition at the cathode $V(0) = 0$ and applying the boundary condition at the anode, the potential for the space charge limited region may be obtained. The ion current density due to space charge limitation (saturation current) is given by the Child–Langmuir law, which, for an electrode defined as x_a, is given by

$$j_{\text{scl}} = \frac{4}{9}\varepsilon_0\sqrt{\frac{2q}{m_i V_0}}\left(\frac{V_0}{x_a}\right)^2. \tag{11.14}$$

For the one-dimensional case, the electric field satisfies

$$\frac{dE(x)}{dx} = \frac{\rho(x)}{\varepsilon_0} = \frac{qn_i(x)}{\varepsilon_0}. \tag{11.15}$$

The corresponding acceleration potential function and electric field, in terms of the accelerating potential voltage V_0 and electrode spacing x_a, are, respectively, given by

$$\phi(x) = -V_0(x/x_a)^{4/3}, E(x) = (4V_0/3x_a)(x/x_a)^{1/3}. \tag{11.16}$$

In this case, the space charge limited current density is again given by Eq. (11.14).

11.2.4 Electro-static/Electro-magnetic Thrust Generation

There is an important consequence of the Child–Langmuir law. From the classical rocket thrust equation $F_T = \dot{m}v_e$, the mass flow rate $\dot{m} = j_{\text{scl}}Am_i/q$, the ion exit velocity $v_e = \sqrt{2qV_0/m_i}$ and the Child–Langmuir law, Eq. (11.14), it follows that

$$\frac{F_T}{A} = \frac{\dot{m}v_e}{A} = \frac{jAm_iv_e}{Aq} = \frac{jm_i}{q}\sqrt{\frac{2qV_0}{m_i}} = \frac{8}{9}\varepsilon_0\left(\frac{V_0}{x_a}\right)^2. \tag{11.17}$$

The operation of an electric thruster can generally be divided into three stages: ionization, acceleration and neutralization. In the first stage or the ionization stage, a propellant such as a noble gas like argon or xenon is most commonly used which is injected into the ionization chamber where it is ionized by electron bombardment from a hollow or heated cathode. Electrons colliding with the propellant's atoms cause it to lose one electron per atom to form positive xenon ions. Electron bombardment can be achieved by various methods. The first method is direct current discharge in which electrons flow back from cathode to anode (as in a gridded ion thruster). The second method is radio frequency discharge where electrons are held circulating in an electro-magnetic radio wave field (e.g. radio frequency ion thruster, RIT). The last method is electron cyclotron resonance (ECR) in which electrons are excited by microwaves circling around an electro-magnetic field [e.g. high-power electric propulsion (HiPEP) thrusters]. The propellant's ions are then extracted from the ionization chamber through a multi-aperture grid extraction system by diffusion. In the second stage or the acceleration stage, the plasma ions enter the grid system (in between the positive and negative grid). The potential difference could be as high as of 1280 V between the positive (screen) and negative (accelerator) molybdenum

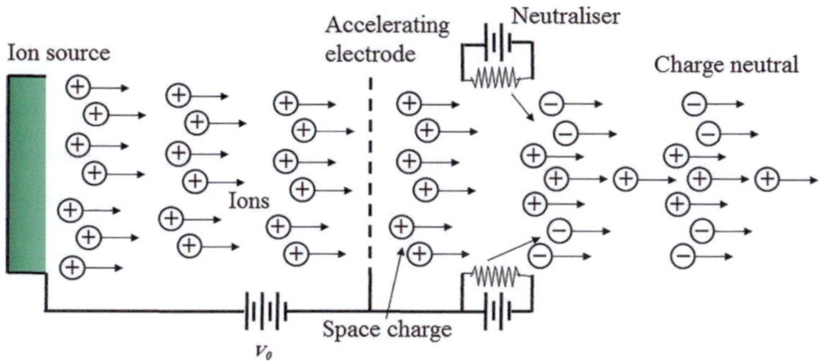

Fig. 11.3 The principle of operation of typical ion thrusters

grid which creates an electric field that accelerates the ions out of the exhaust at over 100,000 km/h. Ion energy typically reaches 1–2 keV for certain thrusters. The reaction force of this jet of ion stream out of the exhaust of the engine generates the thrust of the engine. In the final stage or the neutralization stage, in order to prevent the spacecraft from gaining a net negative charge due to the efflux of positive ions and subsequent accumulation of the charge, a cathode neutralizer injects electrons into the exhaust to neutralize the positive propellant's ions. The principle of operation of typical ion thrusters is illustrated in Fig. 11.3. In general, for a typical *ion thruster*,

$$v_{e_i} = \sqrt{2qV_{\text{ion}}/m_{\text{ion}}},\ T = I_{\text{ion}-\text{beam}}\sqrt{2m_{\text{ion}}V_{\text{ion}}/q}, I_{\text{sp}} = (\gamma_c\eta_m/g) \times \sqrt{qV_{\text{ion}}/m_{\text{ion}}}, \tag{11.18}$$

where q is the ionic charge, V_{ion} is the voltage driving the ions, m_{ion} is the ionic mass, $I_{\text{ion}-\text{beam}}$ is the ionic beam current, is the thruster mass utilization efficiency and γ_c is the total thrust correction factor.

11.3 Typical Plasma Thrusters

There are indeed several types of thrusters. The gridded ion thrusters use electrodes and grids to accelerate the ions, to protect the accelerator grid and to decelerate the ions as required. They also use a neutralizer to compensate for the space charge. The principle of operation is illustrated in Fig. 11.4.

There are also electrode-free designs which use radio frequency antennas if microwaves excite the ions. These are illustrated in Figs. 11.5 and 11.6.

The neutralizer is a critical component of a gridded ion thruster. Its function is to completely neutralize the space charge. Partial neutralization of the space charge could result in a drastic decrease in the thruster's exhaust velocity. At high power

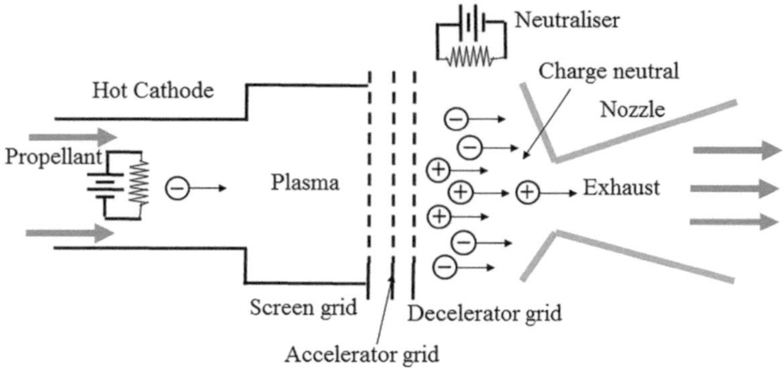

Fig. 11.4 The principle of operation of typical gridded ion thrusters

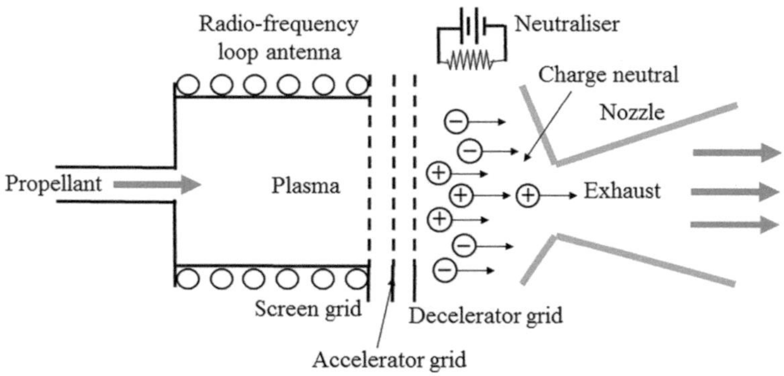

Fig. 11.5 The principle of operation of radio frequency ion thrusters

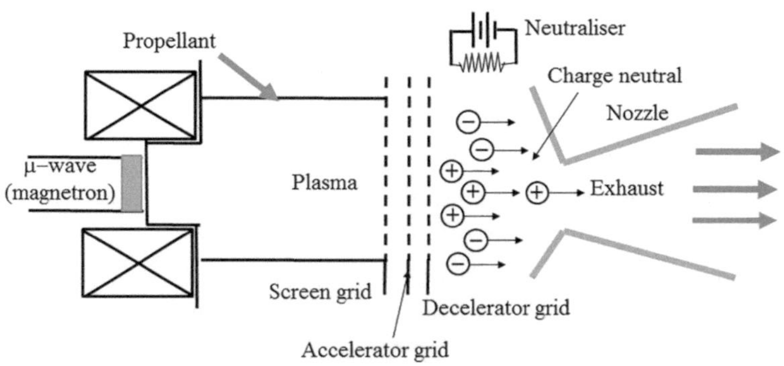

Fig. 11.6 The principle of operation of microwave ion thrusters

levels, electrodes generally wear out over time, resulting in a decrease in the life of the thruster. Electrodeless plasma thrusters also do not require an external neutralizer, in principle, as a quasi-neutral plasma is expelled by the exhaust. They are

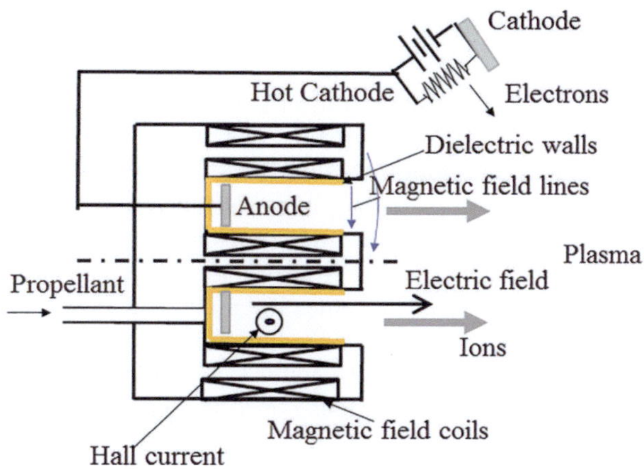

Fig. 11.7 The principle of operation of Hall current thrusters

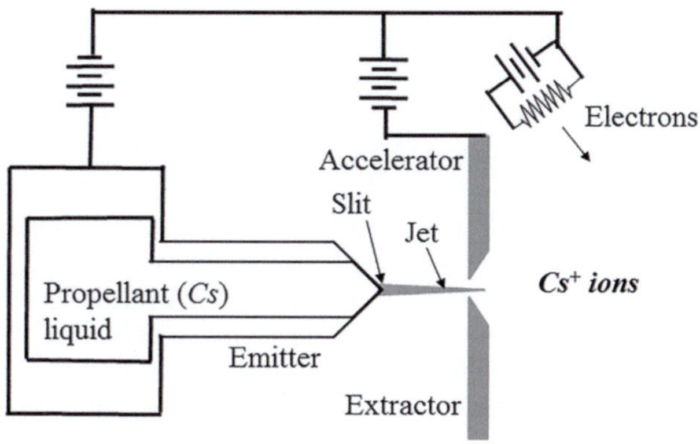

Fig. 11.8 The principle of operation of field emission ion thrusters

generally both anode-less and cathode-less. Radio frequency and microwave thrusters have been built without the use of grids, electrodes and neutralizers.

Hall current thrusters which use electro-magnets and an electric field to generate a Hall current and accelerate the ions are illustrated in Fig. 11.7. Field emission–based electric thrusters are illustrated in Fig. 11.8.

The sequence of events generally for all types of ion or plasma thrusters is (i) charged species (ions give the thrust) need to be generated, (ii) the source potential for the ions must be high relative to the ambient space plasma which is essentially at 0 Volts, (iii) the species are accelerated in an intense electric field region and focussed, and finally (iv) for a charged plasma (with a net positive charge), the ion beam must be neutralized.

11.3.1 Gridded Ion Engines

The typical gridded ion thruster is an archetypal electro-static thruster of the Kaufman type (Mazouffre, 2016). Its operation can be divided into ionization, acceleration and neutralization. Xenon is most commonly used as the propellant in these thrusters. Xenon propellant is first injected into the ionization chamber where it is ionized by electron bombardment from a hollow cathode. Electrons colliding with the xenon atoms cause it to lose one or more electron per atom to form positive xenon ions. Xenon ions enter the multi-grid system (in between the positive and negative electrodes). The potential difference of up to 1280 V between the positive (screen) and negative (accelerator at -180 V) molybdenum grid creates an electric field that accelerates the ions out of the exhaust at over 100,000 km/h. Ion energy typically reaches 1–2 keV. The reaction force of this jet of ion stream out of the exhaust of the engine generates the thrust of the engine. In order to prevent the spacecraft from gaining a net negative charge due to the efflux of positive ions, a cathode neutralizer injects electrons into the exhaust to neutralize the positive xenon ions. Various design variations of the gridded ion thruster have been manufactured. An example of such an engine is NASA's Solar Electric Propulsion Technology Application Readiness (NSTAR) thruster which is completely controllable by digital control and interface unit. The 30 cm NSTAR ion thruster is designed to operate over a large range of thrust levels as required for deep space missions (Bond & Christensen, 1999). NASA's Evolutionary Xenon Thruster (NEXT) is another gridded ion thruster developed for NASA. NEXT combines the best elements of NSTAR to yield a thruster with highly specific impulse, specific mass, throttling range and a much longer life (Fisher, 2020). In these engines, the control of thrust and exhaust velocity is feasible due to ionization of the propellant, followed by acceleration of the ions in a physically separated and isolated region with the thruster.

The chamber design of gridded ion thrusters could be of various shapes including cylindrical, conical, spherical, semi-spherical, spheroidal and other geometries. Gridded ion thrusters could be uniquely different in the way electrical energy is transferred to the propellant. When operated with a direct current supply, ions are created by direct electron impact, with the energetic electrons arising from a hollow cathode into a chamber where they are magnetically confined to increase the degree of ionization. This type of ionization is typical of a Kaufman type engine. Rather than use a direct current supply, an external inductive coil is wound around the thruster body to inductively couple the radio frequency electro-magnetic wave to the ions within the chamber. Although such RF-excited gridded ion engines are slightly less efficient, the internal cathode is not required, and consequently, the electrode erosion is almost halved. Rather than using magnetic confinement, the plasma is heated by a static magnetic field and/or a high-frequency electro-magnetic wave (in the microwave frequency range). When the microwave frequency matches the electron Larmor frequency, the result is electron cyclotron resonance (ECR) which

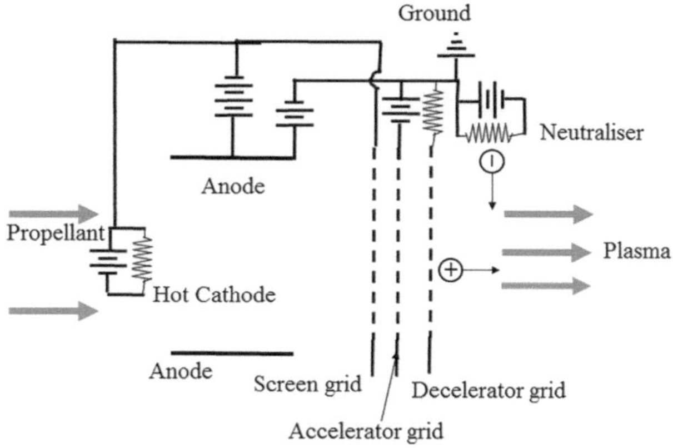

Fig. 11.9 Potentials applied in a gridded ion thruster

tends to increase the energy transfer to the ions. The design of the cavity is critical for microwave-excited engines to avoid cavity resonance which reduces the thrust level.

The inter-grid accelerating potential must be limited in order to avoid excessive penetration of the grid electric field into the discharge chamber containing the plasma, to avoid excessive curvature of the plasma sheath. Direct impingement of energetic ions causes sputtering of the grid material, leading to excessive grid erosion, leading to the widening of the holes and performance degradation and eventually limiting the thruster's life. Hence, a screen grid is used to avoid sputtering and excessive grid erosion. The various potentials applied to the electrodes and grids are illustrated in Fig. 11.9. A negative potential is required to accelerate positive charge: the external surface must be at zero potential. If source is at zero potential with a large negative accelerator, ions would never reach infinity and would return to the spacecraft due to energy losses in the generation process. The collimation of the ion beam is established by the screen grid potential, which is at a high potential, while the acceleration grid which follows is at a lower potential so as to facilitate the acceleration of the ions. The deceleration grid is located further downstream and is at a higher potential. The relative locations of the three grids, the corresponding potentials and typical ion trajectories are illustrated in Fig. 11.10.

During electro-static electric propulsion, a continuous stream of charges is emitted by the spacecraft's thrusters. Consequently, there is an accumulation of charge within the ambient plasma. Let us assume that this results in a net current from the spacecraft, equal to J_i. There will be a net capacitance between the plasma and the spacecraft. The capacitor may be modelled as a spherical capacitor, and the capacitance is assumed to develop in the gap between two spheres. For a sphere, the capacitance is given by $C = 4\pi\varepsilon_0 R_{gap} \sim 10^{-10}$Farads/m. However, since $i = C(dV/dt)$, a voltage is developed across the surfaces of the spheres. If the saturation current is of the order of 40 mA and the saturation voltage change is about 4000 V, the time period dt is of the order of $dt = C(dV/i) = 10\ \mu s$. The ion beam will be arrested by the space charge in 10 μs.

As already mentioned, the neutralizer is essential to balance the polarity of the charges in the plasma, within the space charge region. This is to avoid an

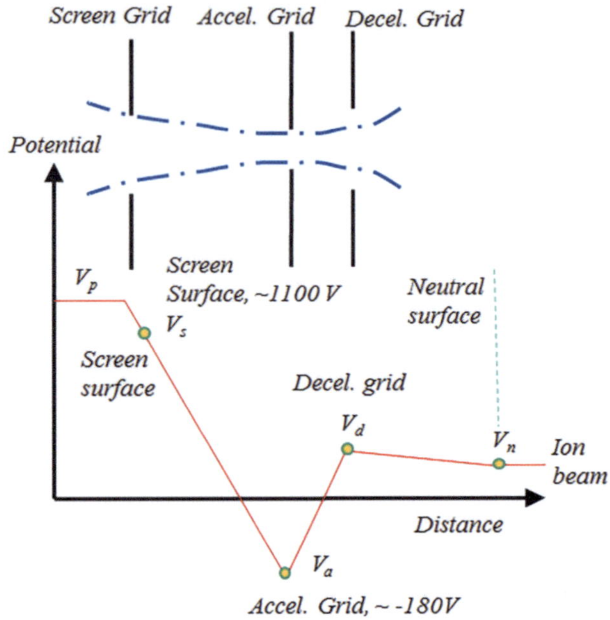

Fig. 11.10 Locations of the three grids, the corresponding potentials and typical ion trajectories

accumulation of charge (of the same polarity) within the space charge region. This in turn will avoid the saturation of the current density within the plasma. Neutralization is achieved by the provision of a flux of electrons, usually from a second hollow cathode discharge, which are mixed effectively within the ion beam.

A model of the idealized case ignores the details of the three grids used. Since the kinetic energy is equal to the potential energy lost by the charge $(1/2)mv^2(x) = q \times V(-x)$, and if the ion source is held at a voltage $V = V_0$ with respect to the accelerating grid, then the velocity of the ions is, $v_e = \sqrt{2qV_0/m_i}$. Using the Child–Langmuir law, the thrust is given by Eq. (11.17), and power/unit area is

$$\frac{P}{A} = \frac{F_T}{A}\frac{v_e}{2} = \frac{4}{9}\varepsilon_0\left(\frac{V_0}{x_a}\right)^2\sqrt{\frac{2qV_0}{m_i}} = \frac{F_T}{A}\sqrt{\frac{qV_0}{2m_i}}. \tag{11.19}$$

It is observed that, for a given source voltage V_0 and electrode distance x_a fixed, the thrust per unit area is independent of the charge-to-mass ratio (*q/m*), while the exhaust velocity and hence the power per unit area do depend on *q/m*. Electrons are generated in the hollow cathode. The electrons are magnetized and set in the motion and hence undergo inelastic collisions with the propellant, say xenon (*Xe*) neutral atoms to form *Xe+*. For a single electronic charge, $1.60217662 \times 10^{-19}$ C, and for a *xenon* ion, atomic mass = 131.293 amu, where the *amu* is an atomic mass units (*u*) (1 *u* is equal to 1/12 the mass of one atom of carbon-12). The mass of a *xenon* atom is $131.3 \times 1.6606 \times 10^{-27}$ kg. Thus, the charge-to-mass ratio is typically equal to $q/m = 7.3486 \times 10^5$ C/kg. When the net acceleration potential is 1100 V, exit velocity of the ions is $v_e = \sqrt{2qV_0/m_i} = 40.2081\text{km/s}$. The thrust developed could also be estimated directly from the mean electric field. Thus, the thrust experienced in an electric field per unit area can be expressed as $F_T/A = (1/2)\varepsilon_0 E_{\text{ave}}^2$, and based on the maximum breakdown field of typical ceramic, ~10^6V/m, the maximum thrust per unit area is of the order of 4.5 N/m^2. The maximum power is therefore of the order of 181 kW/m^2. (In the case of a xenon thruster, conditions must be established to minimize the generation of Xe++ as these lead to increased collisions due to a factor 2 increase in speed, which leads to charge exchange processes and increased erosion. Grids are shaped to minimize erosion and hyperbolically shaped surfaces are typically used.) The efficiency of xenon-based gridded ion thrusters is typically 40–80%, and so, the input power is the order of 450 kW/m^2. Typical maximum power inputs for these thrusters are of the order 4.5 kW, and the specific impulse is of the order of $I_{\text{sp}} = 2500 - 4200\,\text{s}$. Recently, these thrusters (4 × T6 gridded ion thrusters supplied by QinetiQ; Steiger et al., 2019) have been used in the BepiColombo mission, a joint mission between ESA and the Japan Aerospace Exploration Agency (JAXA), to the planet Mercury, launched in 2018. These thrusters require a power input of 10.5 kW and a propellant mass of 580 kg and provide a maximum thrust of 290 mN.

11.3.2 Pulsed Plasma Thrusters

A typical pulsed plasma thruster (PPT) consists of two electrodes (a cathode and an anode); a large capacitor for energy storage; a propellant which could be solid, liquid or gas; and a spark plug to initiate a discharge process and generation of plasma particles (Jahn, 1968). The discharge results in the formation of an arc between the electrodes that ablates and ionizes the propellant to generate plasma made up of ions and electrons following the break-up of the propellant molecules. In the case of a chemically active propellant, energy is also released thermodynamically, and the heat released enhances to the propellant exit velocity and the thrust developed. Energy is stored initially within the main capacitor, with a relatively fast charging time of the order of seconds and a pulsed discharge of the order of 1 Hz. The discharge time of the capacitor is of the order of millisecond, so the instantaneous power delivered by the thruster is typically much more than the average power consumed by it. The plasma velocity as it flows towards the nozzle exit is enhanced by both the self-induced and externally applied magnetic field following the generation of a Lorentz force.

PPTs are classified according to the architecture of electrodes, the type of propellants used and the manner in which the plasma is generated which could be generated by a range of inductive methods without the need for an arc or electrothermally (see, for example, Arrington et al. (1997)), Ziemer (2001), Keidar et al. (2003), Szelecka et al. (2015), Polzin et al. (2020), and Ling et al. (2020)). In all the different types of PPTs, there is the possibility of the pinching of the plasma sheet or the choking of the plasma flow which in turn could result in an unsteady generation of the thrust which is not desirable. It is therefore necessary to ensure that the generation of thrust is relative steady, and this can be ensured if it is regulated.

A typical PPT is an electro-magnetic propulsion device in which the particles of the plasma are accelerated by the force created when the discharge current interacts with a self-induced magnetic field. PPTs utilize the electro-magnetic force and gas–dynamic pressure interactions to accelerate both the ionized and neutral gas by a controlled discharge of a train of pulses. The interaction of the magnetic field with charged particles of the discharge current will generate a Lorentz force that would accelerate the particles further downstream and out of the thruster. Sometimes, an additional external magnetic field could be imposed to increase the strength of the magnetic field (Jahn, 1968; Laperriere, 2005), which would in turn increase the Lorentz force acting on the charged particles in the plasma. The principle of PPTs, illustrated in Fig. 11.11, is to increase exit velocity at the discharge end of the thruster's nozzle, in order to achieve high particle velocities with low propellant consumption. The prefix "pulsed" simply refers to the fact that the discharges are generated by a train of pulses resulting in an almost continuous thrust. Consider a pair of parallel rectangular electrodes and a large capacitor with each of its terminals connected to one of the electrodes. If the capacitor is charged to a high voltage and continuously switched across the electrodes so as to initiate an arc discharge using a spark plug between the electrodes, at the shortest possible distance from the capacitor circuit, it will then propagate down the electrode channel, and eventually, a flow of ionized gases or plasma and the remaining neutral gas will be forced to exit via a nozzle. This flow out of the nozzle's exit generates a net thrust. The ionized gases are generated by the decomposition of a solid polymer propellant (such as Teflon or

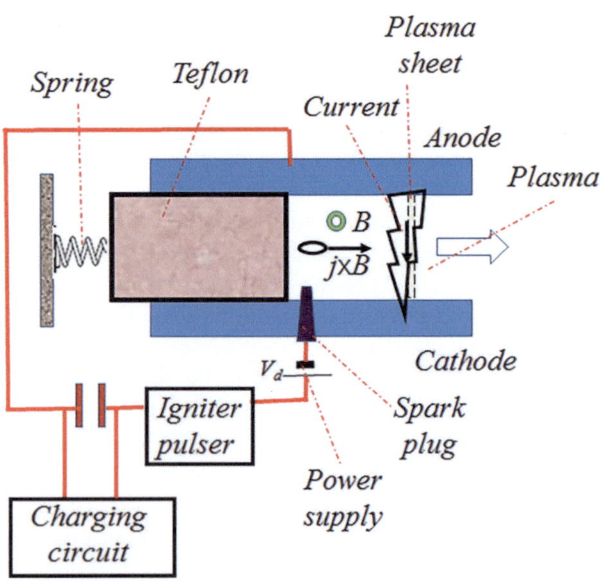

Fig. 11.11 Principle of operation of a solid propellant–based PPT

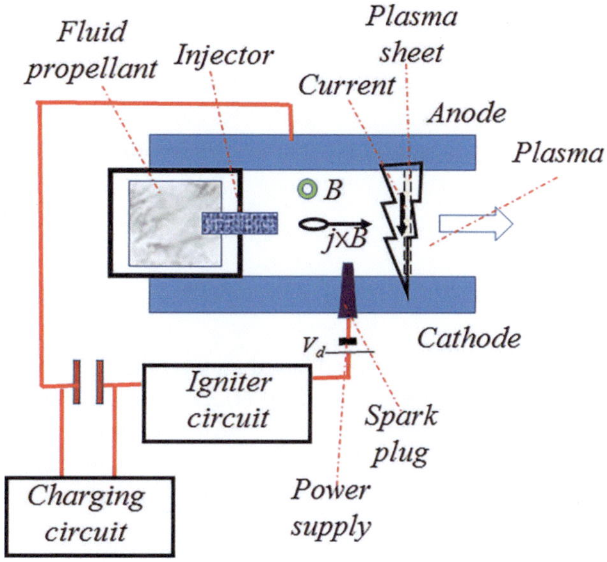

Fig. 11.12 Principle of operation of a liquid or gaseous propellant–based PPT

polytetrafluoroethylene (PTFE)) or by using a liquid or gaseous propellant (Koizumi et al., 2003; Misra, 2018). Figure 11.12 illustrates the working principal of a liquid or gas propellant–fed PPT.

If a thrust of magnitude T is generated by a PPT, the impulse bit is given by

$$I_{\text{bit}} = \int_0^{\Delta t} T d\tau = T\Delta t, \tag{11.20}$$

where Δt is the pulse period. Hence, given the frequency of the pulse train to be f, the thrust is $T = I_{\text{bit}}f$. Similarly, if the energy of the impulse bit is E_{bit}, the power required by the impulse bit is $P = E_{\text{bit}}f$.

Thus, a simple relationship can also be established between the discharge mass of the consumed propellant and the equivalent steady state mass flow rate, $\dot{m}m$, which is given by $\dot{m} = mf$. The specific impulse I_{sp} is given by

$$I_{\text{sp}} = \frac{v_{\text{exit}}}{g} = \frac{T}{\dot{m}g} = \frac{I_{\text{bit}}}{mg}. \tag{11.21}$$

Thus, thrust generation efficiency is given by

$$\eta_t = \frac{1}{2}\frac{\dot{m}v_{\text{exit}}^2}{P} = \frac{1}{2}\frac{\dot{m}}{P}\left(\frac{T}{\dot{m}}\right)^2 = \frac{1}{2}\frac{\dot{m}}{E_{\text{bit}}}\frac{I_{\text{bit}}^2}{\dot{m}m} = \frac{1}{2}\frac{I_{\text{bit}}^2}{mE_{\text{bit}}}. \tag{11.22}$$

The power input efficiency is defined in terms of the bus power P_b supplied as

$$\eta_P = P/P_b. \tag{11.23}$$

The system efficiency is given by

$$\eta = \eta_t \eta_P. \tag{11.24}$$

11.3.3 Hall Current Thrusters

Thrusters that are designed based on the principle of the Hall effect use a cross-field discharge generated by the Hall effect to create and propel the plasma. An electric field in the direction of the motion of the particles is essential as it accelerates the ions to high exhaust velocities as well as a magnetic field, perpendicular to the preceding field, that inhibits electrons out of field motion that would otherwise tend to shorten out the electric field.

In terms of its basic construction, a Hall thruster is a coaxial annular cavity where plasma is generated by an electric field between an anode at the near end and a cathode at the far end, along the axial direction. The propellant is introduced into the cavity by an annular injector around and within the anode which is shaped like an annular disc. A radially distributed magnetic field ensures that the electrons would not travel directly from the cathode to the anode. The existence of the magnetic field results in a circumferential Hall current. The presence of the magnetic field produces that the electrons are not free to accelerate towards the anode but drift azimuthally at a velocity that generates the Hall current as well as a centrifugal force to the magnetic force which is balanced by the force due to the magnetic field. The combined axial and azimuthal motion of the electrons substantially increases the rate of collisions between electrons and propellant atoms, so the ionization would be very efficient. A more appropriate term is electron scattering. The azimuthal velocity of the electrons is given by the product of the cyclotron frequency of angular motion and the radius of the electron's circumferential motion, which is mutually perpendicular to the applied electric and magnetic fields. The exhaust power of the plasma ejected out of the thruster is deterred by the kinetic power of the mass rate of flow of the expelled plasma and the high velocity of it. The efficiency of the thruster is determined by the ratio of the kinetic power of the plasma to the power supplied to the thruster (Fig. 11.13).

The mechanisms for momentum transfer are governed by the Lorentz force. In the absence of a physical wall, the charges deflected by the applied magnetic field accumulate at the two faces of a conductor. There is net flow of electrons. Thus, from Eq. (11.5),

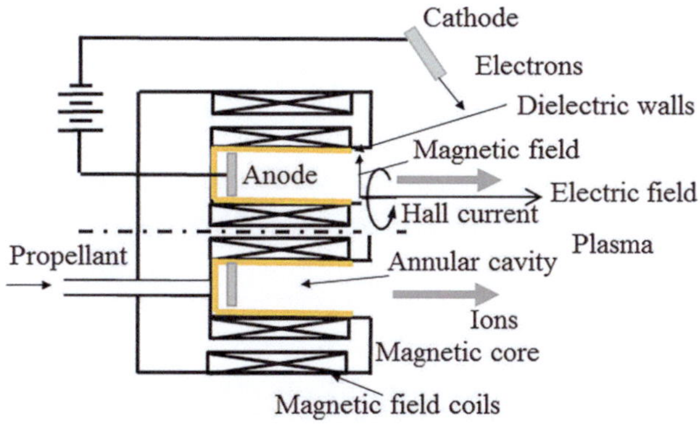

Fig. 11.13 Principle of operation of a Hall current thruster

$$\mathbf{F} = q(\mathbf{E} + \mathbf{v} \times \mathbf{B}) = 0 \Rightarrow \mathbf{E} = -\mathbf{v} \times \mathbf{B}. \tag{11.25}$$

Consequently, only the electrons are given a drift velocity defined by

$$|\mathbf{v}| = \mathbf{E} \times \mathbf{B}/|\mathbf{B}|^2. \tag{11.26}$$

A net azimuthal current density arises, called the Hall current:

$$\vec{j}_\theta = -qn_e \mathbf{E} \times \mathbf{B}/|\mathbf{B}|^2. \tag{11.27}$$

Given this current, the magnetic or Lorentz force and an equal and opposite force is exerted by the electron currents on the magnetic structure. This force is given by

$$\mathbf{F} = \vec{j}_\theta \times \mathbf{B} = -qn_e \mathbf{E} \times \mathbf{B} \times \mathbf{B}/|\mathbf{B}|^2 = qn_e \mathbf{E}. \tag{11.28}$$

The force is the same as the forward electro-static force on the ions in magnitude but is acting on the electrons. The force on the electrons is not electro-statically generated by the electric field but by the magnetically induced Hall current. Thus, the thrusters are aptly referred to as *Hall current thrusters*. Thus, the electro-static Hall current thrusters (HCT) are also similar to the so-called stationary plasma thrusters (SPT). They are simpler electro-mechanical devices than the gridded ion engines. The ion acceleration force generated from the interaction of magnetic and electric fields (Hall current). Ions are generated by electron scattering, and the generated plasma is neutral. The Child–Langmuir law no longer applies, as the generated plasma is always neutral.

A practical construction of a Hall current thruster consists of a coaxial annular cavity where plasma is created by passing current between the annular anode on the upstream end of an otherwise dielectric cavity and the externally located cathode. The propellant enters this plasma cavity via an annular manifold at the anode. A radial magnetic field is applied, either by ring-shaped permanent magnets or through coils and soft iron yokes. The device is also referred to as a cross-field accelerator as

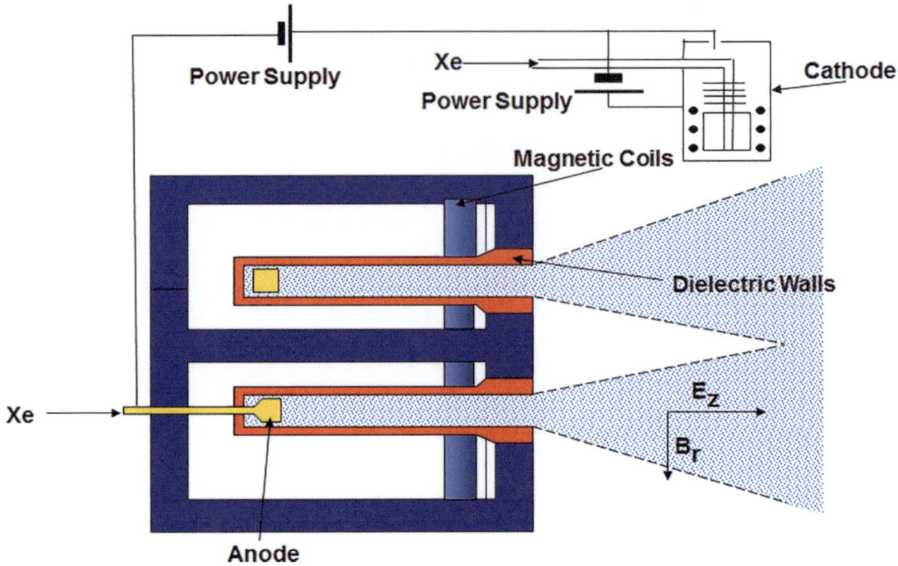

Fig. 11.14 Typical illustration of an annular design of a Hall current thruster. (Adapted from Palaszewski (2014))

the Hall current is mutually perpendicular to the electric **E** and magnetic **B** fields, $\mathbf{j} \sim \mathbf{E} \times \mathbf{B}$. The electric field is axial; the magnetic field is radial. The $\mathbf{E} \times \mathbf{B}$ is normal to both field and hence the current is azimuthal. This is the Hall current which ionizes the propellant such as xenon. Typical values of the axial electric field E_z are on the order of 10^4 Volts/m, and the radial magnetic density B_r is 0.02 T (Tesla), leading to average electron energies >10 eV. The magnetically confined electrons efficiently ionize the propellant gas, and the ions are accelerated towards the exit by the electric field, which produces the thrust. For examples of Hall thrusters with flight heritage, the reader is referred to Levchenko et al. (2018) and Frieman et al. (2022). A typical example of a Hall current thruster is illustrated in Fig. 11.14.

The voltage potentials in an HCT are typically given by anode–cathode voltage ~0.5×10^4 V/m, while the discharge current is of the order of 20 A for a generated thrust of ~200 mN. The variation of the voltage potentials with distance in an HCT is illustrated in Fig. 11.15.

To investigate the feasibility of implementing an HCT, consider the typical Larmor radius r_e for the typical magnetic field of ~0.015 Tesla. For an electron, if v_{th} is defined as the probabilistic mean of the magnitude of the velocity of the atoms or molecules, then consider the mean (thermal) velocity of the electron, given by

$$v_{th} = \sqrt{8k_B T/(\pi m)}. \tag{11.29}$$

Recall that the cyclotron frequency is $\omega_c = Bq/m$. Then, it follows that

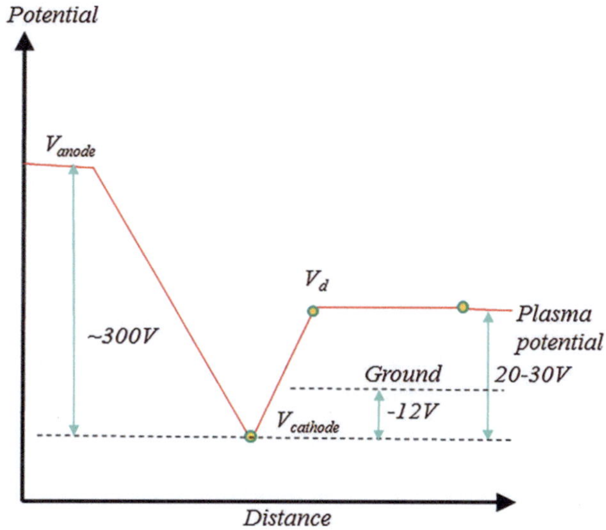

Fig. 11.15 Variation of the voltage potentials with distance in an HCT

$$r_e = \frac{v_{\text{th}}}{\omega_c} = \frac{1}{Be}\sqrt{\frac{8mk_BT_e}{\pi}} = \frac{1}{B}\sqrt{\frac{8}{\pi}\frac{m}{e}T_{eV}}. \tag{11.30}$$

For a xenon ion (atomic mass = 131.293), the axial velocity of the ions, just as in the case of a gridded ion engine, when the ion source is held at voltage $V_0 \sim 800\ V$, is

$$v_i = \sqrt{\frac{2qV_0}{m}} \sim 3.5 \times 10^4 \text{m/s}. \tag{11.31}$$

An estimate of the Larmor radius (from the previous slide) is $r_i = 3$ m (relatively large!) for an ion and $r_e = 0.6$ cm for an electron. In the vicinity of the exit plane, the electric **E** and magnetic **B** fields are approximately of the same order of magnitude. Consequently, it is fair to assume that the magnetic field confines the electrons to within the channel while the ions are not affected by the magnetic field.

One also needs to consider whether the electrons can ionize the gas (propellant). It requires electron/neutral collisions to take place, resulting in electron scattering. Consider the production of ions, I_p per second (function of ionization cross-section), for a given volume,

$$I_p = n_0 n_e \langle \sigma_i v_e \rangle V_0 + n_0 n_p \langle \sigma_i v_p \rangle V_0. \tag{11.32}$$

In Eq. (11.32), σ_i is the ionization cross-section, n_0 is the density of the neutral particles (Loschmidt's number), n_e is the density of the electrons obtained from a Maxwellian distribution, n_p is the density of the positively charged ion, v_e is the velocity of the electrons, v_p is the velocity of the positively charged ion, and V_0 is the applied voltage.

Cross-sections for electron impact ionization of various gases have been measured by Rapp and Englander-Golden (1965) and extensively recorded by Hayashi (2003) in a bibliography; up to 1000 eV are still in use. The direct ionization cross-section of argon have been obtained either theoretically or experimentally, for an energy range that goes from threshold to 1000 eV. Rapp and Englander-Golden (1965) experimentally determined the ionization cross-sections most commonly used in discharge modelling. Based on these studies, it can be concluded that the ionization cross-section for Xenon is dependent on the nature of the ionization process. It reaches a steady state with increasing electron energies and typically between 10^{-20} and 10^{-19} m^2. Thus, it is possible to estimate I_p from Eq. (11.32).

The description of the physics of HCTs cannot be complete without some discussion of the diffusion of electrons towards the anode, forced by its higher electric potential. Eventually, as is the case of the ionization chamber of ion engines, all electrons, primaries and secondaries are collected by the anode. The goal is therefore to maximize the lifetime of electrons to increase the probability of ionizing collisions with neutrals and to maximize the collision frequency.

Unlike the case of a gridded ion engine, there is not a clear-cut upper limit to the obtainable thrust density in a Hall thruster. Increasing the mass flow rate through a given engine would increase thrust by the same factor, provided the same degree of ionization can be maintained. The voltage is presumed to be kept the same, since this controls specific impulse fairly directly. The ion beam current and the electron current leaking into the accelerating channel would increase as the mass flow rate, so that the energetic ability of these electrons to ionize the increased neutral flow would be maintained. Since no space charge develops these increases in flow, current and power could go on indefinitely. But the above considerations have ignored the various effects of collisions.

The collision frequency ν_c affects the ionization rate and hence the Hall current driving the process which needs to be maintained in the channel. Hence, considering the scattering current, one has approximately $\nu_c = \sum n_a \sigma_i v_{th}$. The collision frequency ν_c is a limiting factor in the operation of the Hall thruster. When ν_c is comparable to the cyclotron frequency ω_c, electrons can no longer maintain a pure azimuthal drift, as the Larmor radius is not large enough, relative to the channel size and the collision distance. Thus, electrons are now able to cross the magnetic barrier at a velocity $v_e = (\nu_c/\omega_c) \times (E_z/B_r)$. Thus, the parameter ω_c/ν_c is known as the Hall parameter. When $\omega_c/\nu_c > > 1$, the electron executes many cycles before scattering, and the Hall current remains azimuthal.

So, if $\omega_c/\nu_c = 1.5 \times 10^5$ the Hall current is well established; for a magnetic field of 0.015 Tesla with $\omega_c = 1.8 \times 10^{11}|\mathbf{B}| = 3 \times 10^9 s^{-1}$, $\sigma_i = 5 \times 10^{-20} m^2$, and $v_{th} = 3 \times 10^6$m/s.

Through collisions, some of the electrons gradually make their way to anode, and the resulting current is a loss mechanism amounting to ~30% of the total loss. About 90% of propellant is ionized. Thus, the total efficiency of the HCT is ~60%. There is no space charge limit due to combination of virtual nature of accelerating potential from the cathode plasma, and the propellant remains quasi-neutral. Thrust density is ~10 times higher than gridded ion engines. Typically for an HCT, $I_{sp} = 1000 -$

2400 s and generally closer to about $I_{sp} = 1550$ s, while the generated thrust is ~4 mN–500 mN, and the power requirements are 0.8–16 kW. The thrusters are cylindrical in shape with a diameter from 10 up to 30 cm and ~ 10–25 cm deep. Typically, Hall thrusters with flight heritage use xenon as propellant, with an $I_{sp} = 1500 - 2000$ s requiring an input power of 1.5–4.5 kW, with an efficiency range given by 35–60%. The thrust varies linearly with the discharge power, which is controlled by varying the discharge voltage across the anode and the cathode.

11.3.4 Field Effect Electric, Colloid Propulsion and Electrospray Thrusters

Colloid and field emission–based electric propulsion refers to the process of using a strong electric field to produce a spray of charged ions and/or droplets or an electrospray. The potential difference between an electrode and a liquid surface is balanced by the force of surface tension in the fluid. Consequently, the circular free surface deforms under the action of the electric field to an equilibrium shape that resembles a jet or a cone known as a Taylor cone as it was first discovered by G. I. Taylor.

The field emission electric propulsion (FEEP) thruster is similar in principle to any ion thruster. In an FEEP, the propellant used is a metal in liquid form (such as caesium or indium due to their low ionization energies, low melting points and high mass). The metal is heated and converted into liquid form at an emitter through a tungsten needle/slit of about 1 μm. The propellant flows through the slit to form a free surface at the exit of the slit. The emitter serves as a positive grid and the accelerator plate serves as the negative grid. Thus, an electric field is established.

The surface of the propellant liquid metal at the tip of the needle is in a state of local instability due to surface tension and the electric field. On emission from the slit, the free surface deforms into the shape of a conic, which is a "Taylor cone". When the electric field crosses a certain threshold voltage (of 10^9 V/nm), the metal atoms at the tip are ionized, and the positive ions flow out of it, leaving electrons behind. This results in a stream of ions which resembles a jet and produces the thrust. A neutralizer is used to reduce the net negative charge on the engine by emitting electrons. A typical FEEP is illustrated in Fig. 11.8.

Colloid thrusters are, in principle, similar to FEEP. They are designed to give smaller specific impulses and thrust for high precision control of force. Colloids are organic liquid with high conductivity. Field emission–based thrusters use liquid metal suspensions. These systems can develop precision low thrusts with micro-Newton-level thrusts at high $I_{sp} = 1000 - 6000$ s when an ionic liquid is used as the propellant, particularly suited for some science missions where several satellites are required to operate together as a constellation in order to make measurements—e.g. LISA Pathfinder Mission (Armano et al., 2015) and the Darwin missions (Kaltenegger et al., 2003).

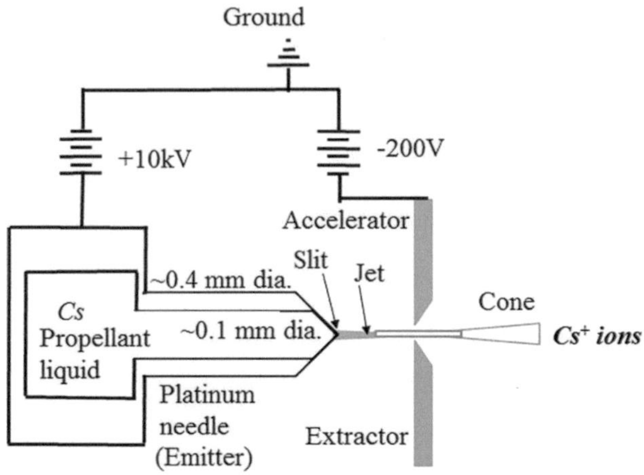

Fig. 11.16 The principle of an electrospray thruster

The underlying process is based on electrospray which is a method for direct ionization of liquids. The primary advantages include simple propellant storage and electronics as well as simple manufacture based around micro-fabrication techniques. The thrusts developed are at the micro-level. The strong field intensity at the tip induces the formation of the jet, composed of ions and droplets. The principle is illustrated in Fig. 11.16.

In an electrospray jet, conductive liquid flows through a small diameter channel at very low flow rates $\sim 10^{-12}$ m^3/s. The applied electric field produces a force leading to fluid motion followed by cone formation. The process involves the phenomenon of ion and droplet evaporation from charged liquid surfaces. For a fluid assumed to be a perfect conductor, the surface of the cone is an equipotential surface. Therefore, to balance the surface tension, the potential gradient must be proportional to $1/\sqrt{r}$. As the potential on the ring electrodes increases, the liquid curvature increases until reaching a half-angle and a Taylor cone forms. However, the space charge effect near an infinitely fine cone point prevents further evaporation in the field.

11.3.5 The Physics of Cone and Droplet Formation

A charged particle within an electric field experiences a force, and as a result, the field is responsible in accelerating the particle. By the definition of an electric field, it is generated or it exists when there is a difference in potential across a region or domain. This applies even when the region is filled with fluid rather than in vacuum. However, the accelerating particle experiences a resistance both due to the viscosity of the fluid and the surface tension. The latter is important if the domain is relatively small. Within the confines of a conical domain, in between the accelerator plates and the platinum needle in Fig. 11.16, the electric field is a maximum at the apex of the conical region due to the relationship between the surface tension and the curvature of the free surface ($1/r$). Surface tension is the elastic cohesive force on the free surface of fluid due to intermolecular forces. As a consequence, energy is required, or work must be done, to increase the free surface area of the liquid. For a cylindrical free surface, given the coefficient of surface tension γ, the pressure differential across the free surface satisfies

$$\Delta p = 2\gamma/r. \tag{11.32}$$

If the pressure results in an increase of the volume dV, the work done $\Delta p dV$ must be equal to the energy supplied by the electric field. Consequently,

$$\Delta p dV = 2\gamma/r dV = (1/2)\varepsilon|\mathbf{E}|^2 dV. \tag{11.33}$$

When the surface tension can no longer balance the pressure due to the applied electro-static field, the free surface breaks to a spray of charged drops or droplets. For a spherical bubble, the corresponding relation given by Eq. (11.33) is

$$\Delta p dV = 4\gamma/r dV = (1/2)\varepsilon|\mathbf{E}|^2 dV. \tag{11.33}$$

Thus, the surface tension within each spherical droplet can still balance the pressure due to the applied electro-static field. These droplets can carry with them one or more of the ionized fluid particles. Like charges repel and the repulsive force between the charges is now balanced by the force of surface tension. Thus, the number of charges as well as the total charge that can be carried in a droplet is limited. Since the surface area of a spherical droplet is in proportion to $V^{2/3}$, it can be shown that the radius of the droplet is proportional to

$$R_{\text{drop}} \propto (1/\sigma)^{2/3}. \tag{11.34}$$

In Eq. (11.34), σ is the conductivity of the fluid. The total charge carried by a droplet is given by

$$q_{\text{drop}} \propto \sqrt{\gamma} R_{\text{drop}}^{3/2} \Rightarrow q_{\text{drop}}/m \propto \sqrt{\gamma}/\sigma. \tag{11.35}$$

While the achievable charge-to-mass ratio of a droplet is significantly lower than a proton (single electrical ion), it is quite large to generate a sufficiently large exhaust velocity in an electrospray-driven thruster. It can be as high as 4000–5000 (Coulombs/kg) for certain propellants. The corresponding velocity of exhaust is

$$v_e = \sqrt{\frac{2qV_0}{m}} \sim \sqrt{2 \times 4000 V_0} = 2000\sqrt{2V_{\text{kV}}}. \tag{11.36}$$

In Eq. 11.36, V_{kV} is the applied potential difference in kilo Volts (kV). The corresponding specific impulse for droplet spray propulsion is $I_{sp} = 200\sqrt{2V_{kV}}$. Thus, when $V_{kV} = 4.5\,\text{kV}$, $I_{sp} = 600\,\text{s}$ which exceeds the value obtained by most chemical propellants. However, for a purely ionic propulsion system, considering a single ion, $q_{ion}/m = 0.95755636 \times 10^8\text{C/kg}$, and $I_{sp} = 1000\sqrt{20V_{kV}}$, which is about $I_{sp} = 9000\,\text{s}$ when $V_{kV} = 4.5\,\text{kV}$.

In field emission tip, both droplets and ions are released as the current across the tip is increased. Associated with them, are two different types of droplets: the Rayleigh droplets that are generated at the tip of the Taylor cone near the region where the fluid is ionized due to instabilities in the jet and the larger-sized Faraday droplets that are generated at the shank of the Taylor cone due to surface wave instabilities. Lord Rayleigh established a criterion required for the droplet to form and subsequently remain stable in terms of both the charge and surface tension. The droplets elongate under the action of an electric field, and the elongating droplets become unstable when the applied electric field reaches a critical limit. The critical electric field limit was established by G. I. Taylor. The optimum flow rate for stable operation, which is very low, both in the droplet spray mode and in the ionic mode, is usually found experimentally.

The flow rates are found experimentally to be directly proportional to the surface tension, the relative permittivity ε_r and the permittivity of free space ε_0 and inversely proportional to the fluid density and conductivity. Thus, the flow rate in m^3/s is

$$\dot{Q} \propto \gamma\varepsilon_r\varepsilon_0/(\rho\sigma). \tag{11.37}$$

Given that for a typical propellant, $\varepsilon_r = 70$, $\sigma = 2\,\text{S/m}$, $\gamma = 0.006\,\text{N/m}$, and $\varepsilon_0 = 8.8542 \times 10^{-12}\text{F/m}$, $\rho\dot{Q} \propto 1.77 \times 10^{-12}$. Thus, the thrust is typically of the order of $T = \dot{m}v_e = \rho\dot{Q}v_e = 3.5 \times 10^{-8}\sqrt{20V_{kV}}$ for an ionic propulsion system. Typically, the thrust is of the order of $T \propto 10^{-7}\text{N}$. In the case of droplet spray system, $\sigma = 0.1\text{S/m}$, and $T = \dot{m}v_e = \rho\dot{Q}v_e = 3.5 \times 10^{-7}\sqrt{2V_{kV}}$. Typically, the thrust is of the order of $T \propto 10^{-6}\text{N}$. Thus, several emitters are generally used to match the thrust of ionic propulsion system.

The droplet spray propulsion system is the basis of a colloid propulsion system, as a conductive colloid is used as a propellant. Its advantage is that it can be miniaturized and manufactured by micro-fabrication and arrays of several thousand emitters can be used to generate thrusts of the orders of milli-Newtons. Furthermore, the polarity of the emitter can be electronically switched rapidly to maintain a neutral plasma, thereby avoiding the formation of space charge and need for a neutralizer. Alexander et al. (2007) have discussed a method for the precise thrust control in colloid electrospray propulsion. Stark et al. (2007), Jhuree et al. (2007), and Grustan-Gutierrez et al. (2017) have investigated the ion/droplet beam efficiencies and applied their results to colloid thrusters. Dandavino et al. (2014) have discussed the micro-fabrication of electrospray emitters and their application to small satellite propulsion. Dandavino et al. (2014) developed a process which allows the wafer-level fabrication of large arrays of emitters, each with integrated extractor and

Table 11.2 Comparison of the performance of various types of thrusters

Thruster type	I_{sp}	Thrust (N) or power required (kW)	Advantages	Disadvantages
Chemical mono-prop	200–250	<40,000 N for launch vehicles, 3000–30 N for satellites	Reliable, suitable both for launch vehicles and satellites	Low I_{sp}, needs propellant management
Chemical bi-prop	300–350	<120,000 N	Reliable, suitable for launch vehicles	Needs propellant management devices
Solid propellant	250–300	<300,000 N	Simple, useful for launch or one-off use	Thrust not controllable
Resistojets	250–700	<2 kW	I_{sp} better than chemical propulsion	I_{sp} still not high enough
Arc jets	600–1000	<30 kW	Simple design, flight heritage	High operating temperatures, voltages, low efficiency
Gridded ion engines	2500–4000	0.3–5 kW	High I_{sp}	Power processing is complex, needs neutralizer, thrust is limited
PPT	1000–1200	<200 W	Can be precisely controlled	Low efficiency, low thrust
HCT	1500–2000	0.5–12 kW	Good I_{sp} range, suitable for small satellites	Complex power processing requirements
Colloid	500–6000	<0.5 mW	High I_{sp} feasible, high precision, controllability	Large number of emitter arrays need

accelerator electrodes. In the method developed by them, the emitter's designs and dimensions are highly repeatable. To improve reliability of the device, the electrodes were wafer-level bonded to the emitters. They have also explained the fabrication processes involved and provided designs of the thruster arrays.

In Table 11.2, the performance of colloid thrusters is compared with all other types of thrusters discussed so far.

Exercises

1. (a) Briefly highlight the principal features and operating principles of an HET or HCT. In your answer, you should provide a diagram of the typical layout for such a thruster, together with a clear identification of the electrical potentials expected in the device at the anode, cathode and the region downstream of the cathode.

Table 11.3 Parameters for the missions to be carried out using a low thrust electric propulsion technology

	Earth gravity mission	LEO/GEO transfer mission
Operational thruster time	1 year	30 days
Required ΔV	2.7 km/s	4 km/s
Power system–specific mass	15 kg/kW	15 kg/kW
Spacecraft mass	1050 kg	7000 kg

(b) For a xenon plasma, estimate the maximum space charge–limited cathode emission current.

(c) Show that the Child–Langmuir equation for the space charge–limited current can be conveniently written as.

$J_e = 2.33 \times 10^{-6} \frac{1}{\sqrt{V}} \left(\frac{V}{d}\right)^2$, for electrons;

$$J_i = 5.45 \times 10^{-8} \frac{1}{\sqrt{M_{i,\mathrm{amu}} V}} \left(\frac{V}{d}\right)^2,$$

for singly charged ions, where $M_{i,\ \mathrm{amu}}$ is the ion atomic mass in atomic mass units; and.

$$J_i = 4.75 \times 10^{-9} \frac{1}{\sqrt{V}} \left(\frac{V}{d}\right)^2, \text{ for Xenon ions.}$$

(d) Consider two missions to be carried out using a low thrust electric propulsion technology: the first is to study the Earth's gravitational field, and the second is to transfer a satellite from low Earth orbit (LEO) into GEO. The parameters for these missions are provided in Table 11.3.

For each mission, there is a choice of thruster: a gridded ion engine that has a specific impulse of 5500 s and a Hall effect thruster that has a specific impulse of 1750s.

Identify which propulsion system would appear to be the most suitable for each mission.

2. (a) Carefully explain why gridded ion engines (GIE) require a neutralizer to operate.

   (b) Highlight the principal features and operating principles of a gridded ion engine (GIE). In your answer, you should provide a diagram of the typical layout for such a thruster, together with clear identification of the electrical potentials expected in the device at the anode, cathode and the region downstream of the cathode.

   (c) A NASA engineer is faced with the problem of choosing between two evolving electric propulsion thrusters for a deep space mission. The characteristics of the two propulsion systems are summarized in Table 11.4. The time frame of the mission is 9 months. The velocity increment (Δv) required for the mission may be assumed to be 16 km/s.

Table 11.4 Summary of the characteristics of the propulsion systems

System Id.	ET1	ET2
Characteristic	Value	Value
Spec. impulse(s)	4000	3200
Thrust (mN)/thruster	600	300
Number of thrusters	3	3
Thruster power efficiency	0.71	0.62
Specific mass per unit power	4.8	6
Dry mass per thruster (kg)	5	11.4
Fuel mass residual	0.01	0.025
Ratio of fuel tank mass to fuel mass	0.1	0.1
Satellite mass, kg (excluding fuel tanks)	40	40

For the two thrusters, estimate (i) the propellant exhaust relative velocity in km/s, (ii) the mass ratio, (iii) the fuel mass including the residual fuel, (iv) the mass of the fuel tanks, (v) the wet mass, (vi) the Earth parking orbit to deep space transfer time in months, (vii) the useful power developed and (ix) the total power needed.

Based on your estimates, which of the two systems would you recommend for the mission? State the reasons for your recommendation.

3. (a) Describe the principal features of a liquid propulsion system which might be used on board a satellite for primary and secondary propulsion. Highlight the way in which these systems differ from liquid propulsion systems used in launch vehicles.
(b) Compare the different levels (or orders of magnitude) of thrust generated and the specific impulse by (i) a solid propellant thruster, (ii) a bi-propellant thruster, (iii) a resistojet thruster, (iv) an arc jet thruster, (v) a stationary plasma thruster (SPT) also known as a Hall current thruster (HCT) or as a Hall effect thruster (HET), (vi) an ion thruster such as a gridded type and (vii) a field emission electric propulsion (FEEP) thruster.
Briefly explain why both performance metrics are important for the selection of an appropriate electric thruster for a specific mission.
(c) With the aid of diagrams, briefly compare the working principles of a resistojet, an arc jet and an FEEP thruster.

4. (a) Describe the mission profile in detail for a satellite attaining geostationary orbit following injection into a geostationary transfer orbit by a launch vehicle such as Ariane V. The satellite's principal propulsion system is a chemical bi-propellant unit that delivers a thrust of 300 N and specific impulse of 310 s. The dry mass of the satellite is 4000 kg. In your answer, you should identify the nature of the orbit elements of the transfer orbit, but you are not required to calculate specific values for these orbit elements. You should then identify three phases in this mission profile prior to final orbit acquisition and provide indicative duration for each phase.

(b) What major differences would occur in this mission profile if the bi-propellant system was replaced by a gridded ion engine having a thrust of 200 mN and specific impulse of 3200 s? Comment on the mass of propellant required for this option relative to that of the configuration of option (a) above.
(c) A third option considered for the propulsion system is a hybrid system consisting of a solid rocket having a thrust of 70 kN with specific impulse of 250 s, together with a bi-propellant rocket thrust of 100 N with specific impulse of 300 s. What would be the transfer orbit strategy in this case? Carefully explain the role of each of the propulsion units and any significant changes to the overall satellite configuration that would be required to operate the solid rocket motor. Again, comment on the mass implication for this configuration relative to the earlier two options discussed.

5. A gridded ion engine propulsion unit requires input from two power systems to operate: one to ionize and accelerate the propellant (the thrust head) and the other to neutralize the ion beam produced with electrons (the neutralizer).
(a) Carefully explain why a neutralizer is required for gridded ion engines.
(b) If the propulsion system operates with argon, a propellant whose atomic weight is 40 and a mass flow rate of 6×10^{-8} kg/s, and the net accelerating potential for the system is 800 V, calculate the thrust available from this system. Each ion in the propellant is known to have a single electric charge of 1.6×10^{-19} C, and the mass of a proton is 1.7×10^{-27} kg.
(c) If the neutralizer is a hot cathode source and has a specific current requirement of 6 mA per Watt, calculate the power required to neutralize the ion beam.
(d) If the thrust head has an efficiency of 50% in converting electrical power to exhaust kinetic power, calculate the total electrical power input to the propulsion unit.
(e) The thrust head operates at the Langmuir–Child law limit which defines the thrust per unit area of the aperture, in terms of the potential difference between source and accelerating electrode, the separation distance between the electrodes and the permittivity of free space having a value of 8.85×10^{-12} F/m. If it is assumed that the maximum sustainable field in the device is 7×10^5 V/m, calculate the minimum diameter of the thrust aperture required to deliver the desired thrust. State all the assumptions you make.

6. An electric propulsion thruster requires input from two power systems to operate: one to ionize and accelerate the propellant at the thrust head and the other to neutralize the generated ion beam with another beam of electrons injected by a neutralizer.
(a) If the thruster uses helium, whose atomic weight is 4.0, as the propellant, with a mass flow rate of 6×10^{-7} kg/s and a net accelerating potential of 1000 Volts, calculate the minimum and maximum thrust generated by the thruster. An electron is known to have an electric charge of 1.6×10^{-19} C, and the mass of a proton is 1.7×10^{-27} kg.
(b) Explain the role of a neutralizer in this type of propulsion system.

(c) If the neutralizer uses a hot cathode source and has a specific current requirement of 5 mA per Watt, calculate the minimum and maximum power required to neutralize the ion beam.
(d) If the thrust head has an efficiency of 40% in converting electrical power to exhaust kinetic power, calculate the minimum and maximum total electrical power input to the propulsion unit.
(e) The generated thrust in Ns per unit area in square metres is directly proportional to the square of the electric field, and the proportionality constant is $8\varepsilon_0/9$, where ε_0 is the permittivity of free space having a value of 8.85×10^{-12} F/m. If it is assumed that the maximum sustainable field in the device is 1.0×106 V/m, calculate the minimum diameter of the thrust aperture required to deliver the desired minimum thrust.

State all the assumptions you make.

7. A gridded ion engine is used to raise the orbit of a satellite in a nearly circular low Earth orbit ($\mu = 398600$ km^3/s^2) with a velocity of 7.72 km/s. It uses krypton as the fuel for generating the ionic flow through a nozzle. (The atomic weight of krypton is 83.798, and the atomic number is 36.) The effective diameter of an ion is 2 mm, and the ions are accelerated by applying a potential difference of 600 Volts across a distance of 5 mm. The total number of ions flowing out of the thruster's nozzle is 20000 ions. The charge-to-mass ratio for the krypton ion is 11.50e+05 Coulombs/Kg. The thruster is operated for a period of 90 days. The thruster efficiency in generating thrust from applied electric field is 0.66. The power plant–specific mass is 10 Kg/kW. The mass of the propellant tanks is specified as 10% of the mass of the propellant. The payload mass is 300 Kg. The thrust generated by a single ion of krypton may be obtained from the Langmuir–Child law:

$$F/A = (8\varepsilon_0/9)(V_0/x_a)^2 N/m^2.$$

It is required to raise the radius of orbit of the satellite by at least 330 km.
(a) Calculate the total thrust generated by the thruster, the mass flow, the total fuel requirements over the burn time, the power required by the thruster, the mass of the power plant, the mass ratio and the total Δv generated over the burn time. Explain all your steps.
(b) Hence, determine the radius of the low Earth orbit and the height by which the satellite could be raised. Does this exceed the required increase in the orbit radius? Explain all your steps.

(You may assume the absolute permittivity of free space = 8.854×10^{-12} $C^2N^{-1}m^{-2}$.)

Appendix 11.1: Useful Atomic Constants

Symbol	Name	Value
A_v	Avogadro's number (atoms/mole)	$6.02214179 \times 10^{23}$
amu	Atomic mass unit	$1.6602176487 \times 10^{-27}$kg
c	Speed of light	299 792 458 m/s
q	Electronic charge	$1.602176487 \times 10^{-19}$C
g	Acceleration due to gravity	9.81205 m/s^2
k	Boltzmann constant	1.3807×10^{-23}J/°K
m	Electronic mass	$9.1093822 \times 10^{-31}$kg
M	Proton mass	$1.67262164 \times 10^{-27}$kg
q/m	Electron-charge-to-mass ratio	1.75822×10^{11}C/kg
M/m	Proton-mass-to-electron-mass ratio	1836.152673
M_{Ar^+}/m	Argon-ion-mass-to-electron-mass ratio	73,440
M_{Xe^+}/m	Xenon-ion-mass-to-electron-mass ratio	241066.8
M_{Kr^+}/m	Krypton-ion-mass-to-electron-mass ratio	153800.62
M_{Xe}	Mass of xenon atom	131.293 amu
M_{Xe}	Mass of xenon atom	2.17975×10^{-25}kg
ε_0	Permittivity of free space	8.8542×10^{-12}F/m
μ_0	Permeability of free space	$4\pi \times 10^{-7}$H/m
πa_0^2	Atomic cross-section	8.7974×10^{-21}m^2
q/k	Temperature associated with 1 electron volt	11604.5 °K
eV	Energy associated with 1 electron volt	$1.602176487 \times 10^{-19}$J
T_0	Standard temperature at 0°C	273.15°K
p_0	Standard atmospheric pressure (760 tor = 1 atm)	1.0133×10^5Pa
n_0	Loschmidt's number (gas density at STP)	2.6868×10^{25}m^{-3}
R	Universal gas constant	8314 J/(kg × mole)
h	Planck's constant	6.626069×10^{-34}J · s
$\hbar$	Planck's constant/2π	1.054572×10^{-34}J · s

References

Alexander, M. S., Smith, K. L., Paine, M. D., & Stark, J. P. W. (2007). Voltage-modulated flow rate for precise thrust control in colloid electrospray propulsion. *Journal of Propulsion and Power, 23*(5), 1042–1048. https://doi.org/10.2514/1.25309

Armano, M., Audley, H., Auger, G., Baird, J., et al. (2015). LISA Pathfinder Mission. *Journal of Physics: Conference Series, 610*, 012005.

Arrington, L. A., Haag, T. W., Pencil, E. J., & Meckel, N. J. (1997). A performance comparison of pulsed plasma thruster electrode configurations, NASA TM 97-206305, IEPC-97-127, 25th International Electric Propulsion Conference sponsored by the Electric Rocket Propulsion Society Cleveland, Ohio, August 24–28, 1997, NASA Lewis Research Center.

Bond, T. A., & Christensen, J. A. (1999). NSTAR Ion Thrusters and Power Processors, NASA/CR-1999-209162.

Dandavino, S., Ataman, Ç., Ryan, C., Chakraborty, S., Courtney, D. G., Stark, J. P. W., & Shea, H. R. (2014). Microfabricated electrospray emitter arrays with integrated extractor and accelerator electrodes for the propulsion of small spacecraft. *Journal of Micromechanics and Microengineering, 24*, 75011.

Fisher, J. (2020). NEXT-C Flight Ion System Status, AIAA 2020-3604, AIAA Propulsion and Energy 2020 Forum, August 24–28, 2020, https://doi.org/10.2514/6.2020-3604

Frieman, J. D., Kamhawi, H., Mackey, J., Haag, T. W., Peterson, R. Y., & Hofer, R. R. (2022). Expanded performance characterization of the NASA HERMeS Hall thruster, AIAA 2022-1353, AIAA SCITECH 2022 Forum, January 3–7, 2022, San Diego, CA, https://doi.org/10.2514/6.2022-1353

Grustan-Gutierrez, E., Jhuree, S., & Stark, J. P. W. (2017) Modelling of colloid thrusters for mission analysis, IEPC-2017-394. Presented at the 35th International Electric Propulsion Conference, Georgia Institute of Technology, Atlanta, Georgia, USA, October 8–12, 2017.

Hayashi, M. (2003). Bibliography of electron and photon cross sections with atoms and molecules published in the 20th century: Argon NIFS-DATA-72, National Institute for Fusion Science (Jpn) ISSN 0915-6364.

Jahn, R. G. (1968). *Physics of Electric Propulsion*. McGraw-Hill.

Jhuree, S., Alexander, M. S., & Stark, J. P. W. (2007). Electrostatic influence on charge to mass spectrum in mixed ion/droplet mode: implications for colloid thruster design. In *Proceedings of the 30th International Electric Propulsion Conference, 2007* (pp. 1–7).

Kaltenegger, L., Karlsson, A., Fridlund, M., & Absil, O. (2003). Overview of the Darwin Mission, Proceedings of the confereence on Towards Other Earths: DARWIN/TPF and the Search for Extrasolar Terrestrial Planets, Heidelberg, Germany, 22–25 April 2003 (ESA SP-539, October 2003).

Keidar, M., Boyd, I. D., & Beilis, I. I. (2003). Model of an electro-thermal pulsed plasma thruster. *Journal of Propulsion and Power, 19*(3), 424–430.

Koizumi, H., Kakami, A., Furuta, Y., Komurasaki, K., & Arakawa, Y. (2003). Liquid propellant pulsed plasma thruster. In *Proceedings of the 28th International Electric Propulsion Conference, Toulouse, France*, 17–21 March 2003, Number 2003-87.

Laperriere, D. D. (2005). Electromechanical modeling and open-loop control of parallel-plate pulsed plasma micro-thrusters with applied magnetic fields. Masters Theses (All Theses, All Years, 898), https://digitalcommons.wpi.edu/etd-theses/898

Levchenko, L., Bazaka, K., Ding, Y., Raitses, Y., Mazouffre, S., Henning, T., Klar, P. J., Shinohara, S., Schein, J., Garrigues, L., Kim, M., Lev, D., Taccogna, F., Boswell, R. W., Charles, C., Koizumi, H., Shen, Y., Scharlemann, C., Keidar, M., & Xu, S. (2018). Space micropropulsion systems for Cubesats and small satellites: From proximate targets to furthermost frontiers. *Applied Physics Reviews, 5*(011104), 011104-1-36. https://doi.org/10.1063/1.5007734

Ling, W. Y. L., Zhang, S., Fu, H., Huang, M., Quansah, J., Liu, X., & Wang, N. (2020). A brief review of alternative propellants and requirements for pulsed plasma thrusters in micropropulsion applications. *Chinese Journal of Aeronautics*. June 2020, https://www.sciencedirect.com/science/article/pii/S1000936120302065, Accessed online, 02-11-2020.

Mazouffre, S. (2016). Electric propulsion for satellites and spacecraft: established technologies and novel approaches. *Plasma Sources Science and Technology, 25*(3), 033002. https://doi.org/10.1088/0963-0252/25/3/033002

Misra, K. (2018). Mathematical Modeling of Liquid-fed Pulsed Plasma Thruster. *Aerospace (MDPI), 5*(13), 16.

Palaszewski, B. (2014). *Electric propulsion for future space missions, Part I.* https://www.grc.nasa.gov/WWW/K-12/DLN/descriptions/presentations/systemsengineering/SystemsEngPart1.ppt

Pedrini, D., Misuri, T., Paganucci, F., & Andrenucci, M. (2017). Development of Hollow Cathodes for Space Electric Propulsion at Sitael. *Aerospace (MDPI), 4*(26), 1–14. https://doi.org/10.3390/aerospace4020026

Polzin, K., Martin, A., Little, J., Promislow, C., Jorns, B., & Woods, J. (2020). State-of-the-Art and Advancement Paths for Inductive Pulsed Plasma Thrusters. *Aerospace (MDPI), 7*(105), 67.

Rapp, D., & Englander-Golden, P. (1965). Total Cross Sections for Ionization and Attachment in Gases by Electron Impact. I. Positive Ionization. *The Journal of Chemical Physics, 43*, 1464–1479.

Stark, J. P. W., Alexander, M. S. Jhuree, S., Paine, M. D., & Smith, K. L. (2007). Electrospray from pulsation to Multi-Jet: High accuracy measurement set new challenges. In *387th Heraeus Seminar Statics and Dynamics of Electrified Liquids: Droplets, Cones and Jets*, 2007.

Steiger, C., Altay, A., Montagnon, E., Budnik, F., Casasco, M., Espeillac, P., Fugger, S., & Strandberg, T. (2019). GNC Operations for the BepiColombo Mission to Mercury: First In-flight Experience, 8th European Conference for Aeronautics and Space Sciences (EUCASS), https://doi.org/10.13009/EUCASS2019-218, 16 Pages.

Szelecka, A., Kurzyna, J., Daniłko, D., & Barral, S. (2015). Liquid micro pulsed plasma thruster. *Nukleonika, 60*(2), 257–261. https://doi.org/10.1515/nuka-2015-0057

Ziemer, J. K. (2001). Performance scaling of gas-fed pulsed plasma thrusters, PhD dissertation. Department of Mechanical and Aerospace Engineering, Princeton University.

Chapter 12
Space Vehicle Electro-dynamic Propulsion

12.1 Electro-dynamic Thrusters

The dynamics of electromagnetic systems in terms of their influence on the generated currents or in the generation of electromagnetic waves can be extremely useful in the generation and acceleration of plasmas for electric propulsion systems. Mature electric propulsion devices such as ion engines, Hall thrusters and magnetoplasmadynamic (MPD) thrusters have electrodes exposed to the plasmas. However, ion sputtering and erosion can damage these exposed electrodes over time. For propulsion systems that are used over a long period, electrodeless propulsion devices have been proposed and rigorously researched as an alternative option. One basic use is in providing a method of injecting energy into the propellant gas that does not require the use of electrodes that are in contact with the gas, in order to avoid lifetime-limiting erosion problems with the use of electrodes. Such a scheme makes use of electromagnetic waves in the radiofrequency (RF) and microwave spectral domains to transfer energy to the plasma. Alternately, the electromagnetic field can shape and regulate the plasma as in a nozzle. In the class of electric thrusters discussed in this chapter, the dynamics of the electromagnetic systems plays a defining role in the generation, in the acceleration or in the ejection of the plasma from the thruster to generate high exhaust velocities. Thus, electromagnetic thrusters produce thrust by applying the principles of the dynamics of the electromagnetic systems to generate either electromagnetic forces acting on the propellant or electromagnetic waves within propellant domain, to induce either plasma ionization or resonant heating. Examples include radiofrequency and microwave thrusters, helicon wave thrusters, magnetic nozzles, certain classes of pulsed plasma thrusters that exploit the dynamics of the induced magnetic fields, magnetohydrodynamic (MHD) thrusters, magnetoplasmadynamic (MPD) thrusters, ponderomotive force thrusters and the Variable Specific Impulse Magnetoplasma Rocket (VASIMR). Electrodeless

R. Vepa, *Space Vehicle Maneuvering, Propulsion, Dynamics and Control*,
https://doi.org/10.1007/978-3-031-65518-0_12

propulsion systems are specifically represented by the VASIMR and the helicon plasma thrusters. Electrodeless plasma thrusters are here defined as devices in which the energy is transferred to the propellant through the action of body forces without direct contact between the plasma and any of the electrodes.

12.2 Role of a Magnetic Field in Plasma Acceleration

It is important to consider the dynamics of charged particle under the influence of both a magnetic and electric field. Of particular importance is the role of the magnetic field.

Let **E** and **H** denote the electric and magnetic field strengths. Initially, it is assumed that the media is isotropic, so there is no ferromagnetism or permanent polarization. Let ε_0 and μ_0 be the free space dielectric constant and free space permeability. Let ε_r and μ_r be the relative dielectric constant (or permittivity) and the relative permeability. Let **D** and **B** denote the dielectric displacement and magnetic induction. They are assumed to be related to the electric and magnetic field strengths, **E** and **H**, respectively, by the constitutive relations

$$\mathbf{D}=\varepsilon_0\varepsilon_r\mathbf{E},\mathbf{B}=\mu_0\mu_r\mathbf{H}. \tag{12.1}$$

Let **j** denote the current density vector, ρ_q the charge density and c the velocity of light (velocity of electromagnetic or EM radiation). Let ρ_r be the resistivity. The equation defining the conservation of charge is

$$\nabla\cdot\mathbf{j}=-\frac{\partial\rho_r}{\partial t}.$$

According to Ampere's law, relating the work done in carrying a unit pole round a closed circuit to the total current in the circuit due to conduction and dielectric displacement,

$$\nabla\times\mathbf{H}=\mathbf{j}+\frac{\partial\mathbf{D}}{\partial t}\cdot\nabla\times\mathbf{B}=\mu_0\mu_r\mathbf{j}+\frac{\mu_r\varepsilon_r}{c^2}\frac{\partial\mathbf{E}}{\partial\mathbf{t}},c^2=\frac{1}{\mu_0\varepsilon_0}. \tag{12.2}$$

The Lenz law of induction or Faraday's law is

$$\nabla\times\mathbf{E}=-\frac{\partial\mathbf{B}}{\partial t}. \tag{12.3}$$

Taking the divergence of the preceding two equations and using the conservation of charge equation, we obtain Gauss's theorem and the fact that isolated magnetic poles cannot exist. Thus, we have

$$\nabla \cdot \mathbf{D} = \rho_c, \nabla \cdot \mathbf{B} = 0. \tag{12.4}$$

The conductivity relation, with σ defined as the conductivity tensor, is

$$\mathbf{j} = \frac{1}{\rho_r}\mathbf{E} = \sigma\mathbf{E}. \tag{12.5}$$

With a magnetic field included, one has

$$\mathbf{j} = \sigma(\mathbf{E} + \mathbf{v} \times \mathbf{B}). \tag{12.6}$$

Note that in Cartesian coordinates and in cylindrical coordinates, respectively,

$$\nabla \equiv \mathbf{e}_x \frac{\partial}{\partial x} + \mathbf{e}_y \frac{\partial}{\partial y} + \mathbf{e}_z \frac{\partial}{\partial z}, \nabla \equiv \mathbf{e}_r \frac{\partial}{\partial r} + \mathbf{e}_\theta \frac{1}{r}\frac{\partial}{\partial \theta} + \mathbf{e}_z \frac{\partial}{\partial z}. \tag{12.7}$$

For a homogeneous medium, in the absence of a residual charge, using the constitutive relations,

$$\nabla \cdot \mathbf{D} = 0 = \nabla \cdot \mathbf{E} = 0, \nabla \cdot \mathbf{B} = 0 = \nabla \cdot \mathbf{H} = 0. \tag{12.8}$$

$$\nabla \times \mathbf{H} = \frac{\partial \mathbf{D}}{\partial t} = \varepsilon_0 \varepsilon_r \frac{\partial \mathbf{E}}{\partial t}, \nabla \times \mathbf{E} = -\frac{\partial \mathbf{B}}{\partial t} = -\mu_0 \mu_r \frac{\partial \mathbf{H}}{\partial t}. \tag{12.9}$$

But $\nabla \times (\nabla \times) = \nabla (\nabla \cdot) - \nabla^2$. Hence,

$$\nabla \times (\nabla \times \mathbf{H}) = \varepsilon_0 \varepsilon_r \nabla \times \frac{\partial \mathbf{E}}{\partial t} = \varepsilon_0 \varepsilon_r \frac{\partial}{\partial t}(\nabla \times \mathbf{E}) = -\varepsilon_0 \mu_0 \varepsilon_r \mu_r \frac{\partial \mathbf{H}}{\partial t} = -\frac{\varepsilon_r \mu_r}{c^2}\frac{\partial \mathbf{H}}{\partial t}. \tag{12.10}$$

It follows that

$$\nabla(\nabla \cdot \mathbf{H}) - \nabla^2 \mathbf{H} = -\nabla^2 \mathbf{H} = -\frac{\varepsilon_r \chi_m}{c^2}\frac{\partial \mathbf{H}}{\partial t}. \tag{12.11}$$

Thus, one has the classic equation for wave propagation which is

$$\nabla^2 \mathbf{H} = \frac{\varepsilon_r \mu_r}{c^2}\frac{\partial \mathbf{H}}{\partial t}. \tag{12.12}$$

There are several applications of the above equation of wave propagation, but our interest is mainly in the applications to plasma thrusters.

The motion of a charged particle in the presence of an electric field, **E**, and magnetic field, **B**, is given by the Lorentz force law together with Newton's second law as

$$m\frac{d\mathbf{v}}{dt} = q(\mathbf{E}(r,\theta,z,t) + \mathbf{v}\times\mathbf{B}(r,\theta,z,t)). \tag{12.13}$$

If one defines

$$\mathbf{V} = \begin{bmatrix} 0 & -v_z & v_\theta \\ v_z & 0 & -v_r \\ -v_\theta & v_r & 0 \end{bmatrix}, \tag{12.14}$$

then one has the matrix equations for the motion of the charged particles:

$$m\frac{d\mathbf{v}}{dt} = q\mathbf{E}(r,\theta,z,t) + q\mathbf{V}\mathbf{B}(r,\theta,z,t). \tag{12.15}$$

In the first instance, it is assumed that the electron thermal pressure, the thermal drift, the electron diamagnetic drift velocity and the frictional resistance due to electron collision with the ions may be neglected. Let $\mathbf{B}(r,\theta,z,t) \equiv \mathbf{e}_r B_0$; hence, one has

$$\begin{aligned}
\nabla\cdot\mathbf{B} &\equiv \mathbf{e}_r\frac{1}{r}\frac{\partial(rB_r)}{\partial r} + \mathbf{e}_\theta\frac{1}{r}\frac{\partial B_\theta}{\partial\theta} + \mathbf{e}_z\frac{\partial B_z}{\partial z},\\
\nabla\times\mathbf{B} &\equiv \mathbf{e}_r\left(\frac{1}{r}\frac{\partial B_z}{\partial\theta} - \frac{\partial B_\theta}{\partial z}\right) + \mathbf{e}_\theta\left(\frac{\partial B_r}{\partial z} - \frac{\partial B_z}{\partial r}\right) + \mathbf{e}_z\frac{1}{r}\left(\frac{\partial(rB_\theta)}{\partial r} - \frac{\partial B_r}{\partial\theta}\right).
\end{aligned} \tag{12.16}$$

With only a radial magnetic field, $B_\theta = B_z = 0$,

$$\begin{aligned}
\nabla\times\mathbf{B} \equiv \mathbf{e}_\theta\frac{\partial B_r}{\partial z} - \mathbf{e}_z\frac{1}{r}\frac{\partial B_r}{\partial\theta},\ \mathbf{v}\times\mathbf{B}(r,\theta,z,t) &= (v_\theta B_z - v_z B_\theta)\mathbf{e}_r\\
&+ (v_z B_r - v_r B_z)\mathbf{e}_\theta + (v_r B_\theta - v_\theta B_r)\mathbf{e}_z.
\end{aligned} \tag{12.17}$$

Hence, when $B_\theta = B_z = 0$,

$$\frac{q\mathbf{v}}{m}\times\mathbf{B}(r,\theta,z,t) = v_z\frac{qB_0}{m}\mathbf{e}_\theta - v_\theta\frac{qB_0}{m}\mathbf{e}_z = v_z\omega_c\mathbf{e}_\theta - v_\theta\omega_c\mathbf{e}_z, \tag{12.18}$$

where $\omega_c = qB_0/m$ is defined as the cyclotron frequency. Since the unit vectors $\mathbf{e}_r$ and $\mathbf{e}_\theta$ are not constant,

$$d\mathbf{v}/dt = \mathbf{e}_r\left(\dot{v}_r - v_\theta\dot{\theta}\right) + \mathbf{e}_\theta\left(\dot{v}_\theta + v_r\dot{\theta}\right) + \mathbf{e}_z\dot{v}_z. \tag{12.19}$$

The velocity components satisfy

$$\dot{v}_r - v_\theta\dot{\theta} = qE_r/m; \dot{v}_\theta + v_r\dot{\theta} = \omega_c v_z + qE_\theta/m; \dot{v}_z = -\omega_c v_\theta + qE_z/m. \tag{12.20}$$

In matrix form,

$$\frac{d}{dt}\begin{bmatrix} v_r \\ v_\theta \\ v_z \end{bmatrix} = -\begin{bmatrix} 0 & -\dot{\theta} & 0 \\ \dot{\theta} & 0 & -\omega_c \\ 0 & \omega_c & 0 \end{bmatrix}\begin{bmatrix} v_r \\ v_\theta \\ v_z \end{bmatrix} + \frac{q}{m}\begin{bmatrix} E_r \\ E_\theta \\ E_z \end{bmatrix}. \tag{12.21}$$

Thus, for the jth charged particle, one has

$$\frac{d}{dt}\begin{bmatrix} v_{rj} \\ v_{\theta j} \\ v_{zj} \end{bmatrix} = -\begin{bmatrix} 0 & -\dot{\theta}_j & 0 \\ \dot{\theta}_j & 0 & -\omega_{cj} \\ 0 & \omega_{cj} & 0 \end{bmatrix}\begin{bmatrix} v_{rj} \\ v_{\theta j} \\ v_{zj} \end{bmatrix} + \frac{q_j}{m_j}\begin{bmatrix} E_r \\ E_\theta \\ E_z \end{bmatrix}. \tag{12.22}$$

The above equation is compactly expressed as

$$d\mathbf{v}_j/dt = -\mathbf{A}_j\mathbf{v}_j + (q_j/m_j)\mathbf{E}_j. \tag{12.23}$$

The associated current vector is $\mathbf{j} = \sum_i q_i n_i \mathbf{v}_i$. Using the operator, $D = d/dt$,

$$(D\mathbf{I} + \mathbf{A}_j)\mathbf{v}_j = (q_j/m_j)\mathbf{E}_j; \mathbf{v}_j = (q_j/m_j)(D\mathbf{I} + \mathbf{A}_j)^{-1}\mathbf{E}_j. \tag{12.24}$$

When the magnetic field vector is given by

$$\mathbf{B}(r,\theta,z,t) = \begin{bmatrix} B_r \\ B_\theta \\ B_z \end{bmatrix}; \mathbf{V}^T\mathbf{B} = \begin{bmatrix} 0 & -v_z & v_\theta \\ v_z & 0 & -v_r \\ -v_\theta & v_r & 0 \end{bmatrix}^T \begin{bmatrix} B_r \\ B_\theta \\ B_z \end{bmatrix},$$
$$\mathbf{V}^T\mathbf{B} = \begin{bmatrix} 0 & -B_z & B_\theta \\ B_z & 0 & -B_r \\ -B_\theta & B_r & 0 \end{bmatrix}\begin{bmatrix} v_r \\ v_\theta \\ v_z \end{bmatrix}. \tag{12.25}$$

For the wave-like motions,

$$\nabla\times\mathbf{B} = \mu_0\mu_r\mathbf{j} + \frac{\varepsilon_r\mu_r}{c^2}\frac{\partial\mathbf{E}}{\partial t}; \nabla\times\mathbf{E} = -\frac{\partial\mathbf{B}}{\partial t}. \tag{12.26}$$

Consequently,

$$\nabla \times \frac{\partial}{\partial t}\mathbf{B} = \mu_0\mu_r\frac{\partial}{\partial t}\mathbf{j} + \frac{\varepsilon_r\mu_r}{c^2}\frac{\partial^2\mathbf{E}}{\partial t^2}; \nabla \times (\nabla \times \mathbf{E}) + \mu_0\mu_r\frac{\partial}{\partial t}\mathbf{j} + \frac{\varepsilon_r\mu_r}{c^2}\frac{\partial^2\mathbf{E}}{\partial t^2} = 0. \tag{12.27}$$

It follows that

$$\nabla \times (\nabla \times \mathbf{E}) + \mu_0\mu_r\sum_i q_i n_i \frac{\partial \mathbf{v}_i}{\partial t} + \frac{\varepsilon_r\mu_r}{c^2}\frac{\partial^2\mathbf{E}}{\partial t^2} = 0. \tag{12.28}$$

But

$$\frac{d\mathbf{v}_j}{dt} + \begin{bmatrix} 0 & -\dot{\theta}_j & 0 \\ \dot{\theta}_j & 0 & 0 \\ 0 & 0 & 0 \end{bmatrix}\mathbf{v}_j = \frac{q_j}{m_j}\mathbf{E} + \frac{q_j}{m_j}\mathbf{V}_j\mathbf{B}, \mathbf{VB} = -\begin{bmatrix} 0 & -B_z & B_\theta \\ B_z & 0 & -B_r \\ -B_\theta & B_r & 0 \end{bmatrix}\begin{bmatrix} v_r \\ v_\theta \\ v_z \end{bmatrix}. \tag{12.29}$$

Hence,

$$\frac{d\mathbf{v}_j}{dt} + \begin{bmatrix} 0 & -\dot{\theta}_j & 0 \\ \dot{\theta}_j & 0 & 0 \\ 0 & 0 & 0 \end{bmatrix}\mathbf{v}_j = \frac{q_j}{m_j}\mathbf{E} - \frac{q_j}{m_j}\begin{bmatrix} 0 & -B_z & B_\theta \\ B_z & 0 & -B_r \\ -B_\theta & B_r & 0 \end{bmatrix}\mathbf{v}_j, \tag{12.30}$$

and

$$\nabla \times (\nabla \times \mathbf{E}) + \mu_0\mu_r\sum_i q_i n_i \frac{\partial \mathbf{v}_i}{\partial t} + \frac{\varepsilon_r\mu_r}{c^2}\frac{\partial^2\mathbf{E}}{\partial t^2} = 0. \tag{12.31}$$

However,

$$\begin{gathered}\nabla \times (\nabla \times \mathbf{E}) = \nabla(\nabla \cdot \mathbf{E}) - \nabla^2\mathbf{E}, \frac{\varepsilon_r\mu_r}{c^2}\frac{\partial^2\mathbf{E}}{\partial t^2} - \nabla^2\mathbf{E} + \nabla(\nabla \cdot \mathbf{E}) \\ +\mu_0\mu_r\sum_i q_i n_i \frac{\partial \mathbf{v}_i}{\partial t} = 0. \end{gathered} \tag{12.32}$$

Thus, the final equation for the velocity dynamics of a charged particle in a combined magnetic and electric field is

$$\frac{d\mathbf{v}_j}{dt} + \frac{q_j}{m_j}\begin{bmatrix} 0 & -\left(B_z + (m_j/q_j)\dot{\theta}_j\right) & B_\theta \\ B_z + (m_j/q_j)\dot{\theta}_j & 0 & -B_r \\ -B_\theta & B_r & 0 \end{bmatrix}\mathbf{v}_j = \frac{q_j}{m_j}\mathbf{E}. \tag{12.33}$$

Spacecraft are propelled by the directed expulsion of high-velocity particles, typically in the form of a gas or plasma. According to Newton's second law, thrust is defined as the instantaneous force on a spacecraft that is proportional to the vector sum of the time rate of change of the momenta of the expelled propellant particles i. Recall that if the instantaneous rate of expulsion of propellant is $\dot{m}_{pi}$ and the propellant is expelled at a velocity v_{ei}, relative to the spacecraft, then the thrust is given by $F_t = \sum_i \dot{m}_{pi} v_{ei}$, where F_t is the thrust in Newtons. The primary metrics in defining the requirements of a plasma thruster are the specific impulse, the thruster efficiency, the (power) specific mass and the lifetime of the plasma thruster. The last metric is the main advantage of electrodeless thrusters. The thruster efficiency is the ratio of the directed kinetic energy in the jet to the energy supplied by the power supply.

12.2.1 Plasma Creation and Acceleration

Plasma for thrusters may be produced by a variety of mechanisms. One method is through the effect of Faraday's law where a changing magnetic flux induces an electric field that can ionize a low-pressure gas by stripping electrons from atoms and molecules. For example, for a coil of N identical turns, the electro-motive force (EMF) generated is given by the rate of change of the magnetic flux Φ_B:

$$\mathcal{E} = = -N d\Phi_B/dt, \Phi_B = \iint_{\sum(t)} \mathbf{B}(t) \cdot d\mathbf{A}. \tag{12.34}$$

The magnetic flux is given by a surface integral over a moving surface area $\Sigma(t)$, of the vector dot product of the magnetic field and the element of area, representing the element of magnetic flux through the surface area element $d\mathbf{A}$, a vector element of the area of the moving surface.

In another method, radiofrequency waves are typically used with various kinds of antenna, and this process is a useful electrodeless mechanism for producing plasma from gaseous propellants.

Plasma for thrusters may be produced by a traditional DC potential to produce a low current density discharge known as a glow discharge. Plasma is produced by the bombardment of a gas by electrons and ions that are accelerated by the potential between electrodes. This mechanism may be used to produce ions for gridded ion thrusters.

Plasma may also be produced by a high current density discharge or arc that may be initiated when the cathode becomes sufficiently hot that it emits electrons thermionically. Otherwise, an arc may be initiated as a glow discharge. Magnetoplasmadynamic, cathodic arc and electro-thermal thrusters are devices with electrodes that produce ions by this means.

Once created, plasma has to be accelerated to produce thrust. There are a number of mechanisms that can be used for this purpose, some more effective than others. The following paragraphs describe the mechanisms that have been used or proposed as a means of accelerating plasma in electrodeless thrusters. The primary mechanisms for plasma acceleration are as follows:

(1) Lorentz force generation
(2) Magnetic nozzles
(3) Electro-thermal acceleration
(4) Rotating magnetic field (RMF) acceleration
(5) Rotating electric field (REF) acceleration
(6) Acceleration by beating electrostatic waves (BEW)
(7) Ponderomotive force generation

The Lorentz force is the force generated in the majority of plasma thrusters by body forces that are described by the Lorentz equation due to the presence of moving charges in an electric and/or magnetic field. The force is normal both to the velocity of the moving charges and to the magnetic field.

A magnetic nozzle is formed whenever a solenoidal field is used to confine plasma and can be used to direct or accelerate plasma by interactions between the plasma and the magnetic field. The processes in an electric thruster can supply more energy to the propellant than is possible in a typical nozzle attached to a chemical rocket. In general, magnetic nozzles may be regarded as electro-thermal devices where the plasma may be produced by different mechanisms by both magnetic and thermal forces together with ambipolar acceleration that balances an electron pressure gradient and drives the plasma through the nozzle. An ambipolar field applies equally to both positively and negatively charged particles. An ambipolar condition means that electrons and ions move together, as otherwise large fields appear and pull the electrons towards the ions. Hence, the fluxes of electrons and ions are equal. Ambipolar diffusion is diffusion of both positive and negative species with opposite electrical charge due to their interaction via an electric field. A magnetic nozzle is illustrated in Fig. 12.1.

In the case of electro-thermal acceleration, a propellant may be heated and ionized by induction, by radiated energy or by an electrical discharge.

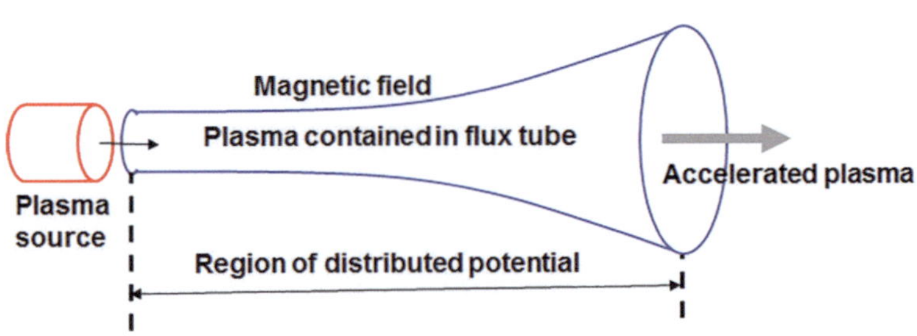

Fig. 12.1 Typical schematic diagram of a magnetic nozzle (adapted from Lafleur (2011))

The other methods of plasma acceleration are rotating magnetic field (RMF) acceleration, with a fixed current density; rotating electric field (REF) acceleration, with fixed magnetic field; and acceleration by beating electrostatic waves (BEW).

The ponderomotive forces represent time-averaged non-linear electromagnetic forces acting on the plasma in the combined presence of both oscillating electric and magnetic fields. The interaction of these fields is non-linear, and time averaging of the non-linear relations results in a net residual force. Ponderomotive force thrusters exploit the non-linear ponderomotive force on a charged particle that is proportional to the negative gradient of the intensity of an inhomogeneous oscillating electromagnetic field and an electric field component in the presence of a magnetic field. Helicon wave electric double-layer acceleration uses ponderomotive forces. With $\rho = nm$, where n is the charged particle's number density and m is the mass of the charged particle, the first two equations in Eq. 9.84, the equation for continuity and conservation of momentum, are

$$\frac{\partial n}{\partial t} + \nabla \cdot (n\mathbf{u}) = 0, m\frac{\partial (n\mathbf{u})}{\partial t} + m\nabla \cdot (n\mathbf{u}\mathbf{u}) = -\nabla p + \sigma\mathbf{E} + \sigma\mathbf{u} \times \mathbf{B} + \nabla \cdot \overline{\overline{\tau}}. \tag{12.35}$$

Assuming wave motion; employing complex variables, so the electric field is of the form $\mathbf{E}$ exp $(i\omega t)$ with $\overline{\overline{\tau}} \equiv 0$; and using Faradays law, $\mathbf{B} = -\nabla \times \mathbf{E}/i\omega$, and the relationship between the current density, the conductivity σ and the electric field given by Eq. (12.5) with the solution to continuity equation, $n = -\nabla \cdot (n\mathbf{u})/i\omega$,

$$i\omega mn\mathbf{u} + m\nabla \cdot (n\mathbf{u}\mathbf{u}) + \nabla p = \sigma\mathbf{E} - \sigma\mathbf{u} \times \nabla \times \mathbf{E}/i\omega = \sigma\mathbf{E} - \mathbf{E} \times \{\nabla \times \mathbf{E}/(i\omega nm)\}. \tag{12.36}$$

The later term may be simplified using the vector identity

$$(1/2)\nabla(\mathbf{E} \cdot \mathbf{E}) = \mathbf{E} \times (\nabla \times \mathbf{E}) + (\mathbf{E} \cdot \nabla)\mathbf{E}. \tag{12.37}$$

Using the vector identity in (12.37), from the right-hand side of Eq. (12.36), which is essentially non-linear in the electric field, the expressions for the ponderomotive force may be simplified for various special cases (see, for example, Dasgupta et al., 2003).

12.2.2 Plasma Magnetic Nozzles, Focussing and Magnetic Mirrors

A magnetic nozzle with its "wall" made up of a convergent–divergent magnetic field created by a combination of electromagnets and permanent magnets to guide the expansion of a hot plasma stream, as it accelerates to supersonic speeds to generate thrust. It functions in ways quite similar in principle to a typical classical convergent–divergent in expanding a neutral gas to very high supersonic speeds. In a magnetic nozzle, however, the walls of the classical nozzle replaced by the magnetic field, while the electromagnetic forces guide the charged particles that make up the plasma. The shape of the additional magnetic field created by the electromagnets that adds to the field generated by the permanent magnets can be shaped, controlled or focussed to create the magnetic field configurations dynamically to match the magnitude or direction of required thrust or the I_{sp} requirements as the mission progresses in time. Although the plasma is quasi-neutral in a holistic sense, it is made up of individual positively charged ions and negatively charged electrons which respond differently to the applied field, particularly the applied magnetic field. The applied magnetic field tends to deflect the light negatively charged electrons more, while the positively charged ions themselves continue almost along the centreline. The electrons follow the magnetic field contour while also moving along the nozzle downstream, aided by their higher velocity as they are both hot and light. The difference in the behaviour of positively charged ions and negatively charged electrons is responsible in creating an ambipolar electric field. The ambipolar electric field is due to charge separation between electrons and ions and yet maintaining quasi-neutrality in an electromagnetic field. The larger thermal velocity of the hot and light electrons than that of the ions leads to the development of a net separation of the charges between the electrons and ions. The resulting ambipolar electric field due to the emerging charge separation acts to accelerate the ions while decelerating the electrons, driving the outward expansion of the two species towards a common single velocity, smaller than the electron thermal speed and also greater than the ion thermal speed. The ambipolar electric field results from the increasing electron pressure with a gradient normal to the magnetic field, resulting from the accumulation of electron charges. Ambipolar electric fields and the associated ambipolar diffusion exist wherever there is a concentration of magnetic lies of force such as the polar regions of the Earth. Martinez (2013) and Little and Choueiri (2013) have discussed the analysis and performance of magnetic nozzles in some detail.

Thus, there is the possibility of dynamically controlling the expansion of a plasma into vacuum with an imposed magnetic field of appropriate geometry and mild intensity, strong enough to have electrons magnetized and yet insufficient to influence the positively charged ions which would require a magnetic field several orders of magnitude stronger. One could then expand the plasma without magnetizing the ions since electrons transport the ions along in the ensuing ambipolar electric field. A magnetic nozzle exhibits a thermoelectric character, as it converts the thermal energy present in the plasma stream provided by a source into directed kinetic energy of ions. In this respect, it somewhat resembles a convergent–divergent nozzle with a supersonic neutral gas flow. The magnetic nozzle is also an electromagnetic device, since the external forces that channel the plasma and convey the generated thrust to the vehicle are magnetically generated. Thus, the forces and hence the thrust could also be directly controlled or throttled electromagnetically.

In a typical magnetically mirror region, a hot plasma is maintained in a steady state inside a shielded cylindrical microwave-excited cavity by injection of particles into the region of the homogeneous magnetic field to effectively balance the loss through the cylinder boundaries where a relatively stronger magnetic field than at the centre is maintained. Mirror-type "magnetic wells" are generally formed by the use of solenoid assemblies in which the magnetic fields at the ends of the solenoid are much stronger than at the centre. Thus, the plasma is confined to the central region. Confinement in these devices is achieved due to constraints on particle motion imposed by the conservation of magnetic moment and energy. However, such open magnetic field line configurations can suffer from magnetohydrodynamic (MHD) plasma instabilities similar to the Rayleigh–Taylor hydrodynamic instability, which are not easily controllable, and consequently, the geometry, shape and structure must be carefully designed to ensure the plasma sustained remains stable and in a steady state.

12.3 Electric Propulsion: Electrodeless and Gridless Systems

Helicon plasma thrusters, pulsed inductive thrusters and radiofrequency ion thrusters (RFITs) that use grids and other means to accelerate plasma employ a radiofrequency-coupled source which is the most commonly used means of producing plasmas in electrodeless plasma thrusters.

Electron cyclotron resonance (ECR)-aided ion sources are suitable electrodeless devices. These sources produce ions by absorption of microwave energy at the cyclotron resonant frequency in an ionized gas in a solenoidal magnetic field. When the time-varying electric field of the microwaves is in phase with the gyration of the electrons at the cyclotron resonant frequency, in the magnetic field and orthogonal to that field, then the electrons absorb energy. The electrons also collide inelastically with neutral gas atoms or molecules, and that process leads to the further ionization of the gas. The physics of electrodeless electric thrusters that use directed plasma to propel spacecraft without employing electrodes subject to plasma erosion is reviewed by Bathgate et al. (2017).

Some helicon thrusters (HT) use an electric double layer, which has two or more peaks of the electric potential and can energize charged particles falling through them and accelerate them without the need of an accelerator grid. These are known as helicon double-layer thrusters (Pottinger et al., 2011). Magnetic-field-aligned electric fields called electric double layers (EDLs) could be productively used to induce particle acceleration. The ion energy distribution function corresponds to the derivative of the measured current–voltage characteristic and shows the presence of two peaks in the EDL region, a low-energy peak which corresponds to a population of cold ions near the probe and a high-energy peak which corresponds to the propellant ion beam (Charles et al., 2006). An EDL could be created by a magnetic mirror constraint by using two or more solenoids to create the multiple peaks in the electric potential along with steep increase in the electric field which then accelerates the ions towards the exhaust. For a plasma constrained by a magnetic mirror (using two solenoids), the position of the EDL, in which an electric field arises between two oppositely charged space charge layers created by the ions and electrons, is directly related to the diverging magnetic field lines. Its location can be clearly identified and controlled.

12.4 Electromagnetic Plasma Thruster Families

Electromagnetic plasma thrusters belong to one of several families. Broadly, they may be classified as follows (Charles, 2009; Levchenko et al., 2020):

Magnetoplasmadynamic (MPDT) thrusters including both self-induced and applied field MPDTs
Pulsed plasma thruster (ablative, gas fed)
Radiofrequency plasma thrusters
Helicon (wave) plasma thrusters
Inertial electrostatic confinement (IEC) thruster
Variable Specific Impulse Magnetoplasma Rocket (VASIMR) which uses a helicon plasma source, ion cyclotron resonance heating and a magnetic nozzle

In the following sections, each of these classes of thrusters would be briefly discussed.

12.4.1 Magnetoplasmadynamic (MPD) Thrusters

A magnetoplasmadynamic (MPD) thruster (MPDT) is a form of electrically powered spacecraft propulsion system which uses the Lorentz force, the force on a charged particle by an electromagnetic field, to generate thrust. The current interacts with induced magnetic field to generate the Lorentz force. It is sometimes referred to as Lorentz force accelerator (LFA) or as an MPD arcjet. A magnetoplasmadynamic thruster (MPDT) is known as a representative high-power electric propulsion device suitable for a long endurance space mission. An MPDT mainly accelerates ionized propellant with an axial (Lorentz) force driven by radial discharge current and induced azimuthal magnetic field. The self-field MPD thruster (MPDT) and applied field (AF) MPDT are high-power electric thrusters that have the potential for further developments, particularly in terms of their controllability and capacity for throttling to enable several applications in future missions. The self-field MPDT operates without additional magnets or electromagnets, using only its self-generated magnetic field for acceleration. AF-MPDTs use additional externally applied magnetic fields typically to increase the thrust, particularly at relatively lower powers within

100–200 kW. In an MPD thruster, plasma acceleration is governed by the magnetohydrodynamic momentum, including small viscous forces due to ion–ion and ion-neutral collisions, and driven primarily by the Lorentz force which contributes substantially at higher currents. While the pressure gradient driving the momentum along the centreline of such thrusters is very important, the electromagnetic accelerating forces for accelerating and constraining the plasma are the axial "blowing" and inward radial "pumping" components of the Lorentz force. With high currents, beyond a critical value of the current, there are instabilities resulting in large-amplitude terminal voltage fluctuations, large transients in plasma properties and growth in anode voltages in the plasma which result in the unstable operation as well as the erosion of the anode at certain spots, due to high current concentrations. For this reason, lower atomic mass propellants (like hydrogen and lithium) are preferred so as to avoid the instabilities. While reasons for the existence of these instabilities have been proposed by Tikhonov and Semenikhin (1993), Tilley et al. (1996), Wagner et al. (1998), Giannelli and Andrenucci (2011), Paganucci et al. (2008), and several others, only passive methods of reducing these instabilities have been proposed, and there is considerable scope for developing methods to completely suppress or eliminate the instabilities, using methods in feedback control.

12.4.2 Self-Induced and Applied Field MPDTs

The principle of operation of self-field MPDT and applied field MPDT is illustrated in Fig. 12.2. A self-field MPD thruster is a plasma accelerator composed of two concentric electrodes between which a very high current discharge is established. In an MPD thruster, a propellant such as argon is first heated by an electric discharge between two electrodes. The discharge current interacts with the self-induced azimuthal magnetic field to generate the main component of the thrust developed by an MPDT. Additionally, there is a thermodynamic component from the expansion of

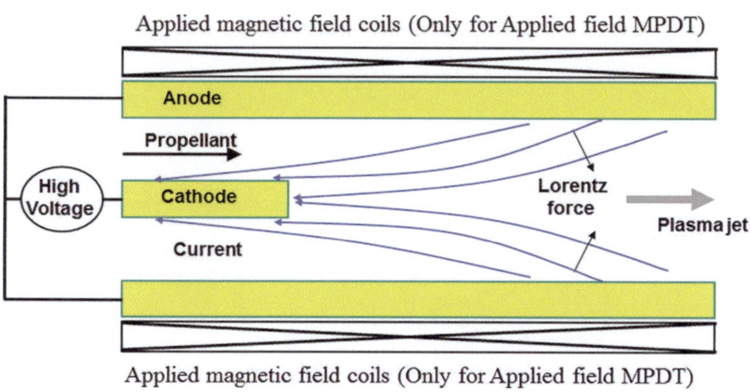

Fig. 12.2 Principle of operation of self-field MPDT and applied field MPDT (nozzle not shown)

the heated propellant in a nozzle. The primary component is proportional to the square of the current, and increasing the primary component's contribution to the total thrust does increase the efficiency. For this reason, the self-induced magnetic field is augmented by the application of an externally applied field. The windings are such that they will enhance the action of the self-magnetic field due to the current flow. However, the total thrust-producing capabilities are limited due to the occurrence of instabilities beyond a critical value of the current. Applied field magnetoplasmadynamic thrusters (AF-MPD thrusters) are hybrid accelerators in which electromagnetic and gas dynamic processes accelerate plasma to high speed. AF-MPD thrusters have considerable potential for future space applications with the significant advantages of highly specific impulse and thrust density.

12.4.3 Air-Breathing Electric Propulsion

The electrostatic and electromagnetic thrusters use the forces generated by electric and magnetic fields which in turn are harnessed for the generation, acceleration and expulsion of an ionized propellant. In contrast, the electro-thermal thrusters use propellants expanded thermodynamically and subsequently accelerated through a nozzle. Thus, the electrostatic and electromagnetic thrusters are eminently suitable for use with an externally supplied propellant such as air drawn from the atmosphere. The principle of operation of an atmosphere-breathing electric propulsion (ABEP) is illustrated in Fig. 12.3. ABEP systems also work on the same principle as the space propulsion systems. Although proposed more than three decades ago (Uematsu et al., 1984), novel ABEP systems were introduced in the early part of the millennium (Nishiyama, 2003; Fujita, 2004). They have been under development for over a decade, as evidenced by several recent publications (Goeksel & Mashek, 2017; Tagawa et al., 2013; Hohman, 2012; Shabshelowitz, 2013; Romano, 2014; Schönherr et al., 2015; Romano et al., 2015; Bozhanov & Rojas, 2017; Cifali

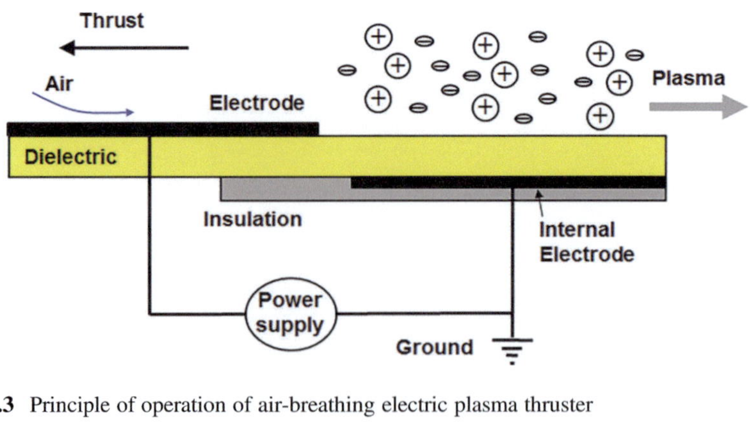

Fig. 12.3 Principle of operation of air-breathing electric plasma thruster

et al., 2011; Garrigues, 2012; Charles et al., 2008) on the subject. ABEP systems can also be classified as electro-thermal, electrostatic and electromagnetic systems. Spacecraft propulsion systems have to store the propellant on-board, while the propellant for ABEP is externally sourced. They use the available gases in the lower atmosphere, which are then accelerated using a nozzle or a nozzle–diffuser combination with or without bypassing some of the airflow. The competing architectures that have been employed in the development of an ABEP thruster are electrostatic or gridded ion thrusters (Nishiyama, 2003; Cifali et al., 2011), Hall effect thrusters (HET) (Shabshelowitz, 2013; Garrigues, 2012), magnetoplasmadynamic (MPD) thrusters (Uematsu et al., 1984) and helicon thrusters (HT) (Charles et al., 2008; Shabshelowitz, 2013).

12.5 Pulsed Plasma Thruster

Pulsed plasma thrusters (PPT) were already briefly discussed in Chap. 11 in Sect. 11.2.2. The principle of operation of a PPT is illustrated in Fig. 12.4. Yet, the dynamics of PPT is to some extent similar to the MPDT, as it also relies on the Lorentz force generated. In Sect. 11.2.2, it was seen that the efficiency and the thrust generated by the thruster was dependent on I_{bit}, as the thrust was given by $T = I_{bit}f$. Next, to obtain an expression for I_{bit}, we consider the mass flow rate. The total mass accelerated per pulse discharge $m_{pd}(t)$ could therefore be given by the sum of the constant neutral mass discharge (m_n) and plasma sheet mass discharge ($m_{ps}(t)$),

$$m_{pd}(t) = m_n + m_{ps}(t). \tag{12.38}$$

The calculation of plasma mass discharge is done by considering the flow of plasma as an idealized, quasi-steady flow. For the magnetohydrodynamic flow, there is a sonic flow speed at some position along the flow discharge channel with a high Reynolds number and pressure much greater than the plasma pressure. The conditions at this point can be used to scale the exhaust velocity by the Alfvén speed and estimate the plasma mass flow rate at the surface of the propellant. Thus, from Burton and Turchi (1998),

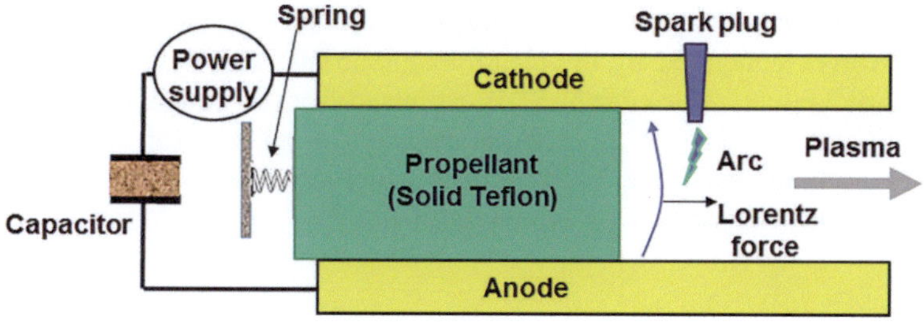

Fig. 12.4 Principle of operation of a PPT (see also Fig. 11.11) (nozzle not shown)

$$\dot{m}_{ps} = \frac{\mu_0}{4.404 V_{crit}} \left(\frac{h}{w}\right) I^2(t), m_{ps} = \frac{\mu_0}{4.404 V_{crit}} \left(\frac{h}{w}\right) \int_0^t I^2(\tau) d\tau, \tag{12.39}$$

where $\mu_0 = 4\pi \times 10^{-7}$ is the magnetic permeability in vacuum, in SI units, and V_{crit} is the Alfven critical speed, given in terms of the ionization potential V_{ion} and molar mass of the propellant M, by

$$V_{crit} = \sqrt{2qV_{ion}/M}. \tag{12.40}$$

The Alfven speed is the speed of magnetic disturbances in a plasma when no electric field is present.

To consider the discharge dynamics of the plasma flow, consider the Lorentz force acting on a plasma particle in an electromagnetic field with a charge density of ρ and a current density of $\mathbf{j} = -I\widehat{\mathbf{z}}/w$ across the electrodes, which is

$$\mathbf{F}_{EM} = \int (\rho \mathbf{E} + \mathbf{j} \times \mathbf{B}) dV. \tag{12.40}$$

It can be shown (Jahn, 1968) that the component of this force along the direction of the plasma thruster $\widehat{\mathbf{x}}$, with $\mathbf{B} = \mu_0 I \widehat{\mathbf{y}}/w$ in the transverse direction $\widehat{\mathbf{y}}$, is

$$\mathbf{F}_{EM}\widehat{\mathbf{x}} = \int_V (\rho \mathbf{E} + \mathbf{j} \times \mathbf{B}) dV \widehat{\mathbf{x}} = \frac{1}{2} L'_{em} I^2(t), \tag{12.41}$$

where the inductance gradient L'_{em} could be simplified to $L'_{em} = \mu_0 h/w$ for $h/w < 1$. For values of $h/w > 1$, Burton and Turchi (1998) have reported that the inductance gradient is $L'_{em} = (\mu_0/\pi) \times (1.5 + \ln(h/(w + \tau)))$, where τ is the thickness of the electrode. The inductance is a function of the position of the current sheet and is given by $L_{em} = L'_{em} x_s + (\mu_0/\pi) \times (0.2235(w + \tau) - h)$. (The corresponding plasma resistance in terms of the plasma conductivity σ_p and current sheet thickness δ is $h/(\sigma_p w \delta)$).

The model of the plasma dynamics assumes that when the plasma is formed, there is no additional mass accumulated in the plasma sheet, while the propellant discharge mass over each cycle is accelerated down the discharge chamber by the combined forces generated by the electromagnetic interactions, the gas dynamics and the inertial effects. The current $I(t)$ is estimated from an equivalent circuit of the thruster. In the presence of an external magnetic field, the force acting on the propellant discharge mass over each cycle now includes a term directly proportional to the current as discussed by Eslava et al. (2014).

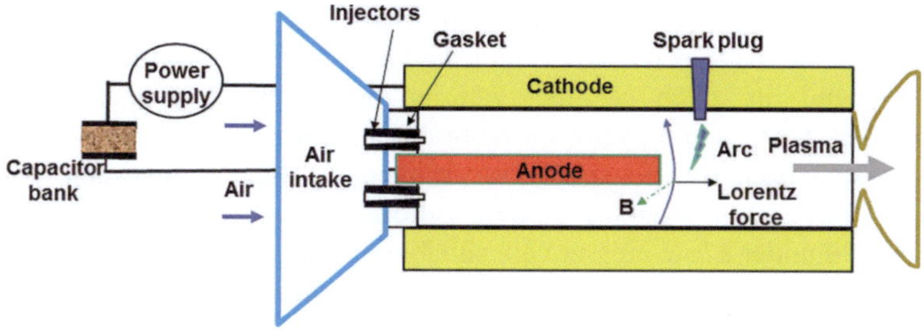

Fig. 12.5 Principle of operation of an air-breathing PPT

12.5.1 Air-Breathing PPTs

A typical air-breathing PPT was developed by Rosales et al. (2019), and it generates a thrust of around 400 mN/kW and is suitable both for aircraft operating above 20–30 km and spacecraft in low Earth orbits. The pulse length was approximately 80 μs, while the maximum current was 4.6 kA. The capacitor bank's capacitance was 40 μf and the power supply was at 2 kV. Switching was achieved by using an insulated gate bipolar transistor. The principle of operation of such an electric plasma thruster is shown in Fig. 12.5.

12.6 RF Excited Plasma Thrusters: Helicon Waves

Helicon waves are clockwise circularly polarized electromagnetic waves that propagate through a metallic chamber in a strong magnetic field with weak collisional damping. Helicon waves are also known as whistler waves since their frequency lies between the ion cyclotron frequency and the electron cyclotron frequency, and they propagate along the magnetic field lines. Helicon waves are characterized by the fact that they propagate in a bounded plasma with wave absorption but not in an infinitely extended plasma. For this reason, they are extremely important for transferring electromagnetic energy to a confined plasma, without the use of electrodes. Chen (1991) (see also Chen, 2015) has provided the theoretical basis to show that the higher than expected gain in plasma acceleration with helicon waves in a noble gas is possible if the ionized electrons are directly accelerated by the wave interaction with the particles.

In radiofrequency discharges, an oscillating electric field is excited by, e.g. an external antenna, causing plasma production in an ionization process by accelerating and heating electrons. Helicon waves are a class of "whistler" waves or propagating radiofrequency waves that can be generated under certain conditions determined by a dispersion relation and which propagate into the plasma with wave absorption leading to plasma generation. Whistler waves propagate at different velocities which are functions of the frequency due to the frequency dependence of the plasma's refractive index. These electromagnetic radiofrequency waves could be picked up by the radio communications receivers and sound like "whistles" as the waves progressively descend in pitch as the various frequency components propagate by. Moreover, the electron cyclotron motion does not influence the solution of the dispersion equation, due to their low frequency. Helicon high-density (up to $\sim 10^{13}$ cm^{-3}) plasma sources driven by a radiofrequency wave in the presence of a magnetic field under a low pressure are suitable plasma sources for plasma thrusters such as HCT and MPD thrusters owing to the flexibility in the choice of operational parameters and their inherent characteristics. In helicon thrusters, high-density plasma could also be produced by accelerating the plasma using RF antennas, with a rotating magnetic field (RMF) coil or even a rotating electric field (REF).

12.6.1 Analysis of Helicon Waves

Helicon waves are known to be right-handed circularly polarized electromagnetic waves, propagating along the magnetic field. The frequency ω is in the range $\omega_{ci} \ll \omega_l \ll \omega \ll \omega_{ce} \ll \omega_{pe}$, where ω_{ci} and ω_{ce} are the ion and electron cyclotron frequencies, ω_l is the lower hybrid frequency and ω_{pe} is the electron plasma frequency. The plasma boundaries cause the helicon wave to develop an eigenmode structure of the form $\exp(i(m\theta + kz - \omega t))$ with the azimuthal mode number m. In this case, when homogeneous plasmas are considered, the dispersion relation simplifies and the wavenumber is directly proportional to the plasma density and inversely proportional to the magnetic field.

Thus, the helicon wave is a fast wave with a high index of refraction and has a frequency greater than the ion cyclotron frequency and much less than the electron cyclotron frequency. Helicons were first studied in connection with the propagation of electromagnetic waves in solid metals at low temperature, where these waves were seen as resonances or standing waves. The term "helicon" was coined in 1960 due to the helical lines of force associated with these waves. Helicon waves in plasmas were then given a mathematical foundation–based Maxwell's EM theory. In a radially unbounded geometry, they correspond to whistler waves and can only be right-hand polarized. In bounded plasmas however, both right-hand polarized and left-hand polarized waves can exist, but the right-hand polarized waves are almost always preferentially excited. Boswell (1970) was the first to use helicon waves to create and sustain a plasma discharge. The physics of the Helicon wave is better understood if one considers its mathematical basis.

Using a set of cylindrical coordinates (r, θ, z) and assuming the magnetic field $\mathbf{B}$, to be in the z direction, the helicon wave in a cylindrical plasma is described by the following physical laws:

(i) Lenz law of induction or Faraday's law is

$$\nabla \times \mathbf{E} = -\partial \mathbf{B}/\partial t. \tag{12.42}$$

(ii) Ampere's law relating the work done in carrying a unit pole round a closed circuit to the total current in the circuit due to conduction and dielectric displacement in general is

$$\nabla \times \mathbf{H} = \mathbf{j} + \frac{\partial \mathbf{D}}{\partial t}, \nabla \times \mathbf{B} = \mu_0 \mu_r \mathbf{j} + \frac{\mu_r \varepsilon_r}{c^2} \frac{\partial \mathbf{E}}{\partial \mathbf{t}}. \tag{12.43}$$

In free space,

$$\nabla \times \mathbf{B} = \mu_0 \mathbf{j} + \frac{1}{c^2} \frac{\partial \mathbf{E}}{\partial \mathbf{t}}. \tag{12.44}$$

In free space in terms of the perturbations,

$$\nabla \times \mathbf{B}_1 = \frac{1}{c_0^2} \boldsymbol{\varepsilon} \frac{\partial \mathbf{E}_1}{\partial \mathbf{t}}. \tag{12.45}$$

Neglecting the displacement current, $\nabla \times \mathbf{H} = \mathbf{j} = \nabla \times \mathbf{B}/\mu_0$.

(iii) Taking the divergence of the preceding two equations and using the conservation of charge equation, Gauss's theorem is obtained and the fact that isolated magnetic poles cannot exist. Thus, one has

$$\nabla \cdot \mathbf{D}_1 = \varepsilon_0 \nabla \cdot \mathbf{E}_1 = \rho_c, \nabla \cdot \mathbf{B}_1 = 0. \tag{12.46}$$

(iv) The conductivity relations are given by Eqs. (12.5) and (12.6). In the above, $\mathbf{E}_1$, $\mathbf{B}_1$ and $\mathbf{j}_1$ are small perturbations of the electric field, the magnetic field and the plasma current, respectively, and n_e is the time-averaged plasma density. The perturbations will be assumed to be of the form $\exp(i(m\theta + kz - \omega t))$. Equilibrium quantities are denoted by the subscript zero. The z axis is aligned with steady-state magnetic field $\mathbf{B} = B_0\hat{z}$ .The electron fluid equation of motion is

$$m_e d\mathbf{v}/dt = -q(\mathbf{E} + \mathbf{v} \times \mathbf{B}_0) - m_e \nu \mathbf{v}. \tag{12.47}$$

In the above, all dissipation mechanisms have been lumped into one phenomenological collision rate. The convective term, the magnetic viscosity force and the pressure gradient term (∇p) have been omitted. The electron viscosity is not important when the electron Larmor radius is well below the radial wavelength of the waves, which is normally >1 cm for helicon waves; thus, for 3 eV electrons, the magnetic viscosity force can be neglected above 0.6 mT. For Trivelpiece–Gould (TG) waves, however, wavelengths shorter than 1 mm are, in principle, possible. As discussed by Chen and Arnush (1997), short-wavelength waves are not likely to exist in practice due to the presence of the finite Larmor radius (FLR) effects; the theory of helicon waves is however inadequate for these waves. Ionizing electrons of 15 eV energy have 1 cm Larmor radii below 1.3 mT, and FLR effects are likely to be important for them at such low fields. Fortunately, their population should be so small that wave propagation would not be affected (Chen & Arnush, 1997).

The zeroth-order pressure term is important under discharge equilibrium conditions but not for wave propagation. The first-order part can be separated into four terms. Since $\mathbf{E} \times \mathbf{B}$ motions are incompressible, n_e vanishes in a uniform plasma whenever FLR effects are negligible. The remaining term can be neglected relative to the $q\mathbf{E}$ term, because fluctuations in KT_e are much less than $KT_e \sim 3$ eV, compared to wave electric potentials of >100 V, typically. Although finite T_e also facilitates drift wave effects in a non-uniform plasma and treating these involves evaluating terms of comparable size, drift frequencies are well below helicon frequencies and should not be important in helicon wave propagation. If we neglect ion motions, the plasma current is $\mathbf{j} = -qn_0\mathbf{v}$. Since

$$m_e\frac{d\mathbf{v}}{dt} = q(\mathbf{E}+\mathbf{v}\times\mathbf{B}) = q\left(\mathbf{E}-\frac{\mathbf{j}}{en_0}\times\mathbf{B}\right) = m_e\frac{d}{dt}\frac{\mathbf{j}}{qn_0}, \mathbf{E} = \frac{d}{dt}\frac{m_e\mathbf{j}}{q^2n_0} - \mathbf{B}\times\frac{\mathbf{j}}{qn_0} = -\frac{i(\omega+i\nu)m_e\mathbf{j}}{q^2n_0} - \mathbf{B}\times\frac{\mathbf{j}}{qn_0}. \tag{12.48}$$

With $\mathbf{B} = B_0\hat{z}$, $\omega_{ce} = qB_0/m_e$, $\delta_r = \omega/\omega_{ce}$ and $\delta = (\omega + i\nu)/\omega_{ce} = (\omega + i\nu)m_e/qB_0$, the electron fluid equation of motion is $\mathbf{E} = -(B_0/qn_0)(i\delta\mathbf{j} + \hat{z}\times\mathbf{j})$. Since $\mathbf{j} = \nabla \times \mathbf{B}/\mu_0$,it follows that

$$\mathbf{E} = -(B_0/(qn_0\mu_0))(i\delta\nabla\times\mathbf{B} + \hat{z}\times\nabla\times\mathbf{B}). \tag{12.49}$$

From

$$\nabla\times\mathbf{E} = -\frac{\partial\mathbf{B}}{\partial t}, \nabla\times\mathbf{E} = -\frac{B_0}{qn_0\mu_0}(i\delta\nabla\times\nabla\times\mathbf{B} + \nabla\times(\hat{z}\times(\nabla\times\mathbf{B}))) = i\omega\mathbf{B}. \tag{12.50}$$

Let $\omega_{pe}^2 = n_0e^2/(\varepsilon_0 m_e)$, $k_s = \omega_{pe}/c$, and $\delta k_s^2 = (\omega + i\nu)n_0e\mu_0/B_0 = (\omega + i\nu)\omega_{pe}^2/c^2\omega_{ce} = k_w^2$. The quantity k_w is simply the wave number of low-frequency whistler waves propagating along $\mathbf{B}$ in free space. Thus,

$$i\delta\nabla\times\nabla\times\mathbf{B}-ik(\nabla\times\mathbf{B})+i\omega(qn_0\mu_0/B_0)\mathbf{B}=\mathbf{0}. \tag{12.51}$$

It follows that

$$\delta\nabla\times\nabla\times\mathbf{B}-k(\nabla\times\mathbf{B})+k_w^2\mathbf{B}=\mathbf{0}. \tag{12.52}$$

It is convenient to factor this equation into

$$(\beta_1-\nabla\times)(\beta_2-\nabla\times)\mathbf{B}=(\beta_1\beta_2-(\beta_1+\beta_2)\nabla\times+\nabla\times\nabla\times)\mathbf{B}=\mathbf{0}, \tag{12.53}$$

where β_1 and β_2 satisfy $\beta_1\beta_2=k_w^2/\delta$ and $\beta_1+\beta_2=k/\delta$. Thus, they are the roots of the characteristic equation

$$\delta\beta^2-k\beta+k_w^2=\mathbf{0}. \tag{12.54}$$

The roots are

$$\beta_{1,2}=\frac{k}{2\delta}\left(1\mp\sqrt{1-4\delta k_w^2/k^2}\right). \tag{12.55}$$

When $m_e\to 0$, $\beta_1=k_w^2/k$, and there is only one finite root. When

$$4\delta k_w^2<<k^2,\beta_{1,2}\approx\frac{k}{2\delta}\left(1\mp 1\pm 2\delta\frac{k_w^2}{k^2}\right). \tag{12.56}$$

Thus, the two roots are approximately given by

$$\beta_1\approx\frac{k_w^2}{k}=\frac{\omega+i\nu}{k}\frac{\omega_{pe}^2}{c^2\omega_{ce}}\text{ and }\beta_2\approx\frac{k}{\delta}\left(1-\delta\frac{k_w^2}{k^2}\right)\approx\frac{k}{\delta}=\frac{k}{\omega+i\nu}\omega_{ce}. \tag{12.57}$$

These correspond to the two asymptotic cases when $\delta\approx 0$, $k\beta_1=k_w^2$, and when $k_w^2\approx 0$, $\delta\beta_2=k$. From the characteristic Eq. (12.54), one obtains $k=\delta\beta+k_w^2/\beta$. Hence, $dk/d\beta=\delta-k_w^2/\beta^2$. Thus, for a maximum or a minimum in the value of k, one has $k=2k_w^2/\beta$. The minimum and maximum are respectively given by

$$k_{\min}=2\frac{\omega_{pe}}{\omega_{ce}}\frac{\omega}{c},\, k_{\max}=\frac{\omega_{pe}}{c}\sqrt{\frac{\omega}{\omega_{ce}-\omega}}. \tag{12.58}$$

If $\mathbf{B}=\mathbf{B}_1+\mathbf{B}_2$, substituting in (12.53), and if $(\beta_1-\nabla\times)\mathbf{B}_1=0$ and $(\beta_2-\nabla\times)\mathbf{B}_2=0$, the equation for $\mathbf{B}$ is satisfied. Taking the curl of the equation $\beta_j\mathbf{B}_j=\nabla\times\mathbf{B}_j$,

$$\beta_j^2 \mathbf{B}_j = \nabla \times \nabla \times \mathbf{B}_j = -\nabla^2 \mathbf{B}_j, \tag{12.59}$$

and since $\nabla \times (\nabla \times) = \nabla (\nabla \cdot) - \nabla^2$, one obtains two vector Helmholtz equations:

$$\beta_j^2 \mathbf{B}_j + \nabla^2 \mathbf{B}_j = 0. \tag{12.60}$$

For the first root, β_1, the solution corresponds to the Helicon wave and the second to the Trivelpiece–Gould wave.

In cylindrical coordinates,

$$\nabla \equiv \mathbf{e}_r \frac{\partial}{\partial r} + \mathbf{e}_\theta \frac{1}{r} \frac{\partial}{\partial \theta} + \mathbf{e}_z \frac{\partial}{\partial z}. \tag{12.61}$$

From the relations for $\nabla \times \mathbf{B}$, $\nabla^2 \mathbf{B}$ and $\nabla \cdot \mathbf{B}$, one obtains

$$\nabla \times \mathbf{B} \equiv \mathbf{e}_r \left(\frac{im}{r} B_z - ikB_\theta \right) + \mathbf{e}_\theta \left(ikB_r - \frac{\partial B_z}{\partial r} \right) + \mathbf{e}_z \frac{1}{r} \left(\frac{\partial (rB_\theta)}{\partial r} - imB_r \right), \nabla \cdot \mathbf{B}$$

$$\equiv \mathbf{e}_r \frac{1}{r} \frac{\partial (rB_r)}{\partial r} + \mathbf{e}_\theta \frac{im}{r} B_\theta + \mathbf{e}_z ikB_z,$$

$$\nabla^2 \mathbf{B} \equiv \frac{1}{r} \frac{\partial}{\partial r} \left(r \frac{\partial}{\partial r} \mathbf{B} \right) - \frac{m^2}{r^2} \mathbf{B} - k^2 \mathbf{B}. \tag{12.62}$$

Hence,

$$\left(\frac{1}{r} \frac{\partial}{\partial r} \left(r \frac{\partial}{\partial r} \right) + \left(\beta_j^2 - k^2 - \frac{m^2}{r^2} \right) \right) \mathbf{B}_j = 0, \mathbf{B}_j \equiv \mathbf{e}_r B_{rj} + \mathbf{e}_\theta B_{\theta j} + \mathbf{e}_z B_{zj}. \tag{12.63}$$

Note that $k = \delta(\beta^2 + k_s^2)/\beta$. In the z direction,

$$\left(\frac{1}{r} \frac{\partial}{\partial r} \left(r \frac{\partial}{\partial r} \right) + \left(T_j^2 - \frac{m^2}{r^2} \right) \right) B_{zj} = 0. \tag{12.64}$$

The solution of the wave equations in the cylindrical frame and geometry can be expressed in terms of Bessel functions $J_n(Tr)$, where the transverse wave number T is defined by $T_j^2 = \beta_j^2 - k^2$, where $j = 1, 2$. The general solution in terms of the Bessel functions of the first and second kinds is

$$\mathbf{B}_j = \mathbf{A}_m J_m(Tr) + \mathbf{C}_m Y_m(Tr). \tag{12.65}$$

The solution that is finite on the axis for either wave in a uniform plasma filling a conducting cylinder of radius a is given by one particular set of solutions, in matrix form, which are

$$\begin{bmatrix} B_{rj} \\ B_{\theta j} \\ B_{zj} \end{bmatrix} = A_j \begin{bmatrix} \left[(\beta_j + k)J_{m-1}(Tr) + (\beta_j - k)J_{m+1}(Tr)\right] \\ i\left[(\beta_j + k)J_{m-1}(Tr) - (\beta_j - k)J_{m+1}(Tr)\right] \\ -2iT_jJ_m(Tr) \end{bmatrix} \\ = -\frac{2iA_j}{rT_j} \begin{bmatrix} i\left(m\beta_jJ_m(Tr) - krJ'_m(Tr)\right) \\ \left(mkJ_m(Tr) + \beta_jrJ'_m(Tr)\right) \\ T_j^2rJ_m(Tr) \end{bmatrix}. \tag{12.66}$$

In the above, the recurrence relations relating the Bessel functions and its derivative were used. These are

$$\frac{2m}{x}J_m(x) = J_{m-1}(x) + J_{m+1}(x), 2\frac{dJ_m(x)}{dx} = J_{m-1}(x) - J_{m+1}(x). \tag{12.67}$$

The relationship between the electric and magnetic field components in the case when the mass of the electron is neglected is

$$\begin{bmatrix} E_{rj} \\ E_{\theta j} \\ E_{zj} \end{bmatrix} = \frac{\omega}{k} \begin{bmatrix} B_{\theta j} \\ -B_{rj} \\ 0 \end{bmatrix}. \tag{12.68}$$

Considering the different modes of helicon waves, they could be obtained from the boundary conditions, at $r = a$. It may be noted that $J_m(Tr)$ is finite at $r = 0$:

$$\mathbf{j}_j = \nabla \times \mathbf{B}_j/\mu_0 = (\beta_j/\mu_0)\mathbf{B}_j. \tag{12.69}$$

12.6.2 Eigenmodes of Helicon Waves

The boundary condition to be satisfied at the walls of an insulator is $j_{rj} = 0$. Thus, $B_{rj} = 0$, and

$$m\beta_jJ_m(Tr) - krJ'_m(Tr) = 0. \tag{12.70}$$

If the cylinder is considered to be a conductor, with the zero electron mass assumption, the boundary condition at the wall is $E_{\theta j} = 0$. Hence, $ikB_{rj} - \partial B_{zj}/\partial r = ikB_{rj} = -i\omega E_{\theta j} = 0$. Thus, in either case, $B_{rj} = 0$, at the walls of the cylinder. For $m = 0$, at $r = a$, this gives

$$J_1(T_1 a) = J_1\left(\sqrt{\beta_1^2 - k^2} a\right) = 0. \tag{12.71}$$

In this case, assuming that $\beta_1^2 \gg k^2$ and $a = 0.01m$,

$$\sqrt{\beta_1^2 - k^2} \approx \beta_1 \approx 0.01 k_w^2 / k = 3.8317. \tag{12.72}$$

It follows that for the lowest mode, $k \approx 0.01 k_w^2/3.83$. For $a = 1m$, $k \approx k_w^2/3.8317$. The particle density is assumed to be $n_0 = 2.0e + 19$. The first of the above two cases is just above $k_{\min}$, while the second is just below k_w over a small frequency range. For the other modes, in the case when $B_{rj} = 0$, which Chen and Arnush (1997) refer to as the 1-BC case, with the assumption of zero electron mass, one has

$$B_{rj} = 0 = (\beta_j + k) J_{m-1}(Ta) + (\beta_j - k) J_{m+1}(Ta) = m\beta_j J_m(Ta) - krJ'_m(Ta) = 0. \tag{12.73}$$

The lowest mode, which was already obtained, is

$$m = 0 \Rightarrow J_1(T_1 a) = J_1\left(\sqrt{\beta_1^2 - k^2} a\right) = 0. \tag{12.74}$$

This mode is axisymmetric and oscillates between being purely electrostatic and purely electromagnetic conditions, having helical field lines. The magnetic field is at one stage purely radial as it changes to a purely azimuthal field and back again. The corresponding electric field can be obtained from the relationship between the electric and magnetic field components, given earlier. The mode is illustrated by Chen and Arnush (1997) (and Arnush & Chen, 1998). It is also a feature of several other modes.

When the mass of the electron cannot be neglected, the general solution is the sum of the both the helicon and TG modal solutions involving both β_1 and β_2. As discussed by Chen and Arnush (1997) (and Arnush & Chen, 1998), the complete boundary condition is

$$\begin{aligned}(\beta_1 + k) J_{m-1}(T_1 a) &+ (\beta_1 - k) J_{m+1}(T_1 a) \\ &= \{(\beta_2 + k) J_{m-1}(T_2 a) + (\beta_2 - k) J_{m+1}(T_2 a)\} \frac{\beta_1 T_1 J_m(T_1 a)}{\beta_2 T_2 J_m(T_2 a)}.\end{aligned} \tag{12.75}$$

The above boundary condition reduces to the 1-BC case, when $\beta_2 T_2 \gg \beta_1 T_1$. For $m = 1$, one has approximately

$$J_1(T_1 a) = J_1\left(\sqrt{\beta_1^2 - k^2} a\right) = \frac{kaT_1}{2\beta_1}\left(J_2\left(\sqrt{\beta_1^2 - k^2} a\right) - J_0\left(\sqrt{\beta_1^2 - k^2} a\right)\right) \approx 0. \tag{12.76}$$

The solution for the wave numbers satisfying the boundary conditions is obtained numerically, using a sensitive root-finding algorithm based on the Newton–Raphson method.

In these modes, the field pattern is stationary in a rotating frame of reference, rotating along the z axis. The field pattern simply rotates in the $\pm\theta$ direction, for $m = \pm 1$, respectively, in time at a fixed position along z, and in the z direction as z increases at a given time. The $m = 1$ mode is right-hand polarized having a straight electrostatic field as one approaches the centre. In the $m = -1$ mode which is left-hand polarized, the field is elliptically shaped at the centre and evolves to a straight electrostatic field as one approaches the boundary. The electric field corresponding to the modes is found from

$$\mathbf{E} = -(B_0/qn_0)(i\delta\mathbf{j} + \hat{z}\times\mathbf{j}). \tag{12.77}$$

Thus, the behaviour of undamped normal modes of helicon waves in a uniform plasma filling an infinitely long conducting or insulating cylinder may be obtained as in Chen and Arnush (1997) (and Arnush & Chen, 1998). By appropriate choice of the helical antenna parameters, the excitation of a helicon wave in a cylindrical magnetoplasma and the consequent production of fully ionized plasma is completely feasible. Helicon sources operate in three distinct modes, and transitions between them are typically accompanied by discontinuous changes in the plasma density. The mode generated by a helicon plasma source depends on both the RF power input and the magnetic field strength. The highest density (helicon) mode, which is one of two inductive modes, is only achievable only if both the magnetic field strength and the RF power are above a certain threshold. Moreover, the power must be supplied to wave-launching antenna to generate magnetic field perturbations within the plasma, by an impedance matching network to maximize the power transfer. A radiofrequency (RF) antenna is usually employed to excite helicon waves and drive the helicon driven plasma discharge.

12.6.3 General Dispersion Relations

To better understand the dispersion relations and to obtain the general dispersion relations, consider Maxwell's equations for the magnetic induction in free space, which are also known as Ampere's and Faraday's laws as

$$\nabla\times\mathbf{B} = \mu_0\mathbf{j} + \frac{1}{c^2}\frac{\partial\mathbf{E}}{\partial \mathrm{t}}, \nabla\times\mathbf{E} = -\frac{\partial\mathbf{B}}{\partial t}. \tag{12.78}$$

We assume that the medium is plasma which is both infinite and uniform. Wave-like solutions are sought for all variables of the form

$$\mathbf{j}(\mathbf{x}, t) = \mathbf{j}_0 \exp(i(\mathbf{k}\cdot\mathbf{x} - \omega t)). \tag{12.79}$$

Hence,

$$i\mathbf{k}\times\mathbf{B}=\mu_0\mathbf{j}-\left(i\omega/c^2\right)\mathbf{E},\, i\mathbf{k}\times\mathbf{E}=i\omega\mathbf{B}. \tag{12.80}$$

From the second of the above pair,

$$i\mathbf{k}\times(i\mathbf{k}\times\mathbf{E})=-\omega\mathbf{k}\times\mathbf{B}=i\omega\mu_0\mathbf{j}+(\omega/c)^2\mathbf{E}. \tag{12.81}$$

The most general linear relationship $\mathbf{j}$ and $\mathbf{E}$ can be expressed as $\mathbf{j}=\sigma\cdot\mathbf{E}$ where σ is the conductivity tensor (a general form of Ohm's law). Hence,

$$-\mathbf{k}\times(\mathbf{k}\times\mathbf{E})=i\omega\mu_0\boldsymbol{\sigma}\cdot\mathbf{E}+(\omega/c)^2\mathbf{E}=\left(i\omega\mu_0\boldsymbol{\sigma}+(\omega/c)^2\mathbf{I}\right)\cdot\mathbf{E}. \tag{12.82}$$

In terms of the dielectric tensor, Ampere's law is

$$\frac{1}{\mu_0}\nabla\times\mathbf{B}=\mathbf{j}+\varepsilon_0\frac{\partial\mathbf{E}}{\partial t}=\boldsymbol{\sigma}\cdot\mathbf{E}+\varepsilon_0\frac{\partial\mathbf{E}}{\partial t}=\boldsymbol{\varepsilon}\varepsilon_0\frac{\partial\mathbf{E}}{\partial t}. \tag{12.83}$$

Thus, one can obtain

$$(\sigma-i\omega\varepsilon_0\mathbf{I})\cdot\mathbf{E}=-\boldsymbol{\varepsilon}i\omega\varepsilon_0\mathbf{E},\,\boldsymbol{\varepsilon}=\mathbf{I}-\boldsymbol{\sigma}/i\omega\varepsilon_0,\,\boldsymbol{\sigma}=i\omega\varepsilon_0(\mathbf{I}-\boldsymbol{\varepsilon}). \tag{12.84}$$

Hence,

$$\mathbf{k}\times(\mathbf{k}\times\mathbf{E})+\left(-(\omega/c)^2(\mathbf{I}-\boldsymbol{\varepsilon})+(\omega/c)^2\mathbf{I}\right)\cdot\mathbf{E}=\mathbf{k}\times(\mathbf{k}\times\mathbf{E})+(\omega/c)^2\boldsymbol{\varepsilon}\cdot\mathbf{E}=\mathbf{0}. \tag{12.85}$$

Recall that $\mathbf{k}\times(\mathbf{k}\times)=\mathbf{k}(\mathbf{k}\cdot)-|\mathbf{k}|^2=\mathbf{kk}-|\mathbf{k}|^2$. Thus,

$$\mathbf{k}\times(\mathbf{k}\times\mathbf{E})+(\omega/c)^2\boldsymbol{\varepsilon}\cdot\mathbf{E}=\left(\mathbf{kk}-|\mathbf{k}|^2+(\omega/c)^2\boldsymbol{\varepsilon}\right)\cdot\mathbf{E}=\mathbf{0}. \tag{12.86}$$

This is a matrix equation, i.e. three simultaneous homogeneous equations for $\mathbf{E}$. Hence, in order that there exists a non-trivial solution,

$$\det\left|\mathbf{kk}-|\mathbf{k}|^2+(\omega/c)^2\boldsymbol{\varepsilon}\right|=0. \tag{12.87}$$

This provides the general dispersion relation for $\mathbf{k}$ which represents the direction of wave propagation.

Considering an isotropic plasma $\boldsymbol{\varepsilon}=\varepsilon\mathrm{I}$ and assuming $\mathbf{k}$ is in the z direction,

$$\det\left|\mathbf{kk}-|\mathbf{k}|^2+(\omega/c)^2\boldsymbol{\varepsilon}\right|=\mathbf{det}\begin{vmatrix}(\omega/c)^2\varepsilon-k^2 & 0 & 0\\ 0 & (\omega/c)^2\varepsilon-k^2 & 0\\ 0 & 0 & (\omega/c)^2\varepsilon\end{vmatrix}=0. \tag{12.88}$$

There are two cases to consider. Let $\mathbf{E} = [E_x \ E_y \ E_z]^T$. When the electric field is transverse to the direction of wave propagation, i.e. if

$$\mathrm{E}_z = 0, \left(\mathbf{kk} - |\mathbf{k}|^2 + (\omega/c)^2\boldsymbol{\varepsilon}\right) \cdot \mathbf{E}$$

$$= \begin{bmatrix} (\omega/c)^2\varepsilon - k^2 & 0 & 0 \\ 0 & (\omega/c)^2\varepsilon - k^2 & 0 \\ 0 & 0 & (\omega/c)^2\varepsilon \end{bmatrix} \begin{bmatrix} E_x \\ E_y \\ 0 \end{bmatrix} = \mathbf{0}. \qquad (12.89)$$

The dispersion relation reduces to

$$\left((\omega/c)^2\varepsilon - k^2\right)^2 = 0. \qquad (12.90)$$

If the electric field is directed towards the z direction, $E_z \neq 0$,, $(\omega^2/c^2)\varepsilon = 0 \Rightarrow \varepsilon = 0$, which is a feasible condition in plasmas.

12.6.4 The Case of Cold Plasma

Consider initially a cold plasma ($\nabla p = 0$) or $T = 0$,, with no applied magnetic field ($\mathbf{B}_0 \equiv 0$) but including an applied electric field. The momentum equation is, considering only electron motions,

$$n_e m_e (d\mathbf{v}/dt + \mathbf{v}\nabla\mathbf{v}) = \nabla p + q n_e (\mathbf{E} + \mathbf{v} \times \mathbf{B}_0) - m_e n_e \nu \mathbf{v} \approx n_e m_e d\mathbf{v}/dt = q n_e \mathbf{E}. \qquad (12.91)$$

Thus,

$$m_e d\mathbf{v}/dt = q\mathbf{E} \Rightarrow \mathbf{v} = -q\mathbf{E}/i\omega m_e. \qquad (12.92)$$

It follows that

$$\mathbf{j} = n_e e\mathbf{v} = -n_e e^2\mathbf{E}/i\omega m_e \Rightarrow \boldsymbol{\sigma} = i n_e e^2/\omega m_e, \boldsymbol{\varepsilon} = 1 - \boldsymbol{\sigma}/i\omega\varepsilon_0 = 1 - n_e e^2/\omega^2\varepsilon_0 m_e. \qquad (12.93)$$

If the electric field is directed towards the z direction, from the dispersion relation,

$$\varepsilon = 0, n_e e^2/\omega^2\varepsilon_0 m_e = 1 \Rightarrow \omega^2 = n_e e^2/\varepsilon_0 m_e. \qquad (12.94)$$

This is the plasma frequency for electrons:

$$\omega^2 = \omega_{pe}^2 = n_e e^2 / \varepsilon_0 m_e. \tag{12.95}$$

If one includes ion motions as well,

$$\begin{aligned}\mathbf{j} = \sum_j n_j q_j \mathbf{v}_j, j = e, i, \boldsymbol{\varepsilon} = 1 - \frac{\omega_{pe}^2 + \omega_{pi}^2}{\omega^2} = 1 - \frac{n_e q_e^2}{\omega^2 \varepsilon_0 m_e} - \frac{n_i q_i^2}{\omega^2 \varepsilon_0 m_i} \cdot \omega^2 \\ = \omega_p^2 = \omega_{pe}^2 + \omega_{pi}^2 = \frac{n_e q_e^2}{\varepsilon_0 m_e} + \frac{n_i q_i^2}{\varepsilon_0 m_i}.\end{aligned} \tag{12.96}$$

The charge on the electron is denoted as q_e to distinguish it from the charge on the ion q_i. When the electric field is transverse to the direction of wave propagation, i.e. if, $E_z = 0$,

$$\begin{aligned}(\omega/c)^2 \varepsilon - k^2 &= (\omega/c)^2 \left(1 - \frac{n_e q_e^2}{\omega^2 \varepsilon_0 m_e} - \frac{n_i q_i^2}{\omega^2 \varepsilon_0 m_i}\right) - k^2 \\ &= (\omega/c)^2 \left(1 - \frac{\omega_p^2}{\omega^2}\right) - k^2 = 0, \\ &\Rightarrow \omega^2 = \omega_p^2 + c^2 k^2.\end{aligned} \tag{12.97}$$

Assuming $\boldsymbol{k}$ is in the z direction, but considering the plasma to be anisotropic,

$$\begin{aligned}\det\left|\mathbf{kk} - |\mathbf{k}|^2 + (\omega/c)^2 \boldsymbol{\varepsilon}\right| = \mathbf{det}\left| -k^2 \begin{bmatrix} 1 & 0 & 0 \\ 0 & 1 & 0 \\ 0 & 0 & 0 \end{bmatrix} + (\omega/c)^2 \boldsymbol{\varepsilon} \right| \\ = (\omega/c)^4 \mathbf{det}\left| \boldsymbol{\varepsilon} - \frac{c^2 k^2}{\omega^2} \begin{bmatrix} 1 & 0 & 0 \\ 0 & 1 & 0 \\ 0 & 0 & 0 \end{bmatrix} \right| = 0.\end{aligned} \tag{12.98}$$

The dielectric tensor for cold plasmas is Hermitian. This means there is no dissipation.

12.6.5 Application to Magnetized Plasma with Two Species

One is primarily interested in the application to waves in magnetised plasmas with two species. The momentum equation is, considering only electron motions, is

$$n_e m_e(d\mathbf{v}/dt + \mathbf{v}\nabla\mathbf{v}) = \nabla p + qn_e(\mathbf{E} + \mathbf{v}\times\mathbf{B}_0) - m_e n_e \nu\mathbf{v} \approx n_e m_e d\mathbf{v}/dt$$
$$= qn_e(\mathbf{E} + \mathbf{v}\times\mathbf{B}_0). \tag{12.99}$$

Considering either an electron or an ion and a unidirectional magnetic field,

$$m\frac{d\mathbf{v}}{dt} = q(\mathbf{E}(x,y,z,t) + \mathbf{v}\times\mathbf{B}(x,y,z,t)).\mathbf{V}$$
$$= \begin{bmatrix} 0 & -v_z & v_y \\ v_z & 0 & -v_x \\ -v_y & v_x & 0 \end{bmatrix}, m\frac{d\mathbf{v}}{dt} = q\mathbf{E}(x,y,z,t) \tag{12.100}$$
$$+q\mathbf{V}\mathbf{B}(x,y,z,t), m\frac{d\mathbf{v}}{dt} = q\mathbf{E}(x,y,z,t) + q\mathbf{V}[0 \quad 0 \quad B_0]^T.$$

Thus, in the frequency domain, with $\omega_c = qB_0/m$, $i = \sqrt{-1}$,

$$-i\omega m\begin{bmatrix} v_x \\ v_y \\ v_z \end{bmatrix} - qB_0\begin{bmatrix} v_y \\ -v_x \\ 0 \end{bmatrix} = q\begin{bmatrix} E_x \\ E_y \\ E_z \end{bmatrix}, -i\omega m\begin{bmatrix} v_x \\ v_y \\ v_z \end{bmatrix}$$
$$-m\omega_c\begin{bmatrix} 0 & 1 & 0 \\ -1 & 0 & 0 \\ 0 & 0 & 0 \end{bmatrix}\begin{bmatrix} v_x \\ v_y \\ v_z \end{bmatrix} = q\begin{bmatrix} E_x \\ E_y \\ E_z \end{bmatrix}. \tag{12.101}$$

Solving for the particle velocity vector,

$$\begin{bmatrix} v_x \\ v_y \\ v_z \end{bmatrix} = \frac{-q}{i\omega m\left((i\omega)^2 + \omega_c^2\right)}\begin{bmatrix} (i\omega)^2 & -i\omega\omega_c & 0 \\ i\omega\omega_c & (i\omega)^2 & 0 \\ 0 & 0 & -i\left((i\omega)^2 + \omega_c^2\right) \end{bmatrix}\begin{bmatrix} E_x \\ E_y \\ E_z \end{bmatrix}. \tag{12.102}$$

Since, $\mathbf{j} = \sum_j n_j q_j \mathbf{v}_j$ and $\mathbf{j} = \boldsymbol{\sigma}\cdot\mathbf{E}$, it follows that

$$\boldsymbol{\sigma}_j = \frac{-n_j q_j^2}{-i\omega m_j\left(-(i\omega)^2 - \omega_{cj}^2\right)}\begin{bmatrix} (i\omega)^2 & -i\omega\omega_{cj} & 0 \\ i\omega\omega_{cj} & (i\omega)^2 & 0 \\ 0 & 0 & \left((i\omega)^2 + \omega_{cj}^2\right) \end{bmatrix}, \tag{12.103}$$

where the subscript $j = e$ for an electron or $j = i$ for an ion. In the above, it may be noted that in the case of an ion, the sign of the charge is different. If one includes the collision frequency, ν_j, with $\overline{\omega}_{cj} = \omega_{cj}/(\nu_j - i\omega)$, the conductivity tensor is

$$\boldsymbol{\sigma}_j = \frac{-n_j q_j^2 (\nu_j - i\omega)^2}{(\nu_j - i\omega) m_j \left(-(\nu_j - i\omega)^2 - \omega_{cj}^2\right)} \begin{bmatrix} 1 & \overline{\omega}_{cj} & 0 \\ -\overline{\omega}_{cj} & 1 & 0 \\ 0 & 0 & 1 + \overline{\omega}_{cj}^2 \end{bmatrix}. \tag{12.104}$$

The total conductivity, due to all species, is the sum of the conductivities for each. Hence, since, $\boldsymbol{\sigma} = \sum \boldsymbol{\sigma}_j$, and with

$$\omega_{pj}^2 = n_j q_j^2 / (\varepsilon_0 m_j), \boldsymbol{\varepsilon} = \mathbf{I} + \boldsymbol{\sigma}/(-i\omega)\varepsilon_0, \tag{12.105}$$

$$\boldsymbol{\varepsilon} = \mathbf{I} + \sum_j \frac{\omega_{pj}^2}{(-i\omega)(\nu_j - i\omega)} \left(1 + \frac{\omega_{cj}^2}{(\nu_j - i\omega)^2}\right)^{-1} \begin{bmatrix} 1 & \overline{\omega}_{cj} & 0 \\ -\overline{\omega}_{cj} & 1 & 0 \\ 0 & 0 & 1 + \overline{\omega}_{cj}^2 \end{bmatrix}. \tag{12.106}$$

The above equation is expressed as

$$\boldsymbol{\varepsilon} = \begin{bmatrix} S & -iD & 0 \\ iD & S & 0 \\ 0 & 0 & P \end{bmatrix}, S = 1 + \sum_j \frac{\omega_{pj}^2/(-i\omega)(\nu_j - i\omega)}{\left(1 + \omega_{cj}^2/(\nu_j - i\omega)^2\right)}, D = -\sum_j \frac{\omega_{pj}^2 \omega_{cj}/\omega(\nu_j - i\omega)^2}{\left(1 + \omega_{cj}^2/(\nu_j - i\omega)^2\right)}, P = 1 - \sum_j \frac{\omega_{pj}^2}{i\omega(\nu_j - i\omega)}. \tag{12.107}$$

With $\nu_j = 0$,

$$S = 1 - \Sigma \frac{\omega_{pj}^2/\omega^2}{\left(1 - \omega_{cj}^2/\omega^2\right)}, D = \sum \frac{\omega_{pj}^2 \omega_{cj}/\omega^3}{\left(1 - \omega_{cj}^2/\omega^2\right)}, P = 1 - \sum \frac{\omega_{pj}^2}{\omega^2}. \tag{12.108}$$

It follows from Eq. (12.98) that

$$\left| \boldsymbol{\varepsilon} - \frac{c^2 k^2}{\omega^2} \begin{bmatrix} 1 & 0 & 0 \\ 0 & 1 & 0 \\ 0 & 0 & 0 \end{bmatrix} \right| = \left| \begin{bmatrix} S & -iD & 0 \\ iD & S & 0 \\ 0 & 0 & P \end{bmatrix} - \frac{c^2 k^2}{\omega^2} \begin{bmatrix} 1 & 0 & 0 \\ 0 & 1 & 0 \\ 0 & 0 & 0 \end{bmatrix} \right| = 0, \tag{12.109}$$

which reduces to

$$P\left(\left(S - (ck/\omega)^2\right)^2 - D^2\right) = 0. \tag{12.110}$$

For an electric field in the z direction, this reduces to

$$P=0\Rightarrow\sum_j\frac{\omega_{pj}^2}{\left(\nu_j-i\omega\right)^2}=0 \tag{12.111}$$

For a transverse electric field, with $\nu_j=0$,

$$\begin{aligned}\left(S-\frac{c^2k^2}{\omega^2}\right)^2=D^2\Rightarrow\left(\mathbf{1}-\frac{c^2k^2}{\omega^2}-\sum\frac{\omega_{pj}^2/\omega^2}{\left(1-\omega_{cj}^2/\omega^2\right)}\right)^2\\=\left(\sum\frac{\left(\omega_{pj}^2/\omega^2\right)\left(\omega_{cj}/\omega\right)}{\left(1-\omega_{cj}^2/\omega^2\right)}\right)^2.\end{aligned} \tag{12.112}$$

To understand this case better, consider the general case, with

$$\begin{aligned}\mathbf{k}\times=\begin{bmatrix}0&-k_z&k_y\\k_z&0&-k_x\\-k_y&k_x&0\end{bmatrix},\mathbf{k}\times\mathbf{k}\times=\begin{bmatrix}0&-k_z&k_y\\k_z&0&-k_x\\-k_y&k_x&0\end{bmatrix}^2\\=\begin{bmatrix}-k_z^2-k_y^2&k_xk_y&k_xk_z\\k_xk_y&-k_z^2-k_x^2&k_zk_y\\k_xk_z&k_zk_y&-k_x^2-k_y^2\end{bmatrix}.\end{aligned} \tag{12.113}$$

Thus,

$$\mathbf{k}\times(\mathbf{k}\times\mathbf{E})+(\omega/c)^2\boldsymbol{\varepsilon}\cdot\mathbf{E}=\left(\mathbf{kk}-|\mathbf{k}|^2+(\omega/c)^2\boldsymbol{\varepsilon}\right)\cdot\mathbf{E}=\mathbf{0}. \tag{12.114}$$

In cylindrical coordinates, the above relation is

$$\left(\begin{bmatrix}-k_\theta^2-k_z^2&k_rk_\theta&k_rk_z\\k_rk_\theta&-k_r^2-k_z^2&k_\theta k_z\\k_rk_z&k_\theta k_z&-k_r^2-k_\theta^2\end{bmatrix}+\frac{\omega^2}{c^2}\begin{bmatrix}S&-iD&0\\iD&S&0\\0&0&P\end{bmatrix}\right)\begin{bmatrix}E_x\\E_y\\E_z\end{bmatrix}=\begin{bmatrix}0\\0\\0\end{bmatrix}. \tag{12.115}$$

Let

$$k_r^2 + k_z^2 = k^2, k_z = k\cos\phi, k_r = k\sin\phi; k_r k_\theta = k_z k_\theta = k_\theta = 0; \tag{12.116}$$

then, it follows that

$$\left(\frac{k^2c^2}{\omega^2}\begin{bmatrix} -\cos^2\phi & 0 & \cos\phi\sin\phi \\ 0 & -1 & 0 \\ \cos\phi\sin\phi & 0 & -\sin^2\phi \end{bmatrix} + \begin{bmatrix} S & -iD & 0 \\ iD & S & 0 \\ 0 & 0 & P \end{bmatrix}\right)\begin{bmatrix} E_x \\ E_y \\ E_z \end{bmatrix} = \begin{bmatrix} 0 \\ 0 \\ 0 \end{bmatrix}. \tag{12.117}$$

12.6.6 *The Case of Helicon Waves*

Let $N = kc/\omega$; then,

$$\begin{aligned}&\left(N^2\cos^2\phi - S\right)\left(N^2 - S\right)\left(N^2\sin^2\phi - P\right) - D^2\left(N^2\sin^2\phi - P\right)\\&\quad - \left(N^2\cos\phi\sin\phi\right)^2\left(N^2 - S\right) = 0,\end{aligned} \tag{12.118}$$

which reduces to

$$N^4\left(S\sin^2\phi + P\cos^2\phi\right) - N^2\left(\left(S^2 - D^2\right)\sin^2\phi + PS\left(1 + \cos^2\phi\right)\right) + P\left(S^2 - D^2\right) = 0. \tag{12.119}$$

The above can be expressed as $N^4A - N^2B + PC = 0$, which has no wave solution as long as $N^2 > 0$. Two cases to consider: $\sin\phi = 0$ and $\cos\phi = 0$. When $\sin\phi = 0$,

$$\begin{aligned}&P\left(N^4 - 2SN^2 + \left(S^2 - D^2\right)\right) = 0 \Rightarrow P\left(\left(S - N^2\right)^2 - D^2\right) = 0.\\&\qquad\Rightarrow P = 0, N^2 = S \pm D.\end{aligned} \tag{12.120}$$

The latter relation reduces to

$$N^2 = 1 - \sum \frac{\omega_{pj}^2/\omega^2}{\left(1 \pm \omega_{cj}/\omega\right)} \simeq 1 - \frac{\omega_{pe}^2/\omega^2}{\left(1 - \omega_{ce}/\omega\right)} - \frac{\omega_{pi}^2/\omega^2}{\left(1 + \omega_{ci}/\omega\right)} = \frac{k^2c^2}{\omega^2}. \tag{12.121}$$

This case corresponds to *helicon or whistler waves*.

Helicon waves can be right-hand (RH) or left-hand (LH) polarized, while whistler waves can only be RH polarized. From the boundary conditions discussed earlier, it can be observed that both the current and the magnetic field vectors are parallel to each other. Thus, when both are RH polarized, both vectors are directed in the same direction, and the polarization propagates in one direction. However, on the LH case, the current and the magnetic field vectors are not directed in the same direction. Yet, helicon waves are bounded due to the interaction between the radiation, the wall and the plasma.

When $\cos\phi = 0$,

$$\begin{aligned} &N^4 S - N^2\left(\left(S^2 - D^2\right) + PS\right) + P\left(S^2 - D^2\right) = 0; \\ &\Rightarrow \left(N^2 - P\right)\left(N^2 S - \left(S^2 - D^2\right)\right) = 0. \end{aligned} \tag{12.122}$$

This case corresponds to *upper and lower hybrid resonances*. The lower hybrid frequency ω_{lh} is a resonant frequency between the electron cyclotron frequency ω_{ce} and the ion cyclotron frequency ω_{ci} and has the form

$$\omega_{lh} = \left(\frac{1}{\omega_{ce}\omega_{ci}} + \frac{1}{\omega_{pi}^2 + \omega_{ci}^2}\right)^{-1/2} \approx \sqrt{\omega_{ce}\omega_{ci}} = \frac{eB_0}{\sqrt{m_e m_i}}. \tag{12.123}$$

The ion motions have a significant effect nearer to the lower hybrid resonance frequencies and cannot be neglected. As one approaches the lower hybrid frequency, the wavenumber becomes infinite due to the effect of the resonance. Furthermore, the plasma densities are known to reach a maximum slightly below the lower hybrid resonance frequencies.

An alternate way to parameterize is to write

$$N^2 = \frac{k^2 c^2}{\omega_{pe}^2}\frac{\omega_{pe}^2}{\omega^2} \equiv \widehat{N}^2 \frac{\omega_{pe}^2}{\omega^2}, \widehat{N}^2 = \frac{k^2 c^2}{\omega_{pe}^2}, \text{and } \left[\widehat{S} \quad \widehat{D} \quad \widehat{P}\right] \equiv \left[S \quad D \quad P\right]\frac{\omega^2}{\omega_{pe}^2}. \tag{12.124}$$

Then,

$$\begin{aligned} &\widehat{N}^4\left(\widehat{S}\sin^2\phi + \widehat{P}\cos^2\phi\right) - \widehat{N}^2\left(\left(\widehat{S}^2 - \widehat{D}^2\right)\sin^2\phi + \widehat{PS}\left(1 + \cos^2\phi\right)\right) \\ &\quad + \widehat{P}\left(\widehat{S}^2 - \widehat{D}^2\right) = 0. \end{aligned} \tag{12.125}$$

Neglecting the ion frequencies, the solution for $\widehat{N}$ may be expressed as

$$\widehat{N} = \frac{kc}{\omega_{pe}} = \frac{\omega}{\omega_{ce}|\cos\phi| - (\omega + i\nu_e)}. \tag{12.126}$$

Hence, waves propagate only for the ϕ angles where the refractive index is dominantly real. It is possible to identify the low ϕ angle waves (long wavelength) as helicon modes and the high ϕ angle waves (short wavelength) as Trivelpiece–Gould (TG) modes.

12.6.7 Thrust from Plasma Exhaust

The motion of a particle within the plasma is described by

$$mn\frac{d\mathbf{v}}{dt} = qn(\mathbf{E}(r,\theta,z,t) + \mathbf{v}\times\mathbf{B}(r,\theta,z,t)) - \nabla\cdot p - mn\nu\left(\mathbf{v}-\mathbf{v}_{eq}\right), \mathbf{v}\times\mathbf{B}(r,\theta,z,t) = \mathbf{V}\mathbf{B}(r,\theta,z,t). \tag{12.127}$$

When

$$\mathbf{B}(r,\theta,z,t) = \begin{bmatrix} B_r \\ B_\theta \\ B_z \end{bmatrix}, -\mathbf{v}\times\mathbf{B} = \mathbf{V}^T\mathbf{B} = \begin{bmatrix} 0 & -v_z & v_\theta \\ v_z & 0 & -v_r \\ -v_\theta & v_r & 0 \end{bmatrix}^T \begin{bmatrix} B_r \\ B_\theta \\ B_z \end{bmatrix}$$
$$= \mathbf{B}\times\mathbf{v}, \mathbf{V}^T\mathbf{B} = \begin{bmatrix} 0 & -B_z & B_\theta \\ B_z & 0 & -B_r \\ -B_\theta & B_r & 0 \end{bmatrix} \begin{bmatrix} v_r \\ v_\theta \\ v_z \end{bmatrix}. \tag{12.128}$$

In a steady state with $d\mathbf{v}/dt = 0$ and $\mathbf{v}_{eq} = \mathbf{0}$,

$$\mathbf{v} = \frac{q}{m\nu}\mathbf{E} - \frac{1}{mn\nu}\nabla\cdot p + \frac{q}{m\nu}\mathbf{v}\times\mathbf{B} \simeq \frac{q}{m\nu}\mathbf{E} - \frac{1}{mn\nu}\nabla\cdot p + \left(\frac{q}{m\nu}\right)^2\left(\mathbf{E}\times\mathbf{B} - \frac{1}{nq}(\nabla\cdot p)\times\mathbf{B}\right), \tag{12.129}$$

or

$$\mathbf{E}(r,\theta,z,t) - \frac{1}{nq}\nabla\cdot p = \frac{m\nu}{q}\mathbf{v} - \mathbf{v}\times\mathbf{B}(r,\theta,z,t) = \left(\frac{m\nu}{q}\mathbf{I} + \begin{bmatrix} 0 & -B_z & B_\theta \\ B_z & 0 & -B_r \\ -B_\theta & B_r & 0 \end{bmatrix}\right)\mathbf{v},$$

$$\mathbf{v} = \left(\frac{m\nu}{q}\mathbf{I} + \begin{bmatrix} 0 & -B_z & B_\theta \\ B_z & 0 & -B_r \\ -B_\theta & B_r & 0 \end{bmatrix}\right)^{-1}\left(\mathbf{E}(r,\theta,z,t) - \frac{1}{nq}\nabla\cdot p\right). \tag{12.130}$$

The pressure is given by the isothermal relation $\nabla p = kT\,\nabla\, n$ and the species particle number density is given by n. Conservation of particles and/or charges in the plasma is described by the continuity equation:

$$\frac{\partial n}{\partial t}+\nabla\cdot(n\mathbf{v})=\frac{\partial n}{\partial t}+\nabla\cdot\boldsymbol{\Gamma}=\dot{n}_s. \tag{12.131}$$

Hence,

$$\boldsymbol{\Gamma}=n\mathbf{v}\approx n\left(\frac{q}{m\nu}\mathbf{E}(r,\theta,z,t)-\frac{1}{mn\nu}\nabla\cdot p+\left(\frac{q}{m\nu}\right)^2\left(\mathbf{E}\times\mathbf{B}-\frac{1}{nq}(\nabla\cdot p)\times\mathbf{B}\right)\right),$$

$$\boldsymbol{\Gamma}=n\mathbf{v}\approx\frac{nq}{m\nu}\left(\mathbf{I}+\left(\frac{q}{m\nu}\right)\begin{bmatrix}0 & -B_z & B_\theta\\ B_z & 0 & -B_r\\ -B_\theta & B_r & 0\end{bmatrix}\right)^{-1}\left(\mathbf{E}(r,\theta,z,t)-\frac{1}{nq}\nabla\cdot p\right). \tag{12.132}$$

The term on the right-hand side represents the time-dependent source or sink term for the species being considered. With the drift diffusion approximation, the species flux vector reduces to

$$\boldsymbol{\Gamma}=\pm n\boldsymbol{\mu}\cdot\mathbf{E}-\boldsymbol{D}\cdot\nabla n. \tag{12.133}$$

A special case of this is called Fick's law, in which the flux of particles for either the electric field or the mobility term being zero is given by

$$\boldsymbol{\Gamma}=-\boldsymbol{D}\cdot\nabla n. \tag{12.134}$$

If one includes the electron energy change caused by electric field, the losses in particle numbers due to collisions and the RF power input, one has for the continuity equation

$$\frac{\partial n}{\partial t}+\nabla\cdot\boldsymbol{\Gamma}+\boldsymbol{\Gamma}\cdot\mathbf{E}=S_c+Q_{RF}/e. \tag{12.135}$$

Furthermore, assuming that the electrons have a Maxwellian distribution in the velocity phase space, the species mobility $\boldsymbol{\mu}$ is a tensor and when a magnetic field is applied, it may be expressed as

$$\boldsymbol{\mu}^{-1}=\frac{m\nu}{q}\mathbf{I}+\begin{bmatrix}0 & -B_z & 0\\ B_z & 0 & 0\\ 0 & 0 & 0\end{bmatrix}. \tag{12.136}$$

The conservation of particles and/or charges in the plasma is described by the continuity equation:

$$\frac{\partial n}{\partial t}+\nabla\cdot\boldsymbol{\Gamma}+\boldsymbol{\Gamma}\cdot\mathbf{E}=S_c+Q_{RF}/e. \tag{12.137}$$

The thrust T is equal in magnitude and opposite in direction to momentum flux exhausted from the system. It could be derived by applying Newton's third law and from the momentum conservation law which results in the reaction thrust being equal to the action of the momentum flux. Thus the total momentum flux of the fluid across a cross section area must be equal to the total pressure integrated over the cross section area A. The integral of the static and dynamic pressure or the momentum flux per unit cross-section in the direction of the exhaust gives

$$T = \int_A \left(p_e + m_i n_i v_{i,z}^2\right) dA \equiv \int_A \tau dA. \tag{12.138}$$

The axial momentum flux is essentially derived from momentum equations of the plasma with two species, the ions and electrons. Assuming that the continuity equation is already satisfied, the momentum equation for each particle species of electrons $k = e$ or ions $k = i$ in a steady state is given by

$$m_k n_k (\mathbf{v}_k \cdot \nabla)\mathbf{v}_k = q_k n_k (\mathbf{E}(r, \theta, z, t) + \mathbf{v}_k \times \mathbf{B}(r, \theta, z, t)) - \nabla p_k. \tag{12.139}$$

In the above, m_k, n_k, $\mathbf{v}_k$, q_k and p_k are the mass, density, velocity, charge and pressure of the particle species k, respectively, while $\mathbf{E}(r, \theta, z, t)$ and $\mathbf{B}(r, \theta, z, t)$ are the electric field and magnetic field vectors, respectively. In component form for a quasi-neutral plasma (so $n_p = n_i = n_e$) with both the ion temperature and the electron mass neglected, for an axisymmetric case in cylindrical coordinates, with radial, azimuthal and axial components, the momentum flux equations are

$$\begin{aligned} &qn_p(E_r + v_{e,\theta}B_z) = -\partial p_e/\partial r, qn_p(E_z - v_{e,\theta}B_r) \\ &= -\partial p_e/\partial z, qn_p(E_r + v_{i,\theta}B_z) = 0, qn_p(E_z - v_{i,\theta}B_r) \\ &= (1/r)\partial\left(rm_i n_p v_{i,r} v_{i,z}\right)/\partial r + \partial\left(m_i n_p v_{i,z}^2\right)/\partial z, \end{aligned} \tag{12.140}$$

By eliminating the electric field component E_z from the second and fourth equations in Eq. (12.140),

$$\begin{aligned} E_z &= -\left(qn_p\right)^{-1}\partial p_e/\partial z + v_{e,\theta}B_r, \quad \partial\left(p_e + m_i n_p v_{i,z}^2\right)/\partial z \\ &\equiv \partial\tau/\partial z = qn_p B_r(v_{e,\theta} - v_{i,\theta}) - (1/r)\partial\left(rm_i n_p v_{i,r} v_{i,z}\right)/\partial r. \end{aligned} \tag{12.141}$$

From the first and third equations in Eq. (12.140),

$$\partial\left(p_e + m_i n_p v_{i,z}^2\right)/\partial z \equiv \partial\tau/\partial z = -(B_r/B_z)\partial p_e/\partial r - (1/r)\partial\left(rm_i n_p v_{i,r} v_{i,z}\right)/\partial r. \tag{12.142}$$

Integrating Eq. (12.142) and including a constant of integration yields the thrust equation (Takahashi, 2019), due to the initial flux, the Lorentz force and the axial momentum flux, and is given by

$$T = 2\pi \int_{r_{source}} r p_{e0} dr + 2\pi \int_{z} \int_{r_p(z)} r \partial \tau / \partial z dr dz \equiv T_0 + T_d, \tag{12.143}$$

where

$$T_d = -2\pi \int_{z} \int_{r_p(z)} (r(B_r/B_z)\partial p_e/\partial r) dr dz - 2\pi \int_{z} \int_{r_p(z)} \left(\partial \left(r m_i n_p v_{i,r} v_{i,z}\right)/\partial r\right) dr dz. \tag{12.144}$$

The total thrust is evaluated numerically.

12.6.8 Antennas for Helicon Waves

In helicon-coupled wave driver, the geometry of an antenna is chosen in a way such that the magnetic fields induced by the antenna current launch plasma waves in the bulk plasma volume which are aided by a weak directional DC magnetic field, to generate the waves at specific wave speeds. The main process by which the plasma gains energy is by the dominant collision mechanisms, as the waves propagate through the plasma, which dissipates energy in the form of heat. Depending on the architecture of the antenna, it can excite the $m = 0$ (single-loop antenna) or the $m = \pm 1$ modes (double saddle type, Nagoya II type or Shoji type antennas), with varying degrees of power transfer. The antennas are illustrated in Fig. 12.6. It appears that generally the $m = 1$ couples well with the plasma as well as transmits a larger share to the power to the plasma. In addition to exciting idealized helicon modes along the axis of wave propagation, helicon antennas also excite the electrostatic slower wave modes known as Trivelpiece–Gould (TG) modes. The key question that arises is how one could tailor a particular antenna to propagate a specific mode.

For the purpose of tailoring an antenna to generate a specific mode, the Birdcage family of antennas developed for magnetic resonance imaging and nuclear magnetic resonance applications are particularly suitable. These are illustrated in Fig. 12.7. The antenna is composed of N identical legs equally distributed on a cylinder of radius R. Each leg is connected at both ends to its two closest neighbours by means of capacitors, usually of equal capacitance C. In its simplest design, the RF input is connected directly across a capacitor in the loop as shown in Fig. 12.8. The birdcage antenna is very efficient for high-density 10^{18}–10^{19} m^{-3} plasma production (in noble gases, typically argon) with moderate injection power at low pressure.

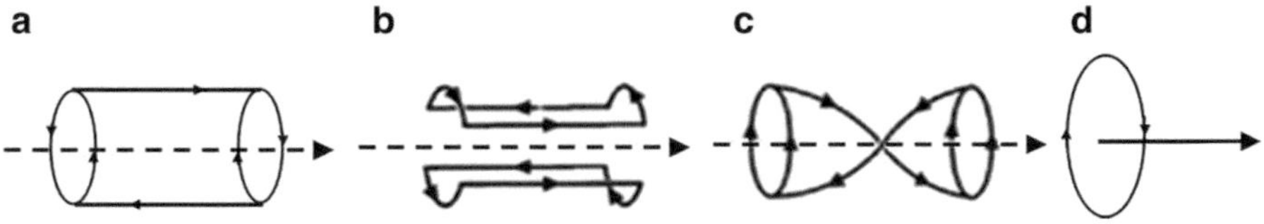

Fig. 12.6 (**a**) Plane-polarized Nagoya III consisting of two horizontal legs connecting two vertical loops; (**b**) Boswell's double saddle with two loops located on either side of a cylindrical surface; (**c**) twisted left-handed half-turn helical Nagoya or Shoji antenna, consisting of two helical legs connecting two loops; and (**d**) single-loop antenna (adapted from Tian (2017))

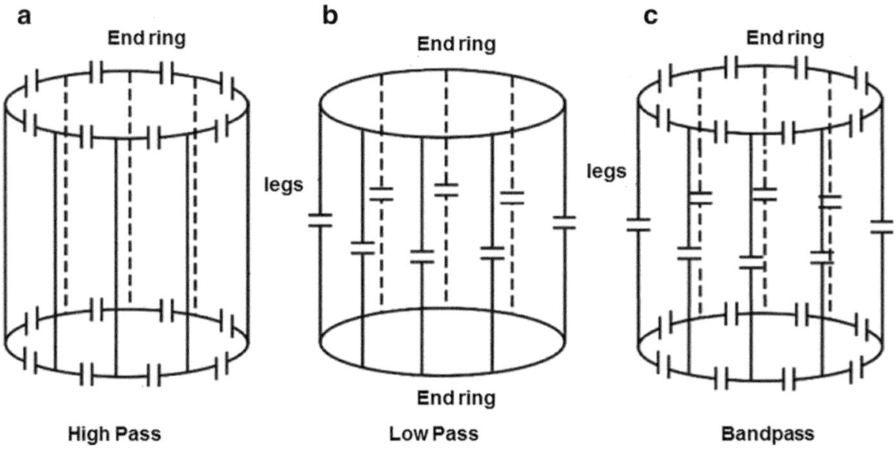

Fig. 12.7 Birdcage antenna: high-pass, low-pass and band-pass (adapted from Romano et al. (2020))

Fig. 12.8 Four-leg birdcage antenna example which serves as a high-pass antenna (adapted from Guittienne et al. (2005))

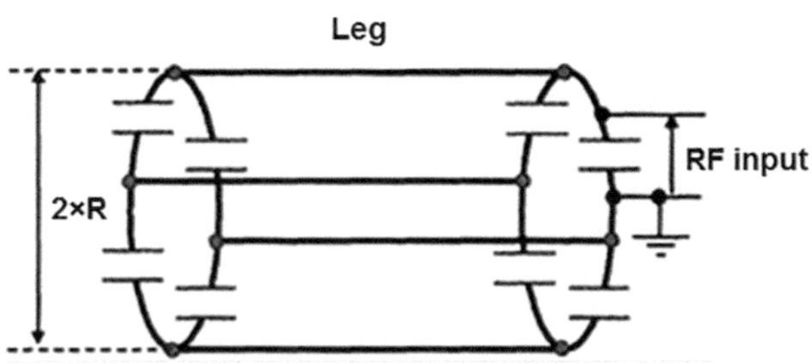

Following Tarakameh (2020), the equivalent circuit of the birdcage antenna may be represented by an equivalent back-to-front connected ladder network, similar to the one shown in Fig. 12.9, where the very last cell in the ladder network is in fact the very first cell.

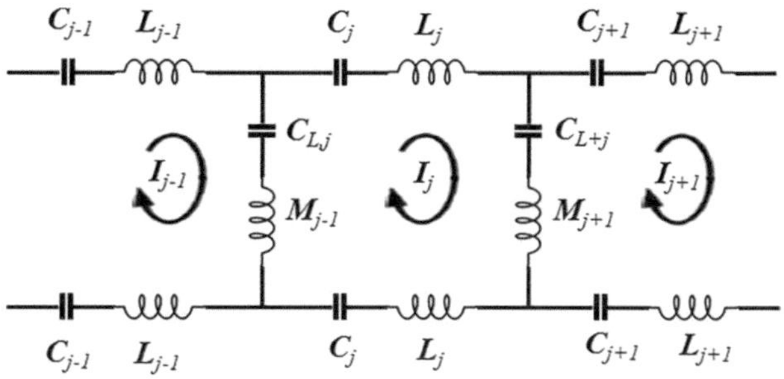

Fig. 12.9 Equivalent circuit representation of the birdcage antenna (adapted from Tarakameh (2020))

The equations for the current dynamics, including the mutual inductances between the rungs in the ladder which represent the legs and the end rings of the birdcage, may be expressed as

$$M_{j,j}\frac{d(I_j - I_{j-1})}{dt} + M_{j+1,j+1}\frac{d(I_j - I_{j+1})}{dt} + 2L_{j,j}\frac{dI_j}{dt} +$$

$$\sum_{\substack{k=1\\k\neq j}}^{N}\left(M_{j,k}\frac{d(I_k - I_{k-1})}{dt} + 2L_{j,k}\frac{dI_k}{dt}\right) - \sum_{k=1}^{N} 2\overline{L}_{j,k}\frac{dI_k}{dt} + \sum_{\substack{k=1\\k\neq j+1}}^{N} M_{j+1,k}\frac{d(I_k - I_{k-1})}{dt}$$

$$+(Q_j - Q_{j-1})/C'_j + (Q_j - Q_{j+1})/C'_{j+1} + 2Q_j/C_j = 0, I_j = dQ_j/dt, j$$
$$= 1, 2, \cdots, N-1, N. \tag{12.145}$$

In Eq. (12.145), N is the number of legs and $N + 1$ is 1, $L_{j,\,j}$ is the self-inductance of the end-ring segment in the jth loop, $L_{j,\,k}$ is the mutual inductance between end-ring segments in j^{th} and k^{th} loops on the same end-ring, $\overline{L}_{j,k}$ is the mutual inductance between end-ring segments in j^{th} and k^{th} loops on the different end-rings, $M_{j,\,j}$ is the self-inductance of the jth rung, $M_{j,\,k}$ is the mutual inductance between j^{th} and k^{th} rungs and I_j Q_j are the current and charge in the jth loop. Thus, the mutual inductance between the segments of the two end rings are also taken into account. The magnitudes of the self and mutual inductances may be found by the methods outlined by Tarakameh (2020), Giovannetti et al. (2002) and Guittienne et al. (2005). The equations for the current dynamics can be reduced to standard eigenvalue problem, and the undamped natural and resonance frequencies may be found by solving the eigenvalue problem. As the birdcage antenna operates as a resonant network, it has the very important property of a large input impedance at resonance, even when coupled to the plasmas. This feature simplifies the problem of impedance matching to the RF source, which is essential not only to maximize the power transfer to the plasma but to minimize the power loss in the matching circuit.

Once the natural frequencies are found, the capacitors could be adjusted to make sure that a resonance is established with the input RF signal. The actual magnetic field is calculated using the Biot–Savart law by numerically integrating the relations

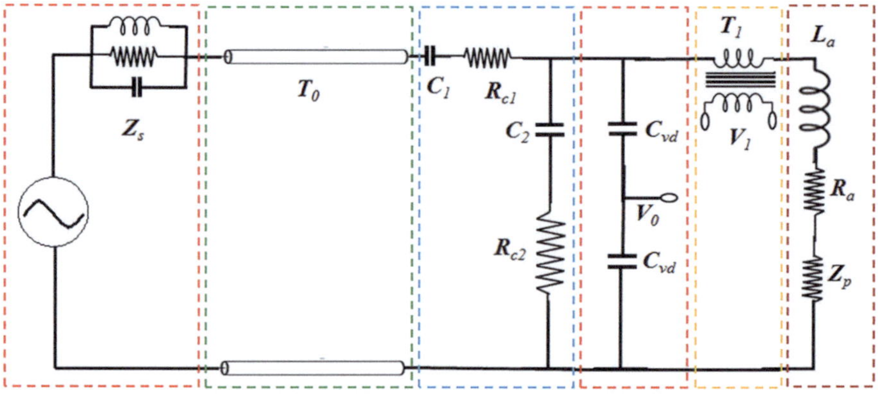

Fig. 12.10 Typical helicon thruster power circuit and equivalents showing the RF power source with known equivalent source impedance Z_s, the matched transmission line, L-type matching network formed by the capacitors C_1 and C_2 with respective resistances Rc_1 and Rc_2 and voltage-sensing divider circuit with C_3 and C_4, current-sensing transformer with T_1, antenna inductance L_A and the combined plasma and antenna resistances $R_A + R_P$ (adapted from N. Sinenian (2008))

arising from the application of the Biot–Savart law and summing them to compute the components of the magnetic field. To connect to an RF circuit, the input is connected across one of the capacitors in an end ring of the birdcage antenna.

The antenna is expected to radiate within the volume it encloses. Thus, in order to prevent radiation travelling outward, it is generally shielded to ensure the radiation does not interfere with other external devices. Then, when a shield is used, it does influence the mutual inductance values between the end rings and the legs of the birdcage. These effects can also be taken into account when evaluating the self and mutual inductances, $L_{j,\ j}$, $L_{j,\ k}$, $\overline{L}_{j,k}$, $M_{j,\ j}$ and $M_{j,\ k}$ within each cell. The shields generally are in the form of two semi-circular metallic shields placed around the antenna on the air side. Not only can the antenna's resonant frequency and consequently its input impedance can be tuned by adjusting the radial position of the screens, the power transfer efficiency to the plasma can be maximized to almost about 90%. The high efficiency coupled with the scalability of the birdcage antenna, even at moderate magnetic fields, is the primary advantage of such a helicon power source.

12.6.9 Helicon Thruster Power Circuit

Figure 12.10 shows a typical Helicon thruster power circuit where a radiofrequency (RF) power source with a known impedance is connected via a transmission line with an equal matching impedance. The transmission line is connected to an antenna via an impedance matching circuit. Impedance matching is essential to ensure that an input system delivers the maximum power to its load.

The voltage and current supplied to the antenna are measured using a voltage divider circuit and a current transformer. The load acting on the antenna output is the equivalent resistance and inductance of the plasma in the plasma chamber.

12.7 Helicon Plasma Thrusters

In the basic helicon plasma thruster concept, the charged particles in a high-density helicon plasma source are guided to the open source exit and accelerated by the magnetic nozzle via a magnetic expansion process. Various gain and loss processes of the particle momentum occur in the thruster, significantly affecting the propulsive performance, where the thrust force is generated by the momentum exhausted from the system. The major momentum loss occurs at the source, near the lateral wall, where the "radial" momentum is transferred to the wall via an electrostatic ion acceleration in the plasma sheath. Although the loss of the axial momentum there has been treated as negligible, data gathered from thrusters clearly shows the presence of the "axial" momentum lost to the lateral wall, which is transferred by the radially lost ions. Such a loss is not desirable as it directly impacts on the magnitude of the generated thrust, which would reduce significantly. The significant axial momentum loss seems to have originated from the internal axial electric field in the plasma core, which appears to be more enhanced by the highly ionized plasmas for the future high-power operation of the helicon plasma thruster. More detailed understanding of the plasma dynamics led to further development of the advanced high-power and electrodeless electric propulsion device and a more advanced helicon plasma thruster. The dynamics of plasmas in a helicon thruster is discussed by Ahedo (2013) in some detail.

The helicon plasma thruster (HPT) was developed in the preceding decades and has emerged as a novel electromagnetic propulsion device and is an electro-dynamic device. The attractiveness of these devices is that in comparison with other electric propulsion devices, such as Hall thrusters, ion engines, MPDs or arc jets, this concept does not need any electrodes, grids or neutralizers. The lack of electrodes, grids or neutralizers suggests that the HPT is simple and robust. A long lifetime is also expected, since the limited plasma–wall interaction due to the magnetic confinement reduces contamination or sputtering of sensitive components, e.g. the cathode in ion or Hall thrusters.

The principles of operation of an HPT are also quite different from that of a MPDT, when considered in some detail. Different physical processes are facilitated in an HPT (Harle, 2015; Williams, 2013) which are briefly summarized as

(i) The emission and propagation of the waves from the antenna which are used to excite the plasma
(ii) The absorption of the RF wave energy, which excites mainly on the electrons

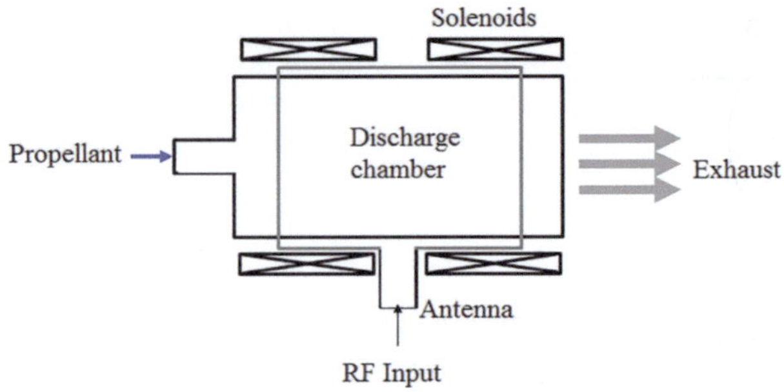

Fig. 12.11 A typical plasma source for a helicon thruster

(iii) The energized electrons bombard the neutral gas, producing a high-density plasma
(iv) The generated plasma is confined and guided by the magnetic field
(v) Forward acceleration of ions is driven by an ambipolar electric field which naturally develops within the plasma to sustain quasi-neutrality
(vi) The magnetic nozzle is used where the plasma continues expanding supersonically
(vii) Thrust is generated due to the increase in the momentum of the supersonic beam
(viii) The thrust is developed due to the interaction of plasma currents with the applied magnetic field

The HPT is composed of the following parts:

(i) A cylindrical chamber, where plasma is produced (plasma source), typically slender and made of a dielectric material, i.e., Pyrex glass. A typical plasma source is illustrated in Fig. 12.11.
(ii) A radiofrequency (RF) antenna wrapped around the chamber that emits within the electromagnetic frequency range of 1–27 MHz, with a wide assortment of topologies: annular, Nagoya III type, helical, saddle, etc. Typical antenna configurations are illustrated in Fig. 12.12.
(iii) The RF power is supplied to the antenna via the RF subsystem, consisting on a power unit, a wave generator/amplifier and a matching network, which adapts the RF power to the plasma source and an electromagnetic chamber.
(iv) A feed system is commonly attached to the back of the chamber.
(v) A set of several electromagnets and/or permanent magnets surrounding the chamber generates the required magnetic field in both inside the chamber (mainly axial) and in the plasma expansion area, to form a divergent magnetic nozzle (MN) topology.

HPT prototypes (Navarro-Cavalle et al., 2013; Rahman & Uzhinsky, 2018) are often classified according to the magnetic circuit they use and the power range in which operate. Several research groups have developed HPTs that implement permanent magnets, mostly in the low-power range, below 1 kW. These are quite compact HPT designs (Harle, 2015). Examples include the permanent magnet

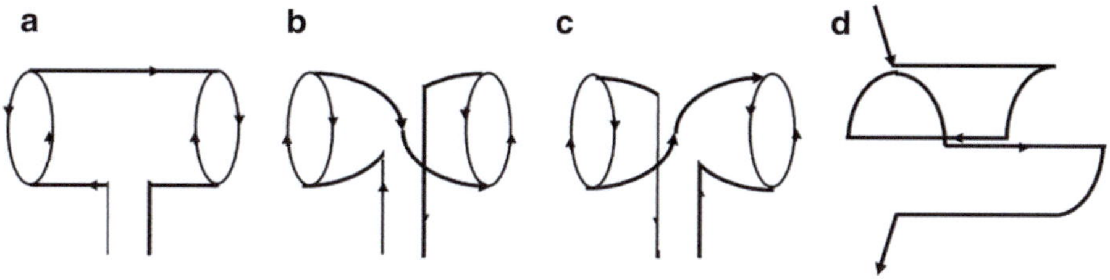

Fig. 12.12 Typical RF antenna configurations for a helicon thruster: (**a**) Nagoya III, (**b**) Nagoya III R, (**c**) Nagoya III L and (**d**) double saddle. (Current paths of the antennas are denoted by the arrows.) (adapted from Williams (2013))

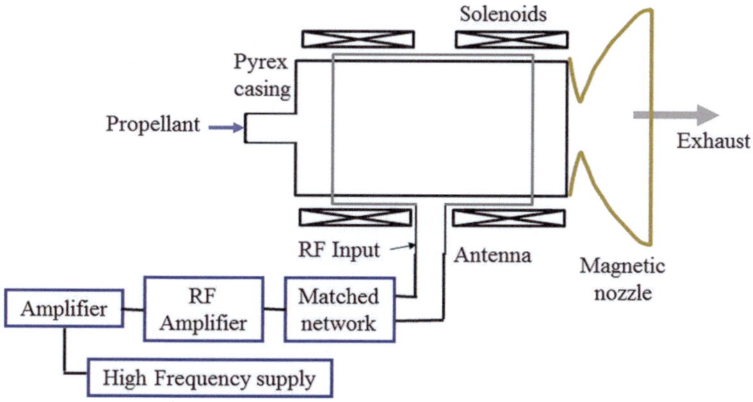

Fig. 12.13 Schematic diagram of a solenoid-based HPT configuration

expanding plasma (PEMP) and the helicon plasma hydrazine combined micro (HPHCOM) thruster. Examples of HPTs that use an electromagnet include helicon double-layer thruster (HDLT) (which relies on the formation of a plasma double layer for ion acceleration), the high-power helicon thruster (HPHT) and the mini helicon thruster experiment (mHTX). A typical schematic diagram of a solenoid-based HPT configuration is illustrated in Fig. 12.13. The principle of operation of a HDLT is illustrated in Fig. 12.14.

12.7.1 Helicon Plasma Thrusters: Design

The inductive plasma thruster is based on an RF-fed cylindrical birdcage antenna combined with an externally applied static magnetic field for the ionization of the propellant, so as to trigger the generation of helicon waves in the plasma, followed by the acceleration and ejection of a quasi-neutral plasma plume. An RF generator is combined with a matching network to provide the RF power to the birdcage antenna. A propellant is injected by the injector/intake into the discharge duct that is engulfed

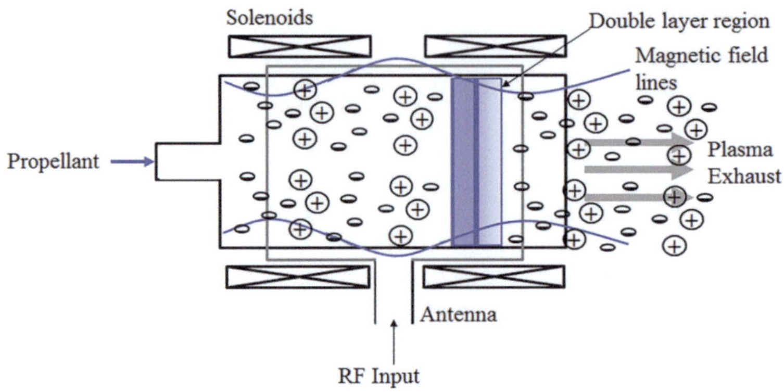

Fig. 12.14 Principle of operation of an HDLT

by the birdcage antenna. An external static magnetic field, generated by a solenoid fed by a direct current (DC) power supply, is applied along the axis of the discharge channel. The EM fields generated by the birdcage antenna ionize the propellant and provide, in conjunction with the applied static magnetic field, the force to accelerate the plasma. An example of an electrodeless RF plasma thruster (Romano et al. 2020) is one with an antenna fed by RF power and an intake to collect atmospheric particles for use as the propellant for the electric thruster.

12.8 Variable Specific Impulse Magnetoplasma Rocket (VASIMR)

In order to produce this gaseous plasma, the Variable Specific Impulse Magnetoplasma Rocket (VASIMR) makes use of helicon radiofrequency waves to ionize the propellant gas within a region enclosed by a magnetic field. The RF waves are produced by a suitable antenna supplied by a matched alternating current source. The waves are tailored to excite the eigenmode of the electrons within the propellant gas. Although it uses the Helicon plasma source, the VASIMR is not really a Helicon device. It is a hybrid device combining an electron RF heating stage with an ion cyclotron resonance heating stage. It is illustrated in Fig. 12.15.

VASIMR has been designed as a high magnetic field and high-power plasma rocket with tens of kilowatts of input power. Various aspects of VASIMR have been discussed by Chang-Diaz et al. (1999), Bering et al. (2005), Arefiev and Breizman (2004) and Squire et al. (2014). The whole VASIMR thruster includes an ion cyclotron resonance heating stage downstream of a HPS, which in fact energizes the plasma that is finally accelerated along a magnetic nozzle, producing thrust. The complete typical engine presents a nominal power of 200 kW, reaching a thrust efficiency over 50%, providing a thrust higher than 3 N and specific impulses above 3000 s.

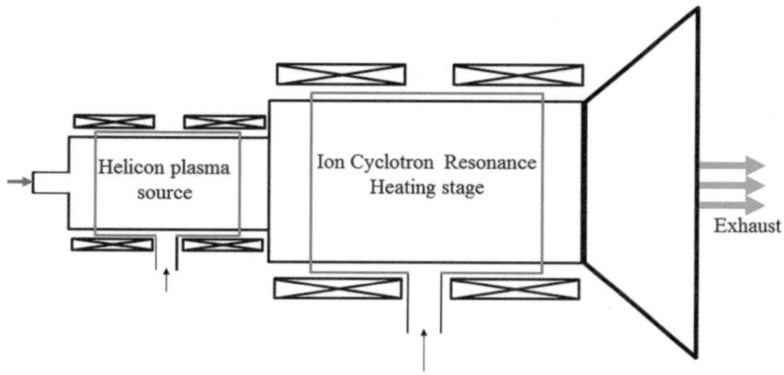

Fig. 12.15 The VASIMR concept

12.9 Inertial Electrostatic Confinement (IEC) Thruster

An inertial electrostatic confinement (IEC) fusion device consists of a vacuum chamber containing two spherical electrode grids, one within the other, designed to accelerate ions inwards towards the centre of the inner spherical electrode from the outer electrode, creating a spherically convergent beam in which ions may be recirculated through the partially transparent grids. The inner cathode grid that is transparent to ions is capable of containing electrons. The purpose is electron confinement at the centre of the device, so as to minimize the energy loss due to electron streaming from the cathode to the anode. Furthermore, the trapped electrons help sustain a large ion density at the centre. Thus, a spherically convergent beam of recirculating ions is generated, by the technique known as inertial electrostatic confinement. The concentric spherical electrodes approximate a radially symmetric electric field, resulting a symmetric potential well for the ions. Negative potentials are applied to the cathode electrode, in the range of 10–100 kV. Neutral gas introduced into the chamber is ionized following collisions with electrons. The resulting ions are accelerated and drawn into the attractive potential well created by the cathode, acquiring a high kinetic energy at the centre where they undergo reactions with other ions. The principle is discussed in detail by Miley and Murali (2013).

12.10 Summary of HPT Thrusters

In this final section, the performance of the HPT thrusters is briefly compared in Table 12.1.

Table 12.1 HPT main prototypes, summary of propulsive performances

Prototype	Power	Thrust	$I_{sp}(s)$	$\eta_u(\%)$	$\eta_t(\%)$
HDLT	<1.5 kW	6 mN	800	25–35	1–3
mHTX	700–1500 W	20 mN	2000	90	1–10
HPHCOM	50 W	1.5 mN	1200	90	13
PMEP	700 W	3 mN	500	<50	1
HPHT	20-50 kW	1–2 N	>2000	~100	55
VASIMIR	200 kW	3 N	>3000	~100	50

Source: Rahman and Uzhinsky (2018)
I_{sp} is the specific impulse (s), η_u the HPS utilization efficiency and η_t the thrust efficiency

Exercises

(1) Considering a radiofrequency ion thruster and given that q_{ion} is the ionic charge, V_{ion} is voltage driving the ions, m_{ion} is the ionic mass, $I_{ion-beam}$ is the ionic beam current, e is the electronic charge, η_m is the thruster mass utilization efficiency and γ_c is the total thrust correction factor, obtain expressions for (i) the ionic velocity, (ii) the specific impulse and (iii) the thrust generated. How do these expressions compare with corresponding expressions for (a) an electro spray thruster, (b) a resistojet, (c) a pulsed plasma thruster, (d) a cold gas propulsion system, (e) a liquid propulsion system and (f) a solid propulsion system.

(2) The concept of low-frequency ($\omega < \omega_{\mathrm{ci}}$) electromagnetic waves travelling through a perfect conductor with parallel phase speed $v_a = B_0/\sqrt{\mu_0 m_i n_i}$ (where B_0 is the background magnetic field, m_{i} is the ion mass and n_{i} is the ion density) was first proposed by Hannes Alfven in 1942.

Considering this case, with only the presence of a magnetic field and with *no applied electric field present*, obtain the dispersion relation for waves in the plasma medium. Hence, obtain the circular frequency corresponding to a transverse shear wave as well as the circular frequency corresponding to a compressive wave in terms of the parallel phase speed given above, the Alfven speed and the speed of magnetic disturbances in a plasma when no electric fields are present.

(3) Collisions are most important for the heating of plasma and cannot be ignored. The anisotropy of the electric and magnetic fields also need to be considered. For this reason, the formulations of the earlier sections are presented again with appropriate modifications both for collisions and for the anisotropy of the plasma.

Furthermore, noble gases which are often used as fuels for plasma are diamagnetic. Super conductors are completely diamagnetic. So, for both these cases, it is important to consider the anisotropy of the plasma. Considering both the collisions and the anisotropy of the fields and the plasma, show that the particle velocity in a Cartesian coordinate frame and the particle velocity in the frequency domain satisfies

$$\mathbf{v}_j = \begin{bmatrix} -j\omega(m_j/q_j) & -B_z & B_y \\ B_z & -j\omega(m_j/q_j) & -B_x \\ -B_y & B_x & -j\omega(m_j/q_j) \end{bmatrix}^{-1} \mathbf{E}.$$

In cylindrical coordinates, show that the corresponding relation is

$$\mathbf{v}_j = \begin{bmatrix} -j\omega(m_j/q_j) & -\left(B_z + (m_j/q_j)\dot{\theta}_j\right) & B_\theta \\ B_z + (m_j/q_j)\dot{\theta}_j & -j\omega(m_j/q_j) & -B_r \\ -B_\theta & B_r & -j\omega(m_j/q_j) \end{bmatrix}^{-1} \mathbf{E}.$$

Hence, generalize the dispersion relations for the Helicon waves, in both sets of reference frames.

Considering the special case when there is only the radial component of the magnetic field is non-zero and given the directional collision frequencies, ν_r, ν_θ and ν_z, and ω_0 the angular velocity of the rotating reference frame, specialize the dispersion relation in terms of the plasma frequency and the cyclotron frequency, for (i) a single species case as well as (ii) a two-species case. State your result as an eigen value problem.

Explain the physical significance of the dispersion relations.

(4) The way pulsed plasma thrusters operate is by rapid release of energy E stored in a capacitor to provide momentum to a small quantity of mass of plasma for high discharge velocity. This propellant mass ejection following a pulse input creates an impulse bit I. Assuming a pulsed plasma thruster is operating at a fixed pulse frequency f, then a steady-state thrust is given by $T = fI$, where I is the impulse bit.

The corresponding steady-state power delivered by the capacitor can be represented by the equation $P = fE$. The characteristics of five different ablative pulsed plasma thrusters are shown in Table 12.2. The power plant–specific mass is 10 Kg/kW. The mass of the propellant tanks is specified as 10% of the mass of the propellant. The payload mass is 300 Kg.

Table 12.2 Characteristics of five different ablative pulsed plasma thrusters

Characteristics	LES8/9	Mighty Sat	APPT-20	APPT-60	Apt-120
Energy in a pulse, J	22.6	43.17	20	60	120
Frequency in Hz	2	2	2	2	2
Impulse bit, $\mu N/s$	300	750	500	1600	3000
Thrust efficiency	7	9.8	14	20	27
Specific impulse	1075	1150	1150	1500	2200

(a) For each of the five thrusters in the table, determine the nozzle exit velocity, the power delivered, the power required, the propulsion system mass, the mass ratio, the propellant mass required, the thrust in Newtons, the mass flow rate at the nozzle exit and the operational mission times required to complete the task, if the total Δv generated over the mission time must be over 60 m/sec.

(b) Based on your calculations, choose an appropriate thruster for a mission, which must be completed under 10 years, stating your reasons for the design choice.

(5) Derive from first principles of circuit theory the equations for the current dynamics, including the mutual inductances between the rungs in the ladder which represent the legs and the end rings of the model of the birdcage antenna shown in Fig. 12.9.

(6) (a) The thrust force equation for a pulsed plasma thruster may be expressed as

$$\sum F_{T_i}(t) = F_{inertial} + F_{gas} + F_{EM}.$$

The first term is given by the rate of change of momentum flux. For the neutral gas, the thermal gas force is

$$F_{gas} = hw \times n_0 \times k_B \times T_e,$$

where h is the height of the electrodes, w is the width of the electrodes, T_e is the plasma electron temperature, n_0 is the particle number density (which could be determined using the Saha approximation) and k_B is the Boltzmann constant. The Lorentz force acting on a plasma particle in an electromagnetic field with a charge density of ρ and a current density of $\mathbf{j}$ across the electrodes is

$$\mathbf{F}_{EM} = \int (\rho \mathbf{E} + \mathbf{j} \times \mathbf{B}) dV.$$

Hence, show that the thrust force, for a self-field–induced pulsed plasma thruster, could be approximated by an equation of the form

$$F_T = -\frac{d}{dt}(m_0 U_0) + hwn_0 k_B T_e + \frac{1}{2}\mu_0 K I^2(t).$$

where F_T is the total thrust acting on the propellant discharge mass and K is a proportionality constant.

(b) Derive the thrust force equation for an applied field pulsed plasma thruster.

(7) The self-induced magnetic field can be obtained as a function of the current density and vice versa, by applying Maxwell's equations (and Ampere's law). Thus,

$$\mathbf{j} = (1/\mu_0)\nabla \times \mathbf{B} \text{ and } \mathbf{F}_{EM} = \int_V (\mathbf{j} \times \mathbf{B}) dV = \int_V \nabla \cdot \mathbf{B} dV.$$

Assuming the thrust generated by a self-field magnetoplasma thruster is dominated by the Lorentz force, derive an approximate expression (known as Maecker's formula) for the thrust generated by such a thruster, given by

$$F_T = \frac{\mu I^2}{4\pi}\left(\ln\left(\frac{r_a}{r_b}\right) + C\right), 0.5 \leq C \leq 1.$$

How is this expression modified by the presence of an applied field?

References

Ahedo, E. (2013). Plasma dynamics in a helicon thruster, Progress in propulsion physics. *EDP Sciences, 4*, 337–354.

Arefiev, A. V., & Briezman, B. N. (2004). *Theoretical components of the VASIMR plasma propulsion concept*. Institute for Fusion Studies, University of Texas Austin.

Arnush, D., & Chen, F. F. (1998). Generalized theory of helicon waves. ii. Excitation and absorption. *Physics of Plasmas (1994–present), 5*(5), 1239–1254.

Bathgate, S. N., Bilek, M. M. M., & Mckenzie, D. R. (2017). Electrodeless plasma thrusters for spacecraft: A review. *Plasma Science and Technology, 19.*, 083001 (24pp). https://doi.org/10.1088/2058-6272/aa71fe

Bering, E. A., III, Brukardt, M., Chang-Diaz, F. R., Squire, J. P., Glover, T. W., Jacobson, V., Tarditi, A., McKaskill, G., & Bengtson, R. D. (2005). *Ion acceleration by single pass ion cyclotron heating in the VASIMR*. 29th International Electric Propulsion Conference, Princeton University.

Boswell, R. W. (1970). Plasma production using a standing helicon wave. *Physics Letters A, 33*(7), 457–458.

Bozhanov, T., & Rojas, A. M. (2017). *Analysis of electric propulsion systems for drag compensation of small satellites in low earth orbits*. In iCubeSat 2017: 6th Interplanetary CubeSat Workshop.

Burton, R., & Turchi, P. (1998). Pulsed plasma thruster. *Journal of Propuls Power, 14*, 716–735.

Chang-Diaz, F. R., Squire, J. P., Ilin, A. V., McGaskill, G. E., Ngyuen, T. X., Winter, D. S., Petro, A. J., Goebel, G. W., Cassady, L. D., Stokke, K. A., Dexter, C. E., Graves, T. P., Amador, L., Jr., George, J. A., Carter, M. D., Baity, F. W., Jr., Barber, G. C., Goulding, R. H., Sparks, D. O., Schwenterly, S. W., Bengston, R. D., Briezman, B. N., Jacobson, V. T., Arefiev, A. V., Sagdeez, R. Z., Karavasilis, K., Novakovski, S. V., Chan, A. A., & Glover, T. W. (1999). *The development of the VASIMR engine*. Proc. of the International Conference on Electromagnetics in Advanced Applications, Torino.

Charles, C. (2009). Plasmas for spacecraft propulsion. *Journal of Physics D: Applied Physics, 42*, 163001. (18pp). https://doi.org/10.1088/0022-3727/42/16/163001

Charles, C., Boswell, R. W., & Lieberman, M. A. (2006). Xenon ion beam characterization in a helicon double layer thruster. *Applied Physics Letters, 89*, 261503(1–3).

Charles, C., Boswell, R. W., Laine, R., & MacLellan, P. (2008). An experimental investigation of alternative propellants for the helicon double layer thruster. *Journal of Physics D: Applied Physics, 41*, 175–213.

Chen, F. F. (1991). Plasma ionisation by helicon waves. *Plasma Physics and Controlled Fusion, 33*(4), 339–354.

Chen, F. F. (2015). Helicon discharges and sources: A review. *Plasma Sources Science and Technology, 24*, 014001.

Chen, F. F., & Arnush, D. (1997). Generalized theory of helicon waves. I. Normal modes. *Physics of Plasmas, 4*(9), 3411–3421. https://doi.org/10.1063/1.872483

Cifali, G., Misuri, T., Rossetti, P., Andrenucci, M., Valentian, D., & Feili, D. (2011). *Preliminary characterization test of HET and RIT with nitrogen and oxygen*. Proceedings of the 47th AIAA/ASME/SAE/ASEE Joint Propulsion Conference and Exhibit, San Diego, AIAA2011-6073. https://doi.org/10.2514/6.2011-6073

Dasgupta, B., Tsurutani, B. T., & Janaki, M. S. (2003). A kinetic approach to the Ponderomotive force. *Geophysical Research Letters, 30*(21), 2128. https://doi.org/10.1029/2003GL017385

Eslava, S., Marchetto, J., & Scougal, E. (2014). *Design of a micro-pulsed plasma thruster for 3U cubesat*. AIAA Conf. Paper, 16 pages. https://www.researchgate.net/publication/261448478_Design_of_a_micro-Pulsed_Plasma_Thruster. Accessed 3 Nov 2023

Fujita, K. (2004). *Air-intake performance estimation of air-breathing ion engines* (pp. 3038–3044). In: Transactions of the Japan Society of Mechanical Engineers. issn: 0(700).

Garrigues, L. (2012). Computational study of hall-effect thruster with ambient atmospheric gas as propellant. *Journal of Propulsion and Power, 28*(2), 344–354.

Giannelli, S., & Andrenucci, M. (2011). *Azimuthal instability of MPD thruster plasmas and inception of critical regimes*. 47th AIAA Joint Propulsion Conference, (AIAA-2011-5887).

Giovannetti, G., Landini, L., Santarelli, M. F., & Positano, V. (2002). A fast and accurate simulator for the design of birdcage coils in MRI. *Magnetic Resonance Materials in Physics, Biology and Medicine, 15*, 36–44.

Goeksel, B., & Mashek, I. (2017). First breakthrough for future air-breathing magneto-plasma propulsion systems. *Journal of Physics: Conference Series, 825*, 012005. http://iopscience.iop.org/article/10.1088/1742-6596/825/1/012005

Guittienne, P., Chevalier, E., & Hollenstein, C. (2005). Towards an optimal antenna for helicon waves excitation. *Journal of Applied Physics, 98*(8), 8, 083304. Accessed 15 Oct 2005.

Harle, T. (2015). Radio frequency plasma thrusters: Performance, evaluation of low magnetic field mode operation through direct thrust measurements, Ph D. In *Dissertation, Surrey Space Centre, Department of Electronic Engineering, Faculty of Engineering and Physical Sciences*. University of Surrey.

Hohman, K. (2012). *Atmospheric breathing electric thruster for planetary exploration*. Presented at the NIAC Spring Symposium.

Jahn, R. G. (1968). *Physics of electric propulsion*. McGraw-Hill.

Lafleur, T. A. (2011). *Helicon wave propagation in low diverging magnetic fields*. Doctor of Philosophy Thesis, The Australian National University.

Levchenko, I., Xu, S., Mazouffre, S., Lev, D., Pedrini, D., Goebel, D., Garrigues, L., Taccogna, F., & Bazaka, K. (2020). Perspectives, frontiers, and new horizons for plasma-based space electric propulsion. *Physics of Plasmas, 27*, 020601. https://doi.org/10.1063/1.5109141

Little, J. M., & Choueiri, E. Y. (2013). Thrust and efficiency model for electron-driven magnetic nozzles. *Physics of Plasmas, 20*, 103501.

Martinez, M. M. (2013). *Analysis of magnetic nozzles for space plasma thrusters*. PhD Thesis, Universidad Politecnica de Madrid.

Miley, G. H., & Murali, S. K. (2013). *Inertial Electrostatic Confinement (IEC) fusion: Fundamentals and applications*. Springer Nature.

Navarro-Cavalle, J., Ahedo, E., Merino, M., Gomez, V., Ruiz, M., & del Amo, J. A. G. (2013). *Helicon plasma thrusters: Prototypes and advances on modeling*. IEPC-2013-285, Presented at the 33rd International Electric Propulsion Conference, The George Washington University, , October 6–10.

Nishiyama, K. (2003). *Air breathing ion engine concept*. Proceedings of the 54th International Astronautical Congress of the International Astronautical Federation, the International Academy of Astronautics, and the International Institute of Space Law, Bremen, IAC-03-S.4.02. https://doi.org/10.2514/6.IAC-03-S.4.02

Paganucci, F., Zuin, M., Agostini, M., Andrenucci, M., Antoni, V., Bagatin, M., Bonomo, F., Cavazzana, R., Franz, P., Marrelli, L., Martin, P., Martines, E., Rossetti, P., Serianni, G., Scarin, P., Signori, M., & Spizzo, G. (2008). MHD instabilities in magneto-plasma-dynamic thrusters. *Plasma Physics and Controlled Fusion, 50*(1–14), 124010. https://doi.org/10.1088/0741-3335/50/12/124010

Pottinger, S., Lappas, V., Charles, C., & Boswell, R. W. (2011). Performance characterization of a helicon double layer thruster using direct thrust measurements. *Journal of Physics D: Applied Physics, 44*(23), 235201. https://doi.org/10.1088/0022-3727/44/23/235201

Rahman, M. M., & Uzhinsky, I. (2018). Radio frequency excited plasma discharge simulation for potential helicon plasma thruster. *IOP Conference Series: Materials Science and Engineering, 449*, 012032.

Romano, F. (2014). *System analysis and test-bed for an air-breathing electric propulsion system.* Master's Thesis, Universit¨at Stuttgart, University of Padova.

Romano, F., Binder, T., Herdrich, G., Fasoulas, S., & Schönherr, T. (2015). *Air-intake design investigation for an air-breathing electric propulsion system* (p. 2015). IEPC-2015-90524/ISTS-2015-b-90524, Presented at Joint Conference of 30th International Symposium on Space Technology and Science, 34th International Electric Propulsion Conference and 6th Nano-satellite Symposium, Hyogo-Kobe, Japan, July 4–10.

Romano, F., Chan, Y.-A., Herdrich, G., Roberts, P. C. E., Traub, C., Fasoulas, S., Crisp, N., Edmondson, S., Haigh, S., Holmes, B. A., Livadiotti, S., Macario-Rojas, A., Abrao-Oiko, V. T., Smith, K., Sinpetru, L., Becedas, J., Dominguez, R. M., Christensen, S., Jensen, T. K., Nielsen, J., Bisgaard, M., Garcia-Almiñana, D., Rodriguez-Donaire, S., Sureda, M., Garcia-Berenguer, M., Kataria, D., Villain, R., Seminari, S., Conte, A., & Belkouchi, B. (2020). *RF helicon-based plasma thruster (IPT): Design, set-up, and first ignition*. 71st International Astronautical Congress (IAC), The CyberSpace Edition, 12–14 October 2020.

Rosales, M. A., Hansen, C., & Winglee, R. M. (2019). Theoretical and experimental analysis for an air-breathing pulsed plasma thruster, AIAA propulsion and energy 2019 forum, 19-22 august 2019. *Indianapolis, IN, USA*. https://doi.org/10.2514/6.2019-3888

Schönherr, T., Komurasaki, K., Romano, F., Massuti-Ballester, B., & Herdrich, G. (2015). Analysis of atmosphere-breathing electric propulsion. *Plasma Science, IEEE Transactions on, 43*(1), 287–294.

Shabshelowitz, A. (2013). *Study of RF plasma technology applied to air-breathing electric propulsion*. University of Michigan.

Sinenian, N. (2008). *Propulsion mechanisms in a helicon plasma thruster*. Master of Science Thesis, Massachusetts Institute of Technology.

Squire, J. P., Carter, M. D., Chang-Diaz, F. R., Giambusso, M., Ilin, A. V., Olsen, C. S., & Bering, E. A., III. (2014). *Development toward a spaceflight capable VASIMR engine and SEP applications*. AIAA Space and Astronautics Forum and Exposition (SPACE 2014), San Diego.

Tagawa, M., Yokota, K., Nishiyama, K., Kuninaka, H., Yoshizawa, Y., Yamamoto, D., & Tsuboi, T. (2013). Experimental study of air breathing ion engine using laser detonation beam source. *Journal of Propulsion and Power, 29*(3), 501–506. https://doi.org/10.2514/1.B34530.IAC-03-S.4.02

Takahashi, K. (2019). Helicon-type radiofrequency plasma thrusters and magnetic plasma nozzles, (in topical collection on recent Progress in physics of plasma-based space propulsion). *Reviews of Modern Plasma Physics, 3*(3), 1–61. https://doi.org/10.1007/s41614-019-0024-2

Tarakameh, A. S. (2020). *Novel techniques and innovative design for the RF chain of magnetic resonance imaging scanners*. Phd thesis, Bilkent University.

Tian, B. (2017). *Modeling of physical processes in radio-frequency plasma thrusters*. Departamento de Bioingenieia Eingenieria Aeroespacial, Universidad Carlos III de Madrid, Leganes, Madrid, July 19, 2017.

Tikhonov, V. B., & Semenikhin, S. A. (1993). *Research of plasma acceleration processes in self-field and applied magnetic field thrusters*. 23rd International Electric Propulsion Conference (IEPC), (IEPC-93-076).

Tilley, D. L., Choueiri, E. Y., Kelley, A. J., & Jahn, R. G. (1996). Microinstabilities in a 10-kilowatt self-field magnetoplasmadynamic thruster. *Journal of Propulsion and Power, 12*(2).

Uematsu, K., Morimoto, S., & Kuriki, K. (1984). MPD thruster performance with various propellants. *Journal of Spacecraft, 22*(4), 412–416. https://doi.org/10.2514/3.25766

Wagner, H. P., Kaeppeler, H. J., & Auweter-Kurtz, M. (1998). Instabilities in MPD thruster flows: 1. Space charge instabilities in unbounded and inhomogeneous plasmas. *Journal of Physics D: Applied Physics, 31*(5), 519.

Williams, L. T. (2013). *Ion acceleration mechanisms of helicon thrusters*. Doctor of Philosophy Dissertation, Georgia Institute of Technology.

Index

R. Vepa, *Space Vehicle Maneuvering, Propulsion, Dynamics and Control*, https://doi.org/10.1007/978-3-031-65518-0

MIX
Papier aus verantwortungsvollen Quellen
Paper from responsible sources
FSC® C105338

If you have any concerns about our products,
you can contact us on
ProductSafety@springernature.com

In case Publisher is established outside the EU,
the EU authorized representative is:
Springer Nature Customer Service Center GmbH
Europaplatz 3, 69115 Heidelberg, Germany

Printed by Libri Plureos GmbH
in Hamburg, Germany